TURBULENT FLUID MOTION

Combustion: An International Series
Norman Chigier, *Editor*

Bayvel and Orzechowski, Liquid Atomization
Chen and Jaw, Fundamentals of Turbulence Modeling
Chigier, Combustion Measurements
Deissler, Turbulent Fluid Motion
Kuznetsov and Sabel'nikov, Turbulence and Combustion
Lefebvre, Atomization and Sprays
Li, Applied Thermodynamics: Availability Method and Energy Conversion
Libby, Introduction to Turbulence
Roy, Propulsion Combustion: Fuels to Emissions

TURBULENT FLUID MOTION

Robert G. Deissler
NASA Lewis Research Center
Cleveland, Ohio

USA	Publishing Office:	Taylor & Francis 325 Chestnut Street, Suite 800 Philadelphia, PA 19106 Tel: (215) 625-8900 Fax: (215) 625-2940
	Distribution Center:	Taylor & Francis 1900 Frost Road, Suite 101 Bristol, PA 19007-1598 Tel: (215) 785-5800 Fax: (215) 785-5515
UK		Taylor & Francis Ltd. 1 Gunpowder Square London EC4A 3DE Tel: 0171 583 0490 Fax: 0171 583 0581

TURBULENT FLUID MOTION

Copyright © 1998 Taylor & Francis. All rights reserved. Printed in the United States of America. Except as permitted under the United States Copyright Act of 1976, no part of this publication may be reproduced or distributed in any form or by any means, or stored in a database or retrieval system, without prior written permission of the publisher.

1 2 3 4 5 6 7 8 9 0   E B E B   9 0 9 8

To June and our family.

A CIP catalog record for this book is available from the British Library.
∞ The paper in this publication meets the requirements of the ANSI Standard Z39.48-1984 (Permanence of Paper)

Library of Congress cataloging-in-publication data is available from the publisher.

ISBN 1-56032-753-7 (cloth)
ISSN 1040-2756

CONTENTS

	Preface	ix
1	**THE PHENOMENON OF FLUID TURBULENCE**	1
1-1	What is Turbulence?	3
1-2	Ubiquity of Turbulence	4
1-3	Why Does Turbulence Occur?	6
1-4	Closing Remarks	8
	References	17
2	**SCALARS, VECTORS, AND TENSORS**	19
2-1	Introduction	19
2-2	Rotation of Coordinate Systems	19
2-3	Vectors (First-Order Tensors)	21
2-4	Second-Order Tensors	22
	2-4-1 Definition and Simple Examples	22
	2-4-2 Stress and the Quotient Law	23
	2-4-3 Kronecker Delta, a Tensor	24
2-5	Third- and Higher-Order Tensors	25
	2-5-1 Vorticity and the Alternating Tensor	25
	2-5-2 A More General Quotient Law	26
2-6	Zero-Order Tensors and Contraction	27
2-7	Outer and Inner Products of Tensors of Higher Order	28
2-8	Subscripted Quantities That Are Not Tensors	28
2-9	Closing Remarks	29
	References	29

3 BASIC CONTINUUM EQUATIONS 31

3-1 Justification of the Use of a Continuum Approach for Turbulence 31
3-2 Equation of Continuity (Conservation of Mass) 32
3-3 Navier-Stokes Equations (Conservation of Momentum) 34
    3-3-1 Stress Tensor 34
    3-3-2 Equations of Motion 38
    3-3-3 Dimensionless Form of Constant-Property Fluid-Flow Equations and Dimensionless Correlation of Friction-Factor Data 41
3-4 Heat Transfer or Energy Equation (Conservation of Energy) 42
    3-4-1 Dimensionless Form of Constant-Property Energy Equation and Dimensionless Correlation of Heat-Transfer Data 44
3-5 Rule for Obtaining Additional Dimensionless Parameters as a System Becomes More Complex 45
3-6 Closing Remarks 47
    References 48

4 AVERAGES, REYNOLDS DECOMPOSITION, AND THE CLOSURE PROBLEM 49

4-1 Average Values and Their Properties 49
    4-1-1 Ergodic Theory and the Randomness of Turbulence 51
    4-1-2 Remarks 51
    4-1-3 Properties of Averaged Values 51
4-2 Equations in Terms of Mean and Fluctuating Components 52
4-3 Averaged Equations 55
    4-3-1 Equations for Mean Flow and Mean Temperature 55
    4-3-2 Simple Closures of the Equations for Mean Flow and Temperature 57
    4-3-3 One-Point Correlation Equations 91
    4-3-4 Two-Point Correlation Equations 98
4-4 Closing Remarks 104
    References 106

5 FOURIER ANALYSIS, SPECTRAL FORM OF THE CONTINUUM EQUATIONS, AND HOMOGENEOUS TURBULENCE 109

5-1 Fourier Analysis of the Two-Point Averaged Continuum Equations 110
    5-1-1 Analysis of Two-Point Averaged Quantities 110
    5-1-2 Analysis of the Two-Point Correlation Equations 112
5-2 Fourier Analysis of the Unaveraged (Instantaneous) Continuum Equations 116
    5-2-1 Analysis of Instantaneous Quantities 116
    5-2-2 Analysis of Instantaneous Continuum Equations 118
5-3 Homogeneous Turbulence without Mean Velocity or Temperature (Scalar) Gradients 122
    5-3-1 Basic Equations 122
    5-3-2 Illustrative Solutions of the Basic Equations 131
5-4 Homogeneous Turbulence and Heat Transfer with Uniform Mean-Velocity or -Temperature Gradients 220

	5-4-1 Basic Equations	220
	5-4-2 Cases for Which Mean Gradients Are Large or the Turbulence Is Weak	222
	5-4-3 Uniformly and Steadily Sheared Homogeneous Turbulence If Triple Correlations May Be Important	355
5-5	Closing Remarks	365
	References	365
6	**TURBULENCE, NONLINEAR DYNAMICS, AND DETERMINISTIC CHAOS**	**373**
6-1	Low-Order Nonlinear System	374
6-2	Basic Equations and a Long-Term Turbulent Solution with Steady Forcing	376
6-3	Some Computer Animations of a Turbulent Flow	382
6-4	Some Turbulent and Nonturbulent Navier-Stokes Flows	383
	6-4-1 Time Series	385
	6-4-2 Phase Portraits	386
	6-4-3 Poincaré Sections	392
	6-4-4 Liapunov Exponent	394
	6-4-5 Ergodic Theory Interpretations	398
	6-4-6 Power Spectra	399
	6-4-7 Dimensions of the Attractors	400
6-5	Closing Remarks	402
	References	403
	AFTERWORD	**405**
	INDEX	**407**

PREFACE

Researchers have been active in serious studies of turbulence for more than a century. Today, as it was a century ago, turbulence is ubiquitous. Although it is still an active field of research, there is no general deductive theory of strong turbulence.

The literature on turbulence is now far too voluminous for anything like a full presentation to be given in a moderately-sized volume. Rather, it is attempted here to give a coherent account of one line of development. Part of this has been given in abbreviated form in Chapter 7 of *Handbook of Turbulence, Volume 1* (Plenum, 1977). In particular, the scope of the work, which was somewhat limited by our inability to solve the fundamental nonlinear equations, has been considerably increased by numerical solutions. Moreover, applications of dynamic systems theory in conjunction with numerical solutions have resulted in, among other things, a sharper characterization of turbulence and a deliniation of routes to turbulence.

The present work is based on a series of six NASA Technical Memoranda by the writer. Throughout the book the emphasis is on understanding the physical processes in turbulent flow. This is done to a large extent by obtaining and interpreting analytical or numerical solutions of the equations of fluid motion. No attempt is made to either emphasize or avoid the use of mathematical analysis. Because most of the material is given in some detail, the student or research worker with a modest knowledge of fluid mechanics should not find the text particularly hard to follow. Some familiarity with Cartesian-tensor notation and Fourier analysis may be helpful, although background material in those subjects is given.

Although turbulence, as it occurs, is more often strong than weak, it appears that much can be learned about its nature by considering weak or moderately weak turbulence, as often is done here. In general, the same processes occur in moderately weak turbulence as occur at much higher Reynolds numbers; the differences are quantitative rather than qualitative. The crux of the matter therefore might be accessible through low– and moderate–Reynolds-number studies.

The basis of the present account of turbulence is the Navier-Stokes and other continuum equations for fluids. It is hoped that the book shows how those equations can act as a unifying thread for such an account.

Some introductory material on fluid turbulence is presented in Chapter 1. This includes discussions and illustrations of what turbulence is and how, why, and where turbulence occurs. Then, in Chapter 2, some of the mathematical apparatus used for the representation and study of turbulence is developed.

A derivation of the continuum equations used for the analysis of turbulence is given in Chapter 3. These equations include the continuity equation, the Navier-Stokes equations, and the heat-transfer or energy equation. An experimental justification for using a continuum approach for the study of turbulence is also given.

Ensemble, time, and space averages as applied to turbulent quantities are discussed in Chapter 4, and pertinent properties of the averages are obtained. Those properties, together with Reynolds decomposition, are used to derive the averaged equations of motion and the one- and two-point moment or correlation equations. The terms in the various equations are interpreted. The closure problem of the averaged equations is discussed, and possible closure schemes are considered. Those schemes usually require an input of supplemental information, unless the averaged equations are closed by calculating their terms by a numerical solution of the original unaveraged equations. The law of the wall for velocities and for temperatures, the velocity- and temperature-defect laws, and the logarithmic laws for velocities and for temperatures are derived. Various notions of randomness and their relation to turbulence are considered in the light of modern ergodic theory.

Background material on Fourier analysis and on the spectral form of the continuum equations, both averaged and unaveraged, are given in Chapter 5. The equations are applied to a number of cases of homogeneous turbulence with and without mean gradients. Some turbulent solutions of the full unaveraged continuum equations are obtained numerically. Closure of the averaged equations by specification of sufficient random initial conditions is considered. The gap problem (the problem of bridging the gap between the infinite amount of data required to specify an initial turbulence and the finite amount generally available) is discussed. Then a solution for the evolution of all of the quantities used to specify the initial turbulence is obtained. Spectral transfer of turbulent activity between scales of motion is studied in some detail. The effects of mean shear, heat transfer, normal strain, and buoyancy are included in the analyses.

Finally, in Chapter 6 the unaveraged Navier-Stokes equations are used numerically in conjunction with tools and concepts from nonlinear dynamics, including time series, phase portraits, Poincaré sections, Liapunov exponents, power spectra, and strange attractors. Initially neighboring solutions for a low–Reynolds-number fully developed turbulence are compared, where the turbulence is sustained by a nonrandom time-independent external force. By reducing the Reynolds number (forcing), several nonturbulent solutions are also obtained and contrasted with the turbulent ones.

I should like to thank all those who helped to make this book a reality. Those include, among others, F.B. Molls, for his work on the calculations, and R.J. Deissler, for helpful discussions in connection with the studies in Chapter 6. E. Reshotko, J. Greber, Y. Kamotani, N. Mhuiris, and F. McCaughan also gave helpful input to Chapter 6. Thanks should go to those at Taylor & Francis and TechBooks who worked on the book and to those who reviewed it.

R. G. Deissler

CHAPTER
ONE

THE PHENOMENON OF FLUID TURBULENCE

Some introductory material on fluid turbulence is presented in Chapter 1. This includes discussions and illustrations of what turbulence is and how, why, and where turbulence occurs.

Consider the steady unidirectional flow of a fluid through a smooth pipe of length L. The pipe is long enough that the flow can be considered fully developed (velocity changes along the pipe are negligible). The flow is incompressible and of constant viscosity. We balance the pressure and shear forces on a cylinder of fluid of radius r and length L and assume that the fluid is Newtonian ($\sigma = -\mu dU/dr$, where σ is the shear stress, μ is the viscosity, and U is the velocity), and that $U = 0$ at the wall. It is then easy to show that the pressure drop Δp along a pipe of diameter D and length L is

$$\Delta p = \frac{128\,\mu QL}{\pi D^4}, \tag{1-1}$$

where Q is the volume of fluid passing any cross-section of the pipe in unit time [1].

Equation (1-1) accurately predicts the pressure drop along a pipe provided the Reynolds number $U_a D/\nu$ is less than about 2000 (U_a is the area-averaged or bulk velocity and ν is the kinematic viscosity). For higher Reynolds numbers, equation (1-1) underestimates the pressure drop, unless care is taken to eliminate disturbances. For instance, for Reynolds numbers of 10^4, 10^5, and 10^6, equation (1-1) underestimates Δp by factors of about 5, 30, and 200, respectively. The deviations from equation (1-1), particularly at high Reynolds numbers, are far from being small. Similar deviations are found if a steady-state analysis is used to predict the heat transfer if a temperature difference occurs between the pipe wall and the fluid.

These deviations first were explained by the classical experiment of Reynolds [2]. He injected dye at the smooth entrance of a glass tube through which water flowed. Results similar to those shown in Fig. 1-1 were obtained. At the low Reynolds number the stream of dye remains intact, indicating that the flow is laminar and unidirectional. At the higher Reynolds number the dye is dispersed, showing that the flow is no longer

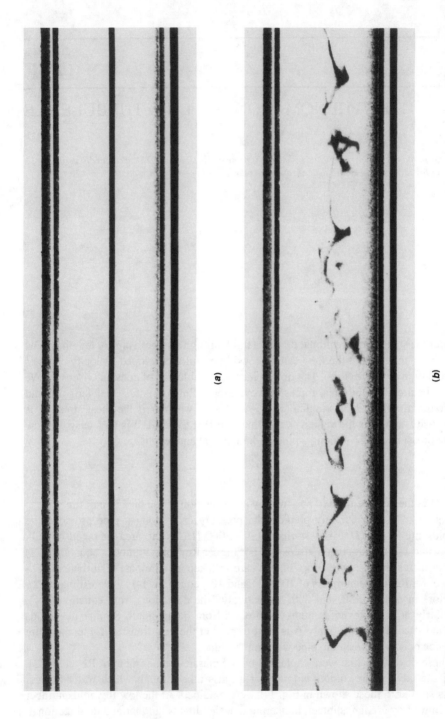

Figure 1-1 Effect of (*a*) low (<2000) and (*b*) higher Reynolds number on ribbon of dye in water flowing through a glass tube. (Photographs by N.H. Johannesen and C. Lowe, reprinted from ref. [3], with permission.)

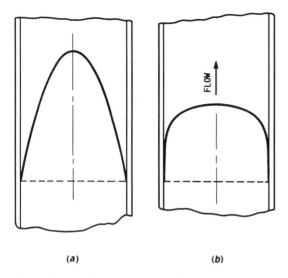

Figure 1-2 Local mean-velocity profiles corresponding to flows in Fig. 1-1: (*a*) low (<2000) and (*b*) higher Reynolds number.

unidirectional. The fluid evidently moves vigorously in the transverse directions, as well as longitudinally, in order to promote the high degree of mixing observed. Thus the high drops in pressure noted in the previous paragraph appear to be caused by lateral momentum transfer, which occurs as the fluid moves transversely. This transverse mixing can produce an effective viscosity that is many times the laminar value, and according to eq. (1-1) (if μ designates an effective viscosity), the pressure drop must increase.

The use of eq. (1-1) to explain the increase in pressure drop at high Reynolds numbers, however, is justified only roughly. Equation (1-1) is derived by assuming that the viscosity (or effective viscosity) is independent of radius. In actuality, the transverse fluid motions must be zero at the solid wall of the pipe, so that the effective viscosity equals the molecular or laminar value at the wall and increases with distance from the wall. Rather than the parabolic constant-viscosity profile shown in Fig. 1-2*a*, the local mean velocity exhibits a profile that is flattened in the center and steep at the wall, as shown in figure 1-2*b*. Because the effective viscosity at the wall is μ, as in laminar flow, the shear stress at the wall is still given by $-\mu \, dU/dr$ but is greater than that for laminar flow, because the velocity gradient at the wall is greater. The pressure drop required to balance the increased shear stress at the wall is then greater than that for laminar flow.

1-1 WHAT IS TURBULENCE?

As suggested by the appearance of the dye in Fig. 1-1*b*, fluid motion at higher Reynolds numbers tends to be haphazard and random. This randomness is seen more clearly in

Fig. 1-3, in which an instantaneous velocity component at a point in a high–Reynolds-number boundary layer is plotted against time. In addition to the random appearance of the plot, as indicated by a visual inspection, it is found that a very small perturbation of the initial velocity fluctuations produces a completely different instantaneous velocity pattern a short time later, although the mean velocity at a point and other mean quantities associated with the flow are not appreciably changed. This is known as *sensitive dependence on initial conditions*. Moreover, a number of scales of motion appear to be present simultaneously in Fig. 1-3, in which small-scale fluctuations are superimposed on the large-scale ones.

Finally, the skewness factor of the velocity derivative of the flow $\overline{(\partial u_1/\partial x_1)^3}/\overline{(\partial u_1/\partial x_1)^2}^{3/2} = S$ is generally negative for nonlinear flows (evidently $0 \geq S > -1$ for turbulent flow [5] (see Fig. 1), where u_1 is the fluctuating velocity component in the x_1 direction and the overbars indicate averaged values. A negative S indicates that, in general, energy is passed from large scales of motion to small ones. The passing of energy among scales of motion could account for the range of length scales in Fig. 1-3. High–Reynolds-number flows with characteristics such as those described here and depicted in Figs. 1-1 through 1-3 are generally called *turbulent*. The apparently random fluctuations that occur in such flows compose what is known as *turbulence*.

Although turbulence generally is characterized as random, it contains a deterministic element. This is because the equations of fluid motion that describe turbulence are deterministic (no random coefficients). Much of the recent work related to turbulence is on something called *deterministic chaos*, which arises in the solutions of deterministic differential equations. The solutions are chaotic because they are extremely sensitive to small changes in initial conditions. In fact, sensitive dependence on initial conditions, which we gave as a characteristic of turbulence, is, strictly speaking, a characteristic of deterministic chaos. However, it has been shown that, at least for typical cases, fluid turbulence is chaotic [6, 7]. It appears reasonable to consider chaoticity as a characteristic of turbulence in general. In order to distinguish between chaos and turbulence, chaos is sometimes designated as having complexity in time and turbulence as having complexity in time and space. More is said about deterministic chaos in a later chapter.

1-2 UBIQUITY OF TURBULENCE

Turbulence phenomena are by no means confined to flows in pipes and boundary layers. In fact, laminar flows in nature, as well as man-made laminar flows, appear to be the exception rather than the rule. For instance the boundary between a column of rising smoke and the surrounding atmosphere is generally irregular and contains a range of scales of motion, thus indicating the presence of turbulence (Fig. 1-4). Similar, perhaps even more striking examples of multiscale turbulence are shown by a volcano erupting into the atmosphere (Fig. 1-5) and by the cloud accompanying a space-shuttle launch (Fig. 1-6). Note the multiscale granular structure of the turbulence in both of those examples. The atmosphere itself is often turbulent, as shown by the irregular appearance of many of the clouds present in it (Figs. 1-7 and 1-8). Jets, wakes, and boundary layers

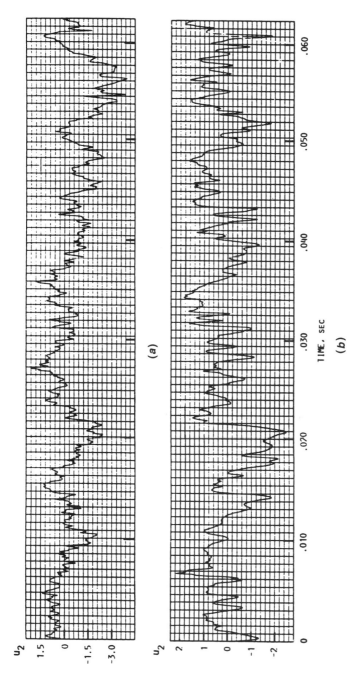

Figure 1-3 Recordings of instantaneous transverse velocity component in a flat-plate turbulent boundary layer at a point (*a*) 0.27 and (*b*) 0.0033 boundary-layer thicknesses from wall. Note that most of the smaller-scale fluctuations are damped out at the point closer to the wall. (Reprinted from ref. [4], with permission.)

6 TURBULENT FLUID MOTION

Figure 1-4 Column of rising turbulent smoke. (Reproduced, with permission, from the Annual Review of Fluid Mechanics, vol. 13, 1981, by Annual Reviews, Inc. (ref. [8]).)

are often turbulent (Figs. 1-9 through 1-12), as are the regions downstream of a grid in a wind tunnel (Fig. 1-13) or downstream of a waterfall. Finally, in order to indicate the all-pervasive character of turbulence, we note that most plasmas and flames, as well as most biological and astrophysical flows (Fig. 1-14), are turbulent.

1-3 WHY DOES TURBULENCE OCCUR?

Before leaving this introductory discussion, it is instructive to attempt to give at least a rough explanation for the occurrence of turbulence. Turbulence occurs as a result of instabilities in a flow. An instability occurs if a perturbation of the flow grows with time. In the examples considered so far, turbulence appears to arise from instabilities produced by layers of fluid sliding over one another or by buoyancy forces.

The former is illustrated in Fig. 1-15. The fluid in the upper part of the sketch moves to the right, that in the lower part to the left. This motion might occur either because of a velocity discontinuity at the central plane or because of a continuous velocity gradient. In the latter case the fluid at the central plane or streamline is stationary. If, for some reason, a portion of fluid moves upward, so that the central streamline becomes convex upward as shown, the velocity of the fluid in the upper part of the figure, as it flows over the curved streamline, tends to be increased (streamlines are closer together), and

Figure 1-5 Turbulent eruption of Mount St. Helens on May 18, 1980. Note the multiscale granular structure of the turbulence in this and the following figure. (Reprinted from ref. [9], with permission.)

that in the lower part decreased (streamlines are farther apart). If we neglect the effect of viscosity, Bernoulli's equation ($U^2/2 + p/\rho = $ const, where U is the velocity and ρ is the density) shows that the pressure above the streamline is reduced and that below it increased, so that the fluid tends to continue moving upward. Similarly, a portion of fluid displaced downward tends to continue downward. That is, the flow is unstable, and conditions favorable for the development of turbulence exist.

Figure 1-16 shows how a density or temperature gradient (density increasing upward) can interact with the downward force of gravity. If a portion of the fluid moves upward (U_3, positive) it will be lighter than the surrounding fluid, and so buoyancy forces tend to cause it to continue moving upward. Similarly, a portion of fluid that moves downward (U_3, negative) tends to continue moving downward, and once again destabilizing forces tend to promote turbulence.

Of course we have not considered the effect of viscosity, and viscous forces may in some cases be stronger than the destabilizing ones. In our earlier example, for instance, we indicated that turbulent flow is sustained in a pipe only for Reynolds numbers greater than about 2000. Thus our simplified analyses do not show that in a particular case turbulence develops. They only point out that there may be destabilizing forces that tend

8 TURBULENT FLUID MOTION

Figure 1-6 Multiscale turbulent cloud at a space shuttle launch. (Reprinted from ref. [10], with permission.)

to produce turbulence. The prediction of whether or not turbulence actually is produced or maintained in a given situation is a much more complex problem and is an active field of research (see, e.g., ref. [15], for transition to turbulence). More is said about the maintenance of turbulence in a later chapter.

1-4 CLOSING REMARKS

Needless to say, the occurrence of turbulence greatly complicates the work of the fluid dynamicist. It appears to be responsible in large measure for maintaining fluid dynamics as an active branch of theoretical physics. In spite of important work in this field carried out over the last century by many eminent scientists (Reynolds [2], Taylor [16],

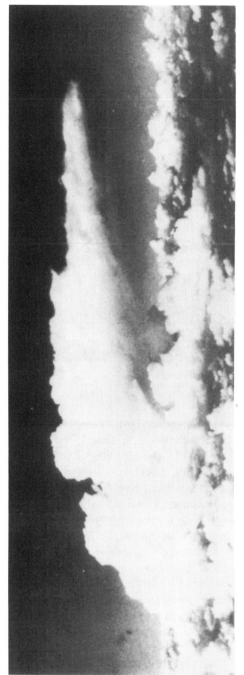

Figure 1.7 Turbulent atmospheric clouds. The shape of the large cloud (anvil) is due to rising air masses with a superimposed wind shear. (Reprinted from ref. [11], with permission.)

10 TURBULENT FLUID MOTION

Figure 1-8 Turbulent Earth (NASA photograph).

Prandtl [17], Heisenberg [18], Von Kármán [19], Kampé DeFeriet [20], and Kolmogorov [21], to name a few), much work remains to be done.

The difficulty in the problem of turbulence lies not so much in determining which equations are appropriate as in using those equations to study turbulence. The required equations of motion, the Navier-Stokes and other fluid equations, are already known (see Chapter 3). Because of their nonlinearity, exact analytical solutions, however, have been obtained only in a few cases (usually linear cases). Nevertheless, as is seen subsequently, those solutions and their interpretation can form the basis for studying many processes occurring in turbulence. For more complicated cases, approximate analyses that are only partially based on the Navier-Stokes equations have been used. Recently, however, the nonlinear problem of turbulence, at least for lower Reynolds numbers, has been studied by obtaining numerical solutions of the full time-dependent equations of motion. It appears that numerical methods can greatly increase the applicability of the Navier-Stokes and other fluid equations to turbulence, and it may be that future progress will come about mainly through the use of those methods.

The present study of turbulence is couched in Cartesian-tensor notation, because the use of that notation greatly simplifies the analysis. Thus tensors and tensor notation are considered in the next chapter.

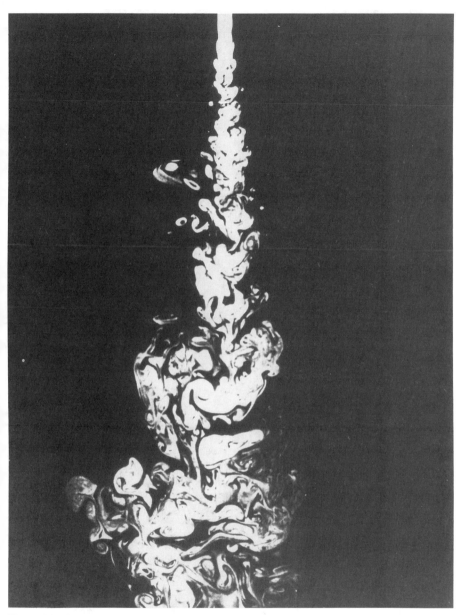

Figure 1-9 Axisymmetric water jet becoming turbulent as it is directed downward into water; Reynolds number ∼ 2000. (Photograph by P.E. Dimotakis, R.C. Lye, and D.Z. Papantoniou; reprinted from ref. [12], with permission.)

Figure 1-10 Axisymmetric jet of air becoming turbulent; Reynolds number ~ 10,000. Note that the turbulence here has a finer structure than that at the lower Reynolds number in the preceding Figure. (Photograph by R. Drubka and H. Nagib; reprinted from ref. [3], with permission.)

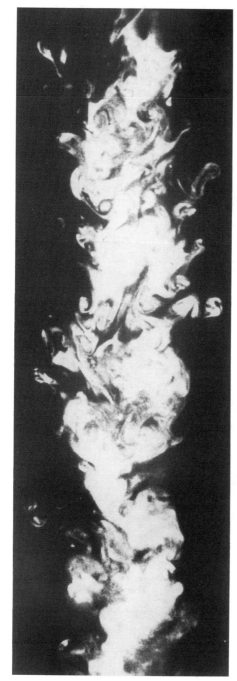

Figure 1-11 Turbulent wake of a cylinder. (Photograph by R. E. Falco; reprinted from ref. [3], with permission.)

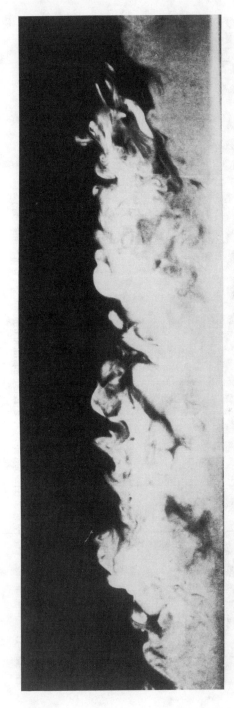

Figure 1-12 Turbulent boundary layer on the floor of a wind tunnel. (Reprinted from ref. [13], with permission.)

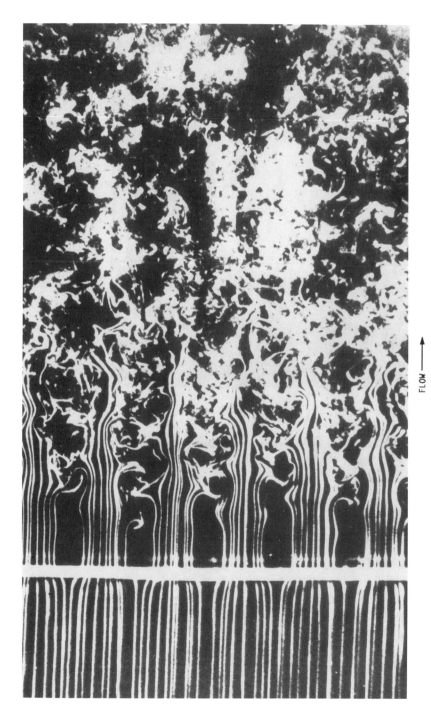

Figure 1-13 Generation of turbulence downstream from a grid. (Photograph by T. Corke and H. Nagib; reprinted from ref. [3], with permission.)

16 TURBULENT FLUID MOTION

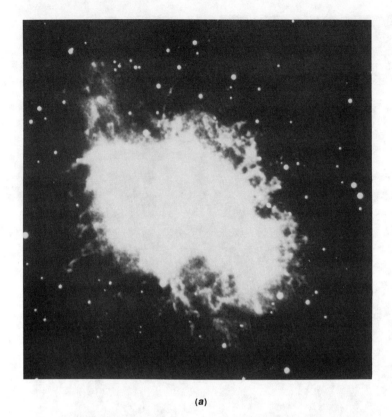

(a)

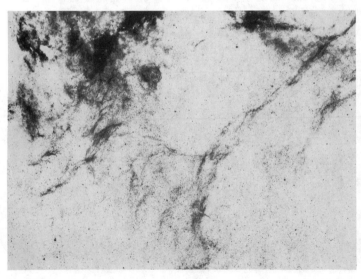

(b)

Figure 1-14 Turbulent astrophysical clouds. (*a*) Crab Nebula (NASA Photograph, NASA EP-167). (*b*) Region in Milky Way (Reprinted from ref. [14], with permission).

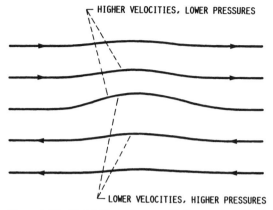

Figure 1-15 Sketch illustrating unstable flow produced by shear. A small perturbation tends to grow.

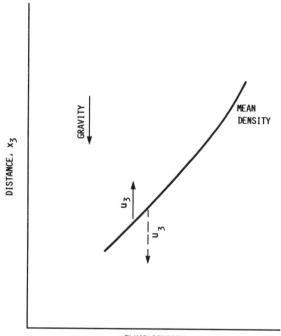

Figure 1-16 Sketch illustrating buoyancy-produced instability. A small perturbation either upward (u_3, positive) or downward (u_3, negative) tends to grow.

REFERENCES

1. Goldstein, S., *Modern Developments in Fluid Dynamics*, vol. 1, pp. 19–21, Clarendon Press, Oxford, England, 1938.
2. Reynolds, O., "An Experimental Investigation of the Circumstances Which Determine Whether the Motion of Water Shall be Direct or Sinuous, and the Law of Resistance in Parallel Channels," *Philos. Trans. R. Soc. London*, vol. 174, pp. 935–982, 1883.

3. Van Dyke, M., *An Album of Fluid Motion*, Parabolic Press, Stanford, CA, 1982.
4. Frenkiel, F.N., and Klebanoff, P.S., "Probability Distributions and Correlations in a Turbulent Boundary Layer," *Phys. Fluids*, vol. 16, no. 6, pp. 725–737, 1973.
5. Tavoularis, S., Bennett, J.C., and Corrsin, S., "Velocity-Derivative Skewness in Small Reynolds Number, Nearly Isotropic Turbulence," *J. Fluid, Mech.*, vol. 88, pt. 1, pp. 63–69, 1978.
6. Deissler, R.G., "Is Navier-Stokes Turbulence Chaotic?" *Phys. Fluids*, vol. 29, no. 5, pp. 1453–1457, 1986.
7. Brandstäter, A., Swift, J., Swinney, H.L., Wolf, A., Farmer, J.D., Jen, E., and Crutchfield, J.P., "Low-Dimensional Chaos in a Hydrodynamic System," *Phys. Rev. Lett.*, vol. 51, no. 16, pp. 1442–1445, 1983.
8. Perry, A.E., Lim, T.T., Chong, M.S., and Teh, E.W., "The Fabric of Turbulence," AIAA Paper 80-1358, July 1980.
9. Broad, W., "Threat to U.S. Airpower: The Dust Factor," *Science*, vol. 213, p. 1476, 1981.
10. McGowan, J., ed., [Cover photograph], *NASA Activities*, vol. 13, no. 5, 1982.
11. Iribarne, J.V., and Cho, H.R., *Atmospheric Physics*, D. Reidel Publishing Company, p. 101, 1980.
12. Fiszdon, W., Kucharczyk, P., and Prosnak, W.J., eds., *Proceedings of the Fifteenth International Symposium on Fluid Dynamics*, Polish Scientific *Publishers*, Warszaw, Poland, 1981.
13. Falco, R.E., "Coherent Motions in the Outer Region of Turbulent Boundary Layers," *Phys. Fluids*, vol. 20, no. 10, pt. II, pp. S124–S132, 1977.
14. Minkowski, R., "The Observational Background of Cosmical Gasdynamics," in *Gasdynamics of Cosmic Clouds*, Interscience Publishers Inc., New York; North Holand Publishing Company, Amsterdam; Elsevier Science Publishing Company, New York, p. 10, 1955.
15. Orszag, S.A., and Kells, L.C., "Transition to Turbulence in Plane Poiseuille and Plane Couette Flow," *J. Fluid Mech.*, vol. 96, pt. 1, pp. 159–205, 1980.
16. Taylor, G.I., "Statistical Theory of Turbulence, Parts I–IV," *Proc. R. Soc. A*, vol. 151, pp. 421–478, 1935.
17. Prandtl, L., "Bericht Über Untersuchungen Zur Ausgebildeten Turbulenz," *Z. Angew. Math. Mech.*, vol. 5, no. 2, pp. 136–139, 1925.
18. Heisenberg, W., "Zur Statistischen Theorie der Turbulenz," *Z. Phys. A.*, vol. 124, pp. 628–657, 1948.
19. Von Kármán, T., "The Fundamentals of the Statistical Theory of Turbulence," *J. Aeronaut. Sci.*, vol. 4, no. 4, pp. 131–138, 1937.
20. Kampé DeFeriet, J., "Some Recent Researches on Turbulence," in *Proceedings of the Fifth International Congress for Applied Mechanics*, edited by J.P. Den Hartog and M. Peters, p. 352, Wiley, 1939.
21. Kolmogorov, A.N., "Local Structure of Turbulence in an Incompressible Fluid at Very High Reynolds Numbers," *Dokl. Akad Nauk SSSR*, vol. 30, no. 4, pp. 299–303, 1941.

CHAPTER
TWO

SCALARS, VECTORS, AND TENSORS

Some of the mathematical apparatus used for the representation and study of turbulence is developed.

2-1 INTRODUCTION

In the study of turbulence one encounters physical quantities such as densities, temperatures, forces, velocities, velocity products, and stresses. We notice that these quantities are not all of the same type. For instance, a density requires only one number to represent it, but a velocity requires three numbers, or it has three components. A quantity that is the product of two velocities, for which each component of one velocity is multiplied in turn by each component of the other, has nine components.

A density is an example of a scalar, or of a *tensor of order zero*. It can be represented by a symbol with no subscripts, say ρ. A velocity is an example of a vector and is called a *tensor of order one*. It can be represented by a symbol with one subscript, say u_i, where $i = 1, 2$, or 3. The three values of i correspond to directions in space parallel to the directions of the three perpendicular coordinate axes x_i ($i = 1, 2, 3$). The x_i form a right-handed coordinate system and are often written as (x, y, z). A quantity that is the product of two velocities is an example of a *second-order tensor*. It can be represented by a symbol with two subscripts, say $T_{ij} = u_i u_j$, where $i = 1, 2, 3$ and $j = 1, 2, 3$. Similarly, products of more than two velocities are tensors of higher order. Thus $u_i u_j u_k$ ($i, j, k = 1, 2$, or 3) is a third-order tensor, and so forth. Averaged values of velocity products, written as $\overline{u_i u_j}$, $\overline{u_i u_j u_k}$, and so forth and called *velocity correlations* (the overbars indicate averaged values), are important quantities in the theory of turbulence.

2-2 ROTATION OF COORDINATE SYSTEMS

It should not be assumed that all quantities represented by symbols with a given number of subscripts are tensors as in the examples of the last paragraph (see, e.g., section 2-8).

To be called a tensor, it is necessary that a quantity obey a certain transformation law if referred to a rotated coordinate system.[1] In considering the transformation laws of tensors, we first note that a rectangular coordinate system transforms under a rotation according to the law

$$x_i^* = a_{ij} x_j, \tag{2-1}$$

where x_i^* is a coordinate of a point in the rotated coordinate system x_i^*, and x_i is a coordinate of the same point in the unrotated system x_i. (Note that x_i^* or x_i can designate either a coordinate system or a coordinate, because a coordinate system is given by its coordinates.) The a_{ij} form a set of nine constants. The conditions under which eq. (2-1) makes sense as a transformation law are considered later in this section.

Equation (2-1) uses the Einstein summation convention, by which a repeated subscript in a term designates a sum of terms, with the subscript successively taking on the values 1, 2, and 3. This convention is used throughout the text, except if otherwise indicated. Note that the symbol used for a repeated subscript is immaterial, so that such a subscript is often called a *dummy subscript*. The expression for x_1^*, written out, is

$$x_1^* = a_{11} x_1 + a_{12} x_2 + a_{13} x_3.$$

The symbols used for the subscripts that are unrepeated in a term of an equation (see, e.g., eq. [2-1]) are also immaterial, so long as the same symbols are used in all terms, and so long as they differ from those used for other subscripts in the equation. Thus, in substituting one equation into another, the symbols used for some of the subscripts must frequently be changed in order to avoid confusion. If a subscript occurs more than twice in a term the equation is generally ambiguous. Also, the same unrepeated subscripts must occur in all terms of an equation. These points, although possibly obvious once they have been mentioned, are important for carrying out meaningful tensor manipulations.

The square of the distance ds between two neighboring points is given in the x_i coordinate system by

$$ds^2 = dx_i dx_i = \delta_{ij} dx_i dx_j \tag{2-2}$$

where δ_{ij} is called the *Kronecker delta*, defined by

$$\begin{aligned} \delta_{ij} &= 1 \quad \text{for} \quad i = j, \\ \delta_{ij} &= 0 \quad \text{for} \quad i \neq j. \end{aligned} \tag{2-3}$$

The truth of eq. (2-2) can be seen by writing out the three terms of $dx_i dx_i$ (summation on i) and the nine terms of $\delta_{ij} dx_i dx_j$ (summation on i and j) and using (2-3). Note that $\delta_{ij} = \delta_{ji}$. In order that the transformation (2-1) make physical or geometrical sense, it is necessary that the distance ds be the same in the x_i^* and the x_i coordinate systems. Thus

$$\begin{aligned} ds^2 = dx_i^* dx_i^* &= a_{ij} a_{ik} dx_j dx_k \\ &= \delta_{jk} dx_j dx_k \end{aligned} \tag{2-4}$$

---

[1] We shall consider here only rectangular Cartesian coordinate systems. Tensors defined in terms of the transformation laws of such coordinate systems are called *Cartesian tensors*. When we use the term "tensor" in this book, "Cartesian tensor" is understood.

where eq. (2-1) in differential form ($dx_i^* = a_{ij}dx_j$) is applied twice, and eq. (2-2) is used. Equation (2-4) gives

$$(a_{ij}a_{ik} - \delta_{jk})dx_j dx_k = 0$$

for all values of the dx_i, or

$$a_{ij}a_{ik} = \delta_{jk}. \tag{2-5}$$

Equation (2-5) gives the nine relations (six of which are different) that must be satisfied by the a_{ij} if eq. (2-1) is to be a sensible transformation law (lengths remain invariant under the transformation). Multiplication of eq. (2-1) by a_{ik} and use of eqs. (2-5) and (2-3) give

$$a_{ik}x_i^* = a_{ik}a_{ij}x_j = \delta_{jk}x_j = x_k, \tag{2-6}$$

where the last step can be verified by writing out the terms and using eq. (2-3). In general, multiplication of a quantity containing, say, a subscript j or a subscript k (or both) by δ_{jk} changes the subscript j to k, or k to j, in that quantity.

Considering only the first and last terms of eq. (2-6), and changing the subscript k to j on both sides of the equation, gives

$$x_j = a_{ij}x_i^*. \tag{2-7}$$

If we differentiate eq. (2-1) with respect to x_j and eq. (2-7) with respect to x_i^*, we get

$$\frac{\partial x_i^*}{\partial x_j} = \frac{\partial x_j}{\partial x_i^*} = a_{ij}. \tag{2-8}$$

A set of equations equivalent to (2-5), but of slightly different form, can be obtained as follows:

$$ds^2 = dx_i dx_i = a_{ji}dx_j^* a_{ki}dx_k^* = dx_i^* dx_i^* = \delta_{jk}dx_j^* dx_k^*$$

where eq. (2-7) in differential form is used. Therefore

$$a_{ji}a_{ki} = \delta_{jk}. \tag{2-5a}$$

2-3 VECTORS (FIRST-ORDER TENSORS)

A quantity u_i is said to be a vector, or a first-order tensor, if it obeys the transformation law

$$u_i^* = a_{ij}u_j, \tag{2-9}$$

where u_i^* is a component of a vector in the rotated coordinate system x_i^* and u_i is a component of the same vector in the unrotated system x_i. (Note that u_i or u_i^*) can designate either a vector component or the vector itself, because a vector can be specified by specifying its components.)

We note by comparing eqs. (2-1) and (2-9) that a vector transforms according to the same law as the coordinates of a point. Thus the coordinates of a point form a vector. They define a directed-line segment drawn from the origin to the point x_i. That vector is usually called a *position* or *displacement* vector. Similarly, any vector defined by eq. (2-9) can be interpreted as a directed-line segment. So the definition given by eq. (2-9) agrees with the perhaps more familiar definition of a vector as a directed-line segment, or as a quantity with both magnitude and direction (displacement, velocity, area, force, etc.).

It is easy to show that the sum or difference of two vectors, say u_i and v_i, is a vector. For, from eq. (2-9),

$$u_i^* + v_i^* = a_{ij}u_j + a_{ij}v_j = a_{ij}(u_j + v_j), \qquad (2\text{-}10)$$

which shows that $u_i + v_i$ obeys the transformation law for a vector. Equation (2-10) is another way of stating the familiar addition law for vectors represented by directed-line segments; instead of adding directed-line segments geometrically, we add corresponding components in either the x_i or x_i^* coordinate system.

2-4 SECOND-ORDER TENSORS

2-4-1 Definition and Simple Examples

A second-order tensor u_{ij} is defined by generalization of eq. (2-9) as a quantity that obeys the transformation law

$$u_{ij}^* = a_{ik}a_{jl}u_{kl} \qquad (2\text{-}11)$$

where, as usual, repeated subscripts (in this case, k and l) indicate summations. Thus eq. (2-11) represents nine equations, each with nine terms on the right side. Tensor notation affords, among other things, considerable economy in writing.

From the definition given by eq. (2-11) it follows that the product of two vectors u_i and v_j is a second-order tensor. For, from eq. (2-9),

$$u_i^* v_j^* = a_{ik}a_{jl}u_k v_l, \qquad (2\text{-}12)$$

which shows that $u_i v_j$ obeys the transformation law for a second-order tensor (eq. [2-11]) The product $u_i v_j$, where the subscript on one vector is not repeated on the other one, is called an *outer product*.

Another example of a second-order tensor is the spatial derivative, or gradient, of a vector $\partial u_i/\partial x_j$. We can show that $\partial u_i/\partial x_j$ is a second-order tensor as follows: Following the rules of partial differentiation, we obtain

$$\frac{\partial u_i^*}{\partial x_j^*} = \frac{\partial u_i^*}{\partial x_k}\frac{\partial x_k}{\partial x_j^*} = \frac{\partial x_k}{\partial x_j^*}\frac{\partial}{\partial x_k}(a_{il}u_l), \qquad (2\text{-}13)$$

where u_i^* is assumed to be a function of x_k, and eq. (2-9) is used. Using eq. (2-8) for $\partial x_k/\partial x_j^*$ we get

$$\frac{\partial u_i^*}{\partial x_j^*} = a_{il}a_{jk}\frac{\partial u_l}{\partial x_k}, \qquad (2\text{-}14)$$

which shows that $\partial u_i/\partial x_j$ obeys the transformation law (2-11) and is thus a second-order tensor.

2-4-2 Stress and the Quotient Law

Still another second-order tensor is the stress σ_{ij} defined by

$$\sigma_{ij}\Delta A_{(i)} = \Delta F_j \quad \text{(no sum on } i\text{)}, \tag{2-15}$$

where ΔF_j is the force component in the x_j direction acting on the small area element ΔA_i, whose normal is in the x_i direction.

To show that σ_{ij} is a tensor, first write a sum of forces of the type ΔF_j:

$$\sigma_{ij}\Delta A_i = \Delta F'_j \quad \text{(sum on } i\text{)} \tag{2-16}$$

A product of two tensors such as that in eq. (2-16), in which a subscript is repeated, is called an *inner product*. But the area ΔA_i (where we designate the area by its components) is a vector, because it can be represented by a directed-line segment normal to the plane of the area (it has magnitude and direction). Similarly the force ΔF_j has magnitude and direction and is thus a vector. The quantity $\Delta F'_j$ is also a vector, because it is a sum of vectors of the type ΔF_j (eq. [2-10]).

Next write eq. (2-16) in the transformed coordinate system x_i^*:

$$\sigma_{ij}^* \Delta A_i^* = \Delta F_j'^*. \tag{2-17}$$

Because ΔA_i and $\Delta F'_j$ are vectors, we have, according to eq. (2-9),

$$\sigma_{ij}^* a_{ik} \Delta A_k = a_{jk} \Delta F'_k$$
$$= a_{jk}\sigma_{ik}\Delta A_i = a_{ji}\sigma_{ki}\Delta A_k,$$

where eq. (2-16) is used in the next-to-last term, and the dummy subscripts i and k are interchanged in the last term. Then, from the first and last terms,

$$\Delta A_k \left(\sigma_{ij}^* a_{ik} - a_{ji}\sigma_{ki}\right) = 0. \tag{2-18}$$

Because eq. (2-18) holds for all values of ΔA_k,

$$a_{ik}\sigma_{ij}^* = a_{ji}\sigma_{ki}. \tag{2-19}$$

To get this equation into the form of eq. (2-11), multiply it by a_{lk}, or

$$a_{ik}a_{lk}\sigma_{ij}^* = a_{ji}a_{lk}\sigma_{ki}.$$

Finally, using eq. (2-5a),

$$\delta_{il}\sigma_{ij}^* = a_{ji}a_{lk}\sigma_{ki}$$

or

$$\sigma_{lj}^* = a_{lk}a_{ji}\sigma_{ki}. \tag{2-20}$$

Comparing eq. (2-20) with (2-11) shows that σ_{ij} is a second-order tensor.

We have shown, starting from eq. (2-16), that if inner multiplication of a quantity σ_{ij} by a vector with arbitrary components gives another vector, then σ_{ij} is a second-order tensor. This is one form of the quotient law. Once it has been established, as has been done here, it provides in some cases a simple test for determining whether a quantity is a tensor. Further discussion of the quotient law is given in section 2-5-2.

2-4-3 Kronecker Delta, a Tensor

In the foregoing paragraphs we showed that a product of two velocities (or vectors), a gradient of a velocity, and a stress, although representing different physical entities, are all alike in that they are second-order tensors. Next we ask whether the Kronecker delta δ_{ij} is a second-order tensor. According to the definition in eq. (2-3) the components of δ_{ij} do not depend on the orientation of coordinate axes. Thus

$$\delta_{ij}^* = \delta_{ij}.$$

Using eq. (2-5a) and (2-3),

$$\delta_{ij}^* = \delta_{ij} = a_{ik}a_{jk} = a_{ik}a_{jl}\delta_{kl}. \tag{2-21}$$

Comparing the first and last members of eqs. (2-21) and (2-11) shows that δ_{ij} is a second-order tensor.

The Kronecker delta δ_{ij} is an example of an *isotropic tensor*. That is, its components remain invariant with rotation of coordinate axes. An isotropic tensor is sometimes called a *numerical tensor*, because its components have the same numerical values for all rotations of the coordinate axes.

We now show that the most general second-order isotropic tensor is $I\delta_{ij}$, or that any second-order isotropic (numerical) tensor can be written as $I\delta_{ij}$, where I is a scalar. The transformation law for a second-order tensor is given by eq. (2-11). Let I_{ij} be any (the most general) second-order isotropic tensor, so that $I_{ij}^* = I_{ij}$. Then eq. (2-11) becomes

$$I_{ij}^* = I_{ij} = a_{ik}a_{jl}I_{kl}.$$

Multiplying the last two members of this equation by a_{jm} and using eq. (2-5) give

$$a_{jm}I_{ij} = a_{ik}a_{jl}a_{jm}I_{kl} = a_{ik}\delta_{lm}I_{kl} = a_{ik}I_{km}.$$

The first and last members of this equation give

$$\delta_{km}a_{jk}I_{ij} = \delta_{ij}a_{jk}I_{km},$$

or

$$a_{jk}(\delta_{km}I_{ij} - \delta_{ij}I_{km}) = 0.$$

Because the relation for I_{ij} cannot depend on the a_{jk} (on the orientation of the coordinate axes), the quantity in parentheses is zero,[2] and we get, after contracting the indices k and m (setting $m = k$),

$$I_{ij} = (I_{kk}/3)\delta_{ij}$$

where I_{kk} is a scalar (see section 2-6). Any value of I_{kk} satisfies this equation, as can be seen by contracting the indices i and j, so that

$$I_{ij} = I\delta_{ij}.$$

That is, any (the most general) second-order isotropic tensor can be written as $I\delta_{ij}$, where I is an arbitrary scaler.

---

[2] If, however, I_{ij} were not an isotropic tensor, then $a_{jk}(\delta_{km}I_{ij}^* - \delta_{ij}I_{km}) = 0$, and we could not set the quantity in parentheses equal to zero; the relation between I_{ij}^* and I_{km} would not be independent of the a_{kj}.

2-5 THIRD- AND HIGHER-ORDER TENSORS

The generalization of eq. (2-11) to tensors of higher order is obvious. For instance a third-order tensor u_{ijk} is defined by

$$u^*_{ijk} = a_{il} a_{jm} a_{kn} u_{lmn} \qquad (2\text{-}22)$$

which represents 27 equations, each with 27 terms on the right side. An example of a third-order tensor is the product of three velocities $u_i u_j u_k$. The product of four velocities forms a fourth-order tensor $u_i u_j u_k u_l$, and so forth.

2-5-1 Vorticity and the Alternating Tensor

An important third-order tensor is the alternating tensor ε_{ijk}, where

$$\varepsilon_{ijk} = \delta_{i1}\delta_{j2}\delta_{k3} + \delta_{i2}\delta_{j3}\delta_{k1} + \delta_{i3}\delta_{j1}\delta_{k2} - \delta_{i1}\delta_{j3}\delta_{k2} - \delta_{i2}\delta_{j1}\delta_{k3} - \delta_{i3}\delta_{j2}\delta_{k1}. \qquad (2\text{-}23)$$

(We call ε_{ijk} a tensor in anticipation of showing that it is such later in this section.) Evaluation of eq. (2-23) shows that

$$\varepsilon_{ijk} = 0 \quad \text{if two subscripts are equal,}$$
$$\varepsilon_{123} = \varepsilon_{231} = \varepsilon_{312} = 1, \qquad (2\text{-}23a)$$
$$\varepsilon_{132} = \varepsilon_{321} = \varepsilon_{213} = -1.$$

The alternating tensor ε_{ijk} is usually defined by eq. (2-23a), but eq. (2-23) is more convenient for our purposes (and may generally be preferable).

We define a quantity ω_i as follows:

$$\omega_i = \varepsilon_{ijk} \frac{\partial u_k}{\partial x_j}, \qquad (2\text{-}24)$$

where u_k is a vector and x_j is a coordinate (also a vector). Equation (2-24) becomes, on using (2-23),

$$\omega_i = (\delta_{i1}\delta_{j2}\delta_{k3} - \delta_{i1}\delta_{j3}\delta_{k2} + \delta_{i2}\delta_{j3}\delta_{k1} - \delta_{i2}\delta_{j1}\delta_{k3} + \delta_{i3}\delta_{j1}\delta_{k2} - \delta_{i3}\delta_{j2}\delta_{k1})\frac{\partial u_k}{\partial x_j}$$

$$= \delta_{i1}\left(\frac{\partial u_3}{\partial x_2} - \frac{\partial u_2}{\partial x_3}\right) + \delta_{i2}\left(\frac{\partial u_1}{\partial x_3} - \frac{\partial u_3}{\partial x_1}\right) + \delta_{i3}\left(\frac{\partial u_2}{\partial x_1} - \frac{\partial u_1}{\partial x_2}\right). \qquad (2\text{-}25)$$

Therefore, the three components of the quantity ω_i are

$$\omega_1 = \frac{\partial u_3}{\partial x_2} - \frac{\partial u_2}{\partial x_3},$$

$$\omega_2 = \frac{\partial u_1}{\partial x_3} - \frac{\partial u_3}{\partial x_1}, \qquad (2\text{-}26)$$

$$\text{and} \quad \omega_3 = \frac{\partial u_2}{\partial x_1} - \frac{\partial u_1}{\partial x_2}.$$

If u_i is the velocity at a point in a fluid, ω_i is called the *vorticity* and is a measure of the local swirl or rotation. Equations (2-26) show that each component ω_i has the magnitude of the rotation of the fluid in an $x_j - x_k$ plane ($j, k \neq i$) and is perpendicular to that plane. Thus the quantity ω_i has both magnitude and direction and so is a vector. The vector ω_i (or $\boldsymbol{\omega}$) is also called the curl of u_i and, in vector notation, is written as $\nabla \times \boldsymbol{u}$. The quantity $\partial u_k/\partial x_j$ in eq. (2-24) is the gradient of a vector and, as shown in section 2-4-1, is a second-order tensor. Because ω_i is a vector, ε_{ijk} is a third-order tensor by a slightly more general form of the quotient law than that in section 2-4-2 (see the next section). By eq. (2-23) it is isotropic (or numerical), because its components remain invariant with rotation of coordinate axes.

2-5-2 A More General Quotient Law

In order to prove the quotient law used in the last section, we note that eq. (2-24) is of the form

$$v_i = v_{ijk} v_{kj}, \tag{2-27}$$

where v_i is a vector and v_{kj} is an arbitrary second-order tensor (its components can have arbitrary values). We want to prove that v_{ijk} is a third-order tensor. In a rotated coordinate system eq. (2-27) becomes

$$v_i^* = v_{ijk}^* v_{kj}^*, \tag{2-28}$$

or, because v_i and v_{kj} are tensors, we can write, using eqs. (2-9) and (2-11),

$$a_{il} v_l = v_{ijk}^* a_{kl} a_{jm} v_{lm}.$$

Substituting for v_l from eq. (2-27),

$$v_{ijk}^* a_{kl} a_{jm} v_{lm} - a_{il} v_{ljk} v_{kj} = 0. \tag{2-29}$$

We factor out v_{lm} after making a change of dummy subscripts in the second term of eq. (2-29). This gives

$$\left(a_{kl} a_{jm} v_{ijk}^* - a_{in} v_{nml}^*\right) v_{lm} = 0.$$

Because v_{lm} is arbitrary, the quantity in parentheses is zero, or

$$a_{kl} a_{jm} v_{ijk}^* = a_{in} v_{nml}.$$

To get this equation into the form of eq. (2-22) we multiply it by $a_{rl} a_{qm}$, or

$$a_{kl} a_{rl} a_{jm} a_{qm} v_{ijk}^* = a_{in} a_{qm} a_{rl} v_{nml}. \tag{2-30}$$

Finally, using eq. (2-5a) on the left side of (2-30) gives

$$\delta_{kr} \delta_{jq} v_{ijk}^* = a_{in} a_{qm} a_{rl} v_{nml}$$

or

$$v_{iqr}^* = a_{in} a_{qm} a_{rl} v_{nml}. \tag{2-31}$$

Comparison of eq. (2-31) with (2-22) shows that v_{ijk} is a third-order tensor. Thus, if in eq. (2-27) v_i is a vector (first-order tensor) and v_{kj} is an arbitrary tensor, then v_{ijk} is a tensor. This proves a rather general form of the quotient law. As mentioned earlier, a product such as that in eq. (2-27), in which a subscript or subscripts in one factor is repeated in the other one, is called an inner product. In general the quotient law states that a quantity is a tensor if an inner multiplication of that quantity with an arbitrary tensor (its components can have arbitrary values) is itself a tensor.

2-6 ZERO-ORDER TENSORS AND CONTRACTION

We notice in the definitions of first-, second-, and third-order tensors (eqs. [2-9], [2-11], and [2-22]), that the number of a_{ij} in the transformation law equals the order of the tensor. Also, of course, the number of subscripts on a tensor equals its order. Thus for a tensor of order zero or a scalar, say u, we should have in place of eq. (2-9) or (2-11),

$$u^* = u. \tag{2-32}$$

So a zero-order tensor, or a scalar, has the same value in the coordinate system x_i^* as in x_i. For that reason it is often called an *invariant*. Because u in eq. (2-32) can be any unsubscripted quantity, we can say that any unsubscripted quantity is a tensor of order zero.

Multiplication of a tensor by a scalar gives a tensor of the same order. For instance multiplication of eq. (2-11) (for a second-order tensor) by a scalar u gives

$$u^* u_{ij}^* = a_{ik} a_{jl} u u_{kl}, \tag{2-33}$$

where eq. (2-32) is used. Thus $u u_{kl}$ transforms as a second-order tensor.

In the second-order tensor u_{ij} we can set $j = i$. That process is called *contraction*. Then, according to eq. (2-11), we have

$$u_{ii}^* = a_{ik} a_{il} u_{kl} = \delta_{kl} u_{kl} = u_{kk} = u_{ii} \tag{2-34}$$

where eq. (2-5) is used. Comparison of eq. (2-34) with (2-32) shows that u_{ii} is a zero-order tensor, or a scalar. Thus, contraction of the subscripts i and j lowered the order of the tensor by two. In general the process of contraction lowers the order of a tensor by two. As another example we contract the second-order tensor $\partial u_i/\partial x_j$ to form the scalar $\partial u_i/\partial x_i$, which is called the divergence of u_i, and which, according to eqs. (2-14) and (2-5) is a scalar. In vector notation the divergence of u_i is written as $\nabla \cdot \mathbf{u}$. As a final example of contraction, contract the second-order tensor $u_i v_j$ to form $u_i v_i$. The quantity $u_i v_i$ is a scalar, because

$$u_i^* v_i^* = a_{ik} a_{il} u_k v_l = \delta_{kl} u_k v_l = u_k v_k = u_i v_i. \tag{2-35}$$

It is called the *dot* or inner product of the vectors $\mathbf{u}$ and $\mathbf{v}$ and is often written as $\mathbf{u} \cdot \mathbf{v}$.

We can show that the gradient of a scalar, say u, is a vector. For, proceeding as in obtaining eqs. (2-13) and (2-14),

$$\frac{\partial u^*}{\partial x_j^*} = \frac{\partial u^*}{\partial x_k} \frac{\partial x_k}{\partial x_j^*} = a_{jk} \frac{\partial u}{\partial x_k}, \tag{2-36}$$

which is the transformation law for a vector. In vector notation the gradient of u is written as ∇u.

2-7 OUTER AND INNER PRODUCTS OF TENSORS OF HIGHER ORDER

It is shown in sections 2-4-1 and 2-6 respectively that outer and inner products of vectors are tensors of some order. It is straightforward to show that outer and inner products of tensors of any order are tensors. For example the outer product $u_i u_{jk}$ is a third-order tensor, because

$$u_i^* u_{jk}^* = a_{il} a_{jm} a_{kn} u_l u_{mn}. \tag{2-37}$$

Also, the inner product $u_i u_{ik}$ is a first-order tensor (vector), because, using eq. (2-5),

$$u_i^* u_{ik}^* = a_{il} a_{im} a_{kn} u_l u_{mn} = \delta_{lm} a_{kn} u_l u_{mn}$$
$$= a_{kn} u_m u_{mn}.$$

2-8 SUBSCRIPTED QUANTITIES THAT ARE NOT TENSORS

In order to give a better understanding of what tensors are, we give here some examples of quantities that, although subscripted, are not tensors. First consider the quantity a_{ij}. Recall that the a_{ij} form a set of nine constants defined by eq. (2-1). Assume first that a_{ij} is a second-order tensor. Then

$$a_{ij}^* = a_{ik} a_{jl} a_{kl} = a_{ik} \delta_{jk} = a_{ij}, \tag{2-38}$$

where eqs. (2-11), (2-5a), and (2-3) are used. Thus if a_{ij} is a tensor, it must be isotropic. We show in section 2-4-3, however, that the most general second-order isotropic tensor is $I\delta_{ij}$, where I is a scalar. So a_{ij} is a tensor only if it is equal to $I\delta_{ij}$. If $a_{ij} = I\delta_{ij}$, eq. (2-1) becomes

$$x_i^* = I\delta_{ij} x_j = I x_i.$$

In contrast to the statement of this equation, however, the x_i^* are not proportional to the x_i for arbitrary rotations, so that $a_{ij} \neq I\delta_{ij}$, and according to our argument a_{ij} is not a tensor.

We show in the previous paragraph that a_{ij} is not a tensor, because $a_{ij} \neq I\delta_{ij}$. More generally we can say that any quantity h_{ij} whose components have the same numerical values in all rectangular coordinate systems is a nontensor if $h_{ij} \neq I\delta_{ij}$, because $I\delta_{ij}$ is the most general isotropic (numerical) tensor (see section 2-4-3).

As another example consider a quantity $w_i = u_{(i)} v_i$ (no sum on i), where u_i and v_i are both vectors. One might imagine that $w_i = u_{(i)} v_i$ is also a vector (first-order tensor), because it has one assignable subscript. However, $u_i^* = a_{ij} u_j$ and $v_i^* = a_{ik} v_k$, so that

$$w_i^* = u_{(i)}^* v_i^* = a_{(i)j} a_{ik} u_j v_k. \tag{2-39}$$

If $w_i = u_{(i)} v_i$ were a first-order tensor it would transform as $w_i^* = a_{ij} w_j = a_{ij} u_{(j)} v_j$, which is considerably different from eq. (2-39). Because both of these expressions cannot be true, $w_i = u_{(i)} v_i$ is not a tensor. Similarly, quantities such as $w_{ij} = u_{(i)} u_{ij}$ and $w_{ijk} = u_{(ij)} u_{ijk}$ are not tensors. Note that all of these quantities are products that are neither inner nor outer.

2-9 CLOSING REMARKS

It is shown (eq. [2-10]) that the sum or difference of two vectors is a vector. Similarly, the sum of any two tensors of the same order is a tensor of that order. No meaning is attached to the sum of tensors of different orders, say $u_i + u_{ij}$;—that is not a tensor.

In general, an equation containing tensors has meaning only if all the terms in the equation are tensors of the same order, and if the same unrepeated subscripts appear in all the terms. These facts are used in obtaining appropriate equations for fluid turbulence.

This explanation of Cartesian tensors should contain what is needed for our purposes. It is hoped that it is reasonably clear. Other treatments of tensors are given, for instance, in books by Jeffreys [1], Spain [2], Arfken [3], Lass [4], Langlois [5], and Goodbody [6].

With the background presented here, the derivation of appropriate continuum equations for turbulence should be straightforward. Before deriving them, however, a justification for calling the fluid a continuum for the study of turbulence is given.

REFERENCES

1. Jeffreys, H., *Cartesian Tensors*, University Press, Cambridge, England, 1952.
2. Spain, B., *Tensor Calculus*, Interscience Publishers, New York, 1953.
3. Arfken, G., *Mathematical Methods for Physicists, ed. 2*, chapters 1–4, Academic Press, New York, 1968.
4. Lass, H., *Elements of Pure and Applied Mathematics*, chapters 1–3, McGraw-Hill, New York, 1957.
5. Langlois, W. E., *Slow Viscous Flow*, chapter 1, Macmillan, New York, 1964.
6. Goodbody, A. M., *Cartesian Tensors: With Applications to Mechanics, Fluid Mechanics and Elasticity*, Halsted Press, New York, 1982.

CHAPTER
THREE

BASIC CONTINUUM EQUATIONS

A derivation of the continuum equations used for the analysis of turbulence is given in Chapter 3. These equations include the continuity equation, the Navier-Stokes equations, and the heat-transfer or energy equation. An experimental justification for using a continuum approach for the study of turbulence is also given.

The Navier-Stokes equations, with the other continuum equations for fluids, form the basis for the analysis of turbulence in this book. One might, in fact, refer to that analysis as a Navier-Stokes theory of turbulence. Therefore a derivation of the Navier-Stokes equations and of the other fluids equations seems appropriate, if the book is to be reasonably self-contained. Moreover, the derivations given here, at least those for the Navier-Stokes equations, proceed from fewer assumptions than do those generally given.

3-1 JUSTIFICATION OF THE USE OF A CONTINUUM APPROACH FOR TURBULENCE

Most workers in fluid dynamics, since the early days of the science, have ignored the molecular structure of fluids. They consider fluids as continua for the purpose of mathematical analysis. This point of view is generally justified by pointing out that the macroscopic lengths in most fluid-dynamic problems are many times as large as the corresponding molecular lengths. This may be true even if the flow is turbulent, the characteristic lengths of the turbulence being much larger than the molecular lengths.

In the case of turbulent flow, however, there may be some question about the size of the smallest important lengths. According to the spectral theory of turbulence, which is discussed subsequently, there is no definite lower limit on eddy size. But the presence of viscosity might be expected to limit the efficacy of the smallest eddies. For that reason most workers feel intuitively that the molecular structure of the fluid is unimportant for turbulence analysis. (An exception is highly rarefied gases.)

However, there is a solid experimental basis for a continuum theory of turbulence. This lies in the excellent macroscopic correlation of data for fluids of widely different

32 TURBULENT FLUID MOTION

molecular structure. For instance, gases and liquids have completely different molecular structures. The molecules in a gas are generally so far apart that each molecule can interact with only one other molecule at a time. In a liquid, on the other hand, each closely spaced molecule interacts simultaneously with many others.[1] In spite of this considerable difference in structure, experimental data for turbulent flow of liquids correlate well macroscopically with those for gases.

Figure 3-1 [1–4] compares experimental fully developed turbulent friction factors at various Reynolds numbers for water and for air flowing at low speeds in smooth pipes. "Fully developed" means that the time-averaged velocity does not vary with time or axial position. The dimensionless friction factor f is the ratio of time-averaged wall-shear stress $\bar{\tau}_W$ to dynamic pressure $(1/2)\rho U_a^2$, U_a is the velocity averaged with respect to time and cross-sectional area, D is the pipe diameter, ρ is the density, and ν is the kinematic viscosity. It can be shown from the equations for continuum fluid flow (to be considered later in this chapter) that f is a function only of Reynolds number for a fully developed low-speed continuum flow (see section 3-3-3). If the molecular structure were important, at least one additional dimensionless group would be required (e.g., the ratio of intermolecular distance to the tube diameter). In spite of the great differences between the molecular structures of water and air, the macroscopic or continuum correlation of the experimental data in Fig. 3-1 is excellent. General experience in correlating data for turbulent flow of liquids and gases indicates results similar to those in Fig. 3-1. It thus appears that a continuum theory of turbulence that ignores differences among the molecular structures of various fluids is realistically based.

3-2 EQUATION OF CONTINUITY (CONSERVATION OF MASS)

The rate of mass flow per unit area in the x_1 direction is

$$m_1 = \rho u_1, \tag{3-1}$$

where ρ is the density and u_1 is the velocity component in the x_1 direction. In the x_i direction this becomes

$$m_i = \rho u_i. \tag{3-2}$$

If m_i is measured at x_i, then at $x_i + \Delta x_i$, m_i is replaced by

$$m_i + \frac{\partial m_i}{\partial x_{(i)}} \Delta x_{(i)} + \cdots \text{(no sum on } i\text{)},$$

where m_i has been expanded in a Taylor series. For small Δx_i only the first two terms need be retained. Then the change in m_i in going from x_i to $x_i + \Delta x_i$ is $(\partial m_i / \partial x_{(i)}) \Delta x_{(i)}$.

---

[1] There is an interesting analogy between the theory of liquids and the theory of strong turbulence. In both theories interactions among many modes or degrees of freedom must be considered.

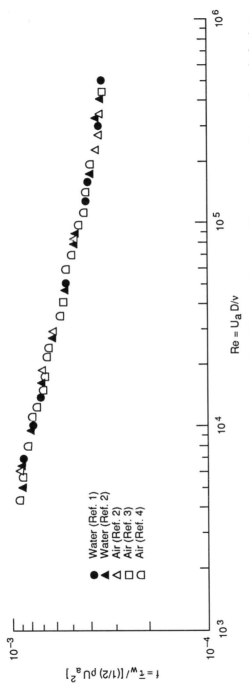

Figure 3-1 Experimental fully developed turbulent friction factors versus Reynolds number for water and for air. Agreement between the results for the two fluids that the differences in molecular structure do not affect the turbulent flow.

34 TURBULENT FLUID MOTION

Consider a small stationary volume element $\Delta x_1 \Delta x_2 \Delta x_3 = \Delta \mathbf{x}$.[2] The net flow of mass into the element through all the faces in a short time Δt is $-(\partial m_i/\partial x_i)\Delta \mathbf{x}\Delta t$ (sum on i). Equating this to the change of mass within the element in time Δt gives

$$\frac{\partial \rho}{\partial t}\Delta t \, \Delta \mathbf{x} = -\frac{\partial m_i}{\partial x_i}\Delta t \, \Delta \mathbf{x}$$

or, using eq. (3-2),

$$\frac{\partial \rho}{\partial t} = -\frac{\partial(\rho u_i)}{\partial x_i}, \tag{3-3}$$

which is the continuity equation for a compressible fluid. If the fluid is incompressible, eq. (3-3) becomes

$$\frac{\partial u_i}{\partial x_i} = 0. \tag{3-4}$$

That is, for an incompressible flow, the net rate of flow of fluid through the faces of a small volume element $\Delta \mathbf{x}$ is zero.

3-3 NAVIER-STOKES EQUATIONS (CONSERVATION OF MOMENTUM)

3-3-1 Stress Tensor

3-3-1-1 Experimental basis and fundamental assumption. We begin with the basic experimental fact concerning the motion of a viscous Newtonian fluid: In a pure shearing motion the shear stress is proportional to the velocity gradient.[3] If, for instance, the velocity is u_1, in the x_1 direction, and the velocity gradient is in the x_2 direction, the shear stress σ_{21} is

$$\sigma_{21} = \mu \frac{du_1}{dx_2}, \tag{3-5}$$

where the scalar μ is called the viscosity and is a property of the fluid but not of its motion. Equation (3-5) is found experimentally to be an excellent assumption for a great many liquids and gases. It gives the shear stress on a nonrotating face 2 (normal to x_2) of a small cube of fluid that is deforming in shear. We can also write eq. (3-5) as

$$\sigma_{21} = -\mu \frac{d\theta_{21}}{dt}, \tag{3-6}$$

---

[2] A volume of arbitrary shape together with Gauss's theorem (divergence theorem) is often used in the derivation of the conservation equations. However, the use of a small rectangular volume element is more direct and gives results as general as does the method that uses Gauss's theorem. Moreover, the derivation of Gauss's theorem generally makes use of small volume elements.

[3] In most derivations of the Navier-Stokes equations the normal components as well as the shear components of the stress are assumed proportional to a velocity gradient. The relations for the normal components, however, are difficult to verify experimentally and are not specified here. Rather they come naturally out of the derivation. We followed this line of development in *Am. J. Phys.*, vol. 44, p. 1128. However, the derivation there appears to be incomplete.

where $d\theta_{21}/dt$ is the time rate of change of the angle between the intersecting faces 1 and 2 (respectively normal to x_1 and x_2) of the deforming cube; the angle θ_{21} is measured from face 2 to face 1. If we change the rate of rotation of the cube as a whole but keep $d\theta_{21}/dt$ constant, σ_{21} does not change; it is the rate of deformation of the cube, not its rotation rate, that determines its state of stress. Both faces 1 and 2 now rotate, however, whereas before (eq. [3-5]) only face 1 rotated. In general, for two intersecting faces i and j of a cube deforming in shear, those faces being initially normal to x_i and x_j, respectively, we can write for the shear stress

$$\sigma_{ji} = -\mu \frac{d\theta_{ji}}{dt}, \quad \text{for } i \neq j, \tag{3-7}$$

in place of eq. (3-6), where $d\theta_{ji}/dt$ is the time rate of change of the angle between faces i and j, and the angle θ_{ji} is originally $\pi/2$, and σ_{ji} is the shear stress on the face initially normal to x_j. For two intersecting faces of a cube deforming in shear, but which are initially normal to x_i^* and x_j^* in a rotated (nonrotating) coordinate system, we write, instead of eq. (3-7),

$$\sigma_{ji}^* = -\mu \frac{d\theta_{ji}^*}{dt}, \quad \text{for } i \neq j, \tag{3-7a}$$

where $d\theta_{ji}^*/dt$ is the time rate of change of the angle between the two faces, which are initially normal to x_i^* and x_j^*, and σ_{ji}^* is the shear stress on the face initially normal to x_j^*. Equation (3-7) or (3-7a) is taken to apply in a general viscous flow, in which all faces of the deforming cube of fluid may rotate and move normally, and for all orientations of the cube. The viscosity μ is a scalar, so that it is invariant with rotations. Our fundamental assumption can be stated somewhat formally as follows: There exists at each point in the fluid a scalar μ called the viscosity, such that for all orientations of the initially normal face pair (i, j), eq. (3-7) or (3-7a) holds.

Because eq. (3-7) applies in a general viscous flow in which all faces of a fluid cube may rotate, $-d\theta_{ji}/dt$ is given, in the x_i coordinate system, by

$$-\frac{d\theta_{ji}}{dt} = \frac{\partial u_i}{\partial x_j} + \frac{\partial u_j}{\partial x_i}, \tag{3-8}$$

or, in the x_i^* system,

$$-\frac{d\theta_{ji}^*}{dt} = \frac{\partial u_i^*}{\partial x_j^*} + \frac{\partial u_j^*}{\partial x_i^*}. \tag{3-8a}$$

Using eqs. (3-7) and (3-8), the expression for the shear stress is

$$\sigma_{ji} = \mu \left(\frac{\partial u_i}{\partial x_j} + \frac{\partial u_j}{\partial x_i} \right), \quad \text{for } i \neq j. \tag{3-9}$$

In the rotated coordinate system x_i^*, eqs. (3-7a) and (3-8a) give

$$\sigma_{ji}^* = \mu \left(\frac{\partial u_i^*}{\partial x_j^*} + \frac{\partial u_j^*}{\partial x_i^*} \right), \quad \text{for } i \neq j. \tag{3-9a}$$

3-3-1-2 Expression for the stress tensor. As shown in section 2-4-2, a stress is a second-order tensor, so that the general expression for σ_{ji} must be such. As it stands, σ_{ji} in eq. (3-9) (or σ_{ji}^* in eq. [3-9a]) is not a tensor because of the qualifying statement $i \neq j$. If we remove that statement, then $\mu(\partial u_i/\partial x_j + \partial u_j/\partial x_i)$ is a second-order tensor; as shown in section 2-4-1, the spatial derivative of a vector is a second-order tensor, as is the sum of second-order tensors and the product of a scalar and a second-order tensor. Because the sum of tensors of the same order is also a tensor, the general expression for σ_{ji} may contain a term (which may in turn be a sum of terms) in addition to $\mu(\partial u_i/\partial x_j + \partial u_j/\partial x_i)$, call it B_{ji}, so that

$$\sigma_{ji} = \mu \left(\frac{\partial u_i}{\partial x_j} + \frac{\partial u_j}{\partial x_i} \right) + B_{ji} \qquad (3\text{-}10)$$

where, in contrast to eq. (3-9), $i, j = 1, 2$, or 3. Using eqs. (2-11) and (2-14), we get, in a rotated coordinate system x_i^*,

$$\sigma_{ji}^* = \mu \left(\frac{\partial u_i^*}{\partial x_j^*} + \frac{\partial u_j^*}{\partial x_i^*} \right) + B_{ji}^*. \qquad (3\text{-}10a)$$

To ensure that the shear components still are given by eq. (3-9) or (3-9a), however, the additional term B_{ji} or B_{ji}^* must be zero for $i \neq j$. However, one cannot know that a particular second-order tensor that is zero for $i \neq j$, say $\psi \delta_{ji}$, is necessarily the complete B_{ji}, unless it can be shown that it is the most general second-order tensor that is zero for $i \neq j$.[4]

Thus we need to determine the form of B_{ji}, that being the most general second-order tensor for which

$$B_{ji} = B_{ji}^* = 0 \quad \text{for} \quad i \neq j. \qquad (3\text{-}11)$$

Note that we have not said anything about B_{ji} for $i = j$; that is considered in what follows.

Because B_{ji} is a second-order tensor, its transformation law is (eq. [2-11]),

$$B_{ji}^* = a_{jk} a_{il} B_{kl}.$$

Multiplying this equation by a_{im} and using eq. (2-5) give

$$a_{im} B_{ji}^* = a_{jk} a_{il} a_{im} B_{kl} = a_{jk} \delta_{lm} B_{kl} = a_{jk} B_{km}.$$

Rewriting the first and last members of this equation,

$$a_{im} B_{ji}^* = a_{jk} B_{km}. \qquad (3\text{-}12)$$

The unreapted (assignable) subscripts in eq. (3-12) are m and j. We set $m = j = 1$, carry out the summations on the repeated subscripts i and k, and, with q and r as general subscripts, let $B_{qr} = B_{qr}^* = 0$ for $q \neq r$ (for shear) (eq. [3-11]). The quantities B_{qr} and B_{qr}^* are both set equal to zero for $q \neq r$ to ensure that for shear, eqs. (3-10) and (3-10a) reduce respectively to (3-9) and (3-9a), as they should. Then eq. (3-12) becomes

$$a_{11}(B_{11}^* - B_{11}) = 0,$$

---

[4]For example, one might imagine that a second-order tensor that is zero for $i \neq j$ depends on the motion of the fluid, as well as on pressure forces that are not necessarily accompanied by motion.

or
$$B^*_{11} = B_{11}.$$

Similarly, by setting $m = 2$, $j = 1$, eq. (3-12) becomes
$$B^*_{11} = B_{22},$$

and by setting $m = 3$, $j = 1$,
$$B^*_{11} = B_{33}.$$

From the last three equations,
$$B_{11} = B_{22} = B_{33}.$$

Thus, starting from eq. (3-11), we have shown that $B_{11} = B_{22} = B_{33}$ if $B_{ij} = B^*_{ij} = 0$ for $i \neq j$. It follows from the definition of the Kronecker delta (eq. [2-3]), that
$$B_{ij} = \psi \delta_{ij},$$

where ψ is a scalar. We have placed no restrictions on B_{ij} other than that it be a second-order tensor with the property that it and B^*_{ij} are zero for $i \neq j$. Therefore, $B_{ij} = \psi \delta_{ij}$ is the most general second-order tensor for which $B_{ij} = B^*_{ij} = 0$ for $i \neq j$. Note that B_{ij} turns out to be an isotropic tensor, although we have not explicitly made that assumption. It is of interest that $B_{ij} = \psi \delta_{ij}$ is the most general second-order isotropic tensor (see section 2-4-3), as well as the most general second-order tensor for which $B_{ij} = B^*_{ij} = 0$ for $i \neq j$.

Thus, the most general expression for the stress tensor that satisfies eq. (3-9) or (3-9a) for the shear components is $\sigma_{ji} = \mu(\partial u_i/\partial x_j + \partial u_j/\partial x_i) + B_{ji}$ (eq. [3-10]), where $B_{ji} = \psi \delta_{ji}$. So
$$\sigma_{ji} = \mu \left(\frac{\partial u_i}{\partial x_j} + \frac{\partial u_j}{\partial x_i} \right) + \psi \delta_{ji}.$$

Note that the above equation, besides reducing to (3-9) for $i \neq j$, reduces to the isotropic form for $u_i = 0$, as it should for a motionless fluid. By contracting that equation, we get
$$\psi = \frac{1}{3}\sigma_{kk} - \frac{2}{3}\mu \frac{\partial u_k}{\partial x_k},$$

so that
$$\sigma_{ji} = \mu \left(\frac{\partial u_i}{\partial x_j} + \frac{\partial u_j}{\partial x_i} - \frac{2}{3}\delta_{ji}\frac{\partial u_k}{\partial x_k} \right) - \sigma \delta_{ji} = \sigma'_{ji} - \sigma \delta_{ji}, \quad (3\text{-}13)$$

where σ'_{ji} is the part of the stress produced by viscous action (friction), and
$$\sigma \equiv -\sigma_{kk}/3. \quad (3\text{-}14)$$

The quantity σ is, by definition, minus the average of the three normal stresses at a point in the fluid and is sometimes called the *mechanical pressure*. Equation (3-13) gives the stress tensor that results from our fundamental assumption, eq. (3-7) (or [3-7a]). The first term on the right side of eq. (3-13) gives the stress produced by viscous or frictional (irreversible) processes; the second term gives the stress produced by pressure forces (nonviscous or reversible processes). For $i = j$, eq. (3-13) gives the normal stresses acting on the faces of a small fluid element; for $i \neq j$ it gives the shear stresses. Inspection of the form of eq. (3-13) shows that $\sigma_{ij} = \sigma_{ji}$.

3-3-2 Equations of Motion

To obtain the equation of motion for a viscous fluid, consider the force $\Delta F_i''$ acting on a small fluid element that moves with the fluid and whose volume is $\Delta x = \Delta x_1 \Delta x_2 \Delta x_3$. Newton's law, applied to the small element, is

$$\Delta F_i'' = \Delta m \frac{du_i}{dt}, \tag{3-15}$$

where t is the time. The quantity Δm is the mass of the element and does not change although the volume Δx can undergo a change of shape in such a way as to continue to enclose the same fluid.

Next obtain the surface force acting on the volume element Δx. We note that if the stress is σ_{ji} at x_j, then at $x_j + \Delta x_j$, σ_{ji} is replaced by

$$\sigma_{ji} + \frac{\partial \sigma_{ji}}{\partial x_{(j)}} \Delta x_{(j)} + \cdots \text{(no sum on } j\text{)}$$

where, as was done for m_i in the last section, σ_{ji} has been expanded in a Taylor series. For small Δx_j only the first two terms need be retained. Then the change in σ_{ji} as one moves from x_j to $x_j + \Delta x_j$ is $(\partial \sigma_{ji}/\partial x_{(j)})\Delta x_{(j)}$. The sum of all the forces in the direction acting on the faces of Δx is $\Delta x\, \partial \sigma_{ji}/\partial x_j$ (sum on j).

In terms of σ_{ji} and an external force, eq. (3-15) becomes

$$\frac{\partial \sigma_{ji}}{\partial x_j} \Delta x + \rho \Delta x g_i = \rho \Delta x \frac{du_i}{dt} \tag{3-16}$$

or

$$\rho \frac{du_i}{dt} = \frac{\partial \sigma_{ji}}{\partial x_j} + \rho g_i, \tag{3-17}$$

where ρ is the density of the fluid and g_i is an external force per unit mass. Introducing the expression for σ_{ji} from eq. (3-13) and the Eulerian derivatives of u_i into eq. (3-17) ($d/dt = \partial/\partial t + u_k \partial/\partial x_k$), we get

$$\rho \frac{\partial u_i}{\partial t} + \rho u_k \frac{\partial u_i}{\partial x_k} = -\frac{\partial \sigma}{\partial x_i} + \frac{\partial}{\partial x_k}\left[\mu\left(\frac{\partial u_i}{\partial x_k} + \frac{\partial u_k}{\partial x_i}\right)\right] - \frac{2}{3}\frac{\partial}{\partial x_i}\left(\mu \frac{\partial u_k}{\partial x_k}\right) + \rho g_i. \tag{3-18}$$

Equation (3-18) is the Navier-Stokes equation for a viscous, compressible fluid.

For incompressible flow and constant viscosity, eq. (3-18) simplifies to

$$\frac{\partial u_i}{\partial t} = -u_k \frac{\partial u_i}{\partial x_k} - \frac{1}{\rho}\frac{\partial \sigma}{\partial x_i} + \nu \frac{\partial^2 u_i}{\partial x_k \partial x_k} + g_i \qquad (3\text{-}19)$$

where the incompressible continuity relation

$$\frac{\partial u_k}{\partial x_k} = 0 \qquad (3\text{-}20)$$

was used, and ν is the kinematic viscosity μ/ρ. Most turbulence studies have been carried out for constant properties for simplicity. The flow is realistic if the turbulence velocities are reasonably small compared with the velocity of sound, and if temperature gradients are not large (small external heat transfer). Although eq. (3-19) is based on a linear stress-strain relationship, the equation is essentially nonlinear because of the presence of the nonlinear convective term $-u_k \partial u_i/\partial x_k$. In fact it is that term that causes most of the difficulties in the turbulence problem; in particular it gives rise to the closure problem, a problem that is considered in later chapters.

In order to interpret the terms in eq. (3-19), it is convenient to multiply it through by ρ and by the stationary volume element $dx_1 dx_2 dx_3$. Then the term on the left side of the equation is the time rate of change of the ith component of momentum $\rho u_i dx_1 dx_2 dx_3$ in the element. This rate of change is contributed to by the terms on the right side of the equation. The first term on the right side, the nonlinear convective or inertia term, is the net rate of flow of the ith component of momentum into the element through its faces. The next term is a (mechanical) pressure–force term and gives the net force acting on the element by virtue of the pressure gradient in the x_i direction. The term containing ν, a linear viscous-force term, gives the net force acting on the element in the x_i direction by virtue of viscous or frictional action. The last term is the external force acting on the element in the x_i direction.

Equations (3-19) and (3-20) constitute a set of four equations in the four unknowns u_i ($i = 1, 2, 3$) and σ. If we take the divergence of eq. (3-19) (differentiate it with respect to x_i) and apply the continuity eq. (3-4), we get

$$\frac{1}{\rho}\frac{\partial^2 \sigma}{\partial x_l \partial x_l} = -\frac{\partial u_i}{\partial x_k}\frac{\partial u_k}{\partial x_i} + \frac{\partial g_i}{\partial x_i}. \qquad (3\text{-}21)$$

Equation (3-21) is a Poisson equation that describes how the scalar σ varies with position under the influence of the source terms on the right side of the equation. Note that it is analogous to the steady-state heat-conduction equation with heat sources, in which σ would be the temperature, $1/\rho$ would be a constant thermal conductivity, and the terms on the right side would be heat sources.

As mentioned earlier, the quantity σ, which is the average of the three normal stresses at a point, is sometimes called the mechanical pressure and is often replaced by the symbol p. Here the symbol σ is retained to distinguish the average of the three normal stresses from the thermodynamic pressure p.

The set of eqs. (3-19) and (3-21) is often more convenient to use than (3-19) and (3-20). This set, which follows from eq. (3-7) (or [3-7a]) with no further assumptions, is complete for incompressible flow with constant viscosity. We are concerned mainly with incompressible flow in this book.

A fundamental turbulence problem for a constant-property fluid is this initial-value problem: given initial values for the u_i as functions of position, a value for ν, and suitable boundary conditions, to calculate u_i and σ/ρ as functions of time and position. Equations (3-19) and (3-21) should be sufficient for doing that. Because of sensitive dependence on initial conditions, however, the initial values of u_i would have to be given with infinite precision. However, even if they are not, as in the real world, the equations should in principle be sufficient for calculating the evolution of averaged values.

Equation (3-18), for variable-property flow, also follows from eq. (3-7) (or [3-7a]) with no further assumptions. However, in order to form a complete set for variable-property flow we need, in addition to the compressible continuity relation (eq. [3-3]), relations giving the variations of properties with temperature and pressure, an expression of the conservation of energy, and finally an assumption that relates the average normal stress σ to the thermodynamic pressure p. The appropriate assumption for σ is, in general, controversial (see, e.g., ref. [5]). However, a sufficiently accurate expression for most fluid-dynamics work is the Stokes hypothesis, obtained by equating σ to the thermodynamic pressure:

$$\sigma \equiv -\sigma_{kk}/3 = p. \tag{3-22}$$

For cases in which buoyancy is important, the velocities and temperature differences are often small, and a good approximation can be obtained by considering properties to be variable only in the buoyancy term (Boussinesq approximation). For doing that, we first write eq. (3-18) for the equilibrium case ($u_i = 0$):

$$0 = -\frac{\partial \sigma_e}{\partial x_i} + \rho_e g_i,$$

where the subscript e designates equilibrium values. Subtracting this equation from eq. (3-18) and assuming that variable properties are important only in the buoyancy term result in

$$\frac{\partial u_i}{\partial t} + u_k \frac{\partial u_i}{\partial x_k} = -\frac{1}{\rho} \frac{\partial (\sigma - \sigma_e)}{\partial x_i} + \nu \frac{\partial^2 u_i}{\partial x_k \partial x_k} + \frac{(\rho - \rho_e)}{\rho} g_i.$$

If $(\rho - \rho_e)/\rho \ll 1$ and is produced mainly by temperature differences, with pressure differences having a negligible effect, we can write this equation as

$$\frac{\partial u_i}{\partial t} = -u_k \frac{\partial u_i}{\partial x_k} - \frac{1}{\rho} \frac{\partial (\sigma - \sigma_e)}{\partial x_i} + \nu \frac{\partial^2 u_i}{\partial x_k \partial x_k} - \beta(T - T_e) g_i, \tag{3-18a}$$

where

$$\beta = -(1/\rho)(\partial \rho/\partial T)_\sigma$$

is the thermal expansion coefficient. Note that the equilibrium temperature T_e is uniform, whereas the equilibrium pressure σ_e is not. Taking the divergence of eq. (3-18a) and using the incompressible continuity equation give the Poisson equation for the (mechanical) pressure difference as

$$\frac{1}{\rho}\frac{\partial^2(\sigma-\sigma_e)}{\partial x_l \partial x_l} = -\frac{\partial u_i}{\partial x_k}\frac{\partial u_k}{\partial x_i} - \beta g_i \frac{\partial T}{\partial x_i}, \qquad (3\text{-}21a)$$

where g_i is constant.

3-3-3 Dimensionless Form of Constant-Property Fluid-Flow Equations and Dimensionless Correlation of Friction-Factor Data

Equations (3-19) and (3-21) can be rescaled or written in dimensionless form for $g_i = 0$ as

$$\frac{\partial u_i}{\partial t} + u_k \frac{\partial u_i}{\partial x_k} = -\frac{\partial \sigma}{\partial x_i} + \frac{1}{Re}\frac{\partial^2 u_i}{\partial x_k \partial x_k} \qquad (3\text{-}23)$$

and

$$\frac{\partial^2 \sigma}{\partial x_l \partial x_l} = -\frac{\partial u_i}{\partial x_k}\frac{\partial u_k}{\partial x_i} \qquad (3\text{-}24)$$

where

$$\frac{u_i}{U_a} \to u_i, \quad \frac{x_k}{L} \to x_k,$$

$$\frac{U_a}{L}t \to t, \quad \frac{\sigma}{\rho U_a^2} \to \sigma,$$

$$\frac{U_a L}{\nu} \to Re$$

and the arrow means "has been replaced by." The quantity U_a is a characteristic average velocity independent of position, L is a characteristic length for the flow, and Re is a Reynolds number. All of the quantities appearing in eqs. (3-23) and (3-24) are dimensionless.

Equations (3-23) and (3-24) can be used to justify the friction-factor correlation for fully developed turbulent flow through a pipe in Fig. 3-1. From eq. (3-9), we get, in dimensional form,

$$(\sigma_{21})_w \equiv \tau_w = \mu \left(\frac{\partial u_1}{\partial x_2}\right)_w \qquad (3\text{-}25)$$

where the subscript w designates values at the wall (τ_w is the shear stress at the wall), u_1 is in the direction along the pipe, and x_2 is normal to the wall. Equation (3-25) can be written in dimensionless (rescaled) form, by using the transformations following eq. (3-24), as

$$\frac{\tau_w}{(1/2)\rho U_a^2} = \frac{2}{Re}\left(\frac{\partial u_1}{\partial x_2}\right)_w, \qquad (3\text{-}26)$$

where the characteristic average velocity U_a is considered to be the axial velocity averaged with respect to time and area over a cross-section of the pipe, and the characteristic

42 TURBULENT FLUID MOTION

length L is taken as the pipe diameter D. (Note that the left side of eq. [3-26] is dimensionless, although the individual quantities there are dimensional, whereas the individual quantities on the right side are dimensionless.)

Consider a long time interval and an axial position for which the flow is fully developed. That is, initial transients have died out and time-averaged velocities $\bar{u}_1$ do not change with increasing time or axial distance. Averaging eq. (3-26) over the long time interval gives, in dimensionless form,

$$\frac{\bar{\tau}_w}{(1/2)\rho U_a^2} \equiv f = \frac{2}{Re}\left(\partial u_1/\partial x_2\right)_w \tag{3-27}$$

where the overbars indicate time-averaged values and f is the friction factor. The time-averaged velocity gradient in eq. (3-27) can be obtained by first computing the velocity field $u_i(x_k, t)$ from eqs. (3-23) and (3-24), starting from a fully developed instantaneous turbulent velocity distribution, and applying appropriate boundary conditions (e.g., $u_i = 0$ at the wall).[5] Then the calculated velocity gradient at the wall is averaged over the time interval at the axial position for which eq. (3-27) was obtained. Thus, for a given value of Reynolds number $Re = u_a D/\nu$ (in eqs. [3-23] and [3-27]), the friction factor $f = 2\bar{\tau}_w/(\rho U_a^2)$ is known, in principle, from eqs. (3-23), (3-24), and (3-27), together with appropriate boundary conditions. That is,

$$f = f(Re) \tag{3-28}$$

for a constant-property fully developed turbulent continuum flow, as was assumed in Fig. 3-1.[6]

3-4 HEAT-TRANSFER OR ENERGY EQUATION (CONSERVATION OF ENERGY)

The heat-transfer rate per unit area by molecular conduction in the x_1 direction is

$$q_1 = -k\frac{\partial T}{\partial x_1},$$

where k is the thermal conductivity and T is the temperature. In any direction x_i,

$$q_i = -k\frac{\partial T}{\partial x_i} \tag{3-29}$$

where q_i is the molecular heat-transfer vector.

To obtain the energy equation for a viscous fluid, consider the energy added in time Δt to a small element that moves with the fluid and whose volume is $\Delta x = \Delta x_1 \Delta x_2 \Delta x_3$. Then the law of conservation of energy states that the change of energy in the element

---

[5]The difficulties involved in actually doing this need not concern us here; this is a thought experiment.

[6]Equation (3-28) also can be obtained by using the method of dimensional analysis, from which eq. (3-28) follows if we assume that $\bar{\tau}/\rho = f(U_a, L, \nu)$. That relation may be obtained intuitively, but in order to be sure that it holds it would seem safer to use the equations of motion for a fluid and an analysis similar to that given here. Consider, however, the argument near the end of section 3-5.

equals the heat added plus the work done on the element. Herein we consider the case in which the only heat added to the element is that transferred into it by molecular conduction. Then an energy balance on the element gives

$$\rho \Delta x \Delta t \frac{d}{dt}(e+(1/2)u_i u_i) = -\Delta x \Delta t \frac{\partial q_i}{\partial x_i} + \Delta x \Delta t \frac{\partial}{\partial x_j}(u_i \sigma_{ji}) + \rho \Delta x \Delta t \, u_i g_i \quad (3\text{-}30)$$

where the energy per unit mass is made up of the internal energy e and the kinetic energy $(1/2)u_i u_i$. The term on the left side of eq. (3-30) gives the change of energy in the element in time Δt; the first term on the right side gives the heat added by molecular conduction. The last two terms are, respectively, the work done on the element by surface forces and by the external body force.

If we substitute for q_i from eq. (3-29) and for $u_i du_i/dt$ and σ_{ji} from eqs. (3-17) and (3-13) respectively, we get, after some manipulation,

$$\rho \frac{de}{dt} = \frac{\partial}{\partial x_j}\left(k \frac{\partial T}{\partial x_j}\right) + \sigma'_{ji} \frac{\partial u_i}{\partial x_j} - \sigma \frac{\partial u_i}{\partial x_i}, \quad (3\text{-}31)$$

where, as in eq. (3-13), σ'_{ji} is the part of the stress produced by viscous action, and σ is the mechanical pressure (eq. [3-14]).

If the fluid is incompressible, $de = c dT$, where c is the specific heat at constant volume, and $\partial u_i/\partial x_i = 0$ (eq. [3-4]). Equation (3-31) then becomes

$$\rho c \left(\frac{\partial T}{\partial t} + u_j \frac{\partial T}{\partial x_j}\right) = \frac{\partial}{\partial x_j}\left(k \frac{\partial T}{\partial x_j}\right) + \sigma'_{ji} \frac{\partial u_i}{\partial x_j}, \quad (3\text{-}32)$$

where the relation between total and Eulerian derivatives $(d/dt = \partial/\partial t + u_k \partial/\partial x_k)$ was used. Equation (3-32) works well for a liquid.

For a perfect gas the term $\sigma \partial u_i/\partial x_i$ in eq. (3-31) cannot be neglected because, even at small velocities compared with the velocity of sound and for small temperature differences (small external heat transfer), that term is not small compared with the term $\rho de/dt$. (As $\sigma \partial u_i/\partial x_i$ becomes small, so also does $\rho de/dt$.)

The relation $de = c dT$ applies to a perfect gas as well as to an incompressible fluid. On using that relation and the perfect-gas relation $c_p dT = c dT + d(p/\rho)$ where c_p is the specific heat at constant pressure, and equating the mechanical pressure σ to the thermodynamic pressure p (eq. [3-22]), eq. (3-31) becomes

$$\rho c_p \frac{dT}{dt} = \frac{dp}{dt} + \frac{\partial}{\partial x_i}\left(k \frac{\partial T}{\partial x_i}\right) + \sigma'_{ji} \frac{\partial u_i}{\partial x_j}. \quad (3\text{-}33)$$

For many flows the velocities are small enough that we can neglect the terms dp/dt and $\sigma'_{ji} \partial u_i/\partial x_j$ in eqs. (3-33) and (3-32) in comparison with the heat-condition term. Moreover, the rate of variation of thermal conductivity with temperature is often small enough that we can consider k to be constant, except for very large temperature differences. Then, to a good approximation, eqs. (3-32) and (3-33) can be written as

$$\frac{\partial T}{\partial t} + u_i \frac{\partial T}{\partial x_i} = \alpha \frac{\partial^2 T}{\partial x_i \partial x_i}, \quad (3\text{-}34)$$

where α is the thermal diffusivity and is given by

$$\alpha = \begin{matrix} k/(\rho c) & \text{for a liquid,} \\ k/(\rho c_p) & \text{for a perfect gas.} \end{matrix} \qquad (3\text{-}35)$$

Equation (3-34) is the form of the energy equation that is used in this book. It can also apply to diffusion processes other than for energy. It is only necessary that the accumulation of some quantity in an element of fluid be balanced by the diffusion of that quantity into the element by a law of the form of eq. (3-29), where q_i is a flux vector for that quantity, k is a constant, and T is a scalar. For instance the scalar T might be a dilute concentration of a foreign substance in the fluid and q_i the flux vector of that substance.

3-4-1 Dimensionless Form of Constant-Property Energy Equation and Dimensionless Correlation of Heat-Transfer Data

Equation (3-34) can be rescaled or written in dimensionless form as

$$\frac{\partial T}{\partial t} + u_i \frac{\partial T}{\partial x_i} = \frac{1}{Pe} \frac{\partial^2 T}{\partial x_i \partial x_i} \qquad (3\text{-}36)$$

where

$$\frac{T_w - T}{T_w - T_a} \to T, \quad \frac{u_i}{U_a} \to u_i,$$

$$\frac{x_k}{L} \to x_k, \quad \frac{U_a}{L} t \to t,$$

$$\frac{U_a L}{\alpha} \to Pe$$

and, as for eqs. (3-23) and (3-24), the arrow means "has been replaced by." In this section we consider the wall temperature T_w in a pipe to be spatially uniform and constant in time. The quantity T_a is an average fluid temperature at a particular axial position, U_a is an average fluid velocity at the same axial position, L is a characteristic length for the flow, Pe is a Peclet number, and α is the thermal diffusivity given by eq. (3-35). All of the quantities appearing in eq. (3-36) are dimensionless. Note that t, x_i, and u_i have been nondimensionalized in the same way as in eqs. (3-23) and (3-24).

From eq. (3-29) we get, in dimensional form,

$$(q_2)_w \equiv q_w = -k \left(\frac{\partial T}{\partial x_2} \right)_w \qquad (3\text{-}37)$$

where the subscript w designates values at the wall, q_w is the heat flux at the wall, and x_2 is normal to the wall. Equation (3-37) can be written in dimensionless (rescaled) form, by using the transformations following eq. (3-36), as

$$\frac{q_w D}{(T_w - T_a)k} = \left(\frac{\partial T}{\partial x_2} \right)_w \qquad (3\text{-}38)$$

where the characteristic average temperature T_a is considered to be the temperature averaged with respect to time and area over a particular cross-section of the pipe, and the characteristic length L is taken as the pipe diameter D. (Note that, as for eq. [3-26], the left side of eq. [3-38] is dimensionless, although the individual quantities there are dimensional, whereas the individual quantities on the right side are dimensionless.)

Consider a long time interval and an axial position for which the flow and heat transfer are fully developed. That is, initial transients have died out and the shapes of the time-averaged velocity and temperature distributions remain similar with increasing time and axial distance. Averaging eq. (3-38) over the long time interval gives, in dimensionless form,

$$\frac{\bar{q}_w D}{(T_w - T_a)k} = \frac{hD}{k} \equiv Nu = \overline{(\partial T/\partial x_2)}_w \qquad (3\text{-}39)$$

where the overbars indicate time-averaged values, $h \equiv \bar{q}_w/(T_w - T_a)$ is the heat-transfer coefficient, and Nu is the Nusselt number. The time-averaged temperature gradient in eq. (3-39) can be obtained by first computing the velocity field $u_i(x_k, t)$ for a given Reynolds number Re from eqs. (3-23) and (3-24). As in section 3-3-1, one starts from a fully developed instantaneous turbulent velocity distribution and applies appropriate boundary conditions (e.g., $u_i = 0$ at the wall). With the velocity field known for a given Reynolds number Re, the temperature field $T(x_k, t)$ for a given Peclet number Pe can be calculated from eq. (3-36), starting from a fully developed instantaneous turbulent temperature distribution, and applying appropriate boundary conditions (e.g., $T = 0$ at the wall; see transformation following eq. [3-36]).[5] The calculated temperature gradient at the wall is averaged over the time interval at the axial position for which eq. (3-39) was obtained. Thus, for a given value of Reynolds number $Re = U_a D/\nu$ (in eq. [3-23]) and of Peclet number $Pe = U_a D/\alpha$ (in eq. [3-36]), the Nusselt number $Nu = hD/k$ is, in principle, known from eqs. (3-23), (3-24), (3-36), and (3-39), together with appropriate boundary conditions. That is,

$$Nu = Nu(Re, Pe) \qquad (3\text{-}40)$$

for a constant-property fully developed turbulent continuum flow and heat transfer. In terms of the Prandtl number $Pr = \nu/\alpha$ and Reynolds number Re, eq. (3-40) can be written in the perhaps more familiar form (because $Pe = RePr$), as

$$Nu = Nu(Re, RePr) = Nu(Re, Pr). \qquad (3\text{-}40a)$$

3-5 RULE FOR OBTAINING ADDITIONAL DIMENSIONLESS PARAMETERS AS A SYSTEM BECOMES MORE COMPLEX

The method that is used in sections 3-3-3 and 3-4-1 to obtain correlating dimensionless parameters from the basic turbulence equations can be extended to more complex systems. However, once a functional relation such as eq. (3-28) or (3-40a) has been obtained for a given system, the functional relation for a system that depends on an additional physical quantity can be written down by the following simple rule: If a system becomes dependent on one additional physical quantity, one additional dimensionless parameter (containing the additonal physical quantity) is added to the functional relation.

46 TURBULENT FLUID MOTION

For example, if the flow or heat transfer in a pipe becomes dependent of the distance from the entrance x_1, as it does if the pipe becomes short, then one dimensionless parameter containing x_1, say x_1/D, where D is the pipe diameter (already included), is added to the functional relation. Thus eqs. (3-28) and (3-40a) become, respectively,

$$f = f(Re, x_1/D) \tag{3-41}$$

and

$$Nu = Nu(Re, Pr, x_1/D). \tag{3-42}$$

If, in addition, the system becomes dependent on the velocity of sound c, as it does for high fluid velocities, then an additional dimensionless parameter, say U_a/c, where U_a is a characteristic velocity (already included in the functional relation), becomes operative, and eqs. (3-41) and (3-42) become, respectively,

$$f = f(Re, x_1/D, U_a/c) \tag{3-43}$$

and

$$Nu = Nu(Re, Pr, x_1/D, U_a/c). \tag{3-44}$$

If the system is also dependent on an external force g_i, then a dimensionless parameter such as $g_i D/U_a^2$ is added in the functional relationship. Finally, although we do not deal with this effect in this book, if intermolecular distances d_m become important, as in a highly rarefied gas, then the functional relationship contains a dimensionless parameter such as d_m/D. If the system is dependent on all the physical quantities mentioned previously, then eqs. (3-28) and (3-40a) become, respectively,

$$f = f\left(Re, x_1/D, U_a/c, g_i D/U_a^2, d_m/D\right) \tag{3-45}$$

and

$$Nu = Nu\left(Re, Pr, x_1/D, U_a/c, g_i D/U_a^2, d_m/D\right). \tag{3-46}$$

Next we ask whether our rule can be applied if we have only one dimensionless parameter to start with. In the dimensional analysis of a system it is often convenient to group the physical variables in such a way as to eliminate as many dimentions as possible. For our case of flow and heat transfer in a pipe we can easily eliminate all dimensions but length and time. Then, assuming that the dimensionless friction factor f is not a function of any variables, that is, starting with only one dimensionless parameter, we have

$$f = \frac{2(\bar{\tau}_w/\rho)}{U_a^2} = f(\text{const}), \tag{3-47}$$

where the dimensions of $(\bar{\tau}_w/\rho)$ are (length)^2/(time)^2, and those of U_a are length/time. In order to obtain a more realistic relation for f, one might expect that the kinematic viscosity ν or the pipe diameter D should be included in the functional relationship. Neither of those quantities, however, combines by itself with quantities already present ($[\bar{\tau}_w/\rho]$ or U_a) to form a new dimensionless parameter; the relationship resulting by adding ν or D separately is not dimensionally correct. However, the quantity (ν/D)

combines with U_a to form $U_a/(\nu/D) = Re$, so that (3-47) becomes

$$f = f(Re),$$

which agrees with eq. (3-28). Thus, adding one physical quantity (ν/D) to the functional relationship in eq. (3-47) adds one dimensionless parameter Re to the relationship, in accordance with our rule.

Similarly, considering the heat transfer and assuming that the dimensionless Nusselt number Nu is not a function of any variables, that is, starting with one dimensionless parameter, we have

$$Nu = \frac{hD}{k} = \frac{[h/(\rho c)]D}{\alpha} = Nu(\text{const}) \qquad (3\text{-}48)$$

where c is the specific heat and α is the thermal diffusivity. As in eq. (3-47), we have grouped the physical variables in such a way as to eliminate all the dimensions but length and time; the dimensions of $[h/(\rho c)]$ are length/time, and those of α are (length)^2/time. In order to obtain a more realistic relation for Nu, one might expect that the kinematic viscosity ν or the mean velocity U_a should be included in the functional relationship. It turns out that each of those can be combined with quantities already present in the functional relationship to form a new dimensionless parameter. Thus ν can be combined with α to form the Prandtl number $Pr = \nu/\alpha$, after which U_a can be combined with ν and D to form the Reynolds number $Re = U_a D/\nu$. Equation (3-48) for the Nusselt number then becomes

$$Nu = Nu(Pr, Re),$$

in agreement with equation (3-40a). Thus, once again a dimensionless parameter is added to the functional relationship each time a physical variable is added.

3-6 CLOSING REMARKS

The analyses given in sections 3-3-3, 3-4-1, and 3-5, which are based on dimensional considerations, whether they proceed in conjunction with the basic fluids equations as in sections 3-3-3 and 3-4-1 or in conjunction with experience with flow situations as in section 3-5 deal mainly with functional relations among dimensionless variables and are primarily an aid to the correlation of experimental data. Thus they do not by themselves, of course, constitute solutions of the fundamental equations.

Theoretically it should be possible to apply the continuum equations obtained in this chapter directly to turbulent flow. In that case the velocities, temperatures, and so forth in the equations are local instantaneous values in the turbulent field. If appropriate initial and boundary conditions are given, the temporal evolution of the turbulent field, in principle, can be calculated from the equations. However, the implementation of such calculations, which is discussed in succeeding chapters, is not easy.

Before considering solutions of the unaveraged equations of this chapter, we consider averaged equations, as well as some equations that contain both averaged and unaveraged quantities.

REFERENCES

1. Nikuradse, J., Gesetzmässigkeiten der turbulenten Strömung in glatten Rohren. VDI-Forschungsheft, no. 356, 1932.
2. Stanton, T.E., and Pannell, J.R., "Similarity of Motion in Relation to the Surface Friction of Fluids." *Philos. Trans. R. Soc. London* A, vol. 214, pp. 199–224, 1914.
3. Laufer, J., "The Structure of Turbulence in Fully Developed Pipe Flow," NACA Report 1174, 1954.
4. Deissler, R.G., "Analytical and Experimental Investigation of Adiabatic Turbulent Flow in Smooth Tubes," NACA TN-2138, 1950.
5. Rosenhead, L., "A Discussion on the First and Second Viscosities of Fluids, Under the Leadership of L. Rosenhead." *Proc. R. Soc. London A*, vol. 226, pp. 1–69, 1954.

CHAPTER
FOUR

AVERAGES, REYNOLDS DECOMPOSITION, AND THE CLOSURE PROBLEM

Ensemble, time, and space averages as applied to turbulent quantities are discussed in Chapter 4, and pertinent properties of the averages are obtained. Those properties, together with Reynolds decomposition, are used to derive the averaged equations of motion and the one- and two-point moment or correlation equations. The terms in the various equations are interpreted. The closure problem of the averaged equations is discussed, and possible closure schemes are considered. Those schemes usually require an input of supplemental information, unless the averaged equations are closed by calculating their terms by a numerical solution of the original unaveraged equations. The law of the wall for velocities and for temperatures, the velocity- and temperature-defect laws, and the logarithmic laws for velocities and for temperatures are derived. Various notions of randomness and their relation to turbulence are considered in the light of modern ergodic theory.

Although the unaveraged equations of the previous chapter can, in principle, be applied directly to turbulence, the historical tendency has been to average out the fluctuations so as to obtain simply varying functions. The idea is that simply varying nonrandom values are easier to deal with than the haphazard motion characteristic of turbulent flow. Physically relevant equations are obtained in this way, but the price to be paid as far as obtaining solutions is concerned (the closure problem) is considerable, as is seen later in this chapter. First we consider various averages of the turbulent quantities.

4-1 AVERAGE VALUES AND THEIR PROPERTIES

For the most general turbulent flows an ensemble average over a large number of macroscopically (but not microscopically) identical flows is appropriate. In all of those flows, the macroscopic determining parameters (e.g., mean velocity, scales, etc.), but not fluctuating quantities, are the same. The ensemble average of a quantity, say the velocity, at a point x_k and time t is the arithmetic average of that quantity for all the flows. For instance, for a velocity component u_i at point x_k and time t,

$$\overline{u_i(x_k, t)}_n = \lim_{N \to \infty} \sum_{n=1}^{N} u_i(n, x_k, t)/N, \tag{4-1}$$

where n indicates the nth flow, N is the total number of flows over which the average is taken, and the overbar with the subscript n designates the ensemble average (the average over n). Needless to say, this type of average, although often used in theoretical work, is hard to implement experimentally, because a large number of macroscopically identical flows are not likely available.

50 TURBULENT FLUID MOTION

Fortunately, in most cases statistical uniformity or stationarity with respect to one or more coordinates or with respect to time obtains. Then the average is taken with respect to the one or more coordinates or with respect to time. For instance, if the turbulence is statistically stationary with respect to time (if the ensemble average does not vary with time t), the time average of u_i at a point x_k is

$$\overline{u_i(x_k)}_t = \lim_{T \to \infty} \int_0^T u_i(x_k, t)\, dt/T, \tag{4-2}$$

where the time average is designated by the overbar and the subscript t.

If the turbulence is statistically stationary with respect to one or more variable coordinates x_j (if the ensemble average does not vary with the one or more coordinates x_j), then the space average with respect to x_j at a time t and at fixed coordinate or coordinates x_k, where $j \neq k$, is

$$\overline{u_i(x_k, t)}_{x_j} = \lim_{X_j \to \infty} \int_{-X_j}^{X_j} u_i(x_k, t, x_j)\, dx_j/(2X_j), \tag{4-3}$$

where the space average is designated by the overbar and the subscript x_j, and the integration is over the one or more coordinates x_j for which statistical uniformity obtains. (Note that if j can have the values 1, 2, and 3, x_k is absent from eq. (4-3), because $k \neq j$.)

A space average also makes sense for periodic boundary conditions, even if the turbulence is not statistically uniform over one or more coordinates. In that case as in the preceding case, the space-averaged quantities (averaged over a period) do not vary with position, because it makes no difference where the period starts ($\partial \overline{u_i(x_k, t)}_{x_j}/\partial x_j = 0$, $k \neq j$). Then

$$\overline{u_i(x_k, t)}_{x_j} = \int_{-X_j}^{X_j} u_i(x_k, t, x_j)\, dx_j/(2X_j), \tag{4-4}$$

where $2X_j$ is one spatial period.

Finally, it might be useful to take the average over both time and one or more coordinates, if the turbulence is statically stationary with respect to both. Then the average with respect to t and x_j is

$$\overline{u_i(x_k)}_{t,x_j} = \lim_{X_j, T \to \infty} \int_{-X_j}^{X_j} \int_0^T u_i(x_k, t, x_j)\, dt\, dx_j/(2X_j T). \tag{4-5}$$

Of course, if the boundary conditions in eq. (4-5) are periodic in some of the directions x_j, the average in those directions need be taken only over one period, as in eq. (4-4).

The averaging processes considered in this section are illustrated by taking average values of a velocity component u_i. The same processes, of course, can be applied to the mechanical pressure σ (eq. [3-14]) and to functions of velocities and of pressures such as $u_i u_j$, $u_i u_j u_k$, $u_i u_j' u_k''$, and $\sigma u_k'$, where the unprimed, primed, and double-primed quantities refer to values at different spatial points. The u_i in eqs. (4-1) through (4-5) need only be replaced by σ, $u_i u_j$, and so forth. It might be mentioned that the simplest average that is descriptive of turbulence is not that of u_i, which describes the overall flow rather than the turbulence, but that of the second-order tensor $u_i u_j$.

AVERAGES, REYNOLDS DECOMPOSITION, AND THE CLOSURE PROBLEM 51

4-1-1 Ergodic Theory and the Randomness of Turbulence

Turbulence is generally taken to be ergodic, in which case the ensemble, time, and all space averages of a turbulent quantity (say $u_i u_j$) should have the same value, assuming that those averages exist. An ergodic system embodies the weakest notion of randomness in a hierarchy of systems [1, 2]. The so-called mixing systems (those whose variables become uncorrelated as their temporal separation $\Delta t \to \infty$) have a stronger notion of randomness than do those that are only ergodic, and systems that exhibit sensitive dependence on initial conditions, or chaoticity, have a stronger notion of randomness than do those that are only ergodic or only ergodic and mixing. Mixing implies ergodicity, and chaoticity implies both ergodicity and mixing, but the converse is not true. At the top of the hierarchy are the most random systems, those that, although deterministic, may appear in a certain sense to behave as randomly as the numbers produced by a roulette wheel [1].

So in order for a flow to be identified as being random (or apparently random) in some sense, and thus as turbulent, it must be at least ergodic. As mentioned in Chapter 1, turbulent systems appear to be at least as random as chaotic systems. In Chapter 6 it is argued that they are likely more random.

Because of the ergodicity of turbulence a distinction among the various kinds of averages usually is not made, and they are written simply as $\overline{u_i}$, $\overline{u_i u_j}$, $\overline{u_i \sigma'}$, and so forth, the subscripts n, t, and x_j in eqs. (4-1) through (4-5) being omitted. It always is assumed, of course, that the averages taken are of an appropriate type, in line with eqs. (4-1) through (4-5).

4-1-2 Remarks

The averaging considered here, which is known as *Reynolds averaging*, is the type that is used in this book. There are, of course, other types of averaging, which have specific uses. For instance, conditional averaging, in which averages are taken under some specified condition such as the condition that only velocities greater than some value be used in the average, is sometimes useful. However, none of the methods of averaging circumvents the closure problem (considered in section 4-3) that occurs if the nonlinear continuum equations are averaged.

4-1-3 Properties of Averaged Values

Finally, we consider some properties of averaged values that are useful in obtaining the averaged equations of fluid motion. We note first that the derivative of an average equals the average of the derivative. Thus, for example, $\partial \overline{u_i u_j}/\partial x_k = \overline{\partial u_i u_j/\partial x_k}$. This equation can be obtained from eq. (4-2) (with u_i replaced by $u_i u_j$) if the average is taken over time. Thus,

$$\frac{\partial \overline{u_i u_j}}{\partial x_k} = \left(\frac{\partial}{\partial x_k}\right)\left[\lim_{T \to \infty} \int_0^T (u_i u_j)\, dt/T\right]$$

$$= \lim_{T \to \infty} \int_0^T \left[\frac{\partial (u_i u_j)}{\partial x_k}\right] dt/T = \overline{\frac{\partial (u_i u_j)}{\partial x_k}}. \qquad (4\text{-}6)$$

52 TURBULENT FLUID MOTION

The same result is obtained if the average is other than that over time. Also, it is easy to show that the sum of averages equals the average of the sum. Thus, again using eq. (4-2),

$$\overline{u_i u'_j} + \overline{u'_i u_j} = \lim_{T \to \infty} \int_0^T (u_i u'_j)\, dt/T + \lim_{T \to \infty} \int_0^T (u'_i u_j)\, dt/T$$

$$= \lim_{T \to \infty} \int_0^T (u_i u'_j + u'_i u_j)\, dt/T = \overline{u_i u'_j + u'_i u_j}. \quad (4\text{-}7)$$

Moreover, taking an average of an average (designated by a double bar) does not change its value, because, for example,

$$\overline{\overline{u_i u_j}} = \lim_{T \to \infty} \int_0^T \overline{(u_i u_j)}\, dt/T = \overline{u_i u_j} \lim_{T \to \infty} \int_0^T dt/T = \overline{u_i u_j}. \quad (4\text{-}8)$$

Finally, the average of the product of an averaged and an unaveraged quantity is the product of the first (averaged) quantity by the second quantity averaged. For example,

$$\overline{\overline{u_i}(u_j u_k)} = \lim_{T \to \infty} \int_0^T \overline{u_i}(u_k u_j)\, dt/T = \overline{u_i} \lim_{T \to \infty} \int_0^T (u_k u_j)\, dt/T$$

$$= \overline{u_i}\, \overline{u_j u_k}. \quad (4\text{-}9)$$

As an example of a relation for averages that does not hold, we note that

$$\overline{u_i u_j} \neq \overline{u_i}\, \overline{u_j},$$

because

$$\lim_{T \to \infty} \int_0^T (u_i u_j)\, dt/T \neq \lim_{T \to \infty} \int_0^T u_i\, dt/T \lim_{T \to \infty} \int_0^T u_j\, dt/T.$$

All of the preceding relations that were shown to hold for time averages hold, of course, for the other kinds of averages. Thus, for instance, if we consider ensemble averages and use eq. (4-1), we find that

$$\overline{\overline{u_i u_j}} = \lim_{N \to \infty} \sum_{n=1}^N \overline{u_i u_j}/N = \overline{u_i u_j}, \quad (4\text{-}8a)$$

which is the same result as that obtained in eq. (4-8) for time averages.

4-2 EQUATIONS IN TERMS OF MEAN AND FLUCTUATING COMPONENTS

Equations for mass, momentum, and energy conservation, as well as for the pressure, that are general enough for all of the work in this book are given by eqs. (3-4), (3-18a),

AVERAGES, REYNOLDS DECOMPOSITION, AND THE CLOSURE PROBLEM

(3-34), and (3-21a) respectively. We first rewrite those equations in slightly different forms and notations:

$$\frac{\partial \tilde{u}_i}{\partial x_i} = 0, \tag{4-10}$$

$$\frac{\partial \tilde{u}_i}{\partial t} = -\frac{\partial (\tilde{u}_i \tilde{u}_k)}{\partial x_k} - \frac{1}{\rho}\frac{\partial (\tilde{\sigma} - \sigma_e)}{\partial x_i} + \nu \frac{\partial^2 \tilde{u}_i}{\partial x_k \partial x_k} - \beta(\tilde{T} - T_e)g_i, \tag{4-11}$$

$$\frac{\partial \tilde{T}}{\partial t} = -\frac{\partial (\tilde{T}\tilde{u}_k)}{\partial x_k} + \alpha \frac{\partial^2 \tilde{T}}{\partial x_k \partial x_k}, \tag{4-12}$$

and

$$\frac{1}{\rho}\frac{\partial^2 (\tilde{\sigma} - \sigma_e)}{\partial x_l \partial x_l} = -\frac{\partial^2 (\tilde{u}_i \tilde{u}_k)}{\partial x_i \partial x_k} - \beta g_i \frac{\partial \tilde{T}}{\partial x_i}. \tag{4-13}$$

These equations are the same as eqs. (3-4), (3-18a), (3-34), and (3-21a), except that tildes, (~) have been placed over the instantaneous velocities u_i, mechanical pressures σ, and temperatures T, and the nonlinear terms have been written in the so-called conservative form by using the continuity eq. (4-10).

Following Reynolds [3], one can break the instantaneous quantities in eqs. (4-10) through (4-13) into mean and fluctuating (or turbulent) components. That process is known as *Reynolds decomposition*. Thus, set

$$\tilde{u}_i = U_i + u_i, \tag{4-14}$$
$$\tilde{\sigma} = P + \sigma, \tag{4-15}$$

and

$$\tilde{T} = T + \tau \tag{4-16}$$

where the first and second terms on the right sides of the equations are respectively mean and fluctuating components, and where

$$\overline{u_i} = \overline{\sigma} = \overline{\tau} = 0, \tag{4-17}$$
$$U_i = \overline{\tilde{u}}_i, \tag{4-18}$$
$$P = \overline{\tilde{\sigma}}, \tag{4-19}$$

and

$$T = \overline{\tilde{T}}. \tag{4-20}$$

As usual, the overbars designate averaged values. Equation (4-10) becomes, on using eqs. (4-14), (4-17), (4-18), and the properties of averaged values given in eqs. (4-6) through (4-8),

$$\frac{\partial U_k}{\partial x_k} = \frac{\partial u_k}{\partial x_k} = 0, \tag{4-21}$$

which shows that both the mean and fluctuating velocity components satisfy conservation of mass. Equations (4-11) through (4-13) become, on using eqs. (4-14) through (4-20) and (4-6) through (4-8), taking averages, and subtracting the averaged equations from the unaveraged ones,

$$\frac{\partial u_i}{\partial t} = -\frac{\partial (u_i u_k)}{\partial x_k} - \frac{1}{\rho}\frac{\partial (\sigma - \sigma_e)}{\partial x_i} + \nu \frac{\partial^2 u_i}{\partial x_k \partial x_k} - \beta g_i \tau - u_k \frac{\partial U_i}{\partial x_k} - U_k \frac{\partial u_i}{\partial x_k} + \frac{\partial \overline{u_i u_k}}{\partial x_k}, \tag{4-22}$$

$$\frac{\partial \tau}{\partial t} = -\frac{\partial (\tau u_k)}{\partial x_k} + \alpha \frac{\partial^2 \tau}{\partial x_k \partial x_k} - u_k \frac{\partial T}{\partial x_k} - U_k \frac{\partial \tau}{\partial x_k} + \frac{\partial \overline{\tau u_k}}{\partial x_k}, \tag{4-23}$$

and

$$\frac{1}{\rho}\frac{\partial (\sigma - \sigma_e)}{\partial x_i \partial x_i} = -\frac{\partial^2 (u_i u_k)}{\partial x_i \partial x_k} - \beta g_i \frac{\partial \tau}{\partial x_i} - 2\frac{\partial u_i}{\partial x_k}\frac{\partial U_k}{\partial x_i} + \frac{\partial^2 \overline{u_i u_k}}{\partial x_i \partial x_k}. \tag{4-24}$$

Equations (4-22) through (4-24) are useful for studying turbulence processes and for constructing evolution equations for correlations (e.g., for $\overline{u_i u_j}$). The first five terms of eq. (4-22), the first three of eq. (4-23), and the first three of eq. (4-24) are similar to the terms in eqs. (4-11) through (4-13) respectively, although their meanings are exactly the same only if $U_i = T = P = 0$ (see eqs. [4-14] through [4-16]).

As was the case for eq. (3-19) (or [4-11]), the first three terms on the right side of eq. (4-22), which give contributions to the rate of change of the velocity fluctuation $\partial u_i/\partial t$, can be interpreted as an inertia-force (or turbulence self-interaction) term, a pressure-force term, and a viscous-force term. The remaining terms are, respectively, a buoyancy-force term, a turbulence-production term, a mean-flow convection term, and a mean turbulent stress term, which may appear if the turbulence is statistically inhomogeneous (if mean turbulence quantities such as $\overline{u_i u_k}$ are functions of position). (The reasons for referring to the production and convection terms as such perhaps will become clearer when the equivalent terms in the averaged equations are discussed.) It is seen in the next chapter that if the mean-velocity gradient is not zero, the term $-U_k \partial u_i/\partial x_k$ generates a small-scale structure in the turbulence by vortex stretching or by a breakup of eddies into smaller ones. The nonlinear self-interaction term $-\partial (u_i u_k)/\partial x_k$ also produces a small-scale structure, and in addition produces randomization of the flow. This effect also is considered in the next chapter.

The terms on the right side of eq. (4-23), which give contributions to the rate of change of the temperature fluctuation $\partial \tau/\partial t$, are respectively a nonlinear temperature–velocity interaction term, a molecular diffusion term for temperature fluctuations, a production term for temperature fluctuations, a mean-flow convection term for temperature fluctuations, and a mean turbulent heat-transfer term that may be present if the turbulence is inhomogeneous.

The Poisson equation for the mechanical pressure fluctuations has four source terms—a nonlinear term, a buoyancy term, a mean-velocity-gradient term, and a mean-turbulent-stress term that appears if the turbulence is inhomogeneous.

4-3 AVERAGED EQUATIONS

Although the averaged equations for fluid motion do not form a closed (complete) set, they are very useful for studying the physical processes in turbulence. Moreover, approximate solutions can be obtained by introducing various closure schemes. Alternatively (or preferably), the terms in the averaged equations can often be calculated from a numerical solution of the unaveraged equations.

4-3-1 Equations for Mean Flow and Mean Temperature

First consider the equations obtained by averaging each term in eqs. (4-11) through (4-13) after applying Reynolds decomposition (eqs. [4-14] through [4-20]) and using (4-21). This gives

$$\rho \frac{\partial U_i}{\partial t} = -\rho U_k \frac{\partial U_i}{\partial x_k} - \frac{\partial (P - \sigma_e)}{\partial x_i} + \frac{\partial}{\partial x_k} \left[\rho v \left(\frac{\partial U_i}{\partial x_k} + \frac{\partial U_k}{\partial x_i} \right) - \rho \overline{u_i u_k} \right]$$
$$- \rho \beta (T - T_e) g_i, \tag{4-25}$$

$$\rho c \frac{\partial T}{\partial t} = -\rho c U_k \frac{\partial T}{\partial x_k} - \frac{\partial}{\partial x_k} \left(-\rho c \alpha \frac{\partial T}{\partial x_k} + \rho c \overline{u_k \tau} \right), \tag{4-26}$$

and

$$\frac{\partial^2 (P - \sigma_e)}{\partial x_l \partial x_l} = -\rho \frac{\partial}{\partial x_l} \left(U_k \frac{\partial U_l}{\partial x_k} \right) - \rho \frac{\partial^2 \overline{u_l u_k}}{\partial x_l \partial x_k} - \rho \beta g_i \frac{\partial T}{\partial x_i}, \tag{4-27}$$

where the properties of averages given by eqs. (4-6) through (4-9) are used. One should note that time averages are not appropriate in these equations unless averaged quantities change much more slowly with respect to time (preferably infinitely more slowly) than the unaveraged quantities. Otherwise, averages with respect to space variables in directions for which the turbulence is statistically stationary are preferable. Equation (4-26) is written for a liquid; for a perfect gas c should be replaced by c_p (see eq. [3-35]). Equations (4-25) through (4-27) look like (4-11) through (4-13) with instantaneous values replaced by average values, but with the important difference that an extra term involving $\overline{u_i u_k}$ or $\overline{u_k \tau}$ now appears in each of the equations. These terms arise from the nonlinear velocity and velocity–temperature terms in eqs. (4-11) through (4-13) and are a manifestation of the closure problem in turbulence. If those terms were absent, the set of eqs. (4-25) through (4-27) would contain as many unknowns as equations and so could be solved. In that case turbulent flows would be no more difficult to calculate than laminar ones. Note that barred quantities in eqs. (4-25) through (4-27) that contain lower-case letters are turbulent quantities. The term $-\rho \overline{u_i u_k}$ was discovered by Reynolds [3] and often is called the *Reynolds term* or the *Reynolds stress*.

4-3-1-1 Interpretation of the terms $-\rho \overline{u_i u_k}$ and $\rho c \overline{u_k \tau}$.
The forms of eqs. (4-25) and (4-26) suggest that the quantities $-\rho \overline{u_i u_k}$ and $\rho c \overline{u_k \tau}$ respectively augment the molecular (or viscous) stress tensor $\rho v (\partial U_i / \partial x_k + \partial U_k / \partial x_i)$ and the molecular heat-transfer vector $-\rho c \alpha \partial T / \partial x_k$. Because they involve fluctuating or turbulent components, we interpret

$-\rho\overline{u_i u_k}$ as a turbulent (or Reynolds) stress tensor and $\rho c\overline{u_k \tau}$ as a turbulent heat transfer vector. (The quantity $\overline{u_i u_k}$ is a second-order tensor, because it is the average value of the product of two vectors [the average value of a tensor is a tensor], and $\overline{u_k \tau}$ is a vector, because it is the average value of the product of a vector and a scalar [the average value of a vector is a vector].) Thus, in eqs. (4-25) and (4-26) we write

$$\bar{\tau}_{ki} = \rho \nu \left(\frac{\partial U_i}{\partial x_k} + \frac{\partial U_k}{\partial x_i} \right) - \rho \overline{u_i u_k} \tag{4-25a}$$

and

$$Q_k = -\rho c\alpha \frac{\partial T}{\partial x_k} + \rho c\overline{u_k \tau}, \tag{4-26a}$$

where $\bar{\tau}_{ki}$ is the total stress tensor (the sum of the molecular and turbulent stress tensors), and Q_k is the total heat-transfer vector (the sum of the molecular and turbulent heat-transfer vectors).

We note that the expression for the turbulent stress corresponds exactly to that for the molecular stress obtained in the kinetic theory of gases (see, e.g., ref. [4]). It is only necessary to replace the macroscopic turbulent velocity fluctuations in $-\rho\overline{u_i u_k}$ by random molecular velocities. A similar correspondence exists in the expressions for the turbulent and molecular heat transfer, in which the macroscopic velocity and temperature fluctuations in $\rho c\overline{u_k \tau}$ are respectively replaced by a random molecular velocity and molecular kinetic energy (molecular temperature) [4].

Consider now the term $-\rho\overline{u_i u_k}$ for $i \neq k$. Then, for instance, $-\rho\overline{u_1 u_2}$, in the presence of a mean-velocity gradient $\partial U_1/\partial x_2$, acts like a turbulent shear stress on an $x_1 - x_3$ plane. Similarly, $\rho c\overline{u_2 \tau}$, in the presence of a mean-temperature gradient $\partial T/\partial x_2$, acts like a turbulent heat transfer in the $-x_2$ direction.

To see how these effects come about, consider Fig. 4-1, in which the curve represents either U_1 or T plotted against x_2. Then if u_2 is positive at a particular location, the fluid instantaneously moves into regions of higher U_1 (if $\partial U_1/\partial x_2$ is positive as shown), and so u_1 tends to be negative (the local velocity tends to be less than U_1). Similarly, if u_2 is negative, the fluid instantaneously moves into regions of lower U_1, and so u_1 tends to be positive. In both cases the product $u_1 u_2$ tends to be negative, and so the average value $\overline{u_1 u_2}$ is nonzero and negative. That is, there is negative correlation between u_1 and u_2. The quantity $-\rho\overline{u_1 u_2}$ is the turbulent (Reynolds) shear stress acting on an $x_1 - x_3$ plane and augments the molecular or viscous shear stress $\rho \nu \partial U_1/\partial x_2$ on that plane. Of course if U_1 is uniform, $\overline{u_1 u_2}$ is zero; there is no correlation between u_1 and u_2.

A similar interpretation applies to $\rho c\overline{u_2 \tau}$ if the curve in Fig. 4-1 represents the mean temperature T plotted against x_2. That is, if u_2 is positive at a particular x_i, the fluid instantaneously moves into regions of higher T (if $\partial T/\partial x_2$ is positive as shown), and so τ tends to be negative (the local temperature tends to be less than T). Similarly, if u_2 is negative τ tends to be positive. In both cases the product $u_2 \tau$ tends to be negative. The average value $\overline{u_2 \tau}$ then is nonzero and negative, so that like the molecular conduction term, $\rho c\overline{u_2 \tau}$ produces heat transfer in the $-x_2$ direction. If T is uniform there is, of course, no correlation between u_2 and τ, and $\overline{u_2 \tau}$ is zero.

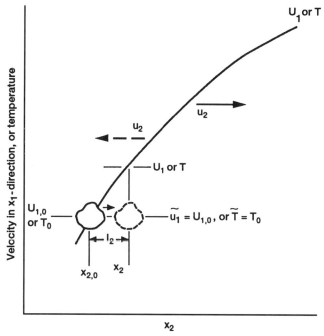

Figure 4-1 Sketch illustrating development of turbulent shear stress $-\rho\overline{u_1 u_2}$ or heat transfer $\rho c \overline{\tau u_2}$ in presence of a mean-velocity or temperature gradient. Regardless of whether the velocity fluctuation u_2 is positive or negative, the product $\overline{u_1 u_2}$ (or $\overline{\tau u_2}$) is negative (see section 4-3-1-1). Also illustrated (in the lower part of the figure) is a mixing-length theory, where l_2 is the mixing length and $x_{2,0}$ is at the virtual origin of an eddy (see section 4-3-2-2).

Finally, we should say something about $-\rho\overline{u_i u_k}$ if $i = k$. In that case $-\rho\overline{u_i u_{(i)}}$ (no sum on i) acts like a normal turbulent stress and augments the normal molecular stress $\rho\nu\partial U_i/\partial x_{(i)}$. For instance $-\rho\overline{u_1^2}$ is the normal stress on an $x_2 - x_3$ plane.

The existence of the turbulent stress tensor $-\rho\overline{u_i u_j}$ and heat-transfer vector $\rho c \overline{\tau u_j}$ is an important and physically significant deduction from the continuum equations for fluids (the Navier-Stokes, energy, and continuity equations). The deduction is obtained from those equations with no approximations by using only Reynolds decomposition and the rules for averaging. Unfortunately the procedure does not provide a way of calculating the values of the turbulent stresses and heat transfer, and so we are left with a closure problem (more unknowns than equations). Next we consider some simple closure schemes.

4-3-2 Simple Closures of the Equations for Mean Flow and Temperature

In order to close the system of eqs. (4-25) through (4-27), that is, to write it in a form in which the number of unknowns equals the number of equations, one must write the quantities $\overline{u_i u_j}$ and $\overline{u_k \tau}$ in terms of the mean velocity U_i, the mean temperature T, and x_i. To do that we necessarily introduce additional information into the equations, so that the theory is not deductive. However, because of the practical importance of obtaining solutions, a great many closure proposals have been made.

Probably the simplest way of closing the system of averaged equations is to assume that $\overline{u_i u_j}$ and $\overline{u_k \tau}$ are respectively proportional to $U_i U_j$ and $U_k T$. One might suppose that this is a reasonable assumption because $U_i U_j$ and $U_k T$ have the same dimensions as $\overline{u_i u_j}$ and $\overline{u_k \tau}$ and are respectively a second-order tensor and a vector (as are $\overline{u_i u_j}$ and $\overline{u_k \tau}$). Consider what happens, however, to the turbulent stress terms in eqs. (4-25) and (4-27) for the simple case in which $U_i = \delta_{i1} U_1$ and U_1 is independent of x_1; that is, the flow is fully developed. In that case the suggested closure assumption gives, in eqs. (4-25) and (4-27), $\partial(\rho \overline{u_i u_k})/\partial x_k = \partial(\rho U_1 U_1)/\partial x_1 = 0$, and $\partial^2 \overline{u_i u_k}/\partial x_l \partial x_k = \partial^2 (U_1 U_1)/\partial x_1^2 = 0$, where ρ is constant. Thus the assumed form for $\overline{u_i u_j}$ does not appear in eqs. (4-25) and (4-27), and there is no effect of turbulence on fully developed flow. Because that result is contrary to experience, the proposed closure assumption cannot be reasonable. Evidently the conditions of correct dimensionality and correct tensor properties, although necessary in a closure assumption, are not sufficient.

4-3-2-1 Eddy diffusivities. Inadequacies such as that noted in the expression for $\overline{u_i u_j}$ that was just considered can be avoided by introducing the so-called eddy diffusivities. This can be done formally by replacing molecular quantities in eqs. (3-13), (3-14), and (3-29) by turbulent quantities. Thus, in those equations, $\sigma_{ij}/\rho \to -\overline{u_i u_j}$, $\mu/\rho \to \varepsilon$, $q_i/(\rho c) \to \overline{\tau u_i}$, and $k/(\rho c) \to \varepsilon_h$, where the arrows are read "becomes." In addition, according to our notation for turbulent flow, the mean velocity is written as U_i. Then, for incompressible flow we get, for $-\overline{u_i u_j}$ and $\overline{\tau u_j}$,

$$-\overline{u_i u_j} = \varepsilon \left(\frac{\partial U_i}{\partial x_j} + \frac{\partial U_j}{\partial x_i} \right) - \frac{1}{3} \overline{u_k u_k} \delta_{ij} \qquad (4\text{-}28)$$

and

$$\overline{\tau u_j} = -\varepsilon_h \frac{\partial T}{\partial x_j}, \qquad (4\text{-}29)$$

where ε is variously called the *turbulent viscosity*, *eddy viscosity*, or *eddy diffusivity for momentum transfer*, and ε_h is the *turbulent conductivity*, *eddy conductivity*, or *eddy diffusivity for heat transfer*.

Unlike the molecular viscosity and conductivity in eqs. (3-13), (3-14), and (3-29), the eddy diffusivities are functions of the character and intensity of the turbulence. We notice that eqs. (4-28) and (4-29) do not suffer from the problems associated with the expressions given near the beginning of section 4-3-2; in general they give nonzero values for both the turbulent stress and heat-transfer terms in eqs. (4-25) through (4-27) for a fully developed flow in a channel or pipe. Moreover they give zero values for $\overline{u_1 u_2}$ and $\overline{\tau u_2}$ if the mean velocity and temperature gradients are zero, as they should be at the center of symmetrical flows (see section 4-3-1-1). For unsymmetrical flows eqs. (4-28) and (4-29) may break down, but we shall not be concerned here with those flows. For statistically homogeneous flows without mean velocity gradients eq. (4-28) gives $\overline{u_i u_j} = \overline{u_k u_k} \delta_{ij}/3$, which is true only for isotropic turbulence. However, for homogeneous turbulence we use other methods. The eddy diffusivities in eqs. (4-28) and (4-29) are scalars. Other expressions, in which they are tensors, have also been proposed [5].

Equations (4-28) and (4-29) do not provide closures for the eqs. (4-25) through (4-27), because we still do not know ε and ε_h as functions of position or mean velocities.

However, they provide a framework in which $\overline{u_i u_j}$ and $\overline{\tau u_j}$ are respectively a second-order tensor and a vector, as they should be, and in which reasonable expressions for $\overline{u_i u_j}$ and $\overline{\tau u_j}$ might be obtained. In general, expressions for those quantities must be tailored to the particular problem being considered.

4-3-2-2 Mixing length. Before considering specific expressions for the eddy diffusivities, we introduce Prandtl's mixing-length hypothesis [5, 6], which gives a rough estimate of $\varepsilon_h/\varepsilon$ and an approximate picture of how momentum and heat might be transferred in a turbulent shear flow.

If turbulence exists in a flow, eddies or portions of fluid move about in an apparently random fashion. If a mean-velocity gradient or temperature gradient exists in a direction transverse to the main flow, some of the eddies move transversely into layers of different mean velocity or temperature (see Fig. 4-1 and section 4-3-1-1). Consider the turbulent fluctuations u_1 and τ at a point x_i (see eqs. [4-14] and [4-16]). The mean velocity and temperature at that point are respectively U_1 and T, and the mean gradients are in the x_2 direction, as in Fig. 4-1.

According to mixing-length theory the fluctuations u_1 and τ at x_i are produced by an eddy or turbulent particle that originates from an instability at another point. We define the virtual origin of the eddy that produced the fluctuations u_1 and τ at x_i as the point $x_{i,0}$ where the eddy would have been born with the mean velocity $U_{1,0}$ and temperature T_0 if the x_1 momentum and temperature of the eddy were conserved as it travels from $x_{i,0}$ to x_i (see Fig. 4-1). Note that actual conservation need not occur, because the virtual origin of the eddy can differ from its actual origin.

From eqs. (4-14) and (4-16), which define Reynolds decomposition, the fluctuations u_1 and τ at x_i are given by

$$u_1 = \tilde{u}_1 - U_1 \qquad (4\text{-}30)$$

and

$$\tau = \tilde{T} - T, \qquad (4\text{-}31)$$

where the tildes designate total instantaneous quantities at x_i, or mean values plus fluctuations. But mean values plus fluctuations in eqs. (4-30) and (4-31), designated by $\tilde{u}_1$ and $\tilde{T}$ respectively, are just the total x_1 velocity and temperature of the eddy considered in Fig. 4-1 and in the definition of virtual origin in the last paragraph. According to that definition, if the eddy is born at the virtual origin $x_{i,0}$ with the mean x_1 velocity $U_{1,0}$ and temperature T_0, then the x_1 momentum and temperature of that eddy effectively are conserved as it moves from $x_{i,0}$ to x_i. Thus $\tilde{u}_1$ and $\tilde{T}$ can be replaced by $U_{1,0}$ and T_0 respectively. Equations (4-30) and (4-31) then become

$$u_1 = U_{1,0} - U_1 \qquad (4\text{-}32)$$

and

$$\tau = T_0 - T. \qquad (4\text{-}33)$$

Substituting eqs. (4-32) and (4-33) into $-\overline{u_1 u_2}$ and $\overline{\tau u_2}$, we get

$$-\overline{u_1 u_2} = \overline{(U_1 - U_{1,0})u_2} \qquad (4\text{-}34)$$

and

$$\overline{\tau u_2} = -\overline{(T - T_0)u_2}. \tag{4-35}$$

Note that $U_1 - U_{1,0} = u_1$ and $T - T_0 = \tau$ are not removed from the bar because they vary with the averaging variable or variables (x_1, x_3, or t). If we expand U_1 and T in Taylor series about $U_{1,0}$ and T_0 respectively and retain only the first two terms in each, we get

$$-\overline{u_1 u_2} = \overline{l_2 \frac{dU_1}{dx_2} u_2} = \overline{l_2 u_2} \frac{dU_1}{dx_2} \tag{4-36}$$

and

$$\overline{\tau u_2} = -\overline{l_2 \frac{dT}{dx_2} u_2} = \overline{l_2 u_2} \frac{dT}{dx_2}, \tag{4-37}$$

where $l_2 = x_2 - x_{2,0}$ and is designated the *mixing length*. Let $i = 1$, $j = 2$, and the mean flow be in the x_1 direction in eqs. (4-28) and (4-29). Comparison of those equations with eqs. (4-36) and (4-37) then gives

$$\varepsilon = \varepsilon_h = \overline{l_2 u_2}. \tag{4-38}$$

The relation between ε and ε_h given by eq. (4-38), besides following from mixing length theory, gives good agreement with experiment except at low Prandtl numbers $c\mu/k$. At low Prandtl numbers, as for liquid metals, the thermal conductivity k is generally so high that heat is conducted to or from an eddy as it moves transversely. The temperatures in eq. (4-35) then must be replaced by values that are closer together, and $|\overline{\tau u_2}|$ decreases [7–9]. In this chapter we are not concerned with low–Prandtl-number fluids and use eq. (4-38).

Equations (4-36) and (4-37) have sometimes been criticized (often mercilessly) because they assume that l_2 is small enough that U_1 and T vary linearly over that distance, whereas in reality it may not be. That problem can be overcome by retaining more terms in the Taylor-series expansions for $U_1 - U_{1,0}$ and $T - T_0$. For instance in the expansion for $U_1 - U_{1,0}$, we can retain an additional term $(1/2)l_2^2 \partial^2 U_1/\partial x_2^2$. However, that would complicate the analysis and may not be worth the effort, because one can probably absorb any second-order effects in the expressions that are assumed for the eddy diffusivity.

4-3-2-3 Nonuniformity of turbulent mixing. A fundamental question about the nature of turbulence concerns how turbulent mixing takes place. Here we consider the instantaneous turbulent mixing that occurs in the presence of mean velocity or temperature gradients. We consider it at this point because of its relevance to mixing-length theory. It turns out that nonuniform mixing is a consequence of several known facts about turbulence.

Let us see what the presence of the turbulent stress and heat transfer terms in the equations for mean flow and heat transfer (eqs. [4-25] and [4-26]) implies about the instantaneous turbulent mixing. Instantaneous mixing refers here to the mixing one sees in a snapshot taken at a particular time.

Note first what would happen if the spatial pattern of instantaneous turbulent mixing were uniform, or nearly so. If that were the case, a portion of fluid as it moves transversely in mean velocity and temperature gradients (mean velocity in the x_1 direction) would have a uniform tendency (because of uniform mixing) to assume the mean x_1 momentum and temperature of the surrounding fluid at each point along its path. That tendency would be more pronounced at higher turbulence intensity or Reynolds number, because small-scale motions become excited with increasing turbulence intensity [10], and so the turbulent mixing (average or instantaneous) increases. (Note that turbulent mixing takes place most efficiently by small-scale motions, because those provide the most intimate contact of the fluid entering a region with that already there.)

Thus if the instantaneous turbulent mixing were spatially uniform, the tendency of a portion of fluid to assume the mean x_1 momentum and temperature of the surrounding fluid at each point as it moves transversely would increase with increasing turbulence intensity. That, however, would cause the fluctuations from the mean, u_1 and τ in the turbulent stress $-\rho\overline{u_i u_j}$ and heat transfer $\rho c \overline{\tau u_j}$ to decrease in magnitude with increasing turbulence intensity or Reynolds number. The stress component $\rho \overline{u_1^2}$ then would decrease. But that trend is unphysical and does not occur. In fact, as might be expected, the opposite trend occurs; as turbulence intensity $(\overline{u_i u_i}/3)^{1/2}$ or Reynolds number increases, $\overline{u_1^2}$ increases [11].

The instantaneous turbulent mixing therefore cannot be spatially uniform, or nearly so, as assumed in obtaining the considered unphysical trend. There must be regions of relative quiescence if x_1 momentum and heat are to be transferred turbulently at high turbulence intensities. But in that case there must also be regions in which the instantaneous mixing is relatively intense and localized, because that is the only way the average mixing could be high for high turbulence intensities if regions of quiescence are present. So the only sensible assumption about the instantaneous mixing is that it is small except in localized regions, where it is intense. Then the above unphysical trends do not occur, because the tendency of a portion of fluid, as it travels transversely, to assume the mean x_1 momentum and temperature of the surrounding fluid is sudden and is confined to localized regions. Note that even if the turbulence is statistically homogeneous, the instantaneous turbulent mixing tends to be highly inhomogeneous.

The fact that instantaneous turbulent mixing takes place mainly in localized regions means that the turbulence must be spatially intermittent in the small scales, because mixing, as mentioned before, takes place mainly by small-scale motions. Intermittency in the small scales has been found experimentally [12].

The localness or suddenness of the turbulent transfer considered in this section also seems to be in agreement with the concept of bursting coherent structures in shear flow near a wall. Much work has been done recently on that phenomenon [13].

Our result that instantaneous turbulent mixing is sudden and localized is congruous with mixing-length theory, which requires a certain suddenness in the turbulent mixing for turbulent transfer to take place. The mixing length can be thought of as the effective distance an eddy moves before mixing with the surrounding fluid. If the mixing took place continuously the mixing length would be zero, and the turbulent shear stress and heat transfer would be zero (eqs. [4-36] and [4-37]).

Thus, although fluid turbulence occurs in a continuum, changes in the momentum and temperature of a moving portion of fluid tend to be sudden and localized. In that respect turbulent systems are not unlike the systems considered in the kinetic theory of

62 TURBULENT FLUID MOTION

gases, in which encounters between particles are sudden and localized. It is of interest that it was apparently kinetic theory that originally inspired turbulent mixing-length theory; the mixing length was supposed to be something like the mean free path of kinetic theory [6].

4-3-2-4 Some conditions satisfied by the turbulent shear stress and heat transfer near a wall. It may be helpful in obtaining models for the turbulent shear stress and heat transfer (or for the eddy diffusivities for momentum and heat transfer) to determine conditions that must be satisfied by those quantities in the region near a wall. Moreover the results obtained in this section, so far as they go, are exact, depending only on the continuity equation and the boundary conditions for the velocity and temperature fluctuations. The results for the turbulent shear stress have been given previously, for example in ref. [14].

Consider an incompressible nonrarefied flow in the x_1 direction parallel to a wall. The normal to the wall is in the x_2 direction, with $x_2 = 0$ at the wall. The continuity equation is

$$\frac{\partial u_i}{\partial x_i} = \frac{\partial u_1}{\partial x_1} + \frac{\partial u_2}{\partial x_2} + \frac{\partial u_3}{\partial x_3} = 0. \tag{4-39}$$

Because the flow is nonrarefied, the nonslip condition holds at the wall, so that

$$(u_1)_{x_2=0} = (u_2)_{x_2=0} = (u_3)_{x_2=0} = 0 \tag{4-40}$$

at every point on the wall. Thus

$$\left(\frac{\partial u_1}{\partial x_1}\right)_{x_2=0} = \left(\frac{\partial u_3}{\partial x_3}\right)_{x_2=0} = 0. \tag{4-41}$$

Equation (4-39) then gives, at the wall,

$$\left(\frac{\partial u_2}{\partial x_2}\right)_{x_2=0} = 0. \tag{4-42}$$

Consider now the turbulent shear stress $-\rho \overline{u_1 u_2}$. Then

$$-\frac{\partial \overline{u_1 u_2}}{\partial x_2} = -\frac{\overline{\partial u_1 u_2}}{\partial x_2} = -\overline{u_1 \frac{\partial u_2}{\partial x_2}} - \overline{u_2 \frac{\partial u_1}{\partial x_2}}, \tag{4-43}$$

$$-\frac{\partial^2 \overline{u_1 u_2}}{\partial x_2^2} = -\overline{u_1 \frac{\partial^2 u_2}{\partial x_2^2}} - 2\overline{\frac{\partial u_1}{\partial x_2}\frac{\partial u_2}{\partial x_2}} - \overline{u_2 \frac{\partial^2 u_1}{\partial x_2^2}}, \tag{4-44}$$

and

$$-\frac{\partial^3 \overline{u_1 u_2}}{\partial x_2^3} = -\overline{u_1 \frac{\partial^3 u_2}{\partial x_2^3}} - 3\overline{\frac{\partial u_1}{\partial x_2}\frac{\partial^2 u_2}{\partial x_2^2}} - 3\overline{\frac{\partial u_2}{\partial x_2}\frac{\partial^2 u_1}{\partial x_2^2}} - \overline{u_2 \frac{\partial^3 u_1}{\partial x_2^3}}, \tag{4-45}$$

where the properties of averages given by eqs. (4-6) and (4-7) are used. By using eqs. (4-40), (4-42), and (4-43) through (4-45) we get, at the wall,

$$-(\overline{u_1 u_2})_{x_2=0} = 0, \tag{4-46}$$

$$-\left(\frac{\partial \overline{u_1 u_2}}{\partial x_2}\right)_{x_2=0} = 0, \tag{4-47}$$

$$-\left(\frac{\partial^2 \overline{u_1 u_2}}{\partial x_2^2}\right)_{x_2=0} = 0, \tag{4-48}$$

and

$$-\left(\frac{\partial^3 \overline{u_1 u_2}}{\partial x_2^3}\right)_{x_2=0} = -3\left(\overline{\frac{\partial u_1}{\partial x_2} \frac{\partial^2 u_2}{\partial x_2^2}}\right)_{x_2=0} \tag{4-49}$$

Because the zero, first, and second derivatives are zero at the wall, $-\overline{u_1 u_2}$, starting at zero, increases very slowly with x_2 near the wall. The third derivative (see eq. [4-49]) gives no information, because at least at this stage we do not know how to evaluate $-(\partial u_1/\partial x_2 \partial^2 u_2/\partial x_2^2)_{x_2=0}$. Thus,

$$0 \leq \left|-\frac{\partial^3 \overline{u_1 u_2}}{\partial x_2^3}\right|_{x_2=0} \leq \infty. \tag{4-50}$$

A similar result holds for higher-order derivatives.

As an example, determine whether

$$-\overline{u_1 u_2} \propto x_2^{18/5} \tag{4-51}$$

is in agreement with eqs. (4-46) through (4-50). Equation (4-51) and its derivatives give, at the wall, $-(\overline{u_1 u_2})_{x_2=0} = -(\partial \overline{u_1 u_2}/\partial x_2)_{x_2=0} = -(\partial^2 \overline{u_1 u_2}/\partial x_2^2)_{x_2=0} = 0$ and $(\partial^3 \overline{u_1 u_2}/\partial x_2^3)_{x_2=0} = 0$. Because these evaluations are in agreement with eqs. (4-46) through (4-50), eq. (4-51) satisfies continuity and the nonslip boundary conditions for u_i at the wall. In section 4-3-2-9 we see that it also gives results in agreement with experiment.

Next consider the turbulent heat transfer $\rho c \overline{\tau u_2}$. Although the procedure is similar to that for the turbulent shear stress, the results can be different because the temperature fluctuation τ at the wall may not be zero if the thermal conductivity of the wall material is not sufficiently high. Thus consider two cases,

$$\tau_{x_2=0} \neq 0 \tag{4-52}$$

for low thermal conductivity of the wall material, and

$$\tau_{x_2=0} = 0 \tag{4-53}$$

for high thermal conductivity of the wall material. Then, proceeding as for $\overline{u_1 u_2}$, we have

$$\frac{\partial \overline{\tau u_2}}{\partial x_2} = \frac{\overline{\partial \tau u_2}}{\partial x_2} = \overline{\tau \frac{\partial u_2}{\partial x_2}} + \overline{u_2 \frac{\partial \tau}{\partial x_2}}, \tag{4-54}$$

$$\frac{\partial^2 \overline{\tau u_2}}{\partial x_2^2} = \overline{\tau \frac{\partial^2 u_2}{\partial x_2^2}} + 2\overline{\frac{\partial \tau}{\partial x_2} \frac{\partial u_2}{\partial x_2}} + \overline{u_2 \frac{\partial^2 \tau}{\partial x_2^2}}, \tag{4-55}$$

and

$$\frac{\partial^3 \overline{\tau u_2}}{\partial x_2^3} = \overline{\tau \frac{\partial^3 u_2}{\partial x_2^3}} + 3\overline{\frac{\partial \tau}{\partial x_2}\frac{\partial^2 u_2}{\partial x_2^2}} + 3\overline{\frac{\partial u_2}{\partial x_2}\frac{\partial^2 \tau}{\partial x_2^2}} + \overline{u_2 \frac{\partial^3 \tau}{\partial x_2^3}}. \tag{4-56}$$

Using eqs. (4-40), (4-42), (4-52), and (4-54) through (4-56), we get, for a wall with low thermal conductivity,

$$(\overline{\tau u_2})_{x_2=0} = 0, \tag{4-57}$$

$$\left(\frac{\partial \overline{\tau u_2}}{\partial x_2}\right)_{x_2=0} = 0, \tag{4-58}$$

$$\left(\frac{\partial^2 \overline{\tau u_2}}{\partial x_2^2}\right)_{x_2=0} = \left(\overline{\tau \frac{\partial^2 u_2}{\partial x_2^2}}\right)_{x_2=0}, \tag{4-59}$$

and

$$\left(\frac{\partial^3 \overline{\tau u_2}}{\partial x_2^3}\right)_{x_2=0} = \left(\overline{\tau \frac{\partial^3 u_2}{\partial x_2^3}}\right)_{x_2=0} + 3\left(\overline{\frac{\partial \tau}{\partial x_2}\frac{\partial^2 u_2}{\partial x_2^2}}\right)_{x_2=0}.$$

Again, because we cannot evaluate the right side of eq. (4-59),

$$0 \le \left|\frac{\partial^2 \overline{\tau u_2}}{\partial x_2^2}\right|_{x_2=0} \le \infty. \tag{4-60}$$

A similar result holds for the third and higher derivatives.

On the other hand, eqs. (4-40), (4-42), (4-53), and (4-54) through (4-56) give, for a wall with high thermal conductivity,

$$-(\overline{\tau u_2})_{x_2=0} = 0, \tag{4-61}$$

$$-\left(\frac{\partial \overline{\tau u_2}}{\partial x_2}\right)_{x_2=0} = 0, \tag{4-62}$$

$$-\left(\frac{\partial^2 \overline{\tau u_2}}{\partial x_2^2}\right)_{x_2=0} = 0, \tag{4-63}$$

$$-\left(\frac{\partial^3 \overline{\tau u_2}}{\partial x_2^3}\right)_{x_2=0} = -3\left(\overline{\frac{\partial \tau}{\partial x_2}\frac{\partial^2 u_2}{\partial x_2^2}}\right)_{x_2=0}, \tag{4-64}$$

and

$$0 \le \left|-\frac{\partial^3 \overline{\tau u_2}}{\partial x_2^3}\right|_{x_2=0} \le \infty. \tag{4-65}$$

A result similar to eq. (4-65) holds for the fourth and higher derivatives.

So, like the turbulent shear stress, the turbulent heat transfer is zero at a wall and increases very slowly with x_2 near a wall. However $|\overline{\tau u_2}|$ may increase slightly more rapidly than $|\overline{u_1 u_2}|$ if the wall thermal conductivity is low, because in that case $(\partial^2 \overline{\tau u_2}/\partial x_2^2)_{x_2=0}$ may be nonzero. Because an effect of wall thermal conductivity on turbulent heat transfer does not seem to have been measured, however, and because equality of eddy diffusivities for heat and momentum transfer gives good results, this possible difference may not be significant in most situations. Of course, new work may show otherwise.

4-3-2-5 Specialization of eqs. (4-25) and (4-26) for fully developed parallel mean flow and fully developed heat transfer without buoyancy.
For simplicity consider a turbulent flow between two infinite parallel walls, which may or may not be moving. The flow is fully developed in space and time and is in the x_1 direction. The direction normal to the walls is x_2. Fully developed flow is here taken to mean that all of the dependent variables in eq. (4-25) except the pressure are independent of time and x_1. The pressure may vary with x_1. If we neglect buoyancy effects by letting $g_i = 0$, eq. (4-25) becomes, for $i = 1$ and 2,

$$0 = -\frac{\partial P}{\partial x_1} + \frac{d}{dx_2}\left(\rho v \frac{dU_1}{dx_2} - \rho\overline{u_1 u_2}\right) \quad (4\text{-}66)$$

and

$$0 = -\frac{\partial P}{\partial x_2} - \rho\frac{d\overline{u_2^2}}{dx_2}. \quad (4\text{-}67)$$

Note that by virtue of the equation following (3-22), σ_e drops out of eq. (4-25) for $g_i = 0$. Differentiating eq. (4-67) with respect to x_1 gives

$$\frac{\partial}{\partial x_2}\left(\frac{\partial P}{\partial x_1}\right) = -\rho\frac{\partial}{\partial x_1}\left(\frac{d\overline{u_2^2}}{dx_2}\right) = 0, \quad (4\text{-}68)$$

because $\overline{u_2^2} = \overline{u_2^2}(x_2)$ does not change with x_1 for fully developed flow. Integrating eq. (4-66) with respect to x_2 (using eqs. [4-68] and [4-25a]) then gives

$$\frac{\partial P}{\partial x_1}x_2 + C_1 = \rho v\frac{dU_1}{dx_2} - \rho\overline{u_1 u_2} = \bar{\tau}_{21}, \quad (4\text{-}69)$$

where C_1 is a constant of integration whose value depends on the boundary conditions, and $\bar{\tau}_{21}$ is the total averaged shear stress (the sum of the molecular and turbulent parts). According to eq. (4-69) the total shear stress varies linearly with x_2.

For the case in which both walls are stationary, $dU_1/dx_2 = \overline{u_1 u_2} = \bar{\tau}_{21} = 0$ at the channel center (at $x_2 = (x_2)_c$), and

$$C_1 = -(x_2)_c\frac{\partial P}{\partial x_1}. \quad (4\text{-}70)$$

Using that value for C_1, eq. (4-69) becomes, at $x_2 = 0$,

$$-(x_2)_c\frac{\partial P}{\partial x_1} = (\bar{\tau}_{21})_w \equiv \bar{\tau}_w$$

and

$$1 - \frac{x_2}{(x_2)_c} = \frac{\rho v}{\bar{\tau}_w}\frac{dU_1}{dx_2} - \frac{\rho}{\bar{\tau}_w}\overline{u_1 u_2} = \frac{\bar{\tau}_{21}}{\bar{\tau}_w}. \quad (4\text{-}71)$$

Thus the total shear stress $\bar{\tau}_{21}$ varies linearly with distance from the wall, being a maximum at the wall and zero at the channel center. Because the turbulent shear stress

$-\rho\overline{u_1u_2}$ is much greater than the molecular shear stress $\rho\nu dU_1/dx_2$, except in the immediate vicinity of the wall, $-\overline{u_1u_2}$ also varies linearly with x_2 over most of the channel cross-section. At the wall, of course, unlike the total shear stress $\bar{\tau}_{21}$, $-\overline{u_1u_2}$ is zero because the velocity fluctuations are zero.

Next consider the energy equation (eq. [4-26]) for the case to which eq. (4-69) applies (a steady, turbulent, unidirectional, fully developed mean flow between infinite parallel walls). The mean temperature is also steady, and the flow is fully developed thermally. Equation (4-26) can then be written using eq. (4-26a), as

$$\rho c U_1 \frac{\partial T}{\partial x_1} = \frac{\partial}{\partial x_1}\left(\rho c \alpha \frac{\partial T}{\partial x_1} - \rho c \overline{u_1\tau}\right) + \frac{\partial}{\partial x_2}\left(\rho c \alpha \frac{\partial T}{\partial x_2} - \rho c \overline{u_2\tau}\right) = -\frac{\partial Q_1}{\partial x_1} - \frac{\partial Q_2}{\partial x_2}, \tag{4-72}$$

where Q_1 and Q_2 are respectively the total (time-averaged) heat transfer (molecular plus turbulent) in the x_1 and x_2 directions. Fully developed thermally is here taken to mean that the molecular and turbulent heat-transfer vectors are independent of x_1. This definition is consistent with eq. (4-72), as can be seen by differentiation of that equation with respect to x_1. (Note that although the left side of eq. (4-72) was not assumed independent of x_1, independence follows from the independence of x_1 of the molecular heat-transfer vector.) Then eq. (4-72) becomes, for fully developed heat transfer,

$$\rho c U_1 \frac{\partial T}{\partial x_1} = \frac{d}{dx_2}\left(\rho c \alpha \frac{\partial T}{\partial x_2} - \rho c \overline{u_2\tau}\right) = -\frac{dQ_2}{dx_2}, \tag{4-73}$$

or

$$\frac{\partial T}{\partial x_1}\int U_1(x_2)dx_2 + C_2 = \alpha\frac{\partial T}{\partial x_2} - \overline{u_2\tau} = -\frac{Q_2}{\rho c}, \tag{4-74}$$

where $\partial T/\partial x_1$ is written outside the integral sign, because the flow is fully developed thermally ($[\partial/\partial x_1][\alpha\partial T/\partial x_2] = 0 = [\partial/\partial x_2][\partial T/\partial x_1]$).

4-3-2-6 Law of the wall. Equations (4-11) and (4-13) can be written in dimensionless form for $g_i = 0$ as

$$\frac{\partial \tilde{u}_i^+}{\partial t^+} = -\frac{\partial(\tilde{u}_i^+\tilde{u}_k^+)}{\partial x_k^+} - \frac{\partial\tilde{\sigma}^+}{\partial x_i^+} + \frac{\partial^2\tilde{u}_i^+}{\partial x_k^+\partial x_k^+} \tag{4-75}$$

and

$$\frac{\partial^2\tilde{\sigma}^+}{\partial x_l^+\partial x_l^+} = -\frac{\partial^2(\tilde{u}_i^+\tilde{u}_k^+)}{\partial x_i^+\partial x_k^+}, \tag{4-76}$$

where a tilde over a quantity indicates a total instantaneous value (mean plus fluctuating) and where

$$\tilde{u}_i^+ \equiv \frac{\tilde{u}_i}{\sqrt{\bar{\tau}_w/\rho}}, \quad x_i^+ \equiv \frac{x_i\sqrt{\bar{\tau}_w/\rho}}{\nu}, \quad \tilde{\sigma}^+ \equiv \frac{\tilde{\sigma}/\rho}{\bar{\tau}_w/\rho}, \quad \text{and} \quad t^+ \equiv \frac{t\bar{\tau}_w}{\rho\nu}. \tag{4-77}$$

(The quantity σ_e drops out of the set of equations because $g_i = 0$; see equation following [3-22].)

The set of eqs. (4-75) and (4-76) contains four equations in the four unknowns $\tilde{u}_1^+$, $\tilde{u}_2^+$, $\tilde{u}_3^+$, and $\tilde{\sigma}^+$. Thus one could do a thought experiment, starting from given initial conditions. Using appropriate boundary conditions at the walls (e.g., $u_i^+ = 0$ for $x_2^+ = 0$, $2(x_2^+)_c$, a solution of eqs. (4-75) and (4-76) in principle could be obtained. For fully developed statistically steady flow one would have to consider a region far from the channel entrance and times large enough for the solution to become independent of initial conditions. The parameter in this system is $(x_2^+)_c$, which appears as one half the maximum value of x_2^+ (the walls are $2[x_2^+]_c$ units apart). The quantity $(x_2^+)_c \equiv (x_2)_c\sqrt{\bar{\tau}_w/\rho}/\nu$ is a kind of Reynolds number, because $\sqrt{\bar{\tau}_w/\rho}$ is a velocity (called the *friction velocity*). The solution of eqs. (4-75) and (4-76), with boundary conditions at $x_2^+ = 0$ and $2(x_2^+)_c$, gives

$$\tilde{u}_i^+ = \tilde{u}_i^+\left[x_k^+, t^+, (x_2^+)_c\right]$$

and

$$\tilde{\sigma}^+ = \tilde{\sigma}^+\left[x_k^+, t^+, (x_2^+)_c\right].$$

However, because mean values are steady and fully developed, $\overline{u_1^+ u_2^+}$ (see eq. [4-14]) and $\bar{u}_1^+ = U_1^+$ (see eq. [4-18]) are not functions of x_1^+, x_3^+, or t^+. Thus,

$$\overline{u_1^+ u_2^+} = \overline{u_1^+ u_2^+}\left[x_2^+, (x_2^+)_c\right], \tag{4-78}$$

and

$$\bar{u}_1^+ = U_1^+ = U_1^+\left[x_2^+, (x_2^+)_c\right]. \tag{4-79}$$

Next let $(x_2^+)_c \to \infty$. Then for the region in which $x_2^+ \ll (x_2^+)_c$, the parameter $(x_2^+)_c$ drops out of the determining boundary conditions. That occurs because the only boundary condition determining the turbulence at $x_2^+ \ll (x_2^+)_c$ is the one at the near wall, the boundary condition at $x_2^+ = 2(x_2^+)_c \to \infty$ having no effect. So eqs. (4-78) and (4-79) become, for $(x_2^+)_c \to \infty$ and $x_2^+ \ll (x_2)_c$,

$$\overline{u_1^+ u_2^+} = \overline{u_1^+ u_2^+}(x_2^+), \tag{4-80}$$

and

$$U_1^+ = U_1^+(x_2^+). \tag{4-81}$$

For the present fully developed unidirectional flow, the eddy viscosity from eq. (4-28)

$$\varepsilon = -\frac{\overline{u_1 u_2}}{dU_1/dx_2}, \tag{4-82}$$

or

$$\frac{\varepsilon}{\nu} = -\frac{\overline{u_1^+ u_2^+}}{dU_1^+/dx_2^+}. \tag{4-83}$$

So, from eqs. (4-78) and (4-79),

$$\frac{\varepsilon}{\nu} = \frac{\varepsilon}{\nu}\left[x_2^+, (x_2^+)_c\right], \tag{4-84}$$

or, from eqs. (4-80) and (4-81),

$$\frac{\varepsilon}{\nu} = \frac{\varepsilon}{\nu}(x_2^+) \qquad (4\text{-}85)$$

for $x_2^+ \ll (x_2^+)_c$ and $(x_2^+)_c \to \infty$.

The fact that eqs. (4-80), (4-81), and (4-85) hold for $x_2^+ \ll (x_2^+)_c$ and $(x_2^+)_c \to \infty$ is often called the *law of the wall*. Although that is usually considered to be an empirical law, we have obtained it here by using the Navier-Stokes equations in a thought experiment. Note that if one considers the region so close to the wall that $\overline{u_1^+ u_2^+} \to 0$, the so-called laminar sublayer, eqs. (4-71) and (4-77) give (for $x_2 \ll [x_2^+]_c$)

$$\frac{dU_1^+}{dx_2^+} = 1, \qquad (4\text{-}86)$$

or

$$U_1^+ = x_2^+. \qquad (4\text{-}87)$$

4-3-2-7 Velocity-defect law. Consider now the central region of the channel, sometimes referred to as the region away from the wall. As in the last section the flow is fully developed and statistically steady. We write eqs. (4-75) and (4-76) as

$$\frac{\partial \tilde{u}_i^+}{\partial t^*} = -\frac{\partial (\tilde{u}_i^+ \tilde{u}_k^+)}{\partial x_k^*} - \frac{\partial \tilde{\sigma}^+}{\partial x_i^*} + \frac{1}{(x_2^+)_c} \frac{\partial^2 \tilde{u}_i^+}{\partial x_k^* \partial x_k^*} \qquad (4\text{-}88)$$

and

$$\frac{\partial^2 \tilde{\sigma}^+}{\partial x_l^* \partial x_l^*} = -\frac{\partial^2 (\tilde{u}_i^+ \tilde{u}_k^+)}{\partial x_i^* \partial x_k^*}, \qquad (4\text{-}89)$$

where

$$x_i^* \equiv \frac{x_i}{(x_2)_c}, \quad (x_2^+)_c \equiv \frac{(x_2)_c \sqrt{\bar{\tau}_w/\rho}}{\nu}, \quad \text{and} \quad t^* \equiv \frac{t\sqrt{\bar{\tau}_w/\rho}}{(x_2)_c}. \qquad (4\text{-}90)$$

The quantities $\tilde{u}_i^+$ and $\tilde{\sigma}^+$ are defined in eq. (4-77). Proceeding as for the law of the wall in the last section, we note that eqs. (4-88) and (4-89) constitute a set of four equations in the four unknowns $\tilde{u}_1^+, \tilde{u}_2^+, \tilde{u}_3^+$, and $\tilde{\sigma}^+$. Here the walls where the boundary conditions are set are at $x_2^* = 0, 2$ (e.g., $\tilde{u}_i^+ = 0$ at $x_2^* = 0, 2$). As in the last section we do a thought experiment, starting with given initial conditions. We consider a cross-section in which the mean flow is fully developed and continue the solution until the flow is statistically steady. Then eqs. (4-88) and (4-89), or their solutions, show that

$$\tilde{u}_i^+ = \tilde{u}_i^+\left[x_k^*, t^*, (x_2^+)_c\right] \qquad (4\text{-}91)$$

and

$$\tilde{\sigma}^+ = \tilde{\sigma}^+\left[x_k^*, t^*, (x_2^+)_c\right]. \qquad (4\text{-}92)$$

Note that whereas in the last section $(x_2^+)_c$ was included in the functional relation because of its use for the location of the boundary condition at $x_2^+ = 0, 2(x_2^+)_c$, here i is included because of its appearance in the differential eq. (4-88); $(x_2^+)_c$ is not needed for the location of the boundary condition at $x_2^* = 2$.

Because mean values are steady and fully developed, $\overline{\tilde{u}_2^+} = U_1^+$ (see eq. [4-18]) is not a function of x_1^*, x_3^*, or t^*, as are the unaveraged quantities in eqs. (4-91) and (4-92). Thus

$$\overline{\tilde{u}_1^+} = U_1^+ = U_1^+\left[x_2^*, (x_2^+)_c\right]. \tag{4-93}$$

Now we confine our attention to the region away from the wall and let $(x_2^+)_c \to \infty$, so that the viscous term, including $(x_2^+)_c$, drops out of eq. (4-88). Batchelor [12] explains this loss of the influence of viscosity in the context of Fourier analysis. He notes that the region of wavenumber space that is affected by the action of viscous forces moves out from the origin toward a wavenumber of infinity as the Reynolds number increases. In the limit of infinite Reynolds number (infinite $[x_2^+]_c$ in our case) the sink of energy is displaced to infinity, and the influence of viscous forces is negligible for wavenumbers (reciprocal eddy sizes) of finite magnitude. (See Chapter 5 for a discussion of Fourier analysis.) Thus, for $(x_2^+)_c \to \infty$, eq. (4-93) can be written, for the region away from the wall, as

$$\left[U_1^+ - (U_1^+)_c\right] = \left[U_1^+ - (U_1^+)_c\right](x_2^*), \tag{4-94}$$

where $(U_1^+)_c$ is the value of U_1^+ at the channel midpoint. Equation (4-94) is written with $U_1^+ - (U_1^+)_c$ as the dependent variable to ensure that curves for different Reynolds numbers collapse at the channel midpoint. According to eq. (4-94) they then also collapse at points *near* the midpoint. The quantity $U_1^+ - (U_1^+)_c$ is called the *dimensionless velocity defect*, and eq. (4-94), which applies in the central region of the channel if viscous forces are negligible there (if $[x_2^+]_c \to \infty$), is known as the *velocity-defect law*.[1] As in the case of the law of the wall, eq. (4-94) usually is considered to be an empirical law, but it is obtained here by using the Navier-Stokes equations in a thought experiment.

4-3-2-8 Logarithmic law. Next we want to determine the form of the mean velocity profile required in a possible overlap region, where eq. (4-81) (the law of the wall) and eq. (4-94) (the velocity-defect law) both apply. From eqs. (4-71), (4-83), and (4-84), which apply over a whole channel cross-section in the fully developed region, we get

$$1 - \frac{x_2^+}{(x_2^+)_c} = \left\{1 + \frac{\varepsilon}{\nu}\left[x_2^+, (x_2^+)_c\right]\right\}\frac{dU_1^+}{dx_2^+}. \tag{4-95}$$

It is known that, except in the immediate vicinity of the wall, where x_2^+ is less than, say, 70, the molecular shear stress is much smaller than the turbulent. We show later in this section that our results are consistent with that statement. Thus, neglecting the molecular

---

[1] Note that letting $(x_2^+)_c \to \infty$ and neglecting the viscous terms in eqs. (4-88) destroys the accuracy of the solution near the wall, because viscous effects are not negligible there. But as already shown, letting $(x_2^+)_c \to \infty$ preserves the accuracy in the region away from the wall, the region in which we are interested here, because viscous effects are negligible away from the wall.

shear stress in eq. (4-95) we have

$$1 - \frac{x_2^+}{(x_2^+)_c} = \frac{\varepsilon}{\nu}[x_2^+, (x_2^+)_c]\frac{dU_1^+}{dx_2^+}. \tag{4-96}$$

In order that eq. (4-81) (the law of the wall) holds, we confine the range of eq. (4-96) to $x_2^+/(x_2^+)_c \ll 1$, but at the same time keep $x_2^+/(x_2^+)_c$ large enough that the turbulent stress is large compared with the molecular stress.[2] For that range we can use eq. (4-85) in eq. (4-96). The latter then becomes

$$1 = \frac{\varepsilon}{\nu}(x_2^+)\frac{dU_1^+}{dx_2^+}. \tag{4-97}$$

Note that eq. (4-97), if integrated from $x_2 = 0$ (at the near wall) to x_2, is in the form of eq. (4-81) and is thus an expression of the law of the wall.

To ensure that eq. (4-97) also satisfies eq. (4-94), the velocity-defect law, we first write the former in terms of $x_2^* \equiv x_2/(x_2)_c = x_2^+/(x_2^+)_c$, $(x_2^+)_c$, and the dimensionless velocity defect $(U_1^+)_c - U_1^+$. That gives

$$1 = -\frac{\varepsilon}{\nu}[(x_2^+)_c x_2^*]\frac{d[(U_1^+)_c - U_1^+]}{(x_2^+)_c dx_2^*}, \tag{4-98}$$

where the subscript c refers to values at the channel center. If eq. (4-98) is to agree with eq. (4-94), then $(x_2^+)_c$ must cancel out of eq. (4-98). For that to happen, ε/ν must be proportional to $(x_2^+)_c$, so that

$$\frac{\varepsilon}{\nu}(x_2^+) = \frac{\varepsilon}{\nu}[(x_2^+)_c x_2^*] = K(x_2^+)_c x_2^* = Kx_2^+, \tag{4-99}$$

where K is a constant (called the Kármán constant). Then eq. (4-98) becomes

$$1 = -Kx_2^*\frac{d[(U_1^+)_c - U_1^+]}{dx_2^*}, \tag{4-100}$$

which, if integrated from $x_2^* = 1$ (at the channel center) to x_2^*, is in agreement with eq. (4-94). Thus, only if ε/ν is given by eq. (4-99) does eq. (4-97) satisfy both eqs. (4-81) and (4-94), as it must in an overlap region.[3] Substituting eq. (4-99) into eq. (4-97) gives

$$1 = Kx_2^+\frac{dU_1^+}{dx_2^+}, \tag{4-101}$$

or, on integration,

$$U_1^+ = \frac{1}{K}\ln x_2^+ + C. \tag{4-102}$$

---

[2] The range of x_2^+ for which that is possible increases as $(u_2^+)_c \to \infty$; the latter was assumed for obtaining the velocity-defect law and the law of the wall.

[3] Actually, there are expressions for ε/ν that satisfy both eqs. (4-81) and (4-94) but that have an appearance which differs from that of eq. (4-99). However, they reduce to eq. (4-99). For example, Prandtl's mixing-length expression [6] and von Kármán's similarity expression [15] are both of the form $(\varepsilon/\nu) = (\varepsilon/\nu)(U_1^+, x_2^+)$ and satisfy eqs. (4-81) and (4-94). But $(\varepsilon/\nu) = (\varepsilon/\nu)(U_1^+, x_2^+) = (\varepsilon/\nu)[U_1^+(x_2^+), x_2^+] = (\varepsilon/\nu)(x_2^+)$. This agrees with eq. (4-85) and leads to eq. (4-99). Equation (4-99) is simpler than the other expressions for ε/ν and, because the other expressions reduce to it, may be more fundamental.

Equation (4-102) is the well-known logarithmic velocity distribution, which applies to the portion of the law-of-the-wall region in which the turbulent shear stress is much greater than the molecular. It can be obtained in a number of ways (e.g., ref. [15]), but is obtained here from eqs. (4-81) and (4-94) by assuming an overlap region. That was first done by Millikan [16], but in a somewhat different way. Besides obtaining the logarithmic law, we have shown that any expression for ε/ν that satisfies both the law of the wall and the velocity-defect law in some overlap region must, according to our analysis, reduce to Kx_2^+ in that region (eq. [4-99]).

We can now see whether the assumption that the turbulent shear stress is much greater than the molecular shear stress except very close to the wall, where x_2^+ is less than about 70, is in agreement with our results. Substituting the expression for ε/ν in eq. (4-99) into eq. (4-95) gives

$$1 - \frac{x_2^+}{(x_2^+)_c} = (1 + Kx_2^+)\frac{dU_1^+}{dx_2^+}, \tag{4-103}$$

so that the ratio of turbulent to molecular shear stress is $Kx_2^+/1$. Because $K \approx 0.4$, the turbulent shear stress is more than an order of magnitude greater than the molecular stress for $x_2^+ > 70$.

4-3-2-9 Expression for ε/ν for fully developed flow that applies at all distances from a wall. We now give a general expression for the eddy diffusivity that satisfies conditions given in sections 4-3-2-4 and 4-3-2-6 through 4-3-2-8, and which applies across the whole channel in the fully developed region. The expression is

$$\frac{\varepsilon}{\nu} = Lx_2^+\left[1 - \frac{x_2^+}{(x_2^+)_c}\right]^{5/4}, \tag{4-104}$$

where

$$L = \begin{cases} a(U_1^+)^{13/5} & \text{for } a(U_1^+)^{13/5} \leq 2/5 \\ 2/5 & \text{for } a(U_1^+)^{13/5} > 2/5 \end{cases} \tag{4-105}$$

and

$$a = 3 \times 10^{-4}.$$

Using eq. (4-104) to close eq. (4-95), a fully developed form of the equation for the mean flow, we get

$$1 - \frac{x_2^+}{(x_2^+)_c} = \left\{1 + Lx_2^+\left[1 - \frac{x_2^+}{(x_2^+)_c}\right]^{5/4}\right\}\frac{dU_1^+}{dx_2^+}, \tag{4-106}$$

where L is given by eq. (4-105).

Consider now whether the conditions obtained in sections 4-3-2-4 and 4-3-2-6 through 4-3-2-8 are satisfied by eqs. (4-104) through (4-106). First note that those equations satisfy eqs. (4-79) and (4-84) for $0 \leq x_2^+ < \infty$. Then for $x_2^+ \ll (x_2^+)_c$ they agree with the law-of-the-wall eqs. (4-81) and (4-85). As $x_2^+ \to 0$, eq. (4-86) or (4-87) is satisfied. That is, as $x_2^+ \to 0$, U_1^+ becomes equal to x_2^+. An additional condition satisfied as $x_2^+ \to 0$ is that $\varepsilon/\nu \to (x_2^+)^{18/5}$, which is equivalent to eq. (4-51). This follows from evaluating eqs. (4-104) and (4-105) for small values of x_2^+ and using eq. (4-87). Then by using eqs. (4-83) and (4-86) we get eq. (4-51), which in turn satisfies eqs. (4-46) through (4-50). These last equations are consequences of continuity and the nonslip boundary condition at the wall, so the expression for ε/ν given by eqs. (4-104) and (4-105) is also in agreement with those consequences.

Next consider whether eq. (4-106) satisfies the velocity-defect law (eq. [4-94]) for the region away from the wall, where the turbulent shear stress is much greater than the molecular (where x_2^+ is greater than say 70). For that region eq. (4-106) can be written as

$$1 = -\frac{2}{5}x_2^*(1 - x_2^*)^{1/4}\frac{d[(U_1^+)_c - U_1^+]}{dx_2^*}, \qquad (4\text{-}107)$$

where $x_2^* = x_2^+/(x_2^+)_c$. But eq. (4-107), if integrated from the channel center to x_2^*, is in the form of eq. (4-94). Thus eq. (4-106), from which eq. (4-107) is obtained for the region away from the wall, satisfies the velocity-defect law in that region. Integration of eq. (4-107) gives

$$(U_1^+)_c - U_1^+ = \frac{5}{2}\ln\left[\frac{1 + (1 - x_2^*)^{1/4}}{1 - (1 - x_2^*)^{1/4}}\right] - 5\tan^{-1}(1 - x_2^*)^{1/4} \qquad (4\text{-}108)$$

for our velocity-defect law.

Finally, for the law-of-the-wall region, where $x_2^+ \ll (x_2^+)_c$, eq. (4-106) becomes

$$1 = \left(1 + \frac{2}{5}x_2^+\right)\frac{dU_1^+}{dx_2^+}. \qquad (4\text{-}109)$$

Then for the part of that region in which the turbulent shear stress is much greater than the molecular (the region away from the wall), we have

$$1 = \frac{2}{5}x_2^+ \frac{dU_1^+}{dx_2^+} \qquad (4\text{-}110)$$

which, if integrated, becomes

$$U_1^+ = \frac{5}{2}\ln x_2^+ + C, \qquad (4\text{-}111)$$

which agrees with the logarithmic distribution in eq. (4-102).

Equation (4-106) is plotted semilogarithmically for several values of $(x_2^+)_c$ in Fig. 4-2. Also included in the plot are some experimental data from refs. [11] and [15], and a numerical solution (direct numerical simulation) of the unaveraged Navier-Stokes equations from reference 17. Although the experimental data are for a pipe, and eq. (4-106) is obtained for a flat channel, a similar equation is obtained for a pipe by letting $(x_2)_c$ be the radius of the pipe. For the region close to a wall, the flow in a pipe should be similar to that in a channel, because for that region the pipe wall can be considered flat. Also, for a pipe, as well as for a channel, the total shear stress varies linearly with wall distance, as can be shown by writing a force balance on an element of fluid for both configurations.

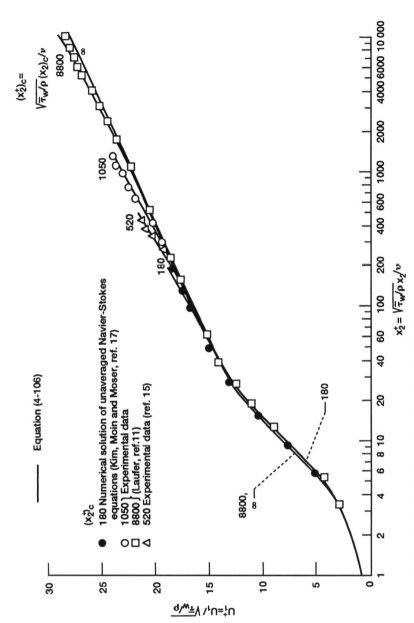

Figure 4-2 Velocity distribution for turbulent flow through a channel or pipe.

It is conceivable that the eddy diffusivity for the central region might be different for the two configurations, but such differences are not apparent in Fig. 4-2. Equation (4-106) appears to represent both the data and the numerical solution quite well.

We can calculate friction factor as a function of Reynolds number by using the velocity distribution from eq. (4-106). Because we are comparing the calculated friction factors with those obtained experimentally for a pipe, the quantity $(x_2)_c$ is the pipe radius (see the previous paragraph). The friction factor f is given by

$$f \equiv \frac{\bar{\tau}_w}{\rho U_a^2/2} = \frac{2}{(U_a^+)^2}, \qquad (4\text{-}112)$$

and the Reynolds number Re is obtained from

$$Re \equiv 2(x_2)_c \frac{U_a}{\nu} = 2(x_2^+)_c U_a^+, \qquad (4\text{-}113)$$

where U_a^+ is a dimensionless bulk or mixed-mean velocity and is given by

$$U_a^+ \equiv \frac{U_a}{\sqrt{\tau_w/\rho}} = [2/(x_2^+)_c] \int_0^{(x_2^+)_c} [(x_2^+)_c - x_2^+] U_1^+ \, dx_2^+ \qquad (4\text{-}114)$$

for a pipe.

A plot of friction factor versus Reynolds number, as obtained from eqs. (4-106), (4-112), and (4-113), is compared with experimental data for turbulent flow in a pipe in Fig. 4-3. As in the case of the velocity distribution in Fig. 4-2 the agreement is quite good.

4-3-2-10 Thermal law of the wall, temperature-defect law, and temperature logarithmic law. Laws analogous to those obtained for the mean velocity distribution in sections 4-3-2-6 through 4-3-2-8 can be obtained for the mean temperature distribution if fully developed heat transfer, in addition to fully developed flow, occurs in a passage. The procedures are similar, and so this section is not detailed. The main difference here is that the energy eq. (4-12) (in appropriate dimensionless forms) is added to the lists of unaveraged equations used in the thought experiments. Thus, in addition to eqs. (4-75) and (4-76) we use

$$\frac{\partial \tilde{T}^+}{\partial t^+} = -\frac{\partial (\tilde{T}^+ \tilde{u}_k^+)}{\partial x_k^+} + \frac{1}{Pr} \frac{\partial^2 \tilde{T}^+}{\partial x_i^+ \partial x_i^+},$$

and in addition to eqs. (4-88) and (4-89) we have

$$\frac{\partial \tilde{T}^+}{\partial t^*} = -\frac{\partial (\tilde{T}^+ \tilde{u}_k^+)}{\partial x_k^*} + \frac{1}{(x_2^*)_c Pr} \frac{\partial^2 \tilde{T}^+}{\partial x_i^* \partial x_i^*},$$

where $\tilde{T}^+ \equiv \frac{(\tilde{T}_w - \tilde{T}) c \bar{\tau}_w}{(Q_2)_w \sqrt{\bar{\tau}_w/\rho}}$, $Pr \equiv \frac{\nu}{\alpha}$ is the Prandtl number, and $(Q_2)_w$ is the time-average normal heat transfer at the wall. The rest of the dimensionless quantities have already been defined in eqs. (4-77) and (4-90). Then one obtains the thermal law of the wall, the temperature-defect law, and the temperature logarithmic law in place of eqs. (4-81

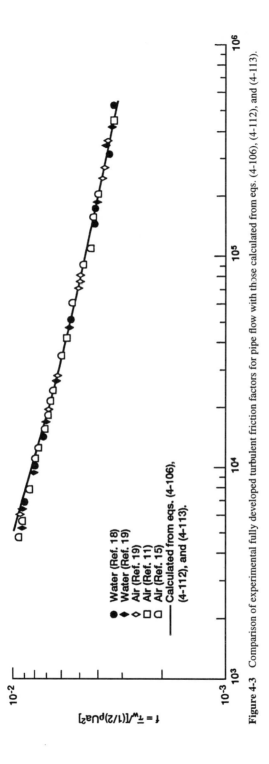

Figure 4-3 Comparison of experimental fully developed turbulent friction factors for pipe flow with those calculated from eqs. (4-106), (4-112), and (4-113).

(4-94), and (4-102), respectively. These are

$$T^+ = T^+(x_2^+, Pr),$$

$$(T_c^+ - T^+) = (T_c^+ - T^+)(x_2^*),$$

and for an overlap region, where both the thermal law of the wall and the temperature defect law apply,

$$T^+ = \frac{1}{K_1} \ln y^+ + C(Pr),$$

where $T^+ = \bar{\bar{T}}^+$ is the time-averaged dimensionless temperature difference, and $K_1 = K$ (see eq. [4-102]) if $\varepsilon_h = \varepsilon$ as in eq. (4-38). Note that the Prandtl number Pr does not appear in the temperature-defect law because the term $\{1/[(x_2^+)Pr]\}\partial^2 \tilde{T}^+/\partial x_i^* \partial x_i^*$ drops out of the dimensionless energy equation used for the temperature-defect law if $(x_2^+)_c \to \infty$. (This loss of the influence of thermal conduction (or of thermal smearing) can be explained by an argument similar to that for the influence of viscosity, as given after eq. (4-93).) On the other hand, Pr appears in the thermal law of the wall because the term $(1/Pr)\partial^2 \tilde{T}^+/\partial x_i^+ \partial x_i^+$ does not drop out of the dimensionless energy equation used for the law of the wall if $(x_2^+)_c \to \infty$. Finally, note that in getting the logarithmic law for the temperature distribution we use, as obtained from eqs. (4-29) and (4-74),

$$\frac{Q_2}{(Q_2)_w} = \left(\frac{1}{Pr} + \frac{\varepsilon_h}{\nu}\right)\frac{dT^+}{dx_2^+} \quad (4\text{-}115)$$

in place of eq. (4-95), where ε_h is the eddy diffusivity for heat transfer. As a result of this change ε is replaced by ε_h or U_1^+ by T^+ in eqs. (4-97) through (4-102). Note that, as for eq. (4-95), and except for very small Prandtl numbers, if one lets $(x_2^+)_c \to \infty$ there is an extensive region in which the turbulent term is much greater than the molecular term and the term on the left side can be replaced by 1. The results in this section, like those in sections 4-3-2-6 through 4-3-2-8, follow from a thought experiment with the unaveraged continuum equations.

4-3-2-11 Calculation of fully developed convective heat transfer. Consider next a treatment of the fully developed heat transfer that can occur if temperature gradients exist in a flowing fluid. If, in accordance with eq. (4-38) we set $\varepsilon_h = \varepsilon$, eq. (4-115) becomes

$$\frac{Q_2}{(Q_2)_w} = \left(\frac{1}{Pr} + \frac{\varepsilon}{\nu}\right)\frac{dT^+}{dx_2^+}, \quad (4\text{-}116)$$

where, again,

$$T^+ \equiv \frac{(T_w - T)c\bar{\tau}_w}{(Q_2)_w \sqrt{\bar{\tau}_w/\rho}}$$

and

$$Pr \equiv \frac{\nu}{\alpha}$$

is the Prandtl number of the fluid. As for the velocity distribution (eq. [4-106]), ε/ν in eq. (4-116) is obtained from eq. (4-104). Equation (4-116) applies to either a channel or a pipe, as does eq. (4-106). However, the quantity $Q_2/(Q_2)_w$ in eq. (4-116) is different for the two configurations, although the effect of the difference on the heat-transfer coefficients is small except for liquid metals (which are not considered here). Because we are comparing our results mainly with experimental heat-transfer (and mass-transfer) data for a pipe, we use $Q_2/(Q_2)_w$ for a pipe [7, 9]:

$$\frac{Q_2}{(Q_2)_w} = \frac{(x_2^+)_c}{(x_2^+)_c - x_2^+} - \frac{2}{[(x_2^+)_c - x_2^+](x_2^+)_c U_a^+} \int_0^{x_2^+} [(x_2^+)_c - \xi] U_1^+(\xi)\, d\xi.$$

(4-117)

The quantity $(x_2^+)_c$ is defined as $(x_2)_c\sqrt{\tau_w/\rho}/\nu$, where $(x_2)_c$ is now the pipe radius, U_1^+ is obtained from eq. (4-106), and U_a^+ is given by eq. (4-114). (Compare eq. [4-117] with that obtained for a channel by using the first and last members of eq. [4-74]).

We want to calculate the Stanton number as a function of Reynolds and Prandtl numbers. The Stanton number is given by

$$St \equiv \frac{h}{\rho c U_a} = \frac{1}{U_a^+ T_a^+},$$

(4-118)

where the heat-transfer coefficient h is defined as

$$h \equiv \frac{(Q_2)_w}{T_w - T_a},$$

(4-119)

and the dimensionless mixed-mean temperature difference T_a^+ is obtained from

$$T_a^+ = \frac{\int_0^{(x_2^+)_c} [(x_2^+)_c - x_2^+] T^+ U_1^+\, dx_2^+}{\int_0^{(x_2^+)_c} [(x_2^+)_c - x_2^+] U_1^+\, dx_2^+},$$

(4-120)

where U_1^+ is calculated from eq. (4-106), and T^+ is obtained from eqs. (4-116) and (4-104). Finally the Reynolds number Re is given by eq. (4-113). By varying the parameter $(x_2^+)_c$, Stanton number can be calculated as a function of Reynolds number and Prandtl number from these equations.

Figure 4-4 shows plots of calculated Stanton number versus Prandtl number for three Reynolds numbers. Included for comparison are experimental data for heat and mass transfer. The mass-transfer data are included by replacing temperatures by concentrations and molecular thermal diffusivities by molecular diffusivities for mass transfer. The agreement of the experimental data with the solid curves calculated by using velocity and temperature distributions from eqs. (4-106), (4-116), and (4-104) is good over the entire range of variables shown.

Of some interest is the fact that good agreement with experiment was obtained by setting the eddy diffusivity for heat transfer (or for mass transfer) ε_h equal to that for momentum transfer ε, as obtained from the mixing-length theory (see eq. [4-38]). By contrast, the agreement obtained by setting $\varepsilon/\varepsilon_h$ equal to the molecular Prandtl number (about 0.7 for air), as is often done, is much poorer. This is indicated by the dashed curve in Fig. 4-4c. These results do not, of course, prove that $\varepsilon/\varepsilon_h = 1$ at all points in the flow, because $\varepsilon/\varepsilon_h$ may vary with position. They only show that the mean effective value of $\varepsilon/\varepsilon_h$ is very close to 1.

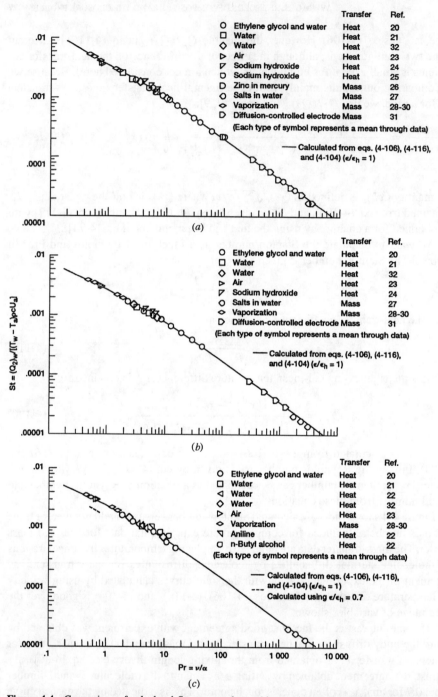

Figure 4-4 Comparison of calculated Stanton numbers with experiment. For the mass-transfer data the temperatures T_w and T_a are replaced by concentrations, and the molecular thermal diffusivity α is replaced by a diffusivity for mass transfer. (a) $Re = 2(x_2)_c U_a/\nu = 10,000$; (b) $Re = 2(x_2)_c U_a/\nu = 25,000$; (c) $Re = 2(x_2)_c U_a/\nu = 50,000$.

4-3-2-12 Some other closure assumptions for fully developed or nearly fully developed (equilibrium) flows.
The results in the previous section utilized an approximate expression for the eddy viscosity that was based partially, however, on exact information. Here we review briefly some earlier approaches.

A reasonable expression for the region away from walls is the von Kármán similarity expression [33]. That expression is most easily obtained by assuming that, away from boundaries, the turbulence at a point is a function only of conditions in the vicinity of the point, in particular, of the first and second derivatives at the point. Then by dimensional analysis we obtain for the eddy viscosity

$$\varepsilon = \varepsilon\left(\frac{\partial U_1}{\partial x_2}, \frac{\partial^2 U_1}{\partial x_2^2}\right) = K^2 \frac{\left(\frac{\partial U_1}{\partial x_2}\right)^3}{\left(\frac{\partial^2 U_1}{\partial x_2^2}\right)^2}, \qquad (4\text{-}121)$$

where K is the Kármán constant.

Von Kármán's hypothesis (eq. [4-121]) has been proposed for the region away from a wall. Close to a wall we assume that the eddy viscosity is a function only of quantities measured relative to a wall, U_1 and x_2, and of ν. The simplest assumption consistent with dimensional analysis and the requirement that the effect of ν should become small for large x_2 is then [33].

$$\varepsilon = \varepsilon(U_1, x_2, \nu) = n^2 U_1 x_2 \left[1 - \exp\left(\frac{-n^2 U_1 x_2}{\nu}\right)\right], \qquad (4\text{-}122)$$

where n is an experimental constant ($n = 0.124$). Equations (4-121) and (4-122) give results for flow and heat (or mass) transfer in tubes that are almost as good as those in Fig. 4-4 (see ref. [33]). The use of those equations also gives good results for the boundary layer on a flat plate [34].

By making a small modification in the derivation of eq. (4-121) we can obtain an expression for ε that should be applicable to a vortex. We assume that the turbulence at a point is dependent only on the shearing deformation rate at the point and in the vicinity of the point; that is, it is a function of the deformation rate and its derivatives. If we exclude derivatives of the deformation rate higher than the first we get, for a parallel flow, eq. (4-121). However, for a circular vortex, we get

$$\varepsilon = \varepsilon\left[\frac{dv}{dr} - \frac{v}{r}, \frac{d}{dr}\left(\frac{dv}{dr} - \frac{v}{r}\right)\right] = \frac{-K^2 \left(\frac{dv}{dr} - \frac{v}{r}\right)^3}{\left[\frac{d}{dr}\left(\frac{dv}{dr} - \frac{v}{r}\right)\right]^2}, \qquad (4\text{-}123)$$

where v is the tangential velocity and r is the radius. For a vortex flow, $v = v_0 r_0/r$, where the subscripts 0 refer to values at some arbitrary radius. Equation (4-123) then becomes

$$\varepsilon = \frac{K^2 v_0 r_0}{2}. \qquad (4\text{-}124)$$

Equation (4-124) has been used profitably for turbulent vortex flows in vortex tubes [35], in astrophysical clouds [36], and in the atmosphere [37].

Finally we mention closures based on Prandtl's mixing-length l_2 [6], where l_2 is defined by

$$\overline{u_1 u_2} = -l_2^2 \left(\frac{\partial U_1}{\partial x_2}\right)^2. \qquad (4\text{-}125)$$

Prandtl assumed that $l_2 = K x_2$, where x_2 is the distance from a wall. Van Driest [38] has modified that assumption by introducing a damping factor that reduces l_2 in the region close to a wall:

$$l_2 = K x_2 \left[1 - \exp\left(\frac{-x_2^+}{A}\right)\right], \qquad (4\text{-}126)$$

where A is an additional experimental constant. Equation (4-126) appears to be reasonably applicable to the regions both close to and away from a wall, as is our eq. (4-104). Equation (4-104) gives, in addition, a velocity profile that is accurate in the so-called wake region, where $x_2 \sim (x_2)_c$ (see Fig. 4-2). Although the use of an equation that is applicable to two or more regions in a flow may be somewhat more convenient than the use of a separate equation for each region, there is basically no reason why one equation should apply to more than one region; the turbulence mechanism is likely different in the different regions.

A word should perhaps be said about the value of the Kármán constant K. If $U_1^+ = U_1^+[x_2^+, (x_2^+)_c]$, as in Fig. 4-2, then a value of K = 0.4, or 2/5, (see eq. [4-105]) gives good results. On the other hand, if it is assumed that $U_1^+ = U_1^+(x_2^+)$, so that a single curve is obtained instead of the multiple curves in Fig. 4-2, then K = 0.36 gives better agreement with the data [33]. Finally, for a vortex (eq. [4-124]), K seems to be close to 0.3 [35]. It is not surprising that the value of K appears to be slightly different for a vortex than for a parallel flow, because, as with all closures, the ones considered here are approximate and correspond only partially to reality. Thus the constants in a closure scheme often must be "fine-tuned" if the flow configuration is changed.

4-3-2-13 Treatment of moderately short highly accelerated turbulent boundary layers.

The previous few sections considered closure schemes for fully developed or nearly fully developed turbulent flows. We turn now to a problem that is in some ways the opposite of those just considered: given a fully developed or nearly fully developed turbulent flow and transverse heat transfer to which the boundary-layer assumptions are applicable, to determine the streamwise evolution if the flow is subjected to a severe streamwise pressure gradient. A boundary layer subjected to a severe pressure gradient can be considered to be the same thing as a highly accelerated boundary layer. According to Bernoulli's equation, which applies outside the boundary layer, where viscous and turbulence stresses are negligible,

$$-\frac{1}{\rho}\frac{dP}{dx_1} = U_\infty \frac{dU_\infty}{dx_1}, \qquad (4\text{-}127)$$

where U_∞ is the velocity just outside the boundary layer. For a thin steady-state two-dimensional boundary layer with constant properties and without buoyancy, eqs. (4-25),

(4-26), and (4-21) become respectively

$$U_1 \frac{\partial U_1}{\partial x_1} = -U_2 \frac{\partial U_1}{\partial x_2} - \frac{1}{\rho}\frac{dP}{dx_1} + \nu \frac{\partial^2 U_1}{\partial x_2^2} - \frac{\partial}{\partial x_2}\overline{u_1 u_2}, \qquad (4\text{-}128)$$

$$U_1 \frac{\partial T}{\partial x_1} = -U_2 \frac{\partial T}{\partial x_2} + \alpha \frac{\partial^2 T}{\partial x_2^2} - \frac{\partial}{\partial x_2}\overline{\tau u_2}, \qquad (4\text{-}129)$$

and

$$\frac{\partial U_2}{\partial x_2} = -\frac{\partial U_1}{\partial x_1}, \qquad (4\text{-}130)$$

where x_1 and x_2 are respectively in the direction along and normal to the wall. The pressure gradient is written as a total derivative dP/dx_1 because P is not a function of x_2 for a thin boundary layer. Equations (4-128) through (4-130) apply to a boundary layer even if the flow is along a curved wall [39].

In order to solve eqs. (4-128) through (4-130) to obtain the evolution of U_1 and T, $\overline{u_1 u_2}$ and $\overline{\tau u_2}$ must be known at each point in the flow. The experiments of Blackwelder and Kovasznay [40] suggest that although severe pressure gradients cause the mean flow in those experiments to change considerably along streamlines, the Reynolds stresses, at least in the important intermediate region of wall distances, are relatively unaffected. This leads us to the hypothesis of frozen Reynolds stresses and turbulent heat transfer along streamlines in a moderately short highly accelerated turbulent boundary layer. We gave theoretical arguments for that hypothesis in refs. [41] and [42], where we showed that it gives good agreement with experiment if applied to eqs. (4-128) through (4-130). Launder [43] had earlier used the concept of a frozen Reynolds stress, but in an approximate integral approach to the boundary layer rather than in the solution of the partial differential eqs. (4-128) and (4-130).

Changes along streamlines, if not zero, should be smaller than, say, changes at constant x_2, because flow along streamlines tends to involve the same fluid or eddies at various streamwise points. In applying the hypothesis of frozen $\overline{u_1 u_2}$ and $\overline{\tau u_2}$ along streamlines it is convenient to transform eqs. (4-128) through (4-130) from (x_1, x_2) to (x_1, ψ) coordinates (von Mises coordinates), where the stream function ψ is given by

$$\frac{\partial \psi}{\partial x_1} = -U_2, \qquad \frac{\partial \psi}{\partial x_2} = U_1, \qquad (4\text{-}131)$$

so that

$$\left(\frac{\partial}{\partial x_1}\right)_{x_2} = \left(\frac{\partial}{\partial x_1}\right)_{\psi} - U_2 \left(\frac{\partial}{\partial \psi}\right)_{x_1}, \qquad \left(\frac{\partial}{\partial x_2}\right)_{x_1} = \left(\frac{\partial}{\partial \psi}\right)_{x_1} U_1, \qquad (4\text{-}132)$$

where the subscripts are quantities held constant. Equations (4-128) through (4-130) then become

$$U_1 \left(\frac{\partial U_1}{\partial x_1}\right)_{\psi} = -\frac{1}{\rho}\frac{dP}{dx_1} + \frac{1}{2}\nu U_1 \frac{\partial^2 U_1^2}{\partial \psi^2} - U_1 \frac{\partial \overline{u_1 u_2}}{\partial \psi} \qquad (4\text{-}133)$$

and

$$\left(\frac{\partial T}{\partial x_1}\right)_\psi = \alpha \frac{\partial}{\partial \psi}\left(U_1 \frac{\partial T}{\partial \psi}\right) - \frac{\partial \overline{\tau u_2}}{\partial \psi}, \qquad (4\text{-}134)$$

where the subscripts ψ indicate changes along a streamline (at constant ψ). By virtue of eq. (4-131), eqs. (4-133) and (4-134) satisfy continuity automatically.

The equations for the evolution of $\overline{u_1 u_2}$ and of $\overline{\tau u_2}$ are obtained later in this chapter (see eqs. [4-140] and [4-141]). For our purpose the important point is that those equations do not contain dP/dx_1, in contrast to eq. (4-133) for the evolution of U_1 or to eq. (4-127) for the evolution of U_∞, the latter of which is a special case of eq. (4-133). Taking that fact into account, as well as the discussion in the paragraph preceding eq. (4-131), one would expect, for large dP/dx_1, that there would be large gradients of U_∞ or of U_1 along streamlines without correspondingly large streamwise gradients of $\overline{u_1 u_2}$ and $\overline{\tau u_2}$. Then if the boundary layer is moderately short, the hypothesis that the turbulent shear stress and transverse heat transfer are frozen at their initial values as one proceeds along a streamline in a flow with large streamwise gradients of U_∞ or P should be a good approximation.

Rescaling the variables in eqs. (4-133) and (4-134) so as to convert them to dimensionless form and introducing the frozen turbulent stress/heat-transfer hypothesis give

$$\frac{1}{2}\left(\frac{\partial U_1^2}{\partial x_1}\right)_\psi = -\frac{dP}{dx_1} + \frac{1}{2}U_1\frac{\partial^2 U_1^2}{\partial \psi^2} - U_1\frac{\partial (\overline{u_1 u_2})_0}{\partial \psi} \qquad (4\text{-}135)$$

and

$$\left(\frac{\partial T}{\partial x_1}\right)_\psi = \frac{1}{Pr}\frac{\partial}{\partial \psi}\left(U_1\frac{\partial T}{\partial \psi}\right) + \frac{\partial (\overline{\tau u_2})_0}{\partial \psi}, \qquad (4\text{-}136)$$

where

$$\frac{U_i}{U_0} \to U_i, \quad \frac{(x_1 - x_0)}{\nu}U_0 \to x_1, \quad \frac{\psi}{\nu} \to \psi,$$

$$\frac{\nu}{\rho U_0^3}\frac{dP}{dx_1} \to \frac{dP}{dx_1}, \quad \frac{(\overline{u_1 u_2})_0}{U_0^2} \to (\overline{u_1 u_2})_0, \qquad (4\text{-}137)$$

$$\frac{T_w - T}{(T_w)_0 - T_\infty} \to T, \quad \frac{(\overline{\tau u_2})_0}{[(T_w)_0 - T_\infty]U_0} \to (\overline{\tau u_2})_0,$$

and the arrow means "has been replaced by." The quantity x_0 is the value of x_1 at the initial station, U_0 is the velocity outside the boundary layer at the initial station, $(T_w)_0$ is the wall temperature at the initial station, T_∞ is the constant temperature outside the thermal boundary layer, and Pr is the Prandtl number. The quantities $(\overline{u_1 u_2})_0$ and $(\overline{\tau u_2})_0$ are, respectively, the values of $\overline{u_1 u_2}$ and $\overline{\tau u_2}$ on the same streamline as $\overline{u_1 u_2}$ and $\overline{\tau u_2}$ but at the initial station.

The frozen turbulent stress/heat-transfer hypothesis can be tested by numerically integrating eqs. (4-135) and (4-136) along streamlines (at constant ψ) and comparing the results with experiments for the same conditions. Upstream of the regions of severe streamwise pressure gradients the pressure gradients were essentially zero.

Velocity profiles (U_1/U_∞ against ψ/ν) are plotted and compared with experiment in Fig. 4-5. (Note the shifted vertical scales.) In all cases the effect of the pressure gradient and the total normal strain parameter U_∞/U_0 is to flatten the profiles. The agreement between theory and experiment is considered good.

Semilogarithmic plots of $U_1/(\bar{\tau}_w/\rho)^{1/2}$ against $(\bar{\tau}_w/\rho)^{1/2}x_2/\nu$ (law-of-the-wall plots) are given in Fig. 4-6. These profiles show the inner region of the boundary layer much better than does Fig. 4-5. The shear stress $\bar{\tau}_w$ at the wall for the theoretical curves was obtained from the slope of the velocity profile at the wall by using points very close to the wall ($[\bar{\tau}_w/\rho]^{1/2}x_2/\nu \ll 1$). Points very close to the wall were necessary because of the nonlinearity of the profile close to the wall in the presence of a severe pressure gradient. It might be pointed out that this nonlinearity makes the experimental determination of the shear stress at the wall extremely difficult. Both theory and experiment indicate that the original logarithmic and wake regions are destroyed by the pressure-gradient and normal-strain effects, although a new logarithmic layer of smaller slope seems to form eventually. Also, the thickness of the sublayer approximately doubles, indicating an apparent "relaminarization," as observed experimentally by many investigators. However, it is not a true relaminarization because, at least in theory, $\overline{u_1 u_2}$ is constant along streamlines. The agreement between theory and experiment appears to be quite good. For values of U_∞/U_0 larger than those shown, the approximation of a frozen Reynolds stress apparently begins to break down.

In order to see how sensitive the development of the mean profile is to the Reynolds shear stress, results were calculated for $\overline{u_1 u_2} = 0$ and are shown dashed in figs. 4-5(b) and 4-7. The effect of $\overline{u_1 u_2}$ on the profiles in Fig. 4-5(b) is slight. The law-of-the-wall plots in Fig. 4-7, on the other hand, show a significant quantitative effect of $\overline{u_1 u_2}$ on the profiles, but qualitatively the curves for $\overline{u_1 u_2} = 0$ and $\overline{u_1 u_2} \neq 0$ are much the same. In both cases the original logarithmic and wake regions are destroyed and the sublayer is thickened. The difference between the indicated quantitative effects of $\overline{u_1 u_2}$ on the profiles in Figs. 4-5(b) and 4-7 is evidently due to the difference in scales and in scaling parameters in the two figures.

Figure 4-8 shows, for a large value of the pressure-gradient parameter, the contributions of various terms in eq. (4-128) to the rate of change of the nondimensional mean kinetic energy $(1/2)\partial(U_1/U_0)^2/\partial(x_1 U_0/\nu)$. (The energy in the transverse velocity component is negligible for a boundary layer.) The contribution of the Reynolds stress term is very large in a narrow region near the wall. However, that tends to be offset by the viscous contribution. Comparison of the curves for $\overline{u_1 u_2} = 0$ and $\overline{u_1 u_2} \neq 0$ shows that the viscous contribution adjusts its value so as to offset the effect of the Reynolds stress. The viscous term is not zero at the wall but balances the pressure-gradient term so that U_1 can remain zero at the wall. Thus, the present velocity profile, in contrast to the case of zero pressure gradient, is nonlinear at the wall. (If it were linear the viscous term in eq. [4-128] would be zero at the wall.)

The pressure-gradient term is independent of wall distance and, for the case shown in Fig. 4-8, becomes dominant for $(\bar{\tau}_w/\rho)^{1/2}x_2/\nu > 40$. Thus, the destruction of the logarithmic and wake regions is due mainly to the pressure-gradient term, rather than

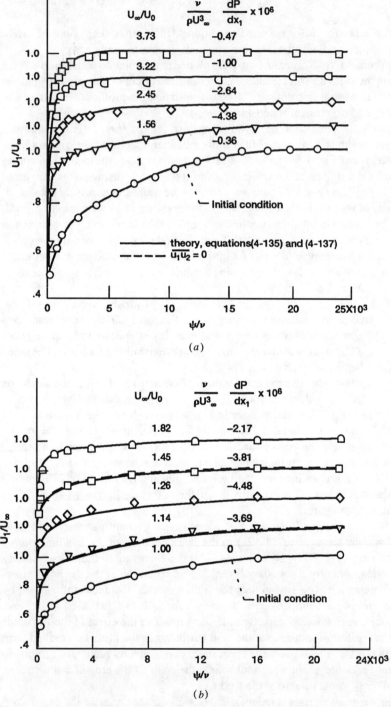

Figure 4-5 Predicted early and intermediate development of mean velocity profile in a turbulent boundary layer with severe favorable pressure gradients and comparison with experiments. Symbols indicate experimental data points. (*a*) Blackwelder and Kovasznay [40]; (*b*) Patel and Head [44] (note shifted vertical scales).

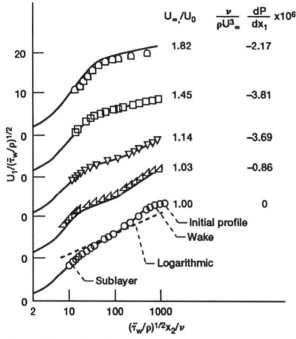

Figure 4-6 Semilogarithmic law-of-the-wall plot of theoretical velocity profiles for severe favorable pressure gradients and comparison with experiment of Patel and Head [44] (note shifted vertical scales). Symbols indicate experimental data points.

to a change in the structure of the turbulence (although some change in structure may occur [45]). Also, the thickening of the sublayer is mostly, although not entirely, due to the pressure-gradient contribution, because as mentioned, viscous effects tend to offset the Reynolds stress contribution.

The results in Fig. 4-8 are, of course, for a large pressure-gradient parameter. For regions of lower pressure gradient, the Reynolds stress has a greater effect, as shown in Fig. 4-7. Also, the velocity profile at any position depends on the whole distribution of pressure gradients up to that position; that is, U_1 is a functional of dP/dx_1, or

$$U_1 = U_1 \left[\frac{dP}{dx_1}(\xi) \right], \tag{4-138}$$

where $0 < \xi < x_1$. Thus, there is a quantitative (but not a qualitative) effect of $\overline{u_1 u_2}$ on the velocity profile, even at those positions at which the pressure-gradient parameter is large.

The analysis can be easily extended to include mass injection at the wall by transforming eq. (4-135) from (x_1, ψ) to (x_1, ψ') coordinates, where $\psi' = \psi - \psi_w$, and ψ_w is the stream function at the wall. The latter varies with x_1 in accordance with

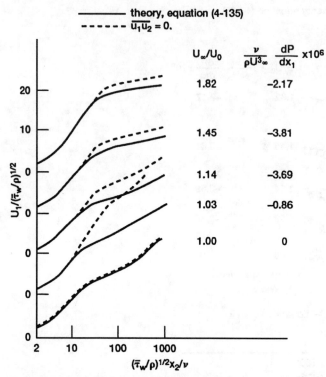

Figure 4-7 Effect of neglecting Reynolds shear stress on theoretical law-of-the-wall plot for experimental conditions of Fig. 4-5 (note shifted vertical scales).

eq. (4-131). Equation (4-135), if written in (x_1, ψ') coordinates, has the additional term $-(1/2)(U_2)_w \partial U_1^2/\partial \psi'$ on the right side, where $(U_2)_w$ is the dimensionless normal velocity at the wall, and where, as in eq. (4-135), all quantities have been nondimensionalized by U_0 and ν (see eq. [4-137]). As before, we use the simplification that $\overline{u_1 u_2}$ remains frozen as it is convected along streamlines (along lines of constant ψ, not constant ψ'). The injected fluid is assumed to be turbulence-free.

To show the effect of mass injection on a boundary layer with severe pressure gradients, mass injection was added in the theoretical calculations in Fig. 4-6. For positive mass injection Fig. 4-9 shows that the normal flow quickly raises the $U_1/(\tau_w/\rho)^{1/2}$ curve, particularly in the wake region, after which the favorable pressure gradient lowers and flattens the curve. The resulting curve still lies above the initial profile. For negative injection, the normal flow and pressure gradient lower and flatten the initial profile. These trends are similar to those observed in the experiments of Julien, Kays, and Moffat [46].

All of the results so far are for favorable pressure gradients, but the analysis should apply as well to severe unfavorable gradients. Figure 4-10 shows a comparison between theory and experiment for the results of Kline et al. [45] for a severe unfavorable pressure gradient (their Fig. 9(a)). The results indicate that the adverse pressure gradients produce an exaggerated wake region, but that the logarithmic and sublayer regions are relatively unaffected. The agreement between theory and experiment is good. It might be mentioned

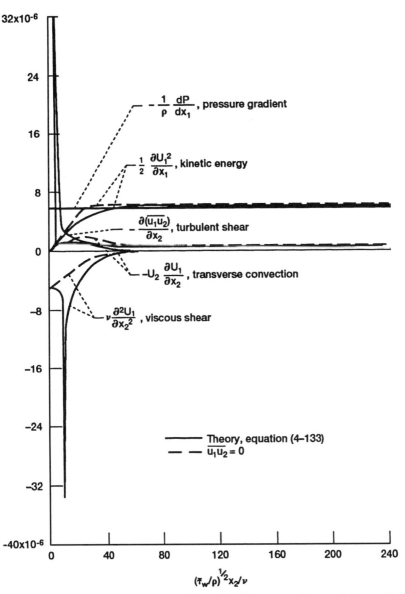

Figure 4-8 Contribution of various terms in eq. (4-128) to streamwise rate of change off dimensionless kinetic energy of mean flow for initial conditions of Fig. 4-6. Terms in eq. (4-128) nondimensionalized by U_0 and ν. $U_\infty/U_0 = 1.14$. Note that transverse component of kinetic energy is negligible for a boundary layer.

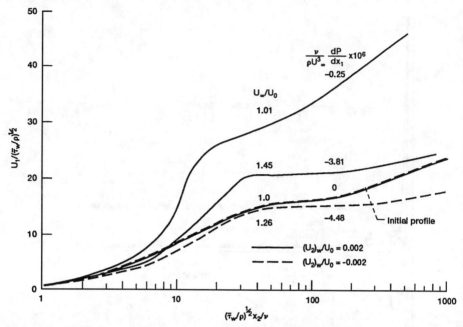

Figure 4-9 Effect of mass injection normal to wall on theoretical velocity profiles for severe favorable pressure gradients. (Initial conditions and pressure gradients correspond to Fig. 4-6.)

that the results for adverse pressure gradients are more sensitive to the $\overline{u_1 u_2}$ distribution than those for favorable gradients. In particular, if $\overline{u_1 u_2}$ is taken as zero, separation occurs upstream for the run shown in Fig. 4-10 for $U_\infty/U_0 = 0.92$. Thus the presence of turbulence appears to delay separation. This is evidently because the Reynolds stress term in eq. (4-128) is positive close to the wall and thus tends to increase U_1 in that region.

Figure 4-11 shows theoretical Stanton-number and skin-friction–coefficient variations with dimensionless longitudinal distance for run 12 from ref. [47]. The shear stress and heat transfer at the wall for the theoretical curves were obtained from the slopes of the velocity and temperature profiles at the wall by using points very close to the wall ($\sqrt{(\tau_w/\rho)}x_2/\nu \ll 1$). Also included in the plot are experimental values for U_∞/U_0. Initial conditions are taken at $x_0 = 4.32$ feet. The difference between the Stanton number and skin-friction–coefficient variations is rather striking and indicates that Reynolds analogy (see, e.g., ref. [33]) does not apply in regions of severe pressure gradients. This difference is also indicated in the experimental results of refs. [40] and [47].

The effect of favorable pressure gradients on velocity and temperature distributions is illustrated in Fig. 4-12, in which theoretical values of T^+ and U_1^+ are plotted against x_2^+ for a low and a high value of pressure-gradient parameter $K = (\nu/\rho U_\infty^3)\, dP/dx_1$. The results are again for run 12 from ref. [47]. The effect of the pressure gradient on the T^+ profile tends to be the opposite of that on the U_1^+ profile. Whereas the pressure gradient flattens the U_1^+ profile in the outer region of the boundary layer, it steepens the T^+ profile in that region. The same trends have been observed experimentally in refs. [44] and [48]. The difference between the velocity and temperature results (or the skin-friction and

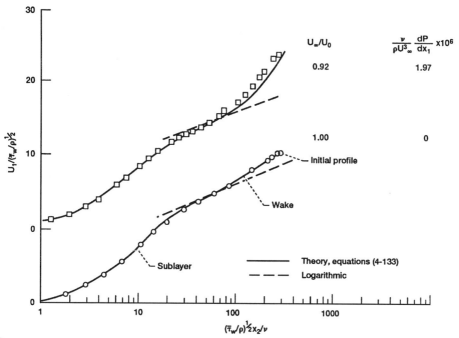

Figure 4-10 Theoretical velocity profiles for severe adverse pressure gradients and comparison with experiment of Kline et al. [45] (note shifted vertical scales).

heat-transfer results) is evidently due to the fact that the equation for the evolution of the mean velocity contains a pressure-gradient term (eq. [4-128]), whereas that for the mean temperature does not (eq. [4-129]). Although the temperature equation does not contain a pressure gradient term, the pressure gradient still can affect the temperature through the mean velocity.

A comparison between theory and experiment for the evolution of Stanton number in severe favorable pressure gradients is presented in Fig. 4-13 for three values of maximum pressure-gradient parameter K_m. The value of x_0 (the initial station for each run was taken at the point at which the local pressure-gradient parameter K_L starts to increase rapidly). For the smaller values of K_m good agreement between theory and experiment is indicated for values of U_∞/U_0 that are not too large. It appears that the range of values of J_∞/U_0 for which the theory applies increases as the pressure-gradient or acceleration parameter increases. For values of $U_0(x_1 - x_0)/\nu$ (or of U_∞/U_0) greater than those shown in Fig. 4-13, the theory appears to break down because the total streamwise strain becomes too great or the local pressure-gradient parameter K_L becomes too small.

Figure 4-13 also shows Stanton numbers calculated for turbulent initial velocity and temperature profiles, but for $\overline{\tau u_2} = \overline{u_1 u_2} = 0$. It is seen that the turbulent stresses and fluxes have a very large effect on the evolution of the Stanton number. The large effect of turbulence on Stanton number perhaps raises questions about calling regions such as those shown in Fig. 4-13 "relaminarization regions," although at the large values of K_m and $U_0(x_1 - x_0)/\nu$ there is some tendency for the zero-turbulence curves to

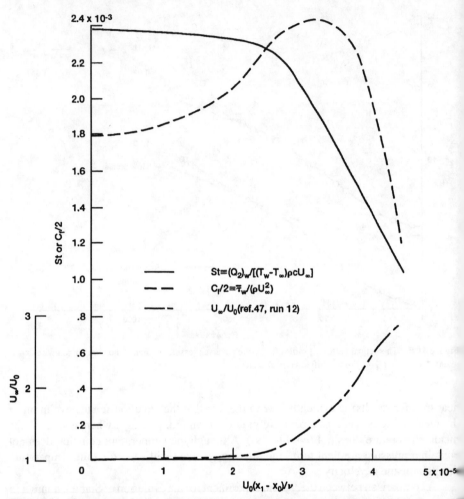

Figure 4-11 Comparison of streamwise evolution of theoretical Stanton number St and skin-friction coefficient C_f in moderately short highly accelerated turbulent boundary layers. x_0 is value of x_1 at initial station. U_0 is value of U_∞ at initial station.

approach those for turbulence. The effects of turbulence on the velocity and temperature distributions are also considerable, although the general trends without turbulence are similar to those in Fig. 4-12.

It should, of course, be remembered that, as is the case for U_1 (eq. [4-138]), Stanton number, as well as C_f, and T at any longitudinal position depend on the whole distribution of pressure gradients up to that position. That is, Stanton number, for instance, is functional of dP/dx_1, or

$$St = St\left[\frac{dP}{dx_1}(\xi)\right], \qquad (4\text{-}139)$$

where $0 < \xi x_1$.

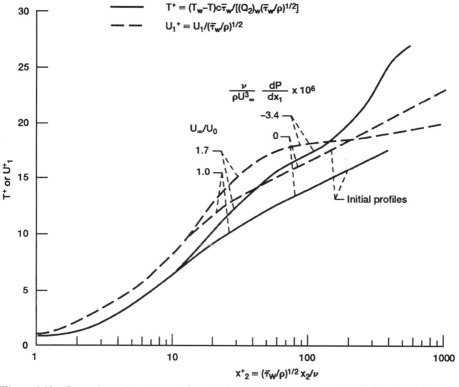

Figure 4-12 Comparison of evolution of theoretical velocity and temperature distributions in moderately short highly accelerated turbulent boundary layers.

4-3-3 One-Point Correlation Equations

Sections 4-3-1 and 4-3-2 consider the equation for the evolution of the mean flow (eq. [4-25]) and that for the mean temperature (eq. [4-26]). Those are obtained by averaging the equations for the instantaneous velocity and temperature (eqs. [4-11] and [4-12]) after applying Reynolds decomposition and contain the important but undetermined turbulent stress tensor $-\rho\overline{u_i u_j}$ and turbulent heat transfer vector $\rho c \overline{u_i \tau}$. This was our first encounter with the closure problem, and sections 4-3-2 through 4-3-2-13 consider some simple closure schemes.

One can construct equations for the evolution of the undetermined quantities $\overline{u_i u_j}$ in eqs. (4-25) and (4-27) from eq. (4-22) for $g_i = 0$ and the following similar equation for the component u_j:

$$\frac{\partial u_j}{\partial t} = -\frac{\partial}{\partial x_k}(u_j u_k) - \frac{1}{\rho}\frac{\partial \sigma}{\partial x_j} + \nu\frac{\partial^2 u_j}{\partial x_k \partial x_k} - u_k\frac{\partial U_j}{\partial x_k} - U_k\frac{\partial u_j}{\partial x_k} + \frac{\partial}{\partial x_k}\overline{u_j u_k}.$$

Note once again that σ_e drops out the equations of motion for $g_i = 0$ by virtue of the equation following eq. (3-22). Next multiply eq. (4-22) by u_j and the above equation for

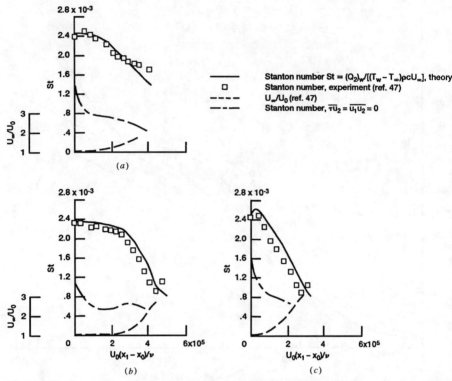

Figure 4-13 Comparison of theory and experiment for streamwise evolution of Stanton number in moderately short highly accelerated turbulent boundary layers. x_0 is value of x_1 at initial station. K_m is maximum value of $\nu/(\rho U_\infty^3)\,dP/dx_1$ for a particular run. (a) $K_m = -1.8 \times 10^{-6}$; (b) $K_m = -3.39 \times 10^{-6}$; (c) $K_m = -3.51 \times 10^{-6}$.

u_j by u_i, add the two equations, and average. This gives, using continuity (eq. [4-21]),

$$\frac{\partial}{\partial t}\overline{u_i u_j} = -\left(\overline{u_j u_k}\frac{\partial U_i}{\partial x_k} + \overline{u_i u_k}\frac{\partial U_j}{\partial x_k}\right) - U_k\frac{\partial}{\partial x_k}\overline{u_i u_j} - \frac{\partial}{\partial x_k}\overline{u_i u_j u_k}$$
$$-\frac{1}{\rho}\left(\frac{\partial}{\partial x_i}\overline{\sigma u_j} + \frac{\partial}{\partial x_j}\overline{\sigma u_i}\right) + \nu\frac{\partial^2 \overline{u_i u_j}}{\partial x_l \partial x_l} + \frac{1}{\rho}\left(\overline{\sigma\frac{\partial u_j}{\partial x_i}} + \overline{\sigma\frac{\partial u_i}{\partial x_j}}\right) - 2\nu\overline{\frac{\partial u_i}{\partial x_l}\frac{\partial u_j}{\partial x_l}}.$$
(4-140)

Similarly, multiplying eq. (4-22) by τ and eq. (4-23) by u_i, adding the two equations and averaging, gives

$$\frac{\partial}{\partial t}\overline{u_i \tau} = -\left(\overline{\tau u_k}\frac{\partial U_i}{\partial x_k} + \overline{u_i u_k}\frac{\partial T}{\partial x_k}\right) - U_k\frac{\partial}{\partial x_k}\overline{u_i \tau} - \frac{\partial}{\partial x_k}\overline{\tau u_i u_k} - \frac{1}{\rho}\frac{\partial}{\partial x_i}\overline{\sigma\tau}$$
$$+\frac{1}{\rho}\overline{\sigma\frac{\partial\tau}{\partial x_i}} + \nu\overline{\tau\frac{\partial^2 u_i}{\partial x_k \partial x_k}} + \alpha\overline{u_i\frac{\partial^2 \tau}{\partial x_k \partial x_k}},$$
(4-141)

AVERAGES, REYNOLDS DECOMPOSITION, AND THE CLOSURE PROBLEM 93

where continuity is again used and buoyancy is neglected. Equations (4-140) and (4-141) are known as *moment* or *correlation* equations. Setting $j = i$ in eq. (4-140) and using continuity, we get, for the rate of change of the kinetic energy per unit mass,

$$\frac{\partial}{\partial t}\left(\frac{\overline{u_i u_i}}{2}\right) = -\overline{u_i u_k}\frac{\partial U_i}{\partial x_k} - U_k\frac{\partial}{\partial x_k}\left(\frac{\overline{u_i u_i}}{2}\right) - \frac{\partial}{\partial x_k}\left(\frac{\overline{u_i u_i}}{2}u_k\right) - \frac{1}{\rho}\frac{\partial}{\partial x_k}\overline{(\sigma u_k)}$$

$$+ \nu\frac{\partial^2\left(\frac{\overline{u_i u_i}}{2}\right)}{\partial x_l \partial x_l} - \nu\overline{\frac{\partial u_i}{\partial x_l}\frac{\partial u_i}{\partial x_l}}. \tag{4-142}$$

As in eqs. (4-25) through (4-27), barred quantities in eqs. (4-140) through (4-142) that contain lowercase letters are turbulent quantities.

The one-point correlation eqs. (4-140) and (4-141) give expressions for the rate of change of $\overline{u_i u_j}$ and $\overline{u_i \tau}$ that might be used in conjunction with eqs. (4-25) through (4-27). The situation with respect to closure, however, is now worse than before. Whereas without eqs. (4-140) and (4-141) we had only to determine $\overline{u_i u_j}$ and $\overline{u_i \tau}$, with them we have to determine quantities like $\overline{u_i u_j u_k}$, $\overline{\tau u_i u_k}$, $\overline{\sigma u_j}$, $\overline{\sigma \tau}$, $\overline{\sigma \partial u_j/\partial x_i}$, $\overline{(\partial u_i/\partial x_l)(\partial u_j/\partial x_l)}$, $\overline{\sigma \partial \tau/\partial x_i}$, and $\overline{\tau \partial^2 u_i/\partial x_k \partial x_k}$. One might use eq. (4-24) to obtain the pressure correlations, but that would only introduce more unknowns. However, eqs. (4-140) through (4-142) are very useful for studying the processes in turbulence, in that most of the terms have clear physical meanings. Moreover, one may be able to calculate terms in those equations from numerical solutions of the unaveraged equations.

An additional one-point equation can be obtained by multiplying eq. (4-23) by τ, averaging, and again neglecting buoyancy. This gives

$$\frac{\partial}{\partial t}\overline{\tau^2} = -2\overline{\tau u_k}\frac{\partial T}{\partial x_k} - U_k\frac{\partial}{\partial x_k}\overline{\tau^2} - \frac{\partial}{\partial x_k}\overline{\tau^2 u_k} + \alpha\frac{\partial^2 \overline{\tau^2}}{\partial x_k \partial x_k} - 2\alpha\overline{\frac{\partial \tau}{\partial x_k}\frac{\partial \tau}{\partial x_k}}, \tag{4-143}$$

where $\overline{\tau^2}$ is the variance of the temperature fluctuation.

4-3-3-1 Physical interpretation of terms in one-point equations. Consider first eqs. (4-140) and (4-142), because physically meaningful interpretations of all the terms in those equations can be given. As in the case of eq. (3-19) it is helpful, for purposes of interpretation, to multiply the terms in eqs. (4-140) and (4-142) through by ρ and by a volume element $dx_1 dx_2 dx_3$. Then the term on the left side of eq. (4-140) or (4-142) gives the time rate of change of $\rho\overline{u_i u_j}$, or of the kinetic energy $\rho\overline{u_i u_i}/2$, within the element. This rate of change is contributed to by the terms on the right sides of the equations. The first of those terms is equal to the net work done on the element by turbulent stresses acting in conjunction with mean-velocity gradients. It is therefore called a *turbulence production* term; it equals the rate of production of $\rho\overline{u_i u_j}$ or of $\rho\overline{u_i u_i}/2$ within the volume element by work done on the element. A somewhat abbreviated interpretation suggested by the form of the term, which often is given, is that it represents work done on the turbulent stress $\rho\overline{u_i u_j}$ by the mean-velocity gradient.

The next term in each of the equations describes the convection or net flow of turbulence or turbulent energy into a volume element by the mean velocity U_k. It moves the turbulence bodily, rather than doing work on it by deforming it, as in the case of the production term. It vanishes either if U_k is zero (no mean flow) or if the turbulence is homogeneous ($\overline{u_i u_j} \neq \overline{u_i u_j}[x_k]$). In the latter case there is no accumulation of turbulence within a volume element, even with a mean flow.

The next three terms in eq. (4-140) and in eq. (4-142) also vanish for homogeneous turbulence. Because they do not contain the mean velocity they do not convect the turbulence bodily or do work on it. Therefore, we interpret them as diffusion terms that diffuse net turbulence from one part of the turbulent field to another by virtue of its inhomogeneity. The pressure-velocity-gradient terms in eq. (4-140) drop out of the contracted eq. (4-142) because of continuity (eq. [4-21]). Therefore, they give zero contribution to the rate of change of the total energy $\overline{u_i u_i}/2$, but they can distribute the energy among the three directional components $u_{(i)}^2/2$ (no sum on i). The last term in eqs. (4-140) and (4-142) is the viscous dissipation term, which dissipates turbulence by the presence of fluctuating velocity gradients.

Consider next eq. (4-141), which on multiplication by ρc and the volume element $dx_1 dx_2 dx_3$ gives the time rate of change of $\rho c \overline{u_i \tau}$ within the element. The first term on the right side gives the production of turbulent heat transfer $\rho c \overline{u_i \tau}$ by the interaction of $\overline{\tau u_k}$ with the mean-velocity gradient and by the interaction of $\overline{u_i u_k}$ with the mean-temperature gradient. The next term gives the net flow of $\rho c \overline{u_i \tau}$ into the volume element by convection. It changes $\rho c \overline{u_i \tau}$ within the element by moving that quantity bodily. The term vanishes for homogeneous turbulence. The next two terms in eq. (4-141) also vanish for homogeneous turbulence, but because they do not contain the mean velocity, they do not convect $\rho c \overline{u_i \tau}$ bodily or produce $\rho c \overline{u_i \tau}$. Therefore, as in the analogous terms in eq. (4-140), we interpret them as diffusion terms that diffuse a turbulence quantity from one part of the turbulence field to another by virtue of its inhomogeneity. The next term changes $\rho c \overline{u_i \tau}$ by the interaction of the pressure fluctuation σ with the temperature-fluctuation gradient. The last two terms change $\rho c \overline{u_i \tau}$ by molecular action. For $\alpha = \nu$ (for a Prandtl number of 1) they can be interpreted by writing them as $\nu \partial^2 \overline{u_i \tau}/\partial x_l \partial x_l - 2\nu \overline{\partial u_i/\partial x_l \partial \tau/\partial x_l}$. Comparing these terms with analogous terms in eq. (4-140), we interpret the first as a diffusion term and the second as a dissipation term.

Finally, consider the equation for the evolution of the temperature variance $\overline{\tau^2}$ (eq. [4-143]). The terms on the right side of that equation are similar to terms already considered in the other evolution equations. They are, respectively, a production term, a convection term, two diffusion terms that diffuse $\overline{\tau^2}$ by virtue of the inhomogeneity of the turbulence field, and a dissipation term that dissipates $\overline{\tau^2}$ by fluctuating temperature gradients.

4-3-3-2 Some direct numerical simulations of terms in one-point moment equations.
In section 4-3-3 we obtain averaged one-point moment (or correlation) equations that give the time evolution of quantities such as $\overline{u_i u_j}$, $\overline{u_i \tau}$, and $\overline{u_i u_i}$. Also, in the previous section we give interpretations of the terms in those equations. However, because of the closure problem that arises in connection with averaged equations, particularly with one-point moment equations, we have not been able to solve those equations or calculate terms. Those terms have, however, been calculated by numerical solution of the unaveraged

AVERAGES, REYNOLDS DECOMPOSITION, AND THE CLOSURE PROBLEM 95

equations (eqs. [4-11] through [4-13]) or by eqs. (4-22) through (4-24) together with eqs. (4-25) through (4-27).

The cases to be considered here are concerned with the evolution of the mean turbulent kinetic energy according to eq. (4-142). The averaged values in that equation vary only in the direction x_2, one of the directions normal to the mean velocity U_1. In that case eq. (4-142) becomes

$$\frac{\partial}{\partial t}\left(\frac{\overline{u_k u_k}}{2}\right) = -\overline{u_1 u_2}\frac{\partial U_1}{\partial x_2} - \frac{1}{\rho}\frac{\partial}{\partial x_2}\overline{\sigma u_2} - \frac{\partial}{\partial x_2}\left(\overline{\frac{u_k u_k}{2}u_2}\right) + \nu\frac{\partial^2}{\partial x_2^2}\left(\frac{\overline{u_k u_k}}{2}\right) - \nu\overline{\frac{\partial u_k}{\partial x_l}\frac{\partial u_k}{\partial x_l}}.$$
(4-144)

Mansour, Kim, and Moin [49] have calculated the terms in eq. (4-144) from a numerical solution of the unaveraged Navier-Stokes equations for a steady fully developed turbulent channel flow. The boundary condition was $u_i = 0$ at the walls. The results, which correspond to the numerical solution plotted in Fig. 4-2, are shown in Fig. 4-14. The corresponding turbulence kinetic energy profile is plotted in Fig. 4-15. Because the turbulence is steady-state, the sum of the terms on the right side of eq. (4-144) is zero.

The production term $-\overline{u_1 u_2}\,dU_1/dx_2$, whose form shows that turbulent energy is produced by work done on the Reynolds shear stress by the mean velocity gradient, is largest in the region close to the wall and peaks at an x_2^+ of about 12. According to Fig. 4-2, the mean velocity profile at that point is just beginning to deviate appreciably from a laminar ($U_1^+ = x_2^+$) profile. Note that the production term $-\overline{u_1 u_2}\,dU_1/dx_2$ in Fig. 4-14 is positive as it should be, because according to the argument in section 4-3-1-1 and Fig. 4-1, if dU_1/dx_2 is positive $\overline{u_1 u_2}$ is negative, and vice versa. Similarly, the dissipation term $-\nu\overline{\partial u_k/\partial x_l \partial u_k/\partial x_l}$ is always negative, because it is the negative of the sum of the squares of velocity gradients. The remaining three terms in eq. (4-144), which are plotted in Fig. 4-14, are diffusion terms. They are the kinetic-energy diffusion term $\partial\overline{u_2 u_k u_k}/2/\partial x_2$, the viscous diffusion term $\nu\partial^2\overline{u_k u_k}/2/\partial x_2^2$, and the pressure diffusion term $-(\partial\overline{\sigma u_2}/\partial x_2)/\rho$. The plots of those terms show that all three of them are positive near $x_2 = 0$, where a large energy sink occurs because of the boundary condition $u_i = 0$ at $x_2 = 0$ (Fig. 4-15). Farther away from the wall they become negative. Thus they remove turbulent energy from the region where the energy is large and deposit it where the energy is smaller. All of the diffusion terms therefore tend to make the turbulence more homogeneous. Far from the wall there is a region of small positive $d\overline{u_2 u_k u_k}/2/dx_2$, apparently because the energy in that region decreases gradually with increasing x_2.

A comparison of the turbulence diffusion processes with spectral transfer processes (discussed in the next chapter) and the directional-transfer processes arising from the pressure velocity-gradient correlations (see section 4-3-3-1) is instructive. The spectral-transfer processes remove energy from wavenumber (or eddy size) regions in which the energy is large and deposit it in regions of smaller energy. The directional-transfer processes remove energy from large-energy directional components and deposit it in a directional component (or components) where the energy is smaller. The turbulence diffusion processes, as shown here, remove energy from regions of space where

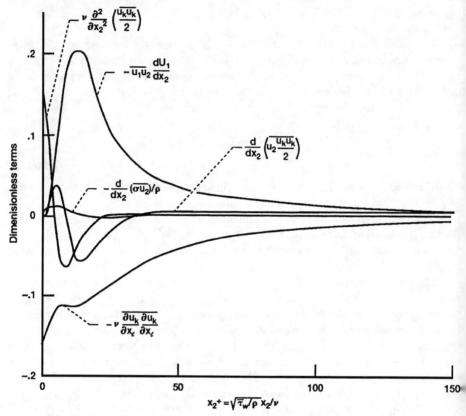

Figure 4-14 Plot of terms in one-point equation for evolution of kinetic energy (eq. [4-144]) for fully developed channel flow. Results obtained by numerical solution of unaveraged Navier-Stokes equations (Mansour, Kim, and Moin [49]). Terms nondimensionalized by $\sqrt{\bar{\tau}_w/\rho}$ and ν. Channel Reynolds number (based on centerline U_1 and channel half-width) is 3200.

the energy is large and deposit it in regions of smaller energy. The spectral transfer, directional transfer, and turbulence diffusion processes tend, respectively, to make the turbulence more uniform in wavenumber space, more isotropic, and more homogeneous in physical space.

Terms in eq. (4-144) have also been calculated for a developing turbulent free shear layer, where the mean quantities were functions only of x_2 and time [50]. The terms were calculated from a numerical solution of the unaveraged equations, where the boundary conditions were periodic. Results were similar to those just considered for a fully developed channel flow, insofar as the kinetic-energy and pressure-diffusion terms diffused turbulence kinetic energy from regions in which the energy was high to regions in which it was lower. As expected (see the discussion for the channel flow) the production term was positive. The viscous dissipation and the viscous diffusion terms were, however, negligibly small, unlike those terms for the channel. That was apparently

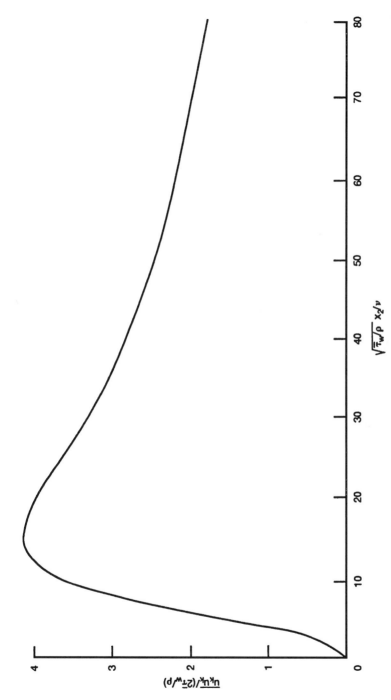

Figure 4-15 Plot of turbulence kinetic energy for fully developed channel flow in Figs. 4-14 and 4-2. Results obtained by numerical solution of unaveraged Navier-Stokes equations (Kim, Moin, and Moser [17]).

because the presence of walls in the channel produced larger mean gradients there than those in the free shear layer.

4-3-4 Two-Point Correlation Equations

As shown in section 4-3-3 the use of one-point correlation or moment equations in conjunction with the equations for the mean flow causes the number of unknowns to go up faster than the number of available equations, so that the closure problem, in effect, gets worse. This leads one to consider the use of two-point equations, in which one might be able to at least eliminate terms containing instantaneous gradients, such as $\overline{\sigma \partial u_j/\partial x_i}$ and $\overline{\partial u_i/\partial x_l \partial u_j/\partial x_l}$. In that case the use of higher-order equations might be expected to increase the number of unknowns at the same rate as the number of available equations. Because an assumption for still undetermined higher-order quantities (e.g., triple correlations) might be expected to affect the flow to a lesser extent than an assumption for lower-order quantities (e.g., $\overline{u_i u_j}$), there might be some advantage in this procedure. Moreover, the use of two-point equations enables us to consider the important spectral-transfer problem of turbulence, as we see in the next chapter.

Before obtaining the two-point correlation equations, consider the structure of the two-point correlations. Some of those correlations have been obtained by Kim, Moin, and Moser [17] in a numerical solution of the unaveraged Navier-Stokes equations and are plotted in Fig. 4-16. As the separation r_k between the defining points increases, the correlation between the velocities at those points goes to zero (as $r_k \to \infty$). This might be considered a consequence of the mixing property of turbulence; although mixing is generally associated with temporal separation (see section 4-1-1), the effects of temporal and spatial separations appear to be similar. In connection with the spanwise separations, note that the correlations go negative before becoming asymptotically zero as r_2 increases, in contrast to the correlations with point separation in the streamwise direction. More is said about that phenomenon in the next chapter, in which it is predicted theoretically.

In this section we consider only two-point equations obtained from the Navier-Stokes equations and neglect heat-transfer and buoyancy effects. Then eq. (4-22) for u_i at a point P and a similar equation for u'_j at a point P' can be written as

$$\frac{\partial u_i}{\partial t} + u_k \frac{\partial U_i}{\partial x_k} + U_k \frac{\partial u_i}{\partial x_k} + \frac{\partial}{\partial x_k}(u_i u_k) - \frac{\partial}{\partial x_k}\overline{u_i u_k} = -\frac{1}{\rho}\frac{\partial \sigma}{\partial x_i} + \nu \frac{\partial^2 u_i}{\partial x_k \partial x_k} \quad (4\text{-}145)$$

and

$$\frac{\partial u'_j}{\partial t} + u'_k \frac{\partial U'_j}{\partial x'_k} + U'_k \frac{\partial u'_j}{\partial x'_k} + \frac{\partial}{\partial x'_k}(u'_j u'_k) - \frac{\partial}{\partial x'_k}\overline{u'_j u'_k} = -\frac{1}{\rho}\frac{\partial \sigma'}{\partial x'_j} + \nu \frac{\partial^2 u'_j}{\partial x'_k \partial x'_k}. \quad (4\text{-}146)$$

Multiplying eq. (4-145) by u'_j and eq. (4-146) by u_i, adding, taking averages, using the fact that quantities at one point are independent of the position of the other point, and introducing the new variables $r_k \equiv x'_k - x_k$ and $(x_k)_n \equiv x_k + (x'_k - x_k)n (0 \le n \le 1$

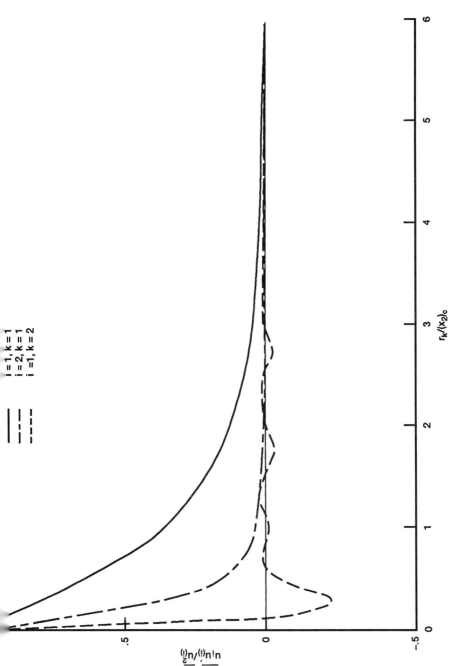

Figure 4-16 Variation of two-point correlation coefficients with point seperation as obtained by Kim, Moin, and Moser [17] from a direct numerical simulation of unaveraged Navier-Stokes equations. $x_2/(x_2)_c = 0.030, x_2^+ = 5.39$.

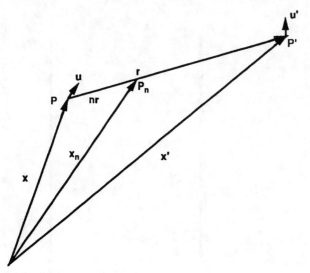

Figure 4-17 Vector configuration for two-point correlation equations.

(Fig. 4-17), results in

$$\frac{\partial}{\partial t}\overline{u_i u'_j} + \overline{u_k u'_j}\frac{\partial U_i}{\partial x_k} + \overline{u_i u'_k}\frac{\partial U'_j}{\partial x'_k} + (U'_k - U_k)\frac{\partial}{\partial r_k}\overline{u_i u'_j}$$

$$+ [(1-n)U_k + nU'_k]\frac{\partial}{\partial (x_k)_n}\overline{u_i u'_j} + \frac{1}{\rho}\left[\frac{\partial}{\partial (x_i)_n}\overline{\sigma u'_j} + \frac{\partial}{\partial (x_j)_n}\overline{u_i \sigma'}\right]$$

$$+ \frac{\partial}{\partial (x_k)_n}\left[n\overline{u_i u'_j u'_k} + (1-n)\overline{u_i u_k u'_j}\right] + \frac{\partial}{\partial r_k}\left(\overline{u_i u'_j u'_k} - \overline{u_i u_k u'_j}\right)$$

$$= \frac{1}{\rho}\left[n\frac{\partial}{\partial (x_i)_n}\overline{\sigma u'_j} + (1-n)\frac{\partial}{\partial (x_j)_n}\overline{u_i \sigma'} - \frac{\partial}{\partial r_j}\overline{u_i \sigma'} + \frac{\partial}{\partial r_i}\overline{\sigma u'_j}\right]$$

$$+ (1 - 2n + 2n^2)\nu\frac{\partial^2 \overline{u_i u'_j}}{\partial (x_k)_n \partial (x_k)_n} - 2(1-2n)\nu\frac{\partial^2 \overline{u_i u'_j}}{\partial (x_k)_n \partial r_k} + 2\nu\frac{\partial^2 \overline{u_i u'_j}}{\partial r_k \partial r_k},$$

(4-147)

where the following transformations are used:

$$\frac{\partial}{\partial x_k} = (1-n)\frac{\partial}{\partial (x_k)_n} - \frac{\partial}{\partial r_k},$$

$$\frac{\partial}{\partial x'_k} = n\frac{\partial}{\partial (x_k)_n} + \frac{\partial}{\partial r_k},$$

$$\frac{\partial^2}{\partial x_k \partial x_k} + \frac{\partial^2}{\partial x'_k \partial x'_k} = (1 - 2n + 2n^2)\frac{\partial^2}{\partial (x_k)_n \partial (x_k)_n} - 2(1-2n)\frac{\partial^2}{(x_k)_n \partial r_k} + 2\frac{\partial^2}{\partial r_k \partial r_k}.$$

AVERAGES, REYNOLDS DECOMPOSITION, AND THE CLOSURE PROBLEM

To obtain the general two-point equations for the pressure-velocity correlations, take the divergence of eqs. (4-145) and (4-146) and use the continuity equation. Thus, from eq. (4-146), one obtains

$$\frac{1}{\rho}\frac{\partial^2 \sigma'}{\partial x'_j \partial x'_j} = -2\frac{\partial u'_k}{\partial x'_j}\frac{\partial U'_j}{\partial x'_k} - \frac{\partial^2 u'_j u'_k}{\partial x'_j \partial x'_k} + \frac{\partial^2 \overline{u'_j u'_k}}{\partial x'_j \partial x'_k}. \qquad (4\text{-}148)$$

Multiplying by u_i, taking averages, and introducing the variables r_k and $(x_k)_n$, as in eq. (4-147), gives

$$\frac{1}{\rho}\left[n^2 \frac{\partial^2 \overline{u_i \sigma'}}{\partial (x_j)_n \partial (x_j)_n} + 2n\frac{\partial^2 \overline{u_i \sigma'}}{\partial (x_j)_n \partial r_j} + \frac{\partial^2 \overline{u_i \sigma'}}{\partial r_j \partial r_j}\right] = -2\frac{\partial U'_j}{\partial x'_k}\left[n\frac{\partial \overline{u_i u'_k}}{\partial (x_j)_n} + \frac{\partial \overline{u_i u'_k}}{\partial r_j}\right]$$

$$- n^2 \frac{\partial^2 \overline{u_i u'_j u'_k}}{\partial (x_j)_n \partial (x_k)_n} - n\frac{\partial^2 \overline{u_i u'_j u'_k}}{\partial (x_j)_n \partial r_k} - n\frac{\partial^2 \overline{u_i u'_j u'_k}}{\partial (x_k)_n \partial r_j} - \frac{\partial^2 \overline{u_i u'_j u'_k}}{\partial r_j \partial r_k}. \qquad (4\text{-}149)$$

Similarly, from eq. (4-145),

$$\frac{1}{\rho}\left[(1-n)^2 \frac{\partial^2 \overline{\sigma u'_j}}{\partial (x_i)_n \partial (x_i)_n} - 2(1-n)\frac{\partial^2 \overline{\sigma u'_j}}{\partial (x_i)_n \partial r_i} + \frac{\partial^2 \overline{\sigma u'_j}}{\partial r_i \partial r_i}\right] = -2\frac{\partial U_i}{\partial x_k}\left[(1-n)\frac{\partial \overline{u_k u'_j}}{\partial (x_i)_n} - \frac{\partial \overline{u_k u'_j}}{\partial r_i}\right]$$

$$- (1-n)^2 \frac{\partial^2 \overline{u_i u_k u'_j}}{\partial (x_i)_n \partial (x_k)_n} + (1-n)\frac{\partial^2 \overline{u_i u_k u'_j}}{\partial (x_i)_n \partial r_k} + (1-n)\frac{\partial^2 \overline{u_i u_k u'_j}}{\partial (x_k)_n \partial r_i} - \frac{\partial^2 \overline{u_i u_k u'_j}}{\partial r_i \partial r_k}.$$

$$(4\text{-}150)$$

The equation for the mean velocity (eq. [4-25] with buoyancy neglected) is

$$\frac{\partial U_i}{\partial t} + U_k \frac{\partial U_i}{\partial x_k} = -\frac{1}{\rho}\frac{\partial P}{\partial x_i} + \frac{\partial}{\partial x_k}\left(\nu \frac{\partial U_i}{\partial x_k} - \overline{u_i u_k}\right) \qquad (4\text{-}151)$$

Equations (4-147), (4-149), and (4-150) constitute the two-point correlation equations for inhomogeneous turbulence with mean velocity gradients. A possible solution for these equations is that all the correlations are zero. In that case no turbulence exists, and the flow is laminar. We are mainly interested here in nontrivial turbulent solutions.

The terms on the right side of the Poisson eqs. (4-149) and (4-150) are source terms associated with the mean velocity and triple correlations. Most of the terms in eqs. (4-147) are similar to terms in the one-point eq. (4-140). Thus eq. (4-147) contains turbulence-production, convection, viscous, and diffusion terms. The sum of the pressure-velocity correlation terms on the right side of the equation transfers energy among directional components. That can be seen by using the relations following eq. (4-147) (the first for $\partial \overline{u_i \sigma'}/\partial r_j$ and the second for $\partial \overline{\sigma u'_j}/\partial r_i$), and using continuity, because the sum of the terms is zero for $i = j$; the terms give zero contribution to the rate of change of energy $\overline{u_i u_i}/2$, but they can transfer energy among directional components. The remaining terms in eq. (4-147), $\partial(\overline{u_i u'_j u'_k} - \overline{u_i u_k u'_j})/\partial r_k$ and $(U'_k - U_k)\partial \overline{u_i u'_j}/\partial r_k$, are new terms that do not have counterparts in the one-point equations. One may interpret them by converting eq. (4-147) to spectral form by taking its Fourier transform. We postpone doing that until the next chapter, in which the spectral form of the equations is discussed.

102 TURBULENT FLUID MOTION

Note that no averaged quantities containing instantaneous gradients appear in those equations, as they do in the one-point equations, so that the number of unknowns, in comparison with the number of available equations, is reduced. However, the two-point correlation equations, together with the equation for the mean velocity (eq. [4-151]) and the continuity equation (eq. [4-21]) still do not form a determinate set for turbulent flow because of the unknown triple correlations.

If the turbulence is weak (low Reynolds number), or if the mean-velocity gradients are generally very large, it may be possible to neglect terms containing triple correlations. Higher Reynolds numbers might be considered by constructing general three- and four-point equations. In each case the set of equations is made determinate by neglecting terms containing the highest-order correlations. However, even for the case of a fully developed flow at low Reynolds number, the difficulties of solution are extremely great. For that reason modelers of inhomogeneous turbulence generally have preferred to use the one-point equations (see, e.g., refs. [51] and [52]).[4]

Still, it appears that the present scheme should, formally, constitute a solution to the turbulent shear-flow problem. But the direct numerical solution of the unaveraged equations, as in refs. [17], [49], and [50], may be simpler. One can, however, get an approximate solution for sustained inhomogeneous turbulence and study, to some extent, the sustaining mechanism [53].

Turbulence is essentially a nonlinear phenomenon.[5] The nonlinear character of the two-point correlation eqs. (4-147), (4-149), and (4-150), even if the triple correlation terms are neglected, is made evident if the mean velocities in those equations are eliminated by introducing eq. (4-151) into them. A simplified problem is considered, in which the mean velocity is in the x_1 direction and mean quantities are steady and can change only in the x_2 direction [53]. Then if the mean velocity is eliminated by eq. (4-151) and the turbulence is weak enough (or the mean-velocity gradients are generally large enough) for triple-correlation terms to be neglected, the set of simultaneous correlation equations that must be considered consists of four nonlinear partial differential equations in four independent and four dependent variables. Although these equations might appear to be too complicated to be of use, some results can be obtained by expanding them in power series in each of the space variables to obtain algebraic expressions for the correlations, some of which are nonlinear (quadratic). The solutions so obtained are accurate only for small values of the space variables.

The reduction of the partial differential equations to a set of algebraic equations, some of which contain quadratic terms, shows the importance of the nonlinear terms in the original equations. The quadratic terms in the algebraic equations come, of course, from the nonlinear terms in the differential equations. If those terms were absent, the algebraic equations all would be linear and homogeneous (no constant terms), because the original equations contained no constant terms. In that case the only solution would be that all the turbulence correlations are zero, so that the flow is laminar. On the other

---

[4] The two-point equations, of course, can be profitably used for statistically homogeneous turbulence, as in the next chapter.
[5] The action in turbulence is somewhat similar to that of a clock, a violin bow, or an electronic oscillator in that in each of these a steady flow of energy is converted into oscillating energy by a nonlinear mechanism.

hand, in the presence of nonlinear terms the system of algebraic equations is quadratic, so that nonzero steady-state turbulence correlations can exist [53].

Reference [53] also shows the importance of the pressure-velocity correlations in sustaining turbulence. The nonlinear production term is absent in the equation for $\overline{u_2 u_2'}$ ($i = j = 2$ in eq. [4-147]), because $U_i = \delta_{i1} U_1$ in our problem. Thus the energy for sustaining $\overline{u_2 u_2'}$ must be transferred from the other components of $\overline{u_i u_j}$. The equations for the other components, of course, do contain nonlinear production terms, and energy can be transferred among the directional components by the pressure-velocity correlations (see discussion following eq. [4-151]).

The set of algebraic equations obtained by expanding the correlation equations in power series in the independent variables has been reduced to a single quadratic equation

$$N_c^2 = \left[N + \overline{(u_1 u_2')}_0^*\right]\left[N + a\overline{(u_1 u_2')}_0^*\right], \qquad (4\text{-}152)$$

where

$$\overline{(u_1 u_2')}_0^* = \frac{l^2 \overline{(u_1 u_2')}_0}{\nu^2}, \qquad N = \frac{\bar{\tau} l^2}{\rho \nu^2},$$

N_c is the critical value of N (at which $\overline{u_1 u_2'} = 0$), the subscript 0 designates the point in (x_2, r_k) space about which the series expansions are made, $\bar{\tau}$ is the shear stress (which is uniform in the present problem), l is a measure of the scale of the turbulence, and a is a constant. Equation (4-152) might be used to estimate the variation of $\overline{u_1 u_2}$ with $\bar{\tau}$ and l, preferably over a limited range; otherwise a might not be constant. For values of N below the critical value the correlation changes sign, according to eq. (4-152). In that case one should use the no-turbulence solution of eqs. (4-147) and (4-149) through (4-151). Note that the no-turbulence solution is valid if N is above its critical value as well as if it is below it. Thus the fluid can be either turbulent or nonturbulent if N is above N_c.

In order to make a rough comparison of our results with experiment we can recast eq. (4-152) so as to obtain an approximate expression for the friction factor for low Reynolds-number fully developed pipe flow. Actually, eq. (4-152) is more nearly applicable to Couette flow than to pipe flow; however, because of the similarity of trends in pipe and Couette flows and the availability of experimental data for transition pipe flow, the comparison is made with pipe-flow data. The approximate equation obtained in ref. [53] is

$$f = \left(16 + B Re_c^{-2}\right) Re^{-1} - B Re^{-3}, \qquad (4\text{-}153)$$

where $f = 2\bar{\tau}_w/(\rho U_a^2)$ is the friction factor, $Re = U_a D/\nu$, U_a is the averaged pipe-flow velocity, D is the pipe diameter, Re_c is the critical Reynolds number, and B is considered constant. For $Re = Re_c$, $f = 16/Re_c$, which is the value of f for laminar flow at Re_c.

Equation (4-153) and the solution for laminar flow, together with experimental data from ref. [54], are plotted in Fig. 4-18. The transition Reynolds number Re_c is taken as 2200, and B is set equal to 0.21×10^9. With these values for Re_c and B, eq. (4-153) is in reasonable agreement with the data for the turbulent region up to a Reynolds number of about 5500. The deviation of the curve from the data at higher Reynolds numbers probably is caused mainly by the neglect of triple correlations. The laminar solution is, of course, part of the present solution, because the case of no turbulence satisfies

104 TURBULENT FLUID MOTION

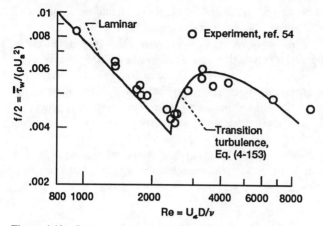

Figure 4-18 Comparison of eq. (4-153) and solution for laminar flow, with pipe-flow data in turbulent transition and laminar regions.

the correlation equations. For $Re < Re_c$ the laminar solution is appropriate because the turbulent solution goes below it and thus becomes unphysical.

4-4 CLOSING REMARKS

We began this chapter by considering kinds of averages and their properties, together with some notions of randomness from ergodic theory. Next, averaging was applied to the continuum equations derived in the previous chapter. It was noted that if the equations are averaged, random chaotically fluctuating variables are replaced by variables which change in a smoother fashion. However, the closure problem then arises. Whereas the original unaveraged equations, together with appropriate initial and boundary conditions, in principle could be completely solved, the solution of the averaged equations is indeterminate. That is, because of the nonlinearity of the basic equations, there are more unknowns than equations in the set of averaged equations. The addition of higher-order moment equations does not alleviate the problem; the number of unknowns goes up as fast as the number of equations if the added moment equations are multipoint, whereas if the added equations are single-point, the number of unknowns goes up faster. However, the use of those equations enables one to study the processes in turbulence. In particular, the numerical solution of the instantaneous (unaveraged) equations enables one to calculate terms in the averaged evolution equations. That procedure, in fact, provides a means of closing the system of averaged equations without introducing information that is supplemental to the basic continuum equations.

Because of the engineering importance of obtaining solutions for turbulent flows at high Reynolds numbers (at which accurate numerical simulations cannot yet be carried out), much effort has been devoted to modeling the undetermined terms in the averaged equations. To this end, we obtained some approximate solutions for a simple fully developed flow and heat transfer that were based, however, partially on exact information. Other approximate solutions and closure schemes, including a frozen-stress model for highly accelerated flows, also were discussed. The basis for a mixing-length theory and the nature of turbulent mixing were considered. Finally, the functional forms of the velocity and temperature profiles near and away from a wall and the logarithmic profiles for the velocity and temperature were derived.

However, the only turbulent solutions considered that can be called completely deductive are numerical solutions of the unaveraged equations. As discussed, it is necessary to use supplementary information if the equations are averaged, unless, of course, they are closed by a numerical solution of the original unaveraged equations. Attempts to obtain analytical solutions of the unaveraged equations have not been successful, mostly because of the nonlinearity of those equations. Sensitive dependence of those equations on initial conditions, in fact, appears to preclude an analytical solution. It is mentioned by Herring [55] that the simplest turbulence theory is just the Navier-Stokes equations; because most turbulence calculations are numerical anyway, no insight is lost by considering direct integration of the Navier-Stokes equations forward in time, starting with some suitable initial data. Because of its relevance and importance in turbulence research, we give here a few additional remarks on direct numerical solution.

Although numerical solution (or numerical simulation) bears some similarity to experiment, there is an important difference. The former uses directly, and attempts to solve, a given set of mathematical equations (say the Navier-Stokes equations). The latter ordinarily does not, although both methods may arrive at the same description of nature. A numerical solution is, in fact, much closer to a theoretical solution than to experiment. A numerical solution differs from a theoretical one only in the tools that are used; numerical methods and high-speed computers may take the place of, or supplement, analytical tools. A significant difference is not, as sometimes stated, that the numerical solution is discrete; as many points as desired can be calculated if a suitable interpolation formula is provided.

Because of the small scale of some of the turbulent eddies, accuracy is a problem in the numerical study of turbulence. Smaller and smaller eddies are generated as the Reynolds number (strength) of the turbulence increases. One can always pick a Reynolds number large enough that the results are quantitatively inaccurate, no matter how fine the numerical mesh. On the other hand, turbulence at any fixed Reynolds number in principle can be calculated, given the availability of a powerful enough computer. Note that there does not seem to be a qualitative difference (no bifurcations) between low- and high–Reynolds-number turbulence; in the latter the turbulent energy is just spread out over a (much) wider range of eddy sizes. The mathematical or computational methods used may be different, of course, for higher Reynolds numbers.

Although we started this section with a discussion of averages and averaged equations, it has been appropriate to close it with a discussion of numerical solutions of the original unaveraged equations, those being the only completely deductive turbulent solutions considered thus far (see, e.g., refs. [17] and [49]). (*Deductive* refers here to solutions of the continuum equations that do not require an input of supplementary information.) Note that, as mentioned earlier, a deductive solution of an unclosed averaged equation (e.g., the evolution equation for the turbulent energy) can be obtained by using it in conjunction with unaveraged equations, the terms in the averaged equation being calculated from the numerical solution of the determinate unaveraged equations. As a final remark, it seems unlikely that a turbulent analytical solution of the unaveraged equations can ever be obtained, because the solution would have to be sensitively dependent on initial conditions (see Chapter 6).

In order to continue with the analysis of turbulence, it is necessary, or at least convenient, to introduce Fourier analysis. That is done in the next chapter.

REFERENCES

1. Lebowitz, J.L., and Penrose, O., "Modern Ergodic Theory," *Phys. Today*, vol. 26, no. 2, pp. 23–29, 1973.
2. Lichtenberg, A.J., and Lieberman, M.A., *Regular and Stochastic Motion*, Springer-Verlag, New York, 1983.
3. Reynolds, O., "On the Dynamical Theory of Incompressible Viscous Fluids and the Determination of the Criterion," *Philos. Trans. R. Soc. London*, vol. 186, pp. 123–164, 1895.
4. Tanenbaum, B.S., *Plasma Physics*, McGraw-Hill, New York, pp. 162–164, 1967.
5. Hinze, J.O., *Turbulence*, McGraw-Hill, New York, 1975.
6. Prandtl, L., "The Mechanics of Viscous Fluids," in *Aerodynamic Theory, A General Review of Progress*, vol. III, edited by W.F. Durand, pp. 34–208, Springer-Verlag, Berlin, 1934.
7. Deissler, R.G., "Analysis of Fully Developed Turbulent Heat Transfer at Low Peclet Numbers in Smooth Tubes with Application to Liquid Metals," National Advisory Committee for Aeronautics, NACA RME52F05, Washington, DC, 1952. Published by NACA (now NASA, National Aeronautics and Space Administration, Washington, DC.)
8. Jenkins, R., "Variation of the Eddy Conductivity with Prandtl Modulus and its Use in Prediction of Turbulent Heat Transfer Coefficients," in *Preprints of Papers—Heat Transfer and Fluid Mechanics Institute*, pp. 147–158, Stanford University Press, Stanford, CA, 1951.
9. Jakob, M., *Heat Transfer, vol. II*, pp. 507–529, John Wiley and Sons, New York, 1957.
10. Deissler, R.G., "On the Decay of Homogeneous Turbulence Before the Final Period," *Phys. Fluids*, vol. 1, no. 2, pp. 111–121, 1958. Published by NACA (now NASA, National Aeronautics and Space Administration, Washington, DC.)
11. Laufer, J., "The Structure of Turbulence in Fully Developed Pipe Flow," National Advisory Committee for Aeronautics, NACA TN-2954, Washington, DC, 1953.
12. Batchelor, G.K., *The Theory of Homogeneous Turbulence*, pp. 107–108, 183–187, Cambridge University Press, New York, 1953.
13. Hussain, A.K.M.F., "Coherent Structures: Reality and Myth." *Phys. Fluids*, vol. 26, no. 10, pp. 2816–2850, 1983.
14. Townsend, A.A., *The Structure of Turbulent Shear Flow*, Cambridge University Press, New York, pp. 218–221, 1956.
15. Deissler, R.G., "Analytical and Experimental Investigation of Adiabatic Turbulent Flow in Smooth Tubes," National Advisory Committee for Aeronautics, NACA TN-2138, Washington, DC, 1950. Published by NACA (now NASA, National Aeronautics and Space Administration, Washington, DC.)
16. Millikan, C.B., "A Critical Discussion of the Turbulent Flows in Channels and Circular Tubes," in *Proceedings of the Fifth International Congress for Applied Mechanics*, edited by J.P. Den Hartog and M. Peters, pp. 386–392, Wiley and Sons, New York, 1938.
17. Kim, J., Moin, P., and Moser, R., "Turbulence Statistics in Fully Developed Channel Flow at Low Reynolds Number," *J. Fluid Mech.*, vol. 177, pp. 133–166, 1987.
18. Nikuradse, J., Gesetzmässigkeiten der Turbulenten Strömung in Glatten Rohren. VDI-Forschungsheft, no. 356, Sept.–Oct. 1932, p. 36.
19. Stanton, T.E., and Pannell, J.R., "Similarity of Motion in Relation to the Surface Friction of Fluids," *Philos. Trans. R. Soc. London A*, vol. 214, pp. 199–224, 1914.
20. Bernardo, E., and Eian, C.S., "Heat Transfer Tests of Aqueous Ethylene Glycol Solutions in an Electrically Heated Tube," National Advisory Committee for Aeronautics, NACA WR E-136, Washington, DC, 1945. Published by NACA (now NASA, National Aeronautics and Space Administration, Washington, DC.)

21. Kaufman, S.J., and Isely, F.D., "Preliminary Investigation of Heat Transfer to Water Flowing in an Electrically Heated Inconel Tube," National Advisory Committee for Aeronautics, NACA RM E50G31, Washington, DC, 1950. Published by NACA (now NASA, National Aeronautics and Space Administration, Washington, DC.)
22. Kreith, F., and Summerfield, M., "Pressure Drop and Convective Heat Transfer with Surface Boiling at High Heat Flux: Data for Aniline and n-Butyl Alcohol," *Trans. ASME*, vol. 72, no. 6, pp. 869–879, 1950.
23. Deissler, R.G., and Eian, C.S., "Analytical and Experimental Investigation of Fully Developed Turbulent Flow of Air in a Smooth Tube with Heat Transfer with Variable Fluid Properties," National Advisory Committee for Aeronautics, NACA TN-2629, Washington, DC, 1952. Published by NACA (now NASA, National Aeronautics and Space Administration, Washington, DC.)
24. Grele, M.D., and Gedeon, L., "Forced-Convection Heat-Transfer Characteristics of Molten Sodium Hydroxide," National Advisory Committee for Aeronautics, NACA RM E53L09, Washington, DC, 1953. Published by NACA (now NASA, National Aeronautics and Space Administration, Washington, DC.)
25. Hoffman, H.W., "Turbulent Forced Convection Heat Transfer in Circular Tubes Containing Molten Sodium Hydroxide," Oak Ridge National Laboratories, ORNL-1370, Oak Ridge, TN, 1952.
26. Bonilla, C.F., "Mass Transfer in Liquid Metal and Fused Salt Systems," *First Quarterly Progress Report*, U.S. Atomic Energy Commission, NYO-3086, Oak Ridge, TN, 1951.
27. Linton, W.H., Jr., and Sherwood, T.K., "Mass Transfer from Solid Shapes to Water in Streamline and Turbulent Flow," *Chem. Eng. Prog.*, vol. 46, no. 5, pp. 258–264, 1950.
28. Barnet, W.I., and Kobe, K.A., "Heat and Vapor Transfer in a Wetted-Wall Tower," *Ind. Eng. Chem.*, vol. 33, no. 4, pp. 436–442, 1941.
29. Chilton, T.H., and Colburn, A.P., "Mass Transfer (Absorption) Coefficients," *Ind. Eng. Chem.*, vol. 26, no. 11, pp. 1183–1187, 1934.
30. Jackson, M.L., and Ceaglske, N.H., "Distillation, Vaporization, and Gas Absorption in a Wetted-Wall Column," *Ind. Eng. Chem.*, vol. 42, no. 6, pp. 1188–1198, 1950.
31. Lin, C.S., Denton, E.B., Gaskill, H.S., and Putnam, G.L., "Diffusion-Controlled Electrode Reactions," *Ind. Eng. Chem.*, vol. 43, no. 9, pp. 2136–2143, 1951.
32. Eagle, A.E., and Ferguson, R.M., "On the Coefficient of Heat Transfer from the Internal Surface of Tube Walls," *Proc. R. Soc. London A*, vol. 127, pp. 540–566, 1930.
33. Deissler, R.G., "Turbulent Heat Transfer and Friction in Smooth Passages," in *Turbulent Flows and Heat Transfer*, edited by C.C. Lin, pp. 288–313, Princeton University Press, Princeton, NJ, 1959.
34. Deissler, R.G., and Loeffler, A.L., "Analysis of Turbulent Flow and Heat Transfer on a Flat Plate at High Mach Numbers with Variable Fluid Properties," National Aeronautics and Space Administration, Report TR R-17, Washington, DC, 1959. From NASA (National Aeronautics and Space Administration, Washington, DC.)
35. Deissler, R.G., and Perlmutter, M., "Analysis of the Flow and Energy Separation in a Turbulent Vortex," *Int. J. Heat Mass Transfer*, vol. 1, nos. 2–3, pp. 173–191, 1960.
36. Deissler, R.G., "Gravitational Collapse of a Turbulent Vortex with Application to Star Formation," *Astrophys. J.*, vol. 209, pp. 190–204, 1976.
37. Deissler, R.G., "Models for Some Aspects of Atmospheric Vortices," *J. Atmos. Sci.*, vol. 34, pp. 1502–1517, 1977.
38. Van Driest, E.R., "On Turbulent Flow Near a Wall," *J. Aeronaut. Sci.*, vol. 23, pp. 1007–1011, 1956.
39. Goldstein, S., *Modern Developments in Fluid Dynamics*, vol. 1, pp. 119–120, Clarendon Press, Oxford, England, 1938.
40. Blackwelder, R.F., and Kovasznay, L.S.G., "Large-Scale Motion of a Turbulent Boundary Layer During Relaminarization," *J. Fluid Mech.*, vol. 53, pp. 61–83, 1972.
41. Deissler, R.G., "Evolution of a Moderately Short Turbulent Boundary Layer in a Severe Pressure Gradient," *J. Fluid Mech.*, vol. 64, pp. 763–774, 1974.
42. Deissler, R.G., "Evolution of the Heat Transfer and Flow in Moderately Short Turbulent Boundary Layers in Severe Pressure Gradients," *Int. J. Heat Mass Transfer*, vol. 17, pp. 1079–1085, 1974.
43. Launder, B.E., "Laminarization of the Turbulent Boundary Layer by Acceleration," Massachusetts Institute of Technology Gas Turbine Laboratory Report 77, 1964.
44. Patel, V.C., and Head, M.R., "Reversion of Turbulent to Laminar Flow," *J. Fluid Mech.*, vol. 34, no. 2, pp. 371–392, 1968.

45. Kline, S.J., Reynolds, W.C., Schraub, F.A., and Runstadler, P.W., "The Structure of Turbulent Boundary Layers," *J. Fluid Mech.*, vol. 30, no. 4, pp. 741–773, 1967.
46. Julien, H.L., Kays, W.M., and Moffat, R.J., "Experimental Hydrodynamics of the Accelerated Turbulent Boundary Layer With and Without Mass Injection," *J. Heat Transfer*, vol. 93, pp. 373–379, 1971.
47. Moretti, P.M., and Kays, W.M., "Heat Transfer to a Turbulent Boundary Layer with Varying Free Stream Temperature: An Experimental Study," *Int. J. Heat Mass Transfer*, vol. 8, pp. 1187–1202, 1965.
48. Back, L.H., and Cuffell, R.F., "Turbulent Boundary Layer and Heat Transfer Measurements Along a Convergent-Divergent Nozzle," *J. Heat Transfer*, vol. 93, pp. 397–407, 1971.
49. Mansour, N.N., Kim, J., and Moin, P., "Reynolds-Stress and Dissipation: Rate Budgets in a Turbulent Channel Flow," *J. Fluid Mech.*, vol. 194, pp. 15–44, 1988.
50. Deissler, R.G., "Turbulent Solutions of the Equations of Fluid Motion," *Rev. Mod. Phys.*, vol. 56, no. 1, pp. 223–254, 1984.
51. Rodi, W., *Turbulence Models for Practical Applications: Introduction to the Modeling of Turbulence*, p. 27. Von Kármán Institute for Fluid Dynamics, Rhode-Saint-Genese, Belgium, 1987.
52. Launder, B.E., Reynolds, W.C., and Rodi, W., *Turbulence Models and Their Applications, vol. 2*, Editions Eyrolles, Paris, France, 1984.
53. Deissler, R.G., "Problem of Steady-State Shear-Flow Turbulence," *Phys. Fluids*, vol. 8, no. 3, pp. 391–398, 1965.
54. Deissler, R.G., Weiland, W.F., and Lowdermilk, W.H., "Analytical and Experimental Investigation of Temperature Recovery Factors for Fully Developed Flow of Air in a Tube," National Advisory Committee for Aeronautics, NACA TN-4376, Washington, DC, 1958. Published by NACA (now NASA, National Aeronautics and Space Administration, Washington, DC.)
55. Herring, J.R., "Statistical Turbulence Theory and Turbulence Phenomenology," in *Free Turbulent Shear Flows, vol. 1: Conference Proceedings*, pp. 41–66, NASA SP-321, National Aeronautics and Space Administration, Washington, DC. 1973.

CHAPTER
FIVE

FOURIER ANALYSIS, SPECTRAL FORM OF THE CONTINUUM EQUATIONS, AND HOMOGENEOUS TURBULENCE

Background material on Fourier analysis and on the spectral form of the continuum equations, both averaged and unaveraged, are given in Chapter 5. The equations are applied to a number of cases of homogeneous turbulence with and without mean gradients. Some turbulent solutions of the full unaveraged continuum equations are obtained numerically. Closure of the averaged equations by specification of sufficient random initial conditions is considered. The gap problem (the problem of bridging the gap between the infinite amount of data required to specify an initial turbulence and the finite amount generally available) is discussed. Then a solution for the evolution of all of the quantities used to specify the initial turbulence is obtained. Spectral transfer of turbulent activity between scales of motion is studied in some detail. The effects of mean shear, heat transfer, normal strain, and buoyancy are included in the analyses.

In chapter 1 it was pointed out that turbulence is multiscaled. That is, it consists of small eddies (small-scale motion) superimposed on larger-scale motion. In order to proceed with our study of the dynamics of turbulence, we should consider how the turbulent activity is distributed among the various scales of motion, together with the interaction, if any, of those scales.

One possible way, the usual way, of representing scales of motion is by means of a series of trigonometric functions, that is, by a Fourier series. For example, one might write a turbulent velocity u as

$$u = \sum_{\kappa=1}^{\infty} (A_\kappa \cos \kappa x + B_\kappa \sin \kappa x),$$

where x is a distance, κ is known as the wavenumber (with dimension 1/length), and the values of the coefficients A_κ and B_κ depend on the turbulence. For simplicity the system is assumed here to be one-dimensional. Larger values of κ correspond to smaller eddies (smaller spatial scales) and vise versa.

Basis functions other than trigonometric are sometimes used to represent turbulence, particularly in numerical simulations with boundary conditions other than periodic [1]. Moreover, attempts are sometimes made to represent the turbulence by functions whose shape is closer to that of a typical eddy (if such can be defined) than are those of trigonometric functions [2, 3]. For instance, the steep instantaneous gradients that are known to occur in turbulent flows, particularly at high Reynolds numbers, might be incorporated into the basis function. However, most of the work on turbulence has used trigonometric basis functions (see, e.g., ref. [4]), and those are adequate for our purposes.

In addition to using Fourier analysis for instantaneous turbulent quantities, one can use it to analyze averaged quantities, such as those in the two-point correlation equations (see section 4-3-4). Because the latter are easier to deal with than unaveraged quantities, which may require the use of generalized functions, they are considered first.

5-1 FOURIER ANALYSIS OF THE TWO-POINT AVERAGED CONTINUUM EQUATIONS

5-1-1 Analysis of Two-Point Averaged Quantities

One can decompose the two-point velocity correlation $\overline{u_i u_j'}(r)$ into a three-dimensional series of trigonometric functions (eddy sizes) as follows:

$$\overline{u_i u_j'}(r) = \sum_{\kappa=-\infty}^{\infty} (\varphi_{ij})_\kappa e^{i\kappa \cdot r}, \tag{5-1}$$

where the trigonometric series has been written in complex notation by using the Euler relation

$$e^{i\kappa \cdot r} = \cos \kappa \cdot r + i \sin \kappa \cdot r, \tag{5-2}$$

and where u_i and u_j' are velocity components at two points that are separated by the vector r, κ is a wavevector, $\kappa \cdot r = \kappa_k r_k = \kappa_1 r_1 + \kappa_2 r_2 + \kappa_3 r_3$ is the dot product of the vectors κ and r, and the $(\varphi_{ij})_\kappa$ are called *Fourier coefficients*. The $(\varphi_{ij})_\kappa$ are complex because of the presence of the complex exponential in eq. (5-1), and because $\overline{u_i u_j'}(r)$ is real. To simplify the notation, possible dependencies on x and time have been omitted. The summation in eq. (5-1) is taken separately over each component of κ.

To determine the coefficients $(\varphi_{ij})_\kappa$, multiply both sides of eq. (5-1) by $e^{-i n \cdot r}$ and by $dr = dr_1 dr_2 dr_3$. Then, integrating over the period 2π for each component of r, we get

$$(\varphi_{ij})_\kappa = \frac{1}{(2\pi)^3} \int_{-\pi}^{\pi} \overline{u_i u_j'}(r) e^{-i\kappa \cdot r} dr, \tag{5-3}$$

because terms for which $n_k \neq \kappa_k$ are zero because of the periodicity of $e^{i(\kappa-n)\cdot r}$. Equation (5-1) gives the three-dimensional Fourier-series expansion of $\overline{u_i u_j'}(r)$, where the $(\varphi_{ij})_\kappa$ are given by eq. (5-3). Because $\overline{u_i u_j'}$ is a second-order tensor (see section 2-4-1), eq. (5-3) has meaning only if $(\varphi_{ij})_\kappa$ is a second-order tensor (see section 2-9).

If the period over which r_k is defined is R_k rather than 2π, where the three R_k are not necessarily equal, eqs. (5-1) and (5-3) become, respectively,

$$\overline{u_i u_j'}(r) = \sum_{\kappa=-\infty}^{\infty} (\varphi_{ij})_\kappa e^{i 2\pi \kappa_k r_k / R_{(k)}}, \tag{5-4}$$

and

$$(\varphi_{ij})_\kappa = \frac{1}{R_1 R_2 R_3} \int_{-R_k/2}^{R_k/2} \overline{u_i u_j'}(v) e^{-i 2\pi \kappa_k v_k / R_{(k)}} dv, \tag{5-5}$$

where the dummy variable r on the right side of eq. (5-3) has been changed to v to avoid later confusion.

Consider now the case in which $R_k \to \infty$. To that end, substitute eq. (5-5) into (5-4). That gives

$$\overline{u_i u'_j}(r) = \sum_{\kappa=-\infty}^{\infty} \frac{1}{R_1 R_2 R_3} e^{i 2\pi \kappa_k r_k / R_{(k)}} \int_{-R_k/2}^{R_k/2} \overline{u_i u'_j}(v) e^{-i 2\pi \kappa_k v_k / R_{(k)}} \, dv, \qquad (5\text{-}6)$$

Next let

$$\frac{2\pi}{R_k} \equiv \Delta s_k, \qquad (5\text{-}7)$$

where s_k is given by

$$s_k \equiv \kappa_k \Delta s_{(k)} \quad (\text{no sum on } k), \qquad (5\text{-}7a)$$

and Δs_k is an increment of s_k. From eqs. (5-7) and (5-7a),

$$s_k = \frac{2\pi \kappa_k}{R_{(k)}}, \qquad (5\text{-}7b)$$

showing that s_k varies linearly with κ_k for a given period R_k. Equation (5-6) then can be written as

$$\overline{u_i u'_j}(r) = \sum_{s=-\infty}^{\infty} \left[\int_{-R_k/2}^{R_k/2} \overline{u_i u'_j}(v) e^{-i s_k v_k} dv \right] e^{i s_k r_k} \frac{\Delta s_1 \Delta s_2 \Delta s_3}{(2\pi)^3}, \qquad (5\text{-}8)$$

where Δs_k is given by eq. (5-7). For $R_k \to \infty$ ($\Delta s_k \to 0$, see eq. [5-7]), eq. (5-8) becomes

$$\overline{u_i u'_j}(r) = \int_{-\infty}^{\infty} \left[\frac{1}{(2\pi)^3} \int_{-\infty}^{\infty} \overline{u_i u'_j}(v) e^{-i s \cdot v} dv \right] e^{i s \cdot r} \, ds, \qquad (5\text{-}9)$$

because, for each value of k ($k = 1, 2, 3$),

$$\lim_{\Delta s_k \to 0} \sum_{s_k = -\infty}^{\infty} f_{s_k} \Delta s_{(k)} = \int_{-\infty}^{\infty} f(s_k) \, ds_{(k)} \quad (\text{no sum on } k). \qquad (5\text{-}10)$$

Equation (5-10) can be considered a definition of the definite integral $\int_{-\infty}^{\infty} f(s_k) ds_{(k)}$. Equation (5-9) gives the *Fourier-integral* representation of $\overline{u_i u'_j}(r)$.

If we let the quantity in brackets in eq. (5-9) be equal to φ_{ij} and replace the dummy variables s and v by κ and r, respectively, we get

$$\overline{u_i u'_j}(r) = \int_{-\infty}^{\infty} \varphi_{ij}(\kappa) e^{i\kappa \cdot r} \, d\kappa, \qquad (5\text{-}11)$$

and

$$\varphi_{ij}(\kappa) = \frac{1}{(2\pi)^3} \int_{-\infty}^{\infty} \overline{u_i u'_j}(r) e^{-i\kappa \cdot r} \, dr. \qquad (5\text{-}12)$$

The quantity $\varphi_{ij}(\kappa)$ is the three-dimensional *Fourier transform* of $\overline{u_i u'_j}(r)$. Similar to $(\varphi_{ij})_\kappa$ in eq. (5-3), it shows how spectral contributions to $\overline{u_i u'_j}(r)$ are distributed in wavenumber space. The difference between the two is that $(\varphi_{ij})_\kappa$ applies if $\overline{u_i u'_j}(r)$ is defined over a finite part of physical space (r space), whereas $\varphi_{ij}(\kappa)$ applies if $\overline{u_i u'_j}(r)$ is defined over all of r space. The quantity $\varphi_{ij}(\kappa)$ exists, of course, only if the integral in eq. (5-12) exists, or if $\overline{u_i u'_j}(r)$ in eq. (5-12) is absolutely integrable over r [5], but that is usually the case (see, e.g., Fig. 4-16). By analogy with $\varphi_{ij}(\kappa)$ (eq. [5-12]), $(\varphi_{ij})_\kappa$ (eq. [5-3]) is sometimes called the *finite Fourier transform* of $\overline{u_i u'_j}(r)$ (applicable if r remains finite [6]). The treatment of the other two-point correlations is similar. For example, the Fourier transform of the two-point (mechanical) pressure–velocity correlation $\overline{\sigma u'_i}(r)$ is given by the vector

$$\lambda_j(\kappa) = \frac{1}{(2\pi)^3} \int_{-\infty}^{\infty} \overline{\sigma u'_j}(r) e^{-i\kappa \cdot r} \, dr \tag{5-13}$$

in place of eq. (5-12), and in place of eq. (5-11) we have

$$\overline{\sigma u'_j}(r) = \int_{-\infty}^{\infty} \lambda_j(\kappa) e^{i\kappa \cdot r} \, d\kappa. \tag{5-14}$$

It is easy to show that taking the derivative of a quantity with respect to r_k multiplies its Fourier transform by $i\kappa_k$. Thus,

$$\frac{\partial}{\partial r_k} \overline{\sigma u'_j}(r) = \int_{-\infty}^{\infty} i\kappa_k \lambda_j(\kappa) e^{i\kappa \cdot r} \, d\kappa, \tag{5-15}$$

or

$$i\kappa_k \lambda_j(\kappa) = \frac{1}{(2\pi)^3} \int_{-\infty}^{\infty} \frac{\partial}{\partial r_k} \overline{\sigma u'_j}(r) e^{-i\kappa \cdot r} \, dr. \tag{5-16}$$

Next we obtain the spectral form of the two-point averaged continuum equations.

5-1-2 Analysis of the Two-Point Correlation Equations

To convert to spectral form the two-point correlation equations obtained in the previous chapter (eqs. [4-147] through [4-150]), we multiply those equations through by $[1/(2\pi)^3] e^{-i\kappa \cdot r} dr$ and integrate from $-\infty$ to $+\infty$. That gives

$$\frac{\partial}{\partial t}\varphi_{ij} + \varphi_{kj}\frac{\partial U_i}{\partial x_k} + \varphi_{ik}\frac{\partial U'_j}{\partial x'_k} + (U'_k - U_k)i\kappa_k\varphi_{ij} + [(1-n)U_k + nU'_k]\frac{\partial}{\partial (x_k)_n}\varphi_{ij}$$

$$+ \frac{1}{\rho}\left[\frac{\partial}{\partial (x_i)_n}\lambda_j + \frac{\partial}{\partial (x_j)_n}\lambda'_i\right] + \frac{\partial}{\partial (x_k)_n}[n\varphi'_{ijk} + (1-n)\varphi_{ikj}] + i\kappa_k(\varphi'_{ijk} - \varphi_{ikj})$$

$$= \frac{1}{\rho}\left[n\frac{\partial}{\partial (x_i)_n}\lambda_j + (1-n)\frac{\partial}{\partial (x_j)_n}\lambda'_i - i\kappa_j\lambda'_i + i\kappa_i\lambda_j\right] + (1 - 2n + 2n^2)\nu$$

$$\times \frac{\partial^2 \varphi_{ij}}{\partial (x_k)_n \partial (x_k)_n} - 2(1-2n)\nu i\kappa_k\frac{\partial \varphi_{ij}}{\partial (x_k)_n} - 2\nu\kappa^2\varphi_{ij}, \tag{5-17}$$

$$\frac{1}{\rho}\left[n^2\frac{\partial^2 \lambda'_i}{\partial (x_j)_n \partial (x_j)_n} + 2n i\kappa_j\frac{\partial \lambda'_i}{\partial (x_j)_n} - \kappa^2\lambda'_i\right]$$

$$= -2\frac{\partial U'_j}{\partial x'_k}\left[n\frac{\partial \varphi_{ik}}{\partial (x_j)_n} + i\kappa_j\varphi_{ik}\right] - n^2\frac{\partial^2 \varphi'_{ijk}}{\partial (x_j)_n \partial (x_k)_n} - n i\kappa_k\frac{\partial \varphi'_{ijk}}{\partial (x_j)_n}$$

$$- n i\kappa_j\frac{\partial \varphi'_{ijk}}{\partial (x_k)_n} + \kappa_j\kappa_k\varphi'_{ijk}, \tag{5-18}$$

and
$$\frac{1}{\rho}\left[(1-n)^2\frac{\partial^2\lambda_j}{\partial(x_i)_n\partial(x_i)_n} - 2(1-n)i\kappa_i\frac{\partial\lambda_j}{\partial(x_i)_n} - \kappa^2\lambda_j\right]$$

$$= -2\frac{\partial U_i}{\partial x_k}\left[(1-n)\frac{\partial\varphi_{kj}}{\partial(x_i)_n} - i\kappa_i\varphi_{kj}\right] - (1-n)^2\frac{\partial^2\varphi_{ikj}}{\partial(x_i)_n\partial(x_k)_n}$$

$$+ (1-n)i\kappa_k\frac{\partial\varphi_{ikj}}{\partial(x_i)_n} + (1-n)i\kappa_i\frac{\partial\varphi_{ikj}}{\partial(x_k)_n} + \kappa_i\kappa_k\varphi_{ikj}, \qquad (5\text{-}19)$$

where φ_{ij} and λ_j are given respectively by eqs. (5-12) and (5-13). Similarly,

$$\lambda'_i(\kappa) = \frac{1}{(2\pi)^3}\int_{-\infty}^{\infty}\overline{u_i\sigma'}(r)e^{-i\kappa\cdot r}\,dr, \qquad (5\text{-}20)$$

$$\varphi_{ikj}(\kappa) = \frac{1}{(2\pi)^3}\int_{-\infty}^{\infty}\overline{u_iu_ku'_j}(r)e^{-i\kappa\cdot r}\,dr, \qquad (5\text{-}21)$$

and

$$\varphi'_{ijk}(\kappa) = \frac{1}{(2\pi)^3}\int_{-\infty}^{\infty}\overline{u_iu'_ju'_k}(r)e^{-i\kappa\cdot r}\,dr, \qquad (5\text{-}22)$$

where n and x_n are defined in Fig. (4-17), and functional dependencies on x_n and t are understood. Finally, the equation for the mean velocity (eq. [4-151]) can be written in spectral form as

$$\frac{\partial U_i}{\partial t} + U_k\frac{\partial U_i}{\partial x_k} = -\frac{1}{\rho}\frac{\partial P}{\partial x_i} + \frac{\partial}{\partial x_k}\left[\nu\frac{\partial U_i}{\partial x_k} - \int_{-\infty}^{\infty}\varphi_{ik}(\kappa)\,d\kappa\right], \qquad (5\text{-}23)$$

where $\overline{u_iu_k}$ has been eliminated by letting $r = 0$ in eq. (5-11).

Equations (5-17) through (5-19), and eq. (5-23), constitute the two-point spectral equations for inhomogeneous turbulence with mean-velocity gradients. Note that these equations have a simpler form than the correlation equations, because derivatives with respect to r or κ are absent. However, the introduction of spectral quantities does not alleviate the closure problem. As was the case for the correlation equations, these spectral equations do not form a complete set; the Fourier transforms of the triple correlations are left undetermined. If the turbulence is sufficiently weak, or some of the mean gradients are sufficiently large, it may be possible to neglect the triple-correlation terms.

As was the case for the Poisson equations (4-149) and (4-150), the terms on the right side of eqs. (5-18) and (5-19) are source terms associated with the mean velocity and triple correlations. The interpretation of most of the terms in eq. (5-17) is also similar to that of terms in the two-point correlation equations. Thus, eq. (5-17) contains turbulence-production, convection, viscous, diffusion, and directional-transfer terms.

5-1-2-1 The spectral-transfer terms in eq. (4-147) or (5-17). It was indicated in section 4-3-4 that the two-point equation (4-147) contains, in addition to the terms mentioned in the last paragraph, the new terms $-\partial(\overline{u_iu'_ju'_k} - \overline{u_iu_ku'_j})/\partial r_k$ and $(U'_k - U_k)\partial\overline{u_iu'_j}/\partial r_k$, which do not have counterparts in the one-point equation (4-140), and which require spectral analysis for their interpretation. The Fourier transforms of those terms, as given in the spectral equation (5-17) are, respectively,

$$-i\kappa_k(\varphi'_{ijk} - \varphi_{ikj}) \equiv T_{ij}(\kappa, x_n) = \frac{1}{(2\pi)^3} \int_{-\infty}^{\infty} [-\partial(\overline{u_i u'_j u'_k} - \overline{u_i u_k u'_j})/\partial r_k] e^{-i\kappa \cdot r} \, dr,$$
(5-24)

and

$$-(U'_k - U_k) i\kappa_k \varphi_{ij} \equiv T'_{ij}(\kappa, x_n) = \frac{1}{(2\pi)^3} \int_{-\infty}^{\infty} [-(U'_k - U_k) \partial \overline{u_i u'_j}/\partial r_k] e^{-i\kappa \cdot r} \, dr,$$
(5-25)

where x is included in the functional designations for T_{ij} and T'_{ij} to emphasize that the turbulence can be inhomogeneous.

Consider first the term $T_{ij}(\kappa, x_n)$, which is the three-dimensional Fourier transform of the turbulence self-interaction term $-\partial(\overline{u_i u'_j u'_k} - \overline{u_i u_k u'_j})/\partial r_k$. That term should be absolutely integrable over r in order for its Fourier transform to exist [5]. Moreover, if a wall is present in the flow, a finite Fourier transform with respect to the component of r normal to the wall is appropriate (eq. [5-3]) (see discussion following eq. [5-12], and ref. [6]).

We want to interpret the term $T_{ij}(\kappa, x_n)$. To that end, referring to Fig. 4-17, one notes that

$$x + r = x', \qquad x + nr = x_n,$$

from which

$$x_n = nx' + (1-n)x.$$

In subscript notation,

$$r_k = x'_k - x_k,$$
(5-26)

and

$$(x_k)_n = nx_k + (1-n)x'_k,$$
(5-27)

where n is a number between 0 and 1. By using eqs. (5-26) and (5-27) and the rules for partial differentiation, one obtains

$$\frac{\partial}{\partial x_k} = (1-n)\frac{\partial}{\partial (x_k)_n} - \frac{\partial}{\partial r_k},$$
(5-28)

and

$$\frac{\partial}{\partial x'_k} = n\frac{\partial}{\partial (x_k)_n} + \frac{\partial}{\partial r_k}.$$
(5-29)

Taking the transform of eq. (5-24), and solving (5-28) and (5-29) for $\partial/\partial r_k$, one can write

$$\int_{-\infty}^{\infty} T_{ij}(\kappa, x_n) e^{i\kappa \cdot r} d\kappa = -\frac{\partial}{\partial r_k} \overline{u_i u'_j u'_k} + \frac{\partial}{\partial r_k} \overline{u_i u_k u'_j}$$

$$= \left(-\frac{\partial}{\partial x'_k} + n\frac{\partial}{\partial (x_k)_n}\right) \overline{u_i u'_j u'_k} + \left(-\frac{\partial}{\partial x_k} + (1-n)\frac{\partial}{\partial (x_k)_n}\right) \overline{u_i u_k u'_j}$$

$$= -\overline{u_i u'_k \frac{\partial u'_j}{\partial x'_k}} - \overline{u'_j \frac{\partial}{\partial x_k} u_i u_k} + \frac{\partial}{\partial (x_k)_n}(n\overline{u_i u'_j u'_k} + (1-n)\overline{u_i u_k u'_j}),$$
(5-30)

where the continuity condition $\partial u_k/\partial x_k = 0$ and the fact that quantities at one point are independent of the position of the other point were used. Equation (5-30) becomes, for $r = 0$,

$$\int_{-\infty}^{\infty} T_{ij}(\kappa, x_n)\, d\kappa = -\left(\overline{u_i u_k}\frac{\partial u_j}{\partial x_k} + u_j\frac{\partial}{\partial x_k}\overline{u_i u_k}\right) + \frac{\partial}{\partial x_k}\overline{u_i u_j u_k}$$

$$= -\frac{\partial}{\partial x_k}\overline{u_i u_j u_k} + \frac{\partial}{\partial x_k}\overline{u_i u_j u_k} = 0, \qquad (5\text{-}31)$$

because, for $r = 0$, $x_k = x'_k = (x'_k)_n$. Therefore, even for a general inhomogeneous turbulence, T_{ij}, if integrated over all wavenumbers, gives zero contribution to the rate of change of $\overline{u_i u_j}$ (see eqs. [4-147] and [5-30]). Thus, $T_{ij}(\kappa, x_n)$ can only transfer Fourier components of $\overline{u_i u_j}$ (energy for $i = j$) from one part of wavenumber space to another.

The quantity $T'_{ij}(\kappa, x_n)$ (see eq. [5-25]), in contrast to $T_{ij}(\kappa, x_n)$ (which produces turbulence self-interaction), is associated with the interaction of the turbulence with the mean flow. However, both terms are related to transfer terms. We can write

$$-(U'_k - U_k)\frac{\partial}{\partial r_k}\overline{u_i u'_j} = \int_{-\infty}^{\infty} T'_{ij}(\kappa, x_n)e^{i\kappa\cdot r}\, d\kappa \qquad (5\text{-}32)$$

where $T'_{ij}(\kappa, x_n)$ is the Fourier transform of $-(U'_k - U_k)(\partial/\partial r_k)\overline{u_i u'_j}$. Letting $r = 0$, eq. (5-32) becomes

$$\int_{-\infty}^{\infty} T'_{ij}(\kappa, x_n)\, d\kappa = 0 \qquad (5\text{-}33)$$

because, for $r = 0$, $U'_k = U_k$. Thus, as in the case of $T_{ij}(\kappa, x_n)$, $T'_{ij}(\kappa, x_n)$ gives zero total contribution to the rate of the change of $\overline{u_i u_j}$ (energy for $i = j$) and can alter only the distribution in wavenumber space of contributions to $\overline{u_i u_j}$. We first interpreted and calculated $T'_{ij}(\kappa)$ as a transfer term for homogeneous turbulence in ref. [7]. (Craya [8] also discusses, in a general way, the modification of homogeneous turbulence by uniform mean gradients, but does not show that T'_{ij} is specifically a spectral-transfer term.)

The quantities T_{ij} and T'_{ij}, defined respectively in eqs. (5-24) and (5-25), appear to be the only terms in the evolution eq. (5-17) that can be interpreted as spectral-transfer terms. As mentioned earlier, the other terms are interpretable as production, convection, directional-transfer, diffusion, and dissipation terms.

The transfer of turbulent activity from one part of wavenumber space to another, or from one eddy size to another, produces a wide range of scales of motion in most turbulent flows. The state of affairs is neatly summarized in a nonmathematical way by a poem written long before eqs. (5-17), (5-31), or (5-33) were known [9]:

> Big whorls have little whorls,
> Which feed on their velocity;
> And little whorls have lesser whorls,
> And so on to viscosity.

As is seen subsequently (e.g., sections 5-3-2-2 and 5-4-2-1), both T_{ij} and T'_{ij} generally (although not always—see Fig. [5-61]) transfer turbulent activity from larger to smaller scales of motion, where it can be dissipated more readily. Thus, T_{ij} and T'_{ij} ordinarily have stabilizing effects.

5-2 FOURIER ANALYSIS OF THE UNAVERAGED (INSTANTANEOUS) CONTINUUM EQUATIONS

5-2-1 Analysis of Instantaneous Quantities

Consider now the Fourier (spectral) analysis of an instantaneous quantity such as the velocity $u_i(x)$. If u_i is defined only for x_k between $-\pi$ and π, one can represent $u_i(x)$ by means of a Fourier series (finite Fourier transform [6]). Then, in place of eqs. (5-1) and (5-3) one has, respectively,

$$u_i(x) = \sum_{\kappa=-\infty}^{\infty} (\varphi_i)_\kappa e^{i\kappa \cdot x} d\kappa, \tag{5-34}$$

and

$$(\varphi_i)_\kappa = \frac{1}{(2\pi)^3} \int_{-\pi}^{\pi} u_i(x) e^{-i\kappa \cdot x} dx. \tag{5-35}$$

If, instead of being defined over a finite portion of physical space, $u_i(x)$ is defined for x_k from $-\infty$ to $+\infty$, one might suppose that eqs. (5-34) and (5-35) become a Fourier-transform pair such as that given by eqs. (5-11) and (5-12). Thus, one might write

$$u_i(x) = \int_{-\infty}^{\infty} \varphi_i(\kappa) e^{i\kappa \cdot x} d\kappa, \tag{5-36}$$

and

$$\varphi_i(\kappa) = \frac{1}{(2\pi)^3} \int_{-\infty}^{\infty} u_i(x) e^{-i\kappa \cdot x} dx, \tag{5-37}$$

where $\varphi_i(\kappa)$ is the Fourier transform of $u_i(x)$. However, $\varphi_i(\kappa)$ as given by eq. (5-37) may not exist in the ordinary sense, because the integral in that equation can be infinite (in contrast to the integral in eq. [5-12]). That happens if the strength of the velocity fluctuation $u_i(x)$ in eq. (5-37) does not approach zero sufficiently fast as $|x| \to \infty$ (for example, if $u_i[x]$ is a stationary random function). The problem is sometimes solved by replacing the integral in eq. (5-36) by a stochastic Fourier–Stieltjes integral, as in refs. [4] and [10]. But it may be simpler to consider $\varphi_i(\kappa)$ as a generalized function [11]–[13].

Probably the simplest generalized function is the Dirac delta (or impulse) function $\delta(\kappa_k - q_k^m)$ which, for our purposes, can be defined as

$$\delta(\kappa_k - q_k^m) \equiv \lim_{g \to \infty} \frac{\sin[g(\kappa_k - q_k^m)]}{\pi(\kappa_k - q_k^m)} \tag{5-38}$$

for $k = 1, 2$, or 3 (no sum on k), and where q_k^m is a particular value of κ_k. Because $\delta(\kappa_k - q_k^m) \to \infty$ as $q_k^m \to \kappa_k$, it has the property that

$$\int_{-\infty}^{\infty} \delta(\kappa_k - q_k^m) a_i(\kappa_k) d\kappa_k = a_i(q_k^m), \tag{5-39}$$

where $a_i(\kappa_k)$ is a continuous function. Note also that (for one dimension),

$$\int_{-\infty}^{\infty} \delta(\kappa) d\kappa = \lim_{g \to \infty} \int_{-\infty}^{\infty} \frac{\sin g\kappa}{\pi \kappa} d\kappa = 1, \tag{5-40}$$

$$\int_{-\infty}^{\infty} e^{i\kappa x} dx = \lim_{g \to \infty} \int_{-g}^{g} (\cos \kappa x + i \sin \kappa x) dx = 2\pi \delta(\kappa), \tag{5-41}$$

and

$$\int_{-\infty}^{\infty} e^{-i\kappa x} dx = 2\pi \delta(\kappa). \tag{5-42}$$

Thus, although the integrals in eqs. (5-41) and (5-42) do not exist in the sense of being ordinary functions, they are equal to the generalized function $2\pi\delta(\kappa)$.

Returning now to the problem of the existence of the Fourier transform $\varphi_i(\kappa)$, as given by eq. (5-37), we try replacing $u_i(x)$ by the trigonometric function $2^3 a_i e^{q^m \cdot x}$. That gives

$$\varphi_i(\kappa) = \frac{1}{\pi^3} \int_{-\infty}^{\infty} a_i e^{i(q^m - \kappa) \cdot x} dx$$

$$= \frac{1}{\pi^3} a_i \lim_{g \to \infty} \frac{\sin[g(\kappa_1 - q_1^m)]}{\kappa_1 - q_1^m} \frac{\sin[g(\kappa_2 - q_2^m)]}{\kappa_2 - q_2^m} \frac{\sin[g(\kappa_3 - q_3^m)]}{\kappa_3 - q_3^m}$$

$$= a_i \delta(\kappa_1 - q_1^m) \delta(\kappa_2 - q_2^m) \delta(\kappa_3 - q_3^m) \equiv a_i \delta(\kappa - q^m), \tag{5-43}$$

where, as before q_k^m is a particular value of κ_k, and where eqs. (5-38) and (5-41) are used. Thus, by replacing $u_i(x)$ in eq. (5-37) by a trigonometric function proportional to $e^{q^m \cdot x}$, we obtained an expression for $\varphi_i(\kappa)$ in terms of generalized functions. The turbulent velocity $u_i(x)$ is, in fact similar to $e^{q^m \cdot x}$, inasmuch as both quantities fluctuate as x varies. Moreover, one can allow for the facts that $u_i(x)$ varies irregularly and that it may include a mean velocity by using a series of trigonometric functions. Thus, let

$$u_i(x) = \sum_{m=1}^{\infty} 2^3 a_i(q_k^m) e^{iq^m \cdot x}, \tag{5-44}$$

where the q^m and $a_i(q^m)$ are random vectors (which may include regular components). Then, instead of eq. (5-43), we have

$$\varphi_i(\kappa) = \frac{1}{\pi^3} \int_{-\infty}^{\infty} \sum_{m=1}^{\infty} a_i(q^m) e^{i(q^m - \kappa) \cdot x} dx,$$

$$= \frac{1}{\pi^3} \sum_{m=1}^{\infty} a_i(q^m) \lim_{g \to \infty} \frac{\sin[g(\kappa_1 - q_1^m)]}{\kappa_1 - q_1^m} \frac{\sin[g(\kappa_2 - q_2^m)]}{\kappa_2 - q_2^m} \frac{\sin[g(\kappa_3 - q_3^m)]}{\kappa_3 - q_3^m}, \tag{5-45}$$

or

$$\varphi_i(\kappa) = \sum_{m=1}^{\infty} a_i(q^m) \delta(\kappa_1 - q_1^m) \delta(\kappa_2 - q_2^m) \delta(\kappa_3 - q_3^m) \equiv \sum_{m=1}^{\infty} a_i(q^m) \delta(\kappa - q^m). \tag{5-46}$$

So although the turbulent velocity $u_i(x)$ can be nonzero over all of x space, its Fourier transform can exist as a generalized function; it can be written as an infinite row of deltas with random coefficients. By virtue of eq. (5-38), the operations of differentiation, integration, multiplication, and so forth can be applied to $\varphi_i(\kappa)$. That is, $\varphi_i(\kappa)$ can be

5-2-2 Analysis of Instantaneous Continuum Equations

The instantaneous eqs. (4-11) through (4-13) can be written, for $g_i = 0$ (no buoyancy), as

$$\frac{\partial \tilde{u}_i}{\partial t} = -\frac{\partial(\tilde{u}_i \tilde{u}_k)}{\partial x_k} - \frac{1}{\rho}\frac{\partial \tilde{\sigma}}{\partial x_i} + \nu \frac{\partial^2 \tilde{u}_i}{\partial x_k \partial x_k}, \qquad (5\text{-}47)$$

$$\frac{\partial \tilde{T}}{\partial t} = -\frac{\partial(\tilde{T}\tilde{u}_k)}{\partial x_k} + \alpha \frac{\partial^2 \tilde{T}}{\partial x_k \partial x_k}, \qquad (5\text{-}48)$$

and

$$\frac{1}{\rho}\frac{\partial^2 \tilde{\sigma}}{\partial x_l \partial x_l} = -\frac{\partial^2(\tilde{u}_i \tilde{u}_k)}{\partial x_i \partial x_k}, \qquad (5\text{-}49)$$

where a tilde over a quantity indicates a total instantaneous value (mean plus fluctuating), and where the quantity σ_e drops out of the set of equations for $g_i = 0$ (see equation following [3-22]).

In order to convert eqs. (5-47) through (5-49) to spectral form, we write the three following Fourier-transform pairs:

$$\tilde{u}_i(x) = \int_{-\infty}^{\infty} \tilde{\varphi}_i(\kappa) e^{i\kappa \cdot x}\, d\kappa, \qquad (5\text{-}50)$$

$$\tilde{\varphi}_i(\kappa) = \frac{1}{(2\pi)^3} \int_{-\infty}^{\infty} \tilde{u}_i(x) e^{-i\kappa \cdot x}\, dx; \qquad (5\text{-}51)$$

$$\tilde{\sigma}(x) = \int_{-\infty}^{\infty} \tilde{\lambda}(\kappa) e^{i\kappa \cdot x}\, d\kappa, \qquad (5\text{-}52)$$

$$\tilde{\lambda}(\kappa) = \frac{1}{(2\pi)^3} \int_{-\infty}^{\infty} \tilde{\sigma}(x) e^{-i\kappa \cdot x}\, dx; \qquad (5\text{-}53)$$

$$\tilde{T}(x) = \int_{-\infty}^{\infty} \tilde{\gamma}(\kappa) e^{i\kappa \cdot x}\, d\kappa, \qquad (5\text{-}54)$$

$$\tilde{\gamma}(\kappa) = \frac{1}{(2\pi)^3} \int_{-\infty}^{\infty} \tilde{T}(x) e^{-i\kappa \cdot x}\, dx. \qquad (5\text{-}55)$$

The κ-space Fourier transforms in eqs. (5-50) through (5-55) are, of course, generalized functions, as discussed in the previous section.

As in the case of averaged quantities, differentiation of an unaveraged quantity in x space with respect to x_k multiplies its Fourier transform in κ space by $i\kappa_k$. For example, from eq. (5-50), one obtains

$$\frac{\partial}{\partial x_k} \tilde{u}_i(x) = \int_{-\infty}^{\infty} i\kappa_k \tilde{\varphi}_i(\kappa) e^{i\kappa \cdot x}\, d\kappa, \qquad (5\text{-}56)$$

or

$$i\kappa_k \tilde{\varphi}_i(\boldsymbol{\kappa}) = \frac{1}{(2\pi)^3} \int_{-\infty}^{\infty} \frac{\partial}{\partial x_k} \tilde{u}_i(\boldsymbol{x}) e^{-i\boldsymbol{\kappa}\cdot\boldsymbol{x}} d\boldsymbol{x}. \tag{5-57}$$

Finally, in order to take the Fourier transforms of eqs. (5-47) through (5-49), one must consider the Fourier transforms of the products $\tilde{u}_i \tilde{u}_k$ and $\tilde{T}\tilde{u}_k$. Thus, writing the Fourier transform of $\tilde{u}_i \tilde{u}_k$ as $\tilde{\zeta}_{ik}(\boldsymbol{\kappa})$, one has

$$\tilde{\zeta}_{ik}(\boldsymbol{\kappa}) = \frac{1}{(2\pi)^3} \int_{-\infty}^{\infty} \tilde{u}_i(\boldsymbol{x}) \tilde{u}_k(\boldsymbol{x}) e^{-i\boldsymbol{\kappa}\cdot\boldsymbol{x}} d\boldsymbol{x} = \frac{1}{(2\pi)^3} \int_{-\infty}^{\infty} \tilde{u}_i(\boldsymbol{x}) \int_{-\infty}^{\infty} \tilde{\varphi}_k(\boldsymbol{\kappa}') e^{i\boldsymbol{\kappa}'\cdot\boldsymbol{x}} d\boldsymbol{\kappa}' e^{-i\boldsymbol{\kappa}\cdot\boldsymbol{x}} d\boldsymbol{x}$$

$$= \frac{1}{(2\pi)^3} \int_{-\infty}^{\infty} \int_{-\infty}^{\infty} \tilde{u}_i(\boldsymbol{x}) \tilde{\varphi}_k(\boldsymbol{\kappa}') e^{-i(\boldsymbol{\kappa}-\boldsymbol{\kappa}')\cdot\boldsymbol{x}} d\boldsymbol{\kappa}' d\boldsymbol{x}$$

$$= \frac{1}{(2\pi)^3} \int_{-\infty}^{\infty} \int_{-\infty}^{\infty} \tilde{u}_i(\boldsymbol{x}) \tilde{\varphi}_k(\boldsymbol{\kappa}') e^{-i(\boldsymbol{\kappa}-\boldsymbol{\kappa}')\cdot\boldsymbol{x}} d\boldsymbol{x} d\boldsymbol{\kappa}'$$

$$= \frac{1}{(2\pi)^3} \int_{-\infty}^{\infty} \tilde{\varphi}_k(\boldsymbol{\kappa}') \int_{-\infty}^{\infty} \tilde{u}_i(\boldsymbol{x}) e^{-i(\boldsymbol{\kappa}-\boldsymbol{\kappa}')\cdot\boldsymbol{x}} d\boldsymbol{x} d\boldsymbol{\kappa}'$$

$$= \int_{-\infty}^{\infty} \tilde{\varphi}_k(\boldsymbol{\kappa}') \tilde{\varphi}_i(\boldsymbol{\kappa} - \boldsymbol{\kappa}') d\boldsymbol{\kappa}', \tag{5-58}$$

where use was made of the relation

$$\int_{-\infty}^{\infty} \int_{-\infty}^{\infty} f(\boldsymbol{x}, \boldsymbol{\kappa}') d\boldsymbol{\kappa}' d\boldsymbol{x} = \int_{-\infty}^{\infty} \int_{-\infty}^{\infty} f(\boldsymbol{x}, \boldsymbol{\kappa}') d\boldsymbol{x} d\boldsymbol{\kappa}'.$$

Inasmuch as $\tilde{\zeta}_{ik}(\boldsymbol{\kappa}) = \tilde{\zeta}_{ki}(\boldsymbol{\kappa})$ (see first equality in eq. [5-58]), one could just as well write the alternate form

$$\tilde{\zeta}_{ik}(\boldsymbol{\kappa}) = \frac{1}{(2\pi)^3} \int_{-\infty}^{\infty} \tilde{u}_k(\boldsymbol{x}) \tilde{u}_i(\boldsymbol{x}) e^{-i\boldsymbol{\kappa}\cdot\boldsymbol{x}} d\boldsymbol{x} = \int_{-\infty}^{\infty} \tilde{\varphi}_i(\boldsymbol{\kappa}') \varphi_k(\boldsymbol{\kappa} - \boldsymbol{\kappa}') d\boldsymbol{\kappa}'. \tag{5-58a}$$

Similarly, the Fourier transform of $\tilde{T}\tilde{u}_k$, call it $\tilde{\eta}_k(\boldsymbol{\kappa})$, can be written as

$$\tilde{\eta}_k(\boldsymbol{\kappa}) = \int_{-\infty}^{\infty} \tilde{\varphi}_k(\boldsymbol{\kappa}') \tilde{\gamma}(\boldsymbol{\kappa} - \boldsymbol{\kappa}') d\boldsymbol{\kappa}'. \tag{5-59}$$

As shown by eqs. (5-56) and (5-57), differentiating a quantity in x space by x_k multiplies its Fourier transform in $\boldsymbol{\kappa}$ space by $i\kappa_k$. Applying this to a product, that in eq. (5-58) for example, shows that

$$i\kappa_k \tilde{\zeta}_{ik} = \frac{1}{(2\pi)^3} \int_{-\infty}^{\infty} [\partial(\tilde{u}_i \tilde{u}_k)/\partial x_k] e^{-i\boldsymbol{\kappa}\cdot\boldsymbol{x}} d\boldsymbol{x} = \int_{-\infty}^{\infty} i\kappa_k \tilde{\varphi}_k(\boldsymbol{\kappa}') \tilde{\varphi}_i(\boldsymbol{\kappa} - \boldsymbol{\kappa}') d\boldsymbol{\kappa}', \tag{5-60}$$

or, using the alternate form for $\tilde{\zeta}_{ik}(\boldsymbol{\kappa})$ (eq. [5-58a]), one obtains

$$i\kappa_k \tilde{\zeta}_{ik} = \frac{1}{(2\pi)^3} \int_{-\infty}^{\infty} [\partial(\tilde{u}_k \tilde{u}_i)/\partial x_k] e^{-i\boldsymbol{\kappa}\cdot\boldsymbol{x}} d\boldsymbol{x} = \int_{-\infty}^{\infty} i\kappa_k \tilde{\varphi}_k(\boldsymbol{\kappa} - \boldsymbol{\kappa}') \tilde{\varphi}_i(\boldsymbol{\kappa}') d\boldsymbol{\kappa}'. \tag{5-60a}$$

The unaveraged eqs. (5-47) through (5-49) now can be written in spectral form by taking their Fourier transforms. Thus, multiplying those equations through by

$[1/(2\pi)^3]e^{-i\boldsymbol{\kappa}\cdot\boldsymbol{x}}d\boldsymbol{x}$, integrating over $\boldsymbol{x}$ from $-\infty$ to $+\infty$, and replacing the subscripts i by j, we get

$$\frac{\partial}{\partial t}\tilde{\varphi}_j(\boldsymbol{\kappa}) = -\int_{-\infty}^{\infty} i\kappa_k \tilde{\varphi}_k(\boldsymbol{\kappa}')\tilde{\varphi}_j(\boldsymbol{\kappa}-\boldsymbol{\kappa}')\,d\boldsymbol{\kappa}' - \frac{1}{\rho}i\kappa_j\tilde{\lambda}(\boldsymbol{\kappa}) - \nu\kappa^2\tilde{\varphi}_j(\boldsymbol{\kappa}), \quad (5\text{-}61)$$

$$\frac{\partial}{\partial t}\tilde{\gamma}(\boldsymbol{\kappa}) = -\int_{-\infty}^{\infty} i\kappa_k \tilde{\varphi}_k(\boldsymbol{\kappa}')\tilde{\gamma}(\boldsymbol{\kappa}-\boldsymbol{\kappa}')\,d\boldsymbol{\kappa}' - \alpha\kappa^2\tilde{\gamma}(\boldsymbol{\kappa}), \quad (5\text{-}62)$$

and

$$-\frac{1}{\rho}\kappa^2\tilde{\lambda}(\boldsymbol{\kappa}) = \int_{-\infty}^{\infty} \kappa_j\kappa_k \tilde{\varphi}_k(\boldsymbol{\kappa}')\tilde{\varphi}_j(\boldsymbol{\kappa}-\boldsymbol{\kappa}')\,d\boldsymbol{\kappa}', \quad (5\text{-}63)$$

where $\kappa^2 = \kappa_l\kappa_l$ and eqs. (5-50) through (5-60) are used. Combining eqs. (5-61) and (5-63) gives

$$\frac{\partial}{\partial t}\tilde{\varphi}_j(\boldsymbol{\kappa}) = -i\kappa_k\left(\delta_{jl} - \frac{\kappa_j\kappa_l}{\kappa^2}\right)\int_{-\infty}^{\infty} \tilde{\varphi}_k(\boldsymbol{\kappa}')\tilde{\varphi}_l(\boldsymbol{\kappa}-\boldsymbol{\kappa}')\,d\boldsymbol{\kappa}' - \nu\kappa^2\tilde{\varphi}_j(\boldsymbol{\kappa}). \quad (5\text{-}64)$$

The evolution equation for $\tilde{\varphi}_j$, if written in the form of (5-64), emphasizes the similarity of the first two terms on the right side of eq. (5-47); the spectral equivalents of those terms are both nonlinear inertia terms that are second degree in $\tilde{\varphi}_k(\boldsymbol{\kappa}')$ and $\tilde{\varphi}_l(\boldsymbol{\kappa}-\boldsymbol{\kappa}')$. As is the case for the unaveraged equations in physical space, eq. (5-64) does not require a closure assumption for its solution. In fact, it often is used in discrete form for numerical solutions of turbulence [1].

Equations (5-62) and (5-64) can be written in terms of complex conjugates by using the relations $\tilde{\varphi}_j(\boldsymbol{\kappa}) = (\tilde{\varphi}_j)_R(\boldsymbol{\kappa}) + i(\tilde{\varphi}_j)_I(\boldsymbol{\kappa})$, $\tilde{\varphi}_i^*(\boldsymbol{\kappa}) = (\tilde{\varphi}_i)_R(\boldsymbol{\kappa}) - i(\tilde{\varphi}_i)_I(\boldsymbol{\kappa})$, $\tilde{\gamma}(\boldsymbol{\kappa}) = \tilde{\gamma}_R(\boldsymbol{\kappa}) + i\tilde{\gamma}_I(\boldsymbol{\kappa})$, and $\tilde{\gamma}^*(\boldsymbol{\kappa}) = \tilde{\gamma}_R(\boldsymbol{\kappa}) - i\tilde{\gamma}_I(\boldsymbol{\kappa})$, where the asterisks designate complex conjugates and the subscripts R and I refer respectively to real and imaginary parts. Thus, we get

$$\frac{\partial}{\partial t}\tilde{\varphi}_i^*(\boldsymbol{\kappa}) = i\kappa_k\left(\delta_{il} - \frac{\kappa_i\kappa_l}{\kappa^2}\right)\int_{-\infty}^{\infty} \tilde{\varphi}_k^*(\boldsymbol{\kappa}')\tilde{\varphi}_l^*(\boldsymbol{\kappa}-\boldsymbol{\kappa}')\,d\boldsymbol{\kappa}' - \nu\kappa^2\tilde{\varphi}_i^*(\boldsymbol{\kappa}), \quad (5\text{-}65)$$

and

$$\frac{\partial}{\partial t}\tilde{\gamma}^*(\boldsymbol{\kappa}) = i\kappa_k\int_{-\infty}^{\infty} \tilde{\varphi}_k^*(\boldsymbol{\kappa}')\tilde{\gamma}^*(\boldsymbol{\kappa}-\boldsymbol{\kappa}')\,d\boldsymbol{\kappa}' - \alpha\kappa^2\tilde{\gamma}^*(\boldsymbol{\kappa}). \quad (5\text{-}66)$$

Multiplying eq. (5-64) by $\tilde{\varphi}_i^*(\boldsymbol{\kappa})$ and eq. (5-65) by $\tilde{\varphi}_j(\boldsymbol{\kappa})$, we get, after adding the two equations,

$$\frac{\partial}{\partial t}\left[\tilde{\varphi}_i^*(\boldsymbol{\kappa})\tilde{\varphi}_j(\boldsymbol{\kappa})\right] = i\kappa_k\left(\delta_{il} - \frac{\kappa_i\kappa_l}{\kappa^2}\right)\int_{-\infty}^{\infty} \tilde{\varphi}_j(\boldsymbol{\kappa})\tilde{\varphi}_k^*(\boldsymbol{\kappa}')\tilde{\varphi}_l^*(\boldsymbol{\kappa}-\boldsymbol{\kappa}')\,d\boldsymbol{\kappa}' - i\kappa_k\left(\delta_{jl} - \frac{\kappa_j\kappa_l}{\kappa^2}\right)$$

$$\times \int_{-\infty}^{\infty} \tilde{\varphi}_i^*(\boldsymbol{\kappa})\tilde{\varphi}_k(\boldsymbol{\kappa}')\tilde{\varphi}_l(\boldsymbol{\kappa}-\boldsymbol{\kappa}')\,d\boldsymbol{\kappa}' - 2\nu\kappa^2\tilde{\varphi}_i^*(\boldsymbol{\kappa})\tilde{\varphi}_j(\boldsymbol{\kappa}). \quad (5\text{-}67)$$

Equation (5-67) gives the evolution of the total (mean plus turbulent) spectral-energy tensor (at a particular $\boldsymbol{\kappa}$). Contracting the indices i and j we get, for the total spectral energy (at a particular κ),

$$\frac{\partial}{\partial t}[\tilde{\varphi}_i^*(\kappa)\tilde{\varphi}_i(\kappa)] = \frac{\partial}{\partial t}[(\tilde{\varphi}_i)_R(\tilde{\varphi}_i)_R + (\tilde{\varphi}_i)_I(\tilde{\varphi}_i)_I] = |\tilde{\varphi}_i(\kappa)\tilde{\varphi}_i(\kappa)|$$

$$= i\kappa_k \int_{-\infty}^{\infty} \left[\tilde{\varphi}_i(\kappa)\tilde{\varphi}_k^*(\kappa')\tilde{\varphi}_i^*(\kappa-\kappa') - \tilde{\varphi}_i^*(\kappa)\tilde{\varphi}_k(\kappa')\tilde{\varphi}_i(\kappa-\kappa')\right]d\kappa'$$

$$- 2\nu\kappa^2 \tilde{\varphi}_i^*(\kappa)\tilde{\varphi}_i(\kappa) \qquad (5\text{-}67a)$$

Note that the spectral-pressure terms (those divided by κ^2) drop out of the contracted equation because of continuity (see eqs. [4-10] and [5-57]). That is, the spectral-pressure terms can transfer total (mean plus turbulent) energy between directional components but do not change the sum of the three directional components at any κ. Similarly, from eqs. (5-62), (5-66), and (5-65), we get

$$\frac{\partial}{\partial t}[\tilde{\gamma}(\kappa)\tilde{\gamma}^*(\kappa)] = \frac{\partial}{\partial t}|\gamma^2(\kappa)| = -i\kappa_k \int_{-\infty}^{\infty} \left[\tilde{\varphi}_k(\kappa')\tilde{\gamma}(\kappa-\kappa')\tilde{\gamma}^*(\kappa)\right.$$

$$\left. - \tilde{\varphi}_k^*(\kappa')\tilde{\gamma}^*(\kappa-\kappa')\tilde{\gamma}(\kappa)\right]d\kappa - 2\alpha\kappa^2\tilde{\gamma}(\kappa)\tilde{\gamma}^*(\kappa) \qquad (5\text{-}68)$$

and

$$\frac{\partial}{\partial t}[\tilde{\gamma}(\kappa)\tilde{\varphi}_i^*(\kappa)] = -i\kappa_k \int_{-\infty}^{\infty} \left[\tilde{\varphi}_k(\kappa')\tilde{\gamma}(\kappa-\kappa')\tilde{\varphi}_i^*(\kappa)\right.$$

$$\left. - \left(\delta_{il} - \frac{\kappa_i\kappa_l}{\kappa^2}\right)\tilde{\varphi}_k^*(\kappa')\tilde{\varphi}_l^*(\kappa-\kappa')\tilde{\gamma}(\kappa)\right]d\kappa - (\alpha+\nu)\kappa^2\tilde{\gamma}(\kappa)\varphi_i^*(\kappa).$$

$$(5\text{-}69)$$

The presence of the integrals in eqs. (5-61) through (5-69) reveals that the Fourier components at each κ depend nonlinearly on Fourier components at every other point in wavenumber space. An important deduction from those equations is the triad nature of the nonlinear interaction of the Fourier components. For instance, eq. (5-64) shows that the evolution of the component $\tilde{\varphi}_j$ at wavevector κ depends on components at wavevectors κ' and $\kappa - \kappa'$. That is, there is an interaction among components at those three wavevectors. Equation (5-67) shows that the evolution of $\tilde{\varphi}_i^*(\kappa)\tilde{\varphi}_j(\kappa)$ also depends on the interaction among components at the three wavevectors κ, κ', and $\kappa - \kappa'$.

The last terms in eqs. (5-64), (5-65), (5-67), and (5-69), which are multiplied by $\nu\kappa^2$, are viscous dissipation terms. The presence of κ^2 in those terms shows that the viscous dissipation occurs at higher wavenumbers than does the bulk of the activity. Similar comments apply to the terms in eqs. (5-62), (5-66), (5-68), and (5-69), which are multiplied by $\alpha\kappa^2$. Those terms cause a reduction of activity by thermal conduction or thermal smearing. Note that eq. (5-69) contains both thermal-smearing and viscous-dissipation terms.

Although eqs. (5-61) through (5-69) have rather compact forms, they are general, applicable to both homogeneous and inhomogeneous turbulence. Because they are written in terms of total instantaneous quantities, however, it is hard to identify the various

122 TURBULENT FLUID MOTION

turbulence processes. For instance, the nonlinear term in eq. (5-64), besides containing spectral-transfer and directional-transfer effects, may contain turbulence production by mean gradients. Thus, further interpretations are postponed to the next section, in which mean velocities are absent (or uniform), and the turbulence is homogeneous.

5-3 HOMOGENEOUS TURBULENCE WITHOUT MEAN VELOCITY OR TEMPERATURE (SCALAR) GRADIENTS

A (statistically) homogeneous turbulence is defined as one in which averaged turbulence quantities are not functions of position. For instance, for homogeneous turbulence

$$\overline{u_i u_j} \neq \overline{u_i u_j}(x),$$
$$\overline{u_i u_j u_k} \neq \overline{u_i u_j u_k}(x)$$

and

$$\overline{\sigma \partial u_i / \partial x_j} \neq \overline{\sigma \partial u_i / \partial x_j}(x),$$

where x is the position vector. Similar statements apply to other averaged turbulence quantities in a homogeneous turbulence.

Homogeneous turbulence without mean gradients is attractive as an area of study because of its conceptual simplicity. Production and diffusion terms, for example, are absent in the equations for that type of turbulence. The absence of those terms, however, may not always be helpful for getting solutions. The presence of large mean-gradient or diffusion effects in fact may be an advantage, because a solution depends to a lesser extent on the difficult-to-determine nonlinear self-interaction terms. In this section we are not interested in mitigating the effects of those terms. Rather we consider homogeneous turbulence without mean gradients mainly as a vehicle for studying the nonlinear self-interaction or dissipation effects.

First we consider the basic equations for homogeneous turbulence without mean velocity or temperature gradients. Then we give some illustrative solutions. In most cases the analytical solutions considered are of the simplest kind, in order to avoid mathematical complexity. Somewhat more widely applicable numerical solutions also are discussed if available and appropriate.

5-3-1 Basic Equations

For homogeneous turbulence without mean velocity gradients, the averaged two-point flow equations in r space (eqs. [4-147] through [4-150] in the last chapter) simplify to

$$\frac{\partial}{\partial t}\overline{u_i u'_j}(r) + \frac{\partial}{\partial r_k}\left(\overline{u_i u'_j u'_k} - \overline{u_i u_k u'_j}\right) = \frac{1}{\rho}\left(\frac{\partial}{\partial r_i}\overline{\sigma u'_j} - \frac{\partial}{\partial r_j}\overline{u_i \sigma'}\right) + 2\nu\frac{\partial^2 \overline{u_i u'_j}}{\partial r_k \partial r_k}, \quad (5\text{-}70)$$

$$\frac{1}{\rho}\frac{\partial^2 \overline{u_i \sigma'}}{\partial r_j \partial r_j} = -\frac{\partial^2 \overline{u_i u'_j u'_k}}{\partial r_j \partial r_k}, \quad (5\text{-}71)$$

and

$$\frac{1}{\rho}\frac{\partial^2 \overline{\sigma u'_j}}{\partial r_i \partial r_i} = -\frac{\partial^2 \overline{u_i u_k u'_j}}{\partial r_i \partial r_k}. \tag{5-72}$$

The corresponding spectral equations in the κ space (eqs. [5-17] through [5-19]) become

$$\frac{\partial}{\partial t}\varphi_{ij}(\kappa) + i\kappa_k\left(\varphi'_{ijk} - \varphi_{ikj}\right) = \frac{1}{\rho}\left(i\kappa_i\lambda_j - i\kappa_j\lambda'_i\right) - 2\nu\kappa^2\varphi_{ij}, \tag{5-73}$$

$$-\frac{1}{\rho}\kappa^2\lambda'_i = \kappa_j\kappa_k\varphi'_{ijk}, \tag{5-74}$$

and

$$-\frac{1}{\rho}\kappa^2\lambda_j = \kappa_i\kappa_k\varphi_{ikj}. \tag{5-75}$$

Combining eqs. (5-73) through (5-75), we get

$$\frac{\partial}{\partial t}\varphi_{ij}(\kappa) = -i\kappa_k\left(\delta_{jl} - \frac{\kappa_j\kappa_l}{\kappa^2}\right)\varphi'_{ilk} + i\kappa_k\left(\delta_{il} - \frac{\kappa_i\kappa_l}{\kappa^2}\right)\varphi_{lkj} - 2\nu\kappa^2\varphi_{ij}. \tag{5-76}$$

One could, of course, obtain eqs. (5-70) through (5-76) directly from the instantaneous eqs. (5-47) and (5-49), and from similar equations written at a point P' with the subscript i replaced by j. The tildes over the instantaneous quantities would be omitted, because mean gradients are absent. So instead of starting with the general eqs. (4-145), (4-146), and (4-148), and then simplifying the final two-point equations, one could start with the simpler equations for homogeneous turbulence without mean gradients. Then, multiplying the unprimed equations by u'_j and the primed equations by u_i, using the fact that quantities at one point are independent of the location of the other point, adding and space-averaging the equations, letting $\partial/\partial x'_k = \partial/\partial r_k$ and $\partial/\partial x_k = -\partial/\partial r_k$, and using the Fourier transforms given by eqs. (5-12), (5-13), and (5-20) through (5-22), one obtains eqs. (5-70) through (5-76).

Equations (5-64), (5-62), and (5-65) through (5-69) become, for homogeneous turbulence without mean gradients,

$$\frac{\partial}{\partial t}\varphi_j(\kappa) = -i\kappa_k\left(\delta_{jl} - \frac{\kappa_j\kappa_l}{\kappa^2}\right)\int_{-\infty}^{\infty}\varphi_k(\kappa')\varphi_l(\kappa - \kappa')\,d\kappa' - \nu\kappa^2\varphi_j(\kappa), \tag{5-77}$$

$$\frac{\partial}{\partial t}\gamma(\kappa) = -\int_{-\infty}^{\infty}i\kappa_k\varphi_k(\kappa')\gamma(\kappa - \kappa')\,d\kappa' - \alpha\kappa^2\gamma(\kappa), \tag{5-78}$$

$$\frac{\partial}{\partial t}\varphi_i^*(\kappa) = i\kappa_k\left(\delta_{il} - \frac{\kappa_i\kappa_l}{\kappa^2}\right)\int_{-\infty}^{\infty}\varphi_k^*(\kappa')\varphi_l^*(\kappa - \kappa')\,d\kappa' - \nu\kappa^2\varphi_i^*(\kappa), \tag{5-79}$$

$$\frac{\partial}{\partial t}\left[\varphi_i^*(\kappa)\varphi_j(\kappa)\right] = -i\kappa_k\left(\delta_{jl} - \frac{\kappa_j\kappa_l}{\kappa^2}\right)\int_{-\infty}^{\infty}\varphi_i^*(\kappa)\varphi_l(\kappa - \kappa')\varphi_k(\kappa')\,d\kappa'$$
$$+ i\kappa_k\left(\delta_{il} - \frac{\kappa_i\kappa_l}{\kappa^2}\right)\int_{-\infty}^{\infty}\varphi_l^*(\kappa - \kappa')\varphi_k^*(\kappa')\varphi_j(\kappa)\,d\kappa'$$
$$- 2\nu\kappa^2\varphi_i^*(\kappa)\varphi_j(\kappa), \tag{5-80}$$

$$\frac{\partial}{\partial t}\left[\varphi_i^*(\kappa)\varphi_i(\kappa)\right] = \frac{\partial}{\partial t}[(\varphi_i)_R(\varphi_i)_R + (\varphi_i)_I(\varphi_i)_I] = |\varphi_i(\kappa)\varphi_i(\kappa)|$$
$$= -i\kappa_k \int_{-\infty}^{\infty}\left[\varphi_i^*(\kappa)\varphi_k(\kappa')\varphi_i(\kappa-\kappa') - \varphi_i(\kappa)\varphi_k^*(\kappa')\varphi_i^*(\kappa-\kappa')\right]d\kappa'$$
$$- \nu\kappa^2\varphi_i^*(\kappa)\varphi_i(\kappa), \qquad (5\text{-}80a)$$

$$\frac{\partial}{\partial t}[\gamma(\kappa)\gamma^*(\kappa)] = \frac{\partial}{\partial t}|\gamma^2(\kappa)|$$
$$= -i\kappa_k \int_{-\infty}^{\infty}\left[\varphi_k(\kappa')\gamma(\kappa-\kappa')\gamma^*(\kappa) - \varphi_k^*(\kappa')\gamma^*(\kappa-\kappa')\gamma(\kappa)\right]d\kappa'$$
$$- 2\alpha\kappa^2\gamma(\kappa)\gamma^*(\kappa), \qquad (5\text{-}81)$$

and

$$\frac{\partial}{\partial t}[\gamma(\kappa)\varphi_i^*(\kappa)] = -i\kappa_k \int_{-\infty}^{\infty}\left[\varphi_k(\kappa')\gamma(\kappa-\kappa')\varphi_i^*(\kappa) - \left(\delta_{il} - \frac{\kappa_i\kappa_l}{\kappa^2}\right)\right.$$
$$\left. \times \varphi_k^*(\kappa')\varphi_l^*(\kappa-\kappa')\gamma(\kappa)\right]d\kappa' - (\alpha+\nu)\kappa^2\gamma(\kappa)\varphi_i^*(\kappa). \qquad (5\text{-}82)$$

Note that eqs. (5-77) through (5-82) are the same as the corresponding equations in the previous section, except that the tildes over the dependent variables have been omitted for homogeneous turbulence without mean velocity or temperature gradients (see eqs. [4-14] through [4-20]).

5-3-1-1 Equivalence of eqs. (5-76) and (5-80). To show the equivalence of the subject equations we first relate $\varphi_i^*(\kappa)\varphi_j(\kappa)$ to $\varphi_{ij}(\kappa)$. To that end, we write

$$\overline{u_i u_j'}(r) = \overline{u_i(x)u_j(x+r)} = \frac{1}{R_1 R_2 R_3} \int_{-R_k/2}^{R_k/2} u_i(x)u_j(x+r)\,dx, \qquad (5\text{-}83)$$

where the overbar designates a space average and R_k is the spatial period in the k direction. Space averages, of course, can be used because the turbulence is homogeneous. Because the period is, at this point, finite, we represent u_i by a Fourier series (finite Fourier transform; see, e.g., eq. [5-4]). Thus,

$$u_i(x) = \sum_{\kappa=-\infty}^{\infty} (\varphi_i)_\kappa e^{i2\pi\kappa_k x_k/R_{(k)}}, \qquad (5\text{-}84)$$

or

$$u_j(x+r) = \sum_{\kappa=-\infty}^{\infty} (\varphi_j)_\kappa e^{i2\pi\kappa_k(x_k+r_k)/R_{(k)}}. \qquad (5\text{-}85)$$

Then eq. (5-83) becomes

$$\overline{u_i(x)u_j(x+r)} = \frac{1}{R_1R_2R_3}\int_{-R_k/2}^{R_k/2} u_i(x)\sum_{\kappa=-\infty}^{\infty}(\varphi_j)_\kappa e^{i2\pi\kappa_k(x_k+r_k)/R_{(k)}}\,dx$$

$$= \frac{1}{R_1R_2R_3}\int_{-R_k/2}^{R_k/2}\sum_{\kappa=-\infty}^{\infty} u_i(x)(\varphi_j)_\kappa e^{i2\pi\kappa_k(x_k+r_k)/R_{(k)}}\,dx$$

$$= \frac{1}{R_1R_2R_3}\sum_{\kappa=-\infty}^{\infty}\int_{-R_k/2}^{R_k/2} u_i(x)(\varphi_j)_\kappa e^{i2\pi\kappa_k(x_k+r_k)/R_{(k)}}\,dx \quad (5\text{-}86)$$

$$= \frac{1}{R_1R_2R_3}\sum_{\kappa=-\infty}^{\infty}(\varphi_j)_\kappa e^{i2\pi\kappa_k r_k/R_{(k)}}\int_{-R_k/2}^{R_k/2} u_i(x)e^{i2\pi\kappa_k x_k/R_{(k)}}\,dx$$

$$= \sum_{s=-\infty}^{\infty}(\varphi_j)_s e^{is_k r_k}\int_{-R_k/2}^{R_k/2} u_i(x)e^{is_k x_k}\,dx\frac{\Delta s_1\Delta s_2\Delta s_3}{(2\pi)^3},$$

where eqs. (5-7) and (5-7b) are used. Note that Δs_k is given by eq. (5-7). Passing to the limit as $R_k \to \infty$ ($\Delta s_k \to 0$), and using eq. (5-10), one obtains

$$\overline{u_i(x)u_j(x+r)} = \frac{1}{(2\pi)^3}\int_{-\infty}^{\infty}\varphi_j(s)e^{is\cdot r}\int_{-\infty}^{\infty} u_i(x)e^{is\cdot x}\,dx\,ds. \quad (5\text{-}87)$$

However,

$$\frac{1}{(2\pi)^3}\int_{-\infty}^{\infty} u_i(x)e^{is\cdot x}\,dx = \varphi_i^*(s), \quad (5\text{-}88)$$

because, for $R_k \to \infty$, eq. (5-84) becomes

$$u_i(x) = \int_{-\infty}^{\infty}\varphi_i(s)e^{is\cdot x}\,ds. \quad (5\text{-}89)$$

Substituting eq. (5-88) into (5-87) and changing the dummy variable from s to κ, one has

$$\overline{u_iu_j'}(r) = \overline{u_i(x)u_j(x+r)} = \int_{-\infty}^{\infty}\varphi_i^*(\kappa)\varphi_j(\kappa)e^{i\kappa\cdot r}\,d\kappa, \quad (5\text{-}90)$$

showing that $\varphi_i^*(\kappa)\varphi_j(\kappa)$ is the Fourier transform (from r space to κ space) of $\overline{u_iu_j'}(r)$. Because $\varphi_{ij}(\kappa)$ is also the Fourier transform of $\overline{u_iu_j'}(r)$ (see eqs. [5-11] and [5-12]),

$$\varphi_{ij}(\kappa) = \varphi_i^*(\kappa)\varphi_j(\kappa). \quad (5\text{-}91)$$

That is, $\varphi_{ij}(\kappa)$ in eq. (5-76) is equal to $\varphi_i^*(\kappa)\varphi_j(\kappa)$ in eq. (5-80). The tensor $\varphi_i^*(\kappa)\varphi_j(\kappa)$ often is written as $\overline{\varphi_i^*(\kappa)\varphi_j(\kappa)}$, where the overbar could represent, say, an ensemble average. However, the overbar does not seem to be strictly necessary because $\varphi_i^*(\kappa)\varphi_j(\kappa)$ is already the Fourier transform of a space-averaged quantity (eq. [5-90]).

Consider next the equivalence of the nonlinear terms in eqs. (5-76) and (5-80). To that end, we relate the tensor $\varphi_{lkj}(\kappa)$ in eq. (5-76) to $\varphi_l^*(\kappa-\kappa')\varphi_k^*(\kappa')\varphi_j(\kappa)$ in eq. (5-80).

Note first that $\varphi_{lkj}(\kappa)$ is the Fourier transform of $\overline{u_l u_k u'_j}(r)$ (see eq. [5-21]), where the overbar again designates a space average. Then, carrying out a development similar to that in the previous paragraph,

$$\overline{u_l u_k u'_j}(r) = \overline{u_l(x) u_k(x) u_j(x+r)} = \frac{1}{R_1 R_2 R_3} \int_{-R_k/2}^{R_k/2} u_l(x) u_k(x) u_j(x+r)\, dx$$

$$= \frac{1}{R_1 R_2 R_3} \int_{-R_k/2}^{R_k/2} u_l(x) u_k(x) \sum_{\kappa=-\infty}^{\infty} (\varphi_j)_\kappa e^{i 2\pi \kappa_k (x_k + r_k)/R_{(k)}}\, dx$$

$$= \frac{1}{R_1 R_2 R_3} \int_{-R_k/2}^{R_k/2} \sum_{\kappa=-\infty}^{\infty} u_l(x) u_k(x) (\varphi_j)_\kappa e^{i 2\pi \kappa_k (x_k + r_k)/R_{(k)}}\, dx \qquad (5\text{-}92)$$

$$= \frac{1}{R_1 R_2 R_3} \sum_{\kappa=-\infty}^{\infty} \int_{-R_k/2}^{R_k/2} u_l(x) u_k(x) (\varphi_j)_\kappa e^{i 2\pi \kappa_k (x_k + r_k)/R_{(k)}}\, dx$$

$$= \frac{1}{R_1 R_2 R_3} \sum_{\kappa=-\infty}^{\infty} (\varphi_j)_\kappa e^{i 2\pi \kappa_k r_k/R_{(k)}} \int_{-R_k/2}^{R_k/2} u_l(x) u_k(x) e^{i 2\pi \kappa_k r_k/R_{(k)}}\, dx.$$

Using eqs. (5-7) through (5-7b), and passing to the limit as $R_k \to \infty$,

$$\overline{u_l u_k u'_j}(r) = \overline{u_l(x) u_k(x) u_j(x+r)} = \frac{1}{(2\pi)^3} \int_{-\infty}^{\infty} \varphi_j(s) e^{i s \cdot r} \int_{-\infty}^{\infty} u_l(x) u_k(x) e^{i s \cdot x}\, dx\, ds$$

$$= \frac{1}{(2\pi)^3} \int_{-\infty}^{\infty} \varphi_j(\kappa) e^{i\kappa \cdot r} \int_{-\infty}^{\infty} u_l(x) u_k(x) e^{i\kappa \cdot r}\, dx\, d\kappa$$

$$= \int_{-\infty}^{\infty} \varphi_j(\kappa) e^{i\kappa \cdot r} \zeta_{lk}^*(\kappa)\, d\kappa = \int_{-\infty}^{\infty} \varphi_j(\kappa) e^{i\kappa \cdot r} \int_{-\infty}^{\infty} \varphi_k^*(\kappa') \varphi_l^*(\kappa - \kappa')\, d\kappa'\, d\kappa$$

$$= \int_{-\infty}^{\infty} \left[\int_{-\infty}^{\infty} \varphi_l^*(\kappa - \kappa') \varphi_k^*(\kappa') \varphi_j(\kappa)\, d\kappa' \right] e^{i\kappa \cdot r}\, d\kappa, \qquad (5\text{-}93)$$

where $\zeta_{lk}^*(\kappa)$ is calculated from eq. (5-58) (tildes omitted). Equation (5-93) shows that $\int_{-\infty}^{\infty} \varphi_l^*(\kappa - \kappa') \varphi_k^*(\kappa') \varphi_j^*(\kappa) d\kappa'$ in eq. [5-80] is the Fourier transform (from r space to κ space) of $\overline{u_l u_k u'_j}(r)$. Because $\varphi_{lkj}(\kappa)$ in eq. (5-76) is also the Fourier transform of $\overline{u_l u_k u'_j}$,

$$\varphi_{lkj}(\kappa) = \int_{-\infty}^{\infty} \varphi_l^*(\kappa - \kappa') \varphi_k^*(\kappa') \varphi_j(\kappa)\, d\kappa'. \qquad (5\text{-}94)$$

Finally, we relate $\varphi'_{ilk}(\kappa)$ in eq. (5-76) to $\varphi_i^*(\kappa) \varphi_l(\kappa - \kappa') \varphi_k(\kappa')$ in eq. (5-80). The Fourier transform of φ'_{ilk} is, according to eq. (5-22), $\overline{u_i u'_l u'_k}(r)$. Carrying out a development similar to those in the previous two paragraphs, with overbars designating space averages,

$$\overline{u_i u'_l u'_k}(r) = \overline{u_i(x) u_l(x+r) u_k(x+r)} = \frac{1}{R_1 R_2 R_3} \int_{-R_k/2}^{R_k/2} u_i(x) u_l(x+r) u_k(x+r) \, dx$$

$$= \frac{1}{R_1 R_2 R_3} \int_{-R_k/2}^{R_k/2} u_i(x) \sum_{\kappa=-\infty}^{\infty} (\zeta_{lk})_\kappa e^{i 2\pi \kappa_k (x_k + r_k)/R_{(k)}} \, dx$$

$$= \frac{1}{R_1 R_2 R_3} \int_{-R_k/2}^{R_k/2} \sum_{\kappa=-\infty}^{\infty} u_i(x) (\zeta_{lk})_\kappa e^{i 2\pi \kappa_k (x_k + r_k)/R_{(k)}} \, dx$$

$$= \frac{1}{R_1 R_2 R_3} \sum_{\kappa=-\infty}^{\infty} \int_{-R_k/2}^{R_k/2} u_i(x) (\zeta_{lk})_\kappa e^{i 2\pi \kappa_k (x_k + r_k)/R_{(k)}} \, dx$$

$$= \frac{1}{R_1 R_2 R_3} \sum_{\kappa=-\infty}^{\infty} (\zeta_{lk})_\kappa e^{i 2\pi \kappa_k r_k/R_{(k)}} \int_{-R_k/2}^{R_k/2} u_i(x) e^{i 2\pi \kappa_k x_k/R_{(k)}} \, dx.$$

Passing to the limit as $R_k \to \infty$ after using eqs. (5-7) through (5-7b), we get

$$\overline{u_i u'_l u'_k}(r) = \overline{u_i(x) u_l(x+r) u_k(x+r)} = \frac{1}{(2\pi)^3} \int_{-\infty}^{\infty} \zeta_{lk}(s) e^{is \cdot r} \int_{-\infty}^{\infty} u_i(x) e^{is \cdot x} \, dx \, ds$$

$$= \frac{1}{(2\pi)^3} \int_{-\infty}^{\infty} \zeta_{lk}(\kappa) e^{i\kappa \cdot r} \int_{-\infty}^{\infty} u_i(x) e^{i\kappa \cdot x} \, dx \, d\kappa$$

$$= \int_{-\infty}^{\infty} \zeta_{lk}(\kappa) e^{i\kappa \cdot r} \varphi_i^*(\kappa) \, d\kappa = \int_{-\infty}^{\infty} \int_{-\infty}^{\infty} \varphi_l(\kappa - \kappa') \varphi_k(\kappa') \, d\kappa' \varphi_i^*(\kappa) e^{i\kappa \cdot r} \, d\kappa$$

$$= \int_{-\infty}^{\infty} \left[\int_{-\infty}^{\infty} \varphi_i^*(\kappa) \varphi_l(\kappa - \kappa') \varphi_k(\kappa') \, d\kappa' \right] e^{i\kappa \cdot r} \, d\kappa, \tag{5-95}$$

where $\zeta_{lk}(\kappa)$ is obtained from eq. (5-58) (tildes omitted). Equation (5-95) shows that $\int_{-\infty}^{\infty} \varphi_i^*(\kappa) \varphi_l(\kappa - \kappa') \varphi_k(\kappa') \, d\kappa'$ in eq. (5-80) is the Fourier transform (from r space to κ space) of $\overline{u_i u'_l u'_k}(r)$. Because φ'_{ilk} in eq. (5-76) is also the Fourier transform of $\overline{u_i u'_l u'_k}$,

$$\varphi'_{ilk}(\kappa) = \int_{-\infty}^{\infty} \varphi_i^*(\kappa) \varphi_l(\kappa - \kappa') \varphi_k(\kappa') \, d\kappa'. \tag{5-96}$$

Using eqs. (5-91), (5-94), and (5-96), one sees that eqs. (5-76) and (5-80) are equivalent, because all the terms in those two equations are equivalent. The strategy for arriving at that conclusion is to show that spectral tensors in eq. (5-80) are Fourier transforms from r to κ space of velocity-correlation tensors (see eqs. [5-90], [5-93], and [5-95]). It already is known that that is the case for the spectral tensors in eq. (5-76); that equation is derived by transforming the evolution equation for $\overline{u_i u'_j}$ from r to κ space. That is not the case for eq. (5-80), which is obtained by transforming from x to κ space (rather than from r to κ space). Hence, the need for the development in the present section.

Because of the equivalence of eqs. (5-76) and (5-80), the latter, like the former, is an averaged equation and thus requires a closure assumption. Note, however, that eq. (5-80)

128 TURBULENT FLUID MOTION

can be closed deductively (without introducing additional information) by calculating the φ_i from a numerical solution of the unaveraged eqs. (5-77) or (5-79).

The demonstration of the equivalence of eqs. (5-76) and (5-80) (at least according to the present method) depends on the homogeneity of the turbulence and the use of space averages in at least one direction. Also, developments similar to those in the present section can be used to show that eqs. (5-81) and (5-82) are Fourier transforms from r to κ space of evolution equations for temperature or temperature–velocity correlations.

5-3-1-2 Further discussion of the equations for homogeneous turbulence without mean gradients.
Consider first the spectral-transfer term. It is shown in section 5-1-2-1 that for a general Navier-Stokes turbulence the spectral terms associated with triple correlations, except for the pressure terms, can be considered as spectral-transfer functions or terms. However, for homogeneous turbulence, the demonstration is much simpler. For that case $\partial/\partial r_k = \partial/\partial x'_k = -\partial/\partial x_k$. Then, in eq. (5-70), the term

$$-\frac{\partial}{\partial r_k}\left(\overline{u_i u'_j u'_k} - \overline{u_i u_k u'_j}\right) = \frac{\partial}{\partial x_k}\overline{u_i u'_j u'_k} + \frac{\partial}{\partial x'_k}\overline{u_i u_k u'_j}$$

$$= \overline{u'_j u'_k \frac{\partial u_i}{\partial x_k}} + \overline{\frac{\partial(u'_j u_k)}{\partial x'_k}u_i} = \int_{-\infty}^{\infty} T_{ij}(\kappa)e^{i\kappa\cdot r}d\kappa, \quad (5\text{-}97)$$

where $T_{ij}(\kappa)$ is the Fourier transform of $-\partial(\overline{u_i u'_j u'_k} - \overline{u_i u_k u'_j})/\partial r_k$.

Then, for $r = 0$,

$$\int_{-\infty}^{\infty} T_{ij}(\kappa)\,d\kappa = \overline{u_j u_k \frac{\partial u_i}{\partial x_k}} + \overline{\frac{\partial(u_j u_k)}{\partial x_k}u_i} = \frac{\partial \overline{(u_i u_j u_k)}}{\partial x_k} = 0 \quad (5\text{-}98)$$

for homogeneous turbulence. Referring to eq. (5-70) and letting $r = 0$ in that equation, we see that T_{ij}, if integrated over all κ, gives zero contribution to the rate of change of $\overline{u_i u_j}$. It can transfer, however, spectral components of $\overline{u_i u_j}$ from one part of wavenumber space to another. So we interpret T_{ij} as a spectral-transfer term associated with turbulence self-interaction. The term $-\partial(\overline{u_i u'_j u'_k} - \overline{u_i u_k u'_j})/\partial r_k$ in eq. (5-70) is therefore the Fourier transform from κ to r space of a self-interaction spectral-transfer term.

Some detail about the spectral transfer of turbulent activity (and about the other turbulence processes) is obtainable from the evolution equation for $\varphi_i^*\varphi_j$ (eq. [5-80]).[1] The spectral energy tensor $\varphi_i^*\varphi_j$ is the Fourier transform from r to κ space of the turbulent-energy tensor $\overline{u_i u'_j}$, and φ_j is the Fourier transform from x to κ space of u_j. Letting $r = 0$ in eq. (5-90) shows that $\varphi_i^*\varphi_j$ gives spectral contributions to $\overline{u_i u_j}$ from various wavevector bands.

The spectral transfer term in eq. (5-80) is

$$T_{ij}(\kappa) = \int_{-\infty}^{\infty} P_{ij}(\kappa,\kappa')\,d\kappa', \quad (5\text{-}99)$$

---

[1] One should note that the discussion preceding section 5-3 is not strictly for turbulence, because the velocity components $\tilde{u}_i$ and their Fourier transforms can include mean components. Thus, the terms in the equations discussed there, unlike those considered in the present section, are not identified as turbulence terms

where

$$P_{ij}(\kappa, \kappa') = i\left[\kappa_k \varphi_i^*(\kappa - \kappa')\varphi_j(\kappa)\varphi_k^*(\kappa') - \kappa_k \varphi_i^*(\kappa)\varphi_j(\kappa - \kappa')\varphi_k(\kappa')\right]. \quad (5\text{-}100)$$

Equations (5-99) and (5-100) show that contributions from various wavevectors κ' between $-\infty$ and $+\infty$ make up the total turbulent transfer at κ. In particular they show that the net transfer into a wavevector band at κ takes place by the interaction of triads of Fourier components at the wavevectors κ, κ', and $\kappa - \kappa'$. That is a hallmark of the turbulence spectral-transfer process and is an important deduction from the Navier-Stokes equations. But it only became evident through the Fourier analysis of the unaveraged velocities $u_i(x)$ in those equations.

By using eq. (5-60a) in place of (5-60) and omitting the tildes for homogeneous turbulence, we get the alternate form

$$P_{ij}(\kappa, \kappa') = i\left[\kappa_k \varphi_k^*(\kappa - \kappa')\varphi_i^*(\kappa')\varphi_j(\kappa) - \kappa_k \varphi_k(\kappa - \kappa')\varphi_i^*(\kappa)\varphi_j(\kappa')\right]. \quad (5\text{-}101)$$

Equation (5-101) becomes, on using continuity in the form $(\kappa_k - \kappa'_k)\varphi_k(\kappa - \kappa') = 0$ (see eqs. [4-21] and [5-36]),

$$P_{ij}(\kappa, \kappa') = i\left[\kappa'_k \varphi_k^*(\kappa - \kappa')\varphi_i^*(\kappa')\varphi_j(\kappa) - \kappa'_k \varphi_k(\kappa - \kappa')\varphi_i^*(\kappa)\varphi_j(\kappa')\right]. \quad (5\text{-}102)$$

Interchanging κ and κ' in eq. (5-101), and noting that $\varphi_k^*(\kappa' - \kappa) = \varphi_k(\kappa - \kappa')$ and $\varphi_i(\kappa' - \kappa) = \varphi_i^*(\kappa - \kappa')$, because u is real, we get

$$P_{ij}(\kappa', \kappa) = i\left[-\kappa'_k \varphi_k^*(\kappa - \kappa')\varphi_i^*(\kappa')\varphi_j(\kappa) + \kappa'_k \varphi_k(\kappa - \kappa')\varphi_i^*(\kappa)\varphi_j(\kappa')\right]. \quad (5\text{-}103)$$

Comparison of eqs. (5-102) and (5-103) shows that

$$P_{ij}(\kappa, \kappa') = -P_{ij}(\kappa', \kappa). \quad (5\text{-}104)$$

Thus, P_{ij} is antisymmetric in κ and κ'. That is a condition that must be satisfied by any expression (assumed or calculated) for P_{ij}. Another condition, obtained by letting $\kappa = \kappa'$ in eq. (5-104), is

$$P_{ij}(\kappa, \kappa) = 0. \quad (5\text{-}105)$$

That is, there is no spectral transfer between Fourier components (eddies) with the same wavevector. This result shows that spectral transfer can take place only between wavevectors that, at least to some extent, are separated.

One can also get the spectral-transfer condition (5-98) from eq. (5-104). Thus, using eq. (5-104) in (5-99) and integrating from $\kappa = -\infty$ to $+\infty$, there results

$$\int_{-\infty}^{\infty} T_{ij}(\kappa) = \int_{-\infty}^{\infty}\int_{-\infty}^{\infty} P_{ij}(\kappa, \kappa')\, d\kappa'\, d\kappa = -\int_{-\infty}^{\infty}\int_{-\infty}^{\infty} P_{ij}(\kappa', \kappa)\, d\kappa\, d\kappa'. \quad (5\text{-}106)$$

Because κ and κ' are dummy variables in the integrals in eq. (5-106), they can be interchanged without changing the values of the integrals. Thus, eq. (5-106) can be true only if

$$\int_{-\infty}^{\infty} T_{ij}(\kappa)\, d\kappa = 0, \quad (5\text{-}107)$$

showing again that $T_{ij}(\kappa)$ gives zero contribution to the rate of change of turbulent-velocity correlation $\overline{u_i u_j}$, but it can transfer turbulent spectral components of $\overline{u_i u_j}$ from one part of wavenumber space to another.

Terms in eqs. (5-81) and (5-82) analogous to $P_{ij}(\kappa, \kappa')$ are, respectively,

$$S(\kappa, \kappa') = -i\left[\kappa_k \varphi_k(\kappa')\gamma(\kappa - \kappa')\gamma^*(\kappa) - \kappa_k \varphi_k^*(\kappa')\gamma^*(\kappa - \kappa')\gamma(\kappa)\right] \quad (5\text{-}108)$$

and

$$V_i(\kappa, \kappa') = -i\left[\kappa_k \varphi_k(\kappa')\gamma(\kappa - \kappa')\varphi_i^*(\kappa) - \kappa_k \varphi_k^*(\kappa')\varphi_i^*(\kappa - \kappa')\gamma(\kappa)\right]. \quad (5\text{-}109)$$

By using developments similar to those for $P_{ij}(\kappa, \kappa')$ and $T_{ij}(\kappa)$ in eqs. (5-99) through (5-107), one can show that

$$S(\kappa, \kappa') = -S(\kappa', \kappa). \quad (5\text{-}110)$$

and

$$V_i(\kappa, \kappa') = -V_i(\kappa', \kappa). \quad (5\text{-}111)$$

For $\kappa' = \kappa$, eqs. (5-110) and (5-111) show that, like eq. (5-105),

$$S(\kappa, \kappa) = V_i(\kappa, \kappa) = 0. \quad (5\text{-}105a)$$

Then, letting

$$W(\kappa) = \int_{-\infty}^{\infty} S(\kappa, \kappa')\, d\kappa' \quad (5\text{-}112)$$

and

$$X_i(\kappa) = \int_{-\infty}^{\infty} V_i(\kappa, \kappa')\, d\kappa', \quad (5\text{-}113)$$

one obtains

$$\int_{-\infty}^{\infty} W(\kappa)\, d\kappa = \int_{-\infty}^{\infty} X_i(\kappa)\, d\kappa = 0. \quad (5\text{-}114)$$

Thus, $W(\kappa)$ and $X_i(\kappa)$ can be given interpretations that are analogous to that given for $T_{ij}(\kappa)$. That is, $W(\kappa)$ gives zero contribution to the rate of change of turbulent temperature correlation $\overline{\tau\tau}$, but it can transfer spectral components $\gamma(\kappa)\gamma^*(\kappa)$ of $\overline{\tau\tau}$ from one part of wavenumber space to another, where τ is a temperature fluctuation. Similarly, $X_i(\kappa)$ gives zero contribution to the rate of change of temperature–velocity correlation $\overline{\tau u_i}$, but it can transfer spectral components $\gamma(\kappa)\varphi_i^*(\kappa)$ of $\overline{\tau u_i}$ from one part of wavenumber space to another.[2]

---

[2] Note, however, that although all of eqs. (5-104) through (5-114) are true, $\overline{\tau u_i}$ (proportional to the turbulent heat transfer) is zero in the present case in which mean gradients are absent. The same can be said about $\overline{u_i u_j}$ for $i \neq j$ (see section 4-3-1-1). (Note that the corresponding spectral quantities $\gamma\varphi_i^*$ and $\varphi_i^*\varphi_j$ for $i \neq j$ are not zero.) Spectral transfer related to $\overline{\tau u_i}$ and to $\overline{u_i u_j}$ for $i \neq j$ for cases in which mean gradients are not absent are considered in section 5-3.

Consider next the turbulent spectral-pressure terms in eqs. (5-80) and (5-82). Those are the terms with κ^2 in the denominator. Like the other inertia terms (the spectral-transfer terms), the spectral-pressure terms contain integrations over wavenumber space from $-\infty$ to $+\infty$. Thus, the spectral-pressure components at κ can be affected by turbulent activity at all parts of wavenumber space. We do not identify them as spectral-transfer terms because, unlike the latter, they can give contributions to the rate of change of $\overline{u_i u_j}$ or of $\overline{u_i \tau}$. The turbulent spectral-pressure terms drop out of the contracted eq. (5-80a) because of continuity, so that they make no contribution to the rate of change of $\varphi_i^*(\kappa)\varphi_i(\kappa)$. But at each κ they can transfer turbulent energy among the three directional components of $\varphi_i^*\varphi_i$. The spectral-pressure term in eq. (5-82), however, does not seem to have a clear physical significance, other than that it can contribute to the rate of change of $\overline{u_i \tau}$ or of $\gamma(\kappa)\varphi_i^*(\kappa)$.

Finally, consider the turbulent dissipation and turbulent thermal-smearing terms in eqs. (5-80) through (5-82). Those are the terms multiplied by κ^2, the turbulent dissipation and turbulent thermal-smearing terms being additionally multiplied respectively by ν and α. They always act in a direction such that the turbulent quantities that evolve according to eqs. (5-80) through (5-82) are brought closer to zero. The fact that the terms are multiplied by κ^2 means that Fourier components at larger wavenumbers (the smaller eddies) are dissipated viscously or thermally smeared more effectively than are those at smaller wavenumbers (the larger eddies). However, just how much more effectively depends on the shapes of the turbulent spectra, as is seen subsequently.[3] To illustrate the turbulence processes and turbulence quantities, we next obtain solutions of the continuum equations for several solvable cases.

5-3-2 Illustrative Solutions of the Basic Equations

Because we are considering a homogeneous turbulent field without mean gradients or external forces, the turbulence decays with time, no energy being added to the system. Thus we have an initial-value problem for which, in lieu of boundary conditions, we specify statistical uniformity.

5-3-2-1 Low turbulence Reynolds number.
The simplest analytical solution of the equations for homogeneous turbulence is that for low Reynolds numbers. Low Reynolds-number turbulence is identified as weak turbulence, the Reynolds number being proportional to a ratio of inertia to viscous forces. It should occur in the final period of decay. The solution might be used to study to some extent the viscous dissipation, that being the only turbulence process accounted for in the analysis.

For homogeneous turbulence without mean gradients the instantaneous equations (5-47) and (5-49) become, respectively,

$$\frac{\partial u_i}{\partial t} = -\frac{\partial (u_i u_k)}{\partial x_k} - \frac{1}{\rho}\frac{\partial \sigma}{\partial x_i} + \nu \frac{\partial^2 u_i}{\partial x_k \partial x_k} \tag{5-115}$$

---

[3] An analogous situation occurs for the transmission of sound in a gas. The shorter higher-frequency waves tend to be attenuated more rapidly by viscous dissipation (and by thermal smearing) than the longer lower-frequency ones.

132 TURBULENT FLUID MOTION

and
$$\frac{\partial^2 \sigma}{\partial x_l \partial x_l} = -\frac{\partial^2 (u_i u_k)}{\partial x_i \partial x_k}, \tag{5-116}$$

where the tildes have been omitted because mean quantities are zero or uniform. Examination of eqs. (5-115) and (5-116) shows that the pressure term in eq. (5-115) and the first term on the right side of that equation are both nonlinear and second-order in u. Thus, as u becomes small, the nonlinear terms in eq. (5-115) approach zero faster than the viscous dissipation term, which is first-order in u. Then, for weak turbulence, eqs. (5-115), (5-77), (5-76), and (5-80) become, respectively,

$$\frac{\partial u_i}{\partial t} = \nu \frac{\partial^2 u_i}{\partial x_k \partial x_k}, \tag{5-117}$$

$$\frac{\partial}{\partial t} \varphi_j(\kappa) = -\nu \kappa^2 \varphi_j(\kappa), \tag{5-118}$$

$$\frac{\partial}{\partial t} \varphi_{ij}(\kappa) = -2\nu \kappa^2 \varphi_{ij}(\kappa), \tag{5-119}$$

and

$$\frac{\partial}{\partial t} \left[\varphi_i^*(\kappa) \varphi_j(\kappa) \right] = -2\nu \kappa^2 \left[\varphi_i^*(\kappa) \varphi_j(\kappa) \right]. \tag{5-119a}$$

Neglecting nonlinear terms at low Reynolds number has resulted in the closed-averaged equations (5-119) and (5-119a). (The unaveraged equations from which (5-117) and (5-118) are obtained of course are already closed.) As shown in section 5-3-1-1, the equations from which (5-119) and (5-119a) are obtained are equivalent, and thus so are eqs. (5-119) and (5-119a).

Equations (5-118) through (5-119a) can be solved to give

$$\varphi_j(\kappa, t) = \varphi_j(\kappa, 0) e^{-\nu \kappa^2 t} \tag{5-120}$$

and

$$\varphi_{ij}(\kappa, t) = \left[\varphi_i^*(\kappa, t)\varphi_j(\kappa, t)\right] = \varphi_{ij}(\kappa, 0) e^{-2\nu \kappa^2 t} = \left[\varphi_i^*(\kappa, 0)\varphi_j(\kappa, 0)\right] e^{-2\nu \kappa^2 t}. \tag{5-121}$$

Equations (5-120) and (5-121) show that Fourier components of $u_j(x, t)$ and of $\overline{u_i u_j'}(r, t)$ are attenuated strongly as (κ) or t becomes large. That attenuation is produced by the viscous dissipation terms in eqs. (5-118) and (5-119), both of which are multiplied by κ^2. Thus, the terms $\partial \varphi_j/\partial t$ and $\partial \varphi_{ij}/\partial t$ in those equations tend to bring φ_j and φ_{ij} close to zero at large κ or t. As a result, with increasing time the turbulent activity (energy for $j = i$) shifts to lower wavenumbers, or to larger eddy sizes. The physical interpretation of this shift is that the smaller eddies die out faster than the larger ones because of the larger velocity gradients (larger shear stresses) between the smaller eddies. The essence of the turbulence dissipation is that it always tends to bring the turbulence activity closer to zero, and that it affects mainly the smaller eddies.

Consider next an illustrative solution for the two-point velocity correlation. To simplify the problem, we look only at the contracted quantity $\overline{u_i u_i'}$, a scalar.

Using eq. (5-121) in (5-11), we get

$$\overline{u_i u_i'}(r) = \int_{-\infty}^{\infty} A\kappa^2 \cos \boldsymbol{\kappa}\cdot\mathbf{r}\, e^{-2\nu\kappa^2 t}\, d\kappa, \qquad (5\text{-}122)$$

where we have replaced $e^{i\boldsymbol{\kappa}\cdot\mathbf{r}}$ by $\cos \boldsymbol{\kappa}\cdot\mathbf{r}$, because $\int_{-\infty}^{\infty} i \sin \boldsymbol{\kappa}\cdot\mathbf{r}\, e^{-2\nu\kappa^2 t}\, d\kappa = 0$. Also, continuing to work toward a one-dimensional scalar solution, we have let $\varphi_{ii}(\kappa, 0) = A\kappa^2$. Introducing spherical coordinates $\binom{r_1}{\kappa_1} = \binom{r}{\kappa}\cos\varphi\sin\theta$, $\binom{r_2}{\kappa_2} = \binom{r}{\kappa}\sin\varphi\sin\theta$, and $\binom{r_3}{\kappa_3} = \binom{r}{\kappa}\cos\theta$, one obtains $\boldsymbol{\kappa}\cdot\mathbf{r} = \kappa r$. Then, integrating over all directions in κ space, eq. (5-122) becomes

$$\overline{u_i u_i'}(r, t) = \int_0^{\infty} A\kappa^2 \cos(\kappa r) e^{-2\nu\kappa^2 t} \int_0^{\pi}\int_0^{2\pi} \kappa^2 \sin\theta\, d\varphi\, d\theta\, d\kappa, \qquad (5\text{-}123)$$

or

$$\overline{u_i u_i'}(r, t) = 4\pi A \int_0^{\infty} \kappa^4 \cos r\kappa\, e^{-2\nu t \kappa^2}\, d\kappa. \qquad (5\text{-}124)$$

Carrying out the integration in eq. (5-123) (see, e.g., ref. [14]) results in

$$\overline{u_i u_i'}(r, t) = \frac{3}{2}\pi^{3/2} A(2\nu t)^{-5/2} \left[1 - (2\nu t)^{-1} r^2 + \frac{1}{12}(2\nu t)^{-2} r^4\right] e^{-r^2/(8\nu t)} \qquad (5\text{-}125)$$

or, evaluating A in terms of $\overline{u_i u_i'}(0, t_0)$ and t_0, where t_0 is an initial time, we get[4]

$$\overline{u_i u_i'}(r, t) = \overline{u_i u_i'}(0, t_0) \left(\frac{t}{t_0}\right)^{-5/2} \left(1 - \frac{r^2}{2\nu t} + \frac{r^4}{48\nu^2 t^2}\right) e^{-r^2/(8\nu t)}. \qquad (5\text{-}126)$$

Equation (5-126) shows that just as φ_{ii} was strongly attenuated as κ increases, so $\overline{u_i u_i'}$ is strongly attenuated with increasing r, as indicated by the negative argument of the exponential. That may seem surprising because at large r, where length scales are large, the viscous effects tend to be small. However, the fact that t appears in the denominator of the argument of the exponential means that $\overline{u_i u_i'}$ decreases less rapidly with time at large r in the region of smaller viscous effects. That is, of course, as it should be. Perhaps the attenuation of $\overline{u_i u_i'}$ with increasing r is best explained by a physical argument. Thus, because a turbulent velocity tends to be a random function of position, the fluid at a point x loses touch, so to speak, with that at $x + r$ as r increases, so that the correlation between u_i and u_i' decreases with increasing r.

One can define characteristic length scales for $\overline{u_i u_i'}(r)$ as

$$\lambda^2 = -\frac{2\overline{u_i u_i'}(0, t)}{\left(\partial^2 \overline{u_i u_i'}(r, t)/\partial r^2\right)_{r=0}}, \qquad (5\text{-}127)$$

and

$$L(t) = \frac{\int_0^{\infty} \overline{u_i u_i'}(r, t)\, dr}{\overline{u_i u_i'}(0, t)}, \qquad (5\text{-}128)$$

---

[4] A more general expression for the two-point velocity correlation in the final period has been obtained by Batchelor and Proudman (ref. [15], eq. [7-7]).

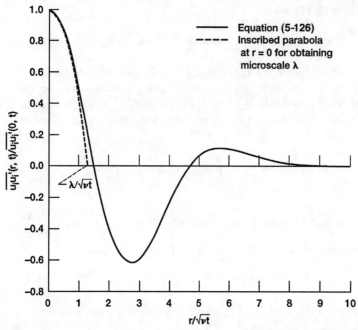

Figure 5-1 Dimensionless plot of a two-point turbulent velocity correlation in the final period of decay.

where λ is a microscale, a measure of the size of the small eddies of the turbulence, and L is a macroscale, a measure of the size of the energy-containing eddies. The former of these is obtained by inscribing a parabola in the curve for $\overline{u_i u_i'}(r)$ versus r at $r = 0$. The microscale λ is then the value of r where the parabola intersects the r axis. If $\overline{u_i u_i'}(r, t)$ is given by eq. (5-126), where $\overline{u_i u_i'}(0, t_0)(t/t_0)^{-5/2} = \overline{u_i u_i'}(0, t_0)$, we have $\lambda/\sqrt{\nu t} = 2\sqrt{2/5}$ and $L/\sqrt{\nu t} = 0$. The perhaps unexpected value for $L/\sqrt{\nu t}$ means that $\overline{u_i u_i'}$ must go negative for some values of r, as confirmed in Fig. 5-1. The value of zero calculated for L means, of course, that the definition of L given by eq. (5-128) is not realistic if $\overline{u_i u_i'}$ goes appreciably negative for some values of r. Possibly a better definition for the macroscale is

$$L' = \frac{\int_0^\infty \left|\overline{u_i u_i'}(r, t)\right| dr}{\overline{u_i u_i'}(0, t)}. \tag{5-129}$$

For $\overline{u_i u_i'}$ given by eq. (5-126), we find $L'/\lambda = 2.34$ for all times in the final period.

One might ask how there can be a negative correlation between u_i and u_i'. Evidently the negative and oscillating correlations shown in Fig. 5-1 might be obtained if there is some nonrandom (possibly periodic) structure in a flow superimposed on random turbulent flow that tends to make the correlation go to zero at large r. Incidentally, there is nothing unusual about correlations in turbulent flow that go negative or oscillate (see, e.g., Fig. 4-16).

5-3-2-2 Turbulence at various times before the final period of decay.

Unless the turbulence level is very low, as in the final period of decay, or the decay times are very short (see section 5-3-2-1), the inertial or spectral-transfer effects are not negligible ($T_{ij} \neq 0$), and so we would like to be able to take them into account in some way. A large number of proposals for calculating the spectral transfer have been given, including those of Heisenberg [16], Kovasznay [17], and Kraichnan [18], to name a few. The number of proposals for calculating the spectral-transfer effects appears, in fact, to approach the number of workers in the field. Reviews of the proposals are given, for instance, in refs. [4], [19], and [20]. Here we consider a simple deductive approach that is essentially a perturbation on the solution for the final period of decay considered in the previous section. That is by far the simplest deductive approach and is sufficient for our purpose, which is to illustrate spectral transfer by a comparatively simple solution.

We consider a correlation-term discard closure. In using that systematic procedure, the infinite set of multipoint correlation equations is made determinate by neglecting the highest-order terms in the highest-order equations considered [21, 22]. The procedure can be shown to be equivalent to a formally exact expansion in powers of Reynolds number (or of time) [23, 24]. It has been speculated that the scheme may be divergent and may give negative spectral energies [10, 25], although the results in refs. [21] and [22] show no such tendencies.[5] Even though the expansion may be divergent, the truncated series should give a reasonable approximation as an asymptotic expansion [24]. That may be the saving feature of the present scheme as well as of other approximations.[6] A more serious problem might be that the calculations are likely to become hopelessly complicated (or impossible) if higher–Reynolds-number turbulence is to be represented.

Several other schemes, such as the cumulant-discard closure and the direct-interaction approximation, which are essentially partial summations of the expansion in powers of Reynolds number [23], have been proposed. Both of those are much more complex than the method considered here. Although at least the latter of those schemes appears to give some realistic results, the problem in their justification is that there seems to be no way of knowing whether the additional terms retained (to all orders in Reynolds number) are more important than those still neglected. In this connection, note that although cumulant-discard approximations retain terms of all orders in Reynolds number or time, the domains of validity of the cumulant-discard approximations and the simpler power-series truncations (correlation-term discard schemes) are apparently the same (small Reynolds number or small time [23, 24]).

---

[5] Perhaps it is too much to expect an approximate (truncated) representation to be good (e.g., produce positive energy) at all Reynolds numbers, times, and wavenumbers. (Of course, the wider the range of applicability, the better.) A more realistic expectation might be that it be good for limited ranges of those parameters. For example, it seems unreasonable to discard a representation solely because the energy goes negative for a range of wavenumbers at large times. It may be perfectly satisfactory at earlier times. (Note that Newtonian mechanics has not been thrown out because it breaks down as the speed of light is approached.)

[6] Other, more sophisticated approximations also appear to work because they are kinds of asymptotic expansions. For example, higher-order expansions related to the direct-interaction approximation appear to give less reasonable results than those of lower-order ones [24, 26].

136 TURBULENT FLUID MOTION

Consider first the two-point correlation equations—equations that involve product mean values or correlations between velocities or between velocities and pressures at two points. Those equations can be obtained by simplification of eqs. (4-147), (4-149), and (4-150) in the previous chapter, but it may be instructive to obtain them directly from the Navier-Stokes equations.

The incompressible Navier–Stokes or momentum equations written for the points P and P' separated by the vector r are

$$\frac{\partial u_i}{\partial t} + \frac{\partial (u_i u_k)}{\partial x_k} = -\frac{1}{\rho}\frac{\partial \sigma}{\partial x_i} + \nu \frac{\partial^2 u_i}{\partial x_k \partial x_k} \tag{5-130}$$

and

$$\frac{\partial u'_j}{\partial t} + \frac{\partial (u'_j u'_k)}{\partial x'_k} = -\frac{1}{\rho}\frac{\partial \sigma'}{\partial x'_j} + \nu \frac{\partial^2 u'_j}{\partial x'_k \partial x'_k}, \tag{5-131}$$

where, as usual, the subscripts can take on the values 1, 2, or 3, and a repeated subscript in a term indicates a summation. The quantities u_i and u'_j are instantaneous velocity components, x_i is a space coordinate, t is the time, ρ is the density, ν is the kinematic viscosity, and σ is the instantaneous mechanical pressure (see eq. [3-14]). Multiplying the first equation by u'_j and the second by u_i, adding, and taking space averages results in

$$\frac{\partial \overline{u_i u'_j}}{\partial t} + \frac{\partial}{\partial x_k}\left(\overline{u_i u'_j u_k}\right) + \frac{\partial}{\partial x'_k}\left(\overline{u_i u'_j u'_k}\right) = -\frac{1}{\rho}\left(\frac{\partial \overline{\sigma u'_j}}{\partial x_i} + \frac{\partial \overline{\sigma' u_i}}{\partial x'_j}\right) + \nu\left(\frac{\partial^2 \overline{u_i u'_j}}{\partial x_k \partial x_k} + \frac{\partial^2 \overline{u_i u'_j}}{\partial x'_k \partial x'_k}\right), \tag{5-132}$$

where the fact that quantities at x_i are independent of x'_i and quantities at x'_i are independent of x_i is used. Equation (5-132) can be written as

$$\frac{\partial \overline{u_i u'_j}}{\partial t} + \frac{\partial}{\partial r_k}\left(\overline{u_i u'_j u'_k} - \overline{u_i u'_j u_k}\right) = -\frac{1}{\rho}\left(\frac{\partial \overline{\sigma' u_i}}{\partial r_j} - \frac{\partial \overline{\sigma u'_j}}{\partial r_i}\right) + 2\nu \frac{\partial^2 \overline{u_i u'_j}}{\partial r_k \partial r_k}, \tag{5-133}$$

where the relations $(\partial/\partial x_i) = -(\partial/\partial r_i)$ and $(\partial/\partial x'_i) = (\partial/\partial r_i)$ are used, and the correlations are functions only of r and t. Equation (5-133) was first obtained by von Kármán and Howarth [27]. It is desirable to write eq. (5-133) in spectral form in order to reduce it to an ordinary differential equation and because of the physical significance of spectral quantities. For this purpose we use the three-dimensional Fourier transforms (from r to κ space):

$$\overline{u_i u'_j}(r) = \int_{-\infty}^{\infty} \varphi_{ij}(\kappa) e^{i\kappa \cdot r} d\kappa, \tag{5-134}$$

$$\overline{u_i u_k u'_j}(r) = \int_{-\infty}^{\infty} \varphi_{ikj}(\kappa) e^{i\kappa \cdot r} d\kappa, \tag{5-135}$$

and

$$\overline{\sigma u'_j}(r) = \int_{-\infty}^{\infty} \lambda_j(\kappa) e^{i\kappa \cdot r} d\kappa, \tag{5-136}$$

where κ is a wavevector and $d\kappa = d\kappa_1 d\kappa_2 d\kappa_3$. The magnitude of κ has the dimension 1/length can be considered to be the reciprocal of an eddy size. From eq. (5-135)

$$\overline{u_i u_k u'_j(-r)} = \int_{-\infty}^{\infty} \varphi_{ikj}(\kappa) e^{-i\kappa \cdot r} d\kappa = \int_{-\infty}^{\infty} \varphi_{ikj}(-\kappa) e^{i\kappa \cdot r} d\kappa,$$

where the last step can be seen more clearly by writing the inverse transform. Interchanging the subscripts i and j and then interchaning the points P and P' gives,

$$\overline{u_i u'_j(r) u'_k(r)} = \overline{u_j u_k u'_i(-r)} = \int_{-\infty}^{\infty} \varphi_{jki}(-\kappa) e^{i\kappa \cdot r} d\kappa. \tag{5-135a}$$

Similarly

$$\overline{u_i \sigma'(r)} = \overline{\sigma u'_i(-r)} = \int_{-\infty}^{\infty} \lambda_i(-\kappa) e^{i\kappa \cdot r} d\kappa. \tag{5-136a}$$

Substituting eqs. (5-134) through (5-136a) into eq. (5-133) gives

$$\frac{d\varphi_{ij}}{dt} + i\kappa_k[\varphi_{jki}(-\kappa) - \varphi_{ikj}] = -\frac{1}{\rho}(i\kappa_j \lambda_i(-\kappa) - i\kappa_i \lambda_j) - 2\nu\kappa^2 \varphi_{ij}, \tag{5-137}$$

where the functional designations (κ) are omitted for brevity. The tensor eq. (5-137) becomes a scalar equation by contraction of the indices i and j. Thus,

$$\frac{d\varphi_{ii}}{dt} + 2\nu\kappa^2 \varphi_{ii} = i\kappa_k \varphi_{iki} - i\kappa_k \varphi_{iki}(-\kappa). \tag{5-138}$$

The pressure terms drop out of eq. (5-138) because of the continuity relation $\partial u_i/\partial x_i = \partial u'_i/\partial x'_i = 0$ (see eq. [5-133]).

The terms on the right side of eq. (5-138) are collectively proportional to what is known as the energy-transfer term. They account for the transfer of energy from one wavenumber to another or from one eddy size to another, but their total contribution to $d\overline{u_i u_i}/dt$ is zero (see section 5-3-1-2). Equation (5-138) applies to homogeneous as well as to isotropic turbulence. However, in order to obtain a solution an expression for the transfer or inertia term on the right side must be obtained. Von Kármán and Howarth [27] neglected that term and obtained a solution applicable in the final period of decay. In the present investigation it is proposed to obtain an expression for the transfer term applicable at times before the final period from the three-point correlation or spectral equations. To obtain the three-point equation, write the Navier-Stokes equation at the points P, P', and P'' separated by the vectors r and r'. The vector configuration is shown in Fig. 5-2. The first two equations are the same as eqs. (5-130) and (5-131), with the dummy subscripts k replaced by l. The third equation is

$$\frac{\partial u''_k}{\partial t} + \frac{\partial}{\partial x''_l}(u''_k u''_l) = -\frac{1}{\rho}\frac{\partial \sigma''}{\partial x''_k} + \nu \frac{\partial^2 u''_k}{\partial x''_l \partial x''_l}. \tag{5-139}$$

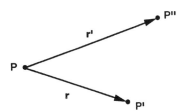

Figure 5-2 Vector configuration for three-point correlation equations.

138 TURBULENT FLUID MOTION

Multiplying the first equation by $u'_j u''_k$, the second by $u_i u''_k$, and the third by $u_i u'_j$, adding the three equations, and taking space averages results in

$$\frac{\partial}{\partial t}\overline{u_i u'_j u''_k} + \frac{\partial}{\partial x_l}\overline{u_i u'_j u''_k u_l} + \frac{\partial}{\partial x'_l}\overline{u_i u'_j u''_k u'_l} + \frac{\partial}{\partial x''_l}\overline{u_i u'_j u''_k u''_l}$$

$$= -\frac{1}{\rho}\left(\frac{\partial \overline{\sigma u'_j u''_k}}{\partial x_i} + \frac{\partial \overline{\sigma' u_i u''_k}}{\partial x'_j} + \frac{\partial \overline{\sigma'' u_i u'_j}}{\partial x''_k}\right) + \nu\left(\frac{\partial^2 \overline{u_i u'_j u''_k}}{\partial x_l \partial x_l} + \frac{\partial^2 \overline{u_i u'_j u''_k}}{\partial x'_l \partial x'_l} + \frac{\partial^2 \overline{u_i u'_j u''_k}}{\partial x''_l \partial x''_l}\right). \tag{5-140}$$

Equation (5-140) can be written in terms of the independent variables r and r' as

$$\frac{\partial}{\partial t}\overline{u_i u'_j u''_k} - \frac{\partial}{\partial r_l}\overline{u_i u'_j u''_k u_l} - \frac{\partial}{\partial r'_l}\overline{u_i u'_j u''_k u_l} + \frac{\partial}{\partial r_l}\overline{u_i u'_j u''_k u'_l} + \frac{\partial}{\partial r'_l}\overline{u_i u'_j u''_k u''_l}$$

$$= -\frac{1}{\rho}\left(-\frac{\partial}{\partial r_i}\overline{\sigma u'_j u''_k} - \frac{\partial}{\partial r'_i}\overline{\sigma u'_j u''_k} + \frac{\partial}{\partial r_j}\overline{\sigma' u_i u''_k} + \frac{\partial}{\partial r'_k}\overline{\sigma'' u_i u'_j}\right)$$

$$+ 2\nu\left(\frac{\partial^2 \overline{u_i u'_j u''_k}}{\partial r_l \partial r_l} + \frac{\partial^2 \overline{u_i u'_j u''_k}}{\partial r'_l \partial r'_l} + \frac{\partial^2 \overline{u_i u'_j u''_k}}{\partial r'_l \partial x'_l}\right), \tag{5-141}$$

where the following relations were used:

$$\frac{\partial}{\partial x'_l} = \frac{\partial}{\partial r_l}, \quad \frac{\partial}{\partial x''_l} = \frac{\partial}{\partial r'_l}, \quad \frac{\partial}{\partial x_l} = -\frac{\partial}{\partial r_l} - \frac{\partial}{\partial r'_l}.$$

In order to convert eq. (5-141) to spectral form, one can define the following six-dimensional Fourier transforms:

$$\overline{u_i u'_j(r) u''_k(r')} = \int_{-\infty}^{\infty}\int_{-\infty}^{\infty} \beta_{ijk}(\boldsymbol{\kappa}, \boldsymbol{\kappa}') e^{i(\boldsymbol{\kappa}\cdot\boldsymbol{r} + \boldsymbol{\kappa}'\cdot\boldsymbol{r}')} d\boldsymbol{\kappa}\, d\boldsymbol{\kappa}', \tag{5-142}$$

$$\overline{u_i u_l u'_j(r) u''_k(r')} = \int_{-\infty}^{\infty}\int_{-\infty}^{\infty} \beta_{iljk}(\boldsymbol{\kappa}, \boldsymbol{\kappa}') e^{i(\boldsymbol{\kappa}\cdot\boldsymbol{r} + \boldsymbol{\kappa}'\cdot\boldsymbol{r}')} d\boldsymbol{\kappa}\, d\boldsymbol{\kappa}', \tag{5-143}$$

$$\overline{\sigma u'_j(r) u''_k(r')} = \int_{-\infty}^{\infty}\int_{-\infty}^{\infty} \alpha_{jk}(\boldsymbol{\kappa}, \boldsymbol{\kappa}') e^{i(\boldsymbol{\kappa}\cdot\boldsymbol{r} + \boldsymbol{\kappa}'\cdot\boldsymbol{r}')} d\boldsymbol{\kappa}\, d\boldsymbol{\kappa}'. \tag{5-144}$$

By using the method for obtaining eq. (5-135a) the following relations result from eqs. (5-143) and (5-144):

$$\overline{u_i u'_l(r) u'_j(r) u''_k(r')} = \overline{u_j u_l u'_i(-r) u''_k(r' - r)}$$

$$= \int_{-\infty}^{\infty}\int_{-\infty}^{\infty} \beta_{ilik}(-\boldsymbol{\kappa} - \boldsymbol{\kappa}', \boldsymbol{\kappa}') e^{i(\boldsymbol{\kappa}\cdot\boldsymbol{r} + \boldsymbol{\kappa}'\cdot\boldsymbol{r}')} d\boldsymbol{\kappa}\, d\boldsymbol{\kappa}', \tag{5-143a}$$

$$\overline{u_i u'_j(r) u''_k(r') u''_l(r')} = \overline{u_k u_l u'_i(-r') u''_j(r - r')}$$

$$= \int_{-\infty}^{\infty}\int_{-\infty}^{\infty} \beta_{klij}(-\boldsymbol{\kappa} - \boldsymbol{\kappa}', \boldsymbol{\kappa}') e^{i(\boldsymbol{\kappa}\cdot\boldsymbol{r} + \boldsymbol{\kappa}'\cdot\boldsymbol{r}')} d\boldsymbol{\kappa}\, d\boldsymbol{\kappa}', \tag{5-143b}$$

where the points P and P' are interchanged to obtain eq. (5-143a). For obtaining (5-143b), P is replaced by P', P' is replaced by P'', and P'' is replaced by P. Similarly,

$$\overline{u_i \sigma'(r) u''_k(r')} = \overline{\sigma u'_i(-r) u''_k(r' - r)} = \int_{-\infty}^{\infty}\int_{-\infty}^{\infty} \alpha_{ik}(-\boldsymbol{\kappa} - \boldsymbol{\kappa}', \boldsymbol{\kappa}') e^{i(\boldsymbol{\kappa}\cdot\boldsymbol{r} + \boldsymbol{\kappa}'\cdot\boldsymbol{r}')} d\boldsymbol{\kappa}\, d\boldsymbol{\kappa}', \tag{5-144a}$$

and

$$\overline{u_i u'_j(r)\sigma''(r')} = \overline{\sigma u'_i(-r)u''_j(r-r')} = \int_{-\infty}^{\infty}\int_{-\infty}^{\infty} \alpha_{ij}(-\kappa-\kappa',\kappa)e^{i(\kappa\cdot r+\kappa'\cdot r')}\,d\kappa\,d\kappa'. \tag{5-145}$$

Substituting the preceeding relations into eq. (5-141) gives

$$\frac{d}{dt}\beta_{ijk} + 2\nu\left(\kappa^2 + \kappa_l\kappa'_l + \kappa'^2\right)\beta_{ijk}$$
$$= \left[i\left(\kappa_l + \kappa'_l\right)\beta_{iljk} - i\kappa_l\beta_{jlik}(-\kappa-\kappa',\kappa') - i\kappa'_l\beta_{klij}(-\kappa-\kappa',\kappa)\right]$$
$$-\frac{1}{\rho}\left[-i\left(\kappa_i + \kappa'_i\right)\alpha_{jk} + i\kappa_j\alpha_{ik}(-\kappa-\kappa',\kappa') + i\kappa'_k\alpha_{ij}(-\kappa-\kappa',\kappa)\right]. \tag{5-146}$$

Equation (5-146) agrees with eq. (2-11) in ref. [28].

The expression in the first bracket on the right side of eq. (5-146) can be interpreted as a transfer term similar to that on the right side of eq. (5-138). Using an argument similar to that given for eq. (5-138) (see section 5-3-1-2), one can obtain an expression for the quadruple correlation terms in eq. (5-141) for r and $r' = 0$ by allowing the points x_i, x'_i, and x''_i in eqs. (5-130), (5-131), and (5-139) to coincide. Then the terms involving quadruple correlations become

$$\overline{u_j u_k \frac{\partial(u_i u_l)}{\partial x_l}} + \overline{u_i u_k \frac{\partial(u_l u_j)}{\partial x_l}} + \overline{u_i u_j \frac{\partial(u_k u_l)}{\partial x_l}} = \overline{u_l u_j u_k \frac{\partial u_i}{\partial x_l}} + \overline{u_l u_i u_k \frac{\partial u_j}{\partial x_l}} + \overline{u_l u_i u_j \frac{\partial u_k}{\partial x_l}}$$
$$= \frac{\partial}{\partial x_l}(\overline{u_i u_j u_k u_l}) = 0, \tag{5-147}$$

where the conditions of homogeneity and continuity are used. From eq. (5-143) for $r = r' = 0$, and eq. (5-147),

$$\int_{-\infty}^{\infty}\int_{-\infty}^{\infty} [\beta_{ijk}]\,d\kappa\,d\kappa' = 0 \tag{5-148}$$

where the expression in the first bracket in eq. (5-146) was set equal to $[\beta_{ijk}]$. Thus $[\beta_{ijk}]$ gives zero total contribution to $d(\overline{u_i u_j u_k})/dt$. However, it can alter the distribution in wavenumber space of contributions to $\overline{u_i u_j u_k}$. It appears that similar transfer terms occur in all higher-order equations.

The tensor eq. (5-146) can be converted to a scalar equation by contraction of the indexes i and j and inner multiplication by κ_k:

$$\frac{d}{dt}(\kappa_k\beta_{iik}) + 2\nu\left(\kappa^2 + \kappa_l\kappa'_l + \kappa'^2\right)\kappa_k\beta_{iik}$$
$$= i\kappa_k\left(\kappa_l + \kappa'_l\right)\beta_{ilik} - i\kappa_k\kappa_l\beta_{ilik}(-\kappa-\kappa',\kappa') - i\kappa_k\kappa'_l\beta_{klii}(-\kappa-\kappa',\kappa)$$
$$-\frac{1}{\rho}\left[-i\kappa_k\left(\kappa_i + \kappa'_i\right)\alpha_{ik} + i\kappa_k\kappa_i\alpha_{ik}(-\kappa-\kappa',\kappa') + i\kappa_k\kappa'_k\alpha_{ii}(-\kappa-\kappa',\kappa)\right]. \tag{5-149}$$

To obtain a relation between the terms on the right side of eq. (5-149), which are derived from the quadruple correlation terms, and from the pressure terms in eq. (5-141), take the divergence of the Navier-Stokes equation and combine the result with the continuity equation to give

$$\frac{1}{\rho}\frac{\partial^2 \sigma}{\partial x_l \partial x_l} = -\frac{\partial^2 (u_l u_m)}{\partial x_l \partial x_m}. \tag{5-150}$$

Multiplying eq. (5-150) by $u'_i u''_k$, taking space averages, and writing the resulting equation in terms of the independent variables r and r', give

$$\frac{1}{\rho}\left(\frac{\partial^2 \overline{\sigma u'_i u''_k}}{\partial r_l \partial r_l} + 2\frac{\partial^2 \overline{\sigma u'_i u''_k}}{\partial r_l \partial r'_l} + \frac{\partial^2 \overline{\sigma u'_i u''_k}}{\partial r'_l \partial r'_l}\right)$$

$$= -\frac{\partial^2 \overline{u_l u_m u'_i u''_k}}{\partial r_m \partial r_l} - \frac{\partial^2 \overline{u_l u_m u'_i u''_k}}{\partial r_m \partial r'_l} - \frac{\partial^2 \overline{u_l u_m u'_i u''_k}}{\partial r'_m \partial r_l} - \frac{\partial^2 \overline{u_l u_m u'_i u''_k}}{\partial r'_m \partial r'_l}. \tag{5-151}$$

The Fourier transform of eq. (5-151) is

$$-\frac{1}{\rho}(\kappa^2 + 2\kappa_l \kappa'_l + \kappa'^2)\alpha_{ik} = (\kappa_l \kappa_m + \kappa'_l \kappa_m + \kappa_l \kappa'_m + \kappa'_l \kappa'_m)\beta_{lmik}$$

or

$$-\frac{1}{\rho}\alpha_{ik} = \frac{\kappa_l \kappa_m + \kappa'_l \kappa_m + \kappa_l \kappa'_m + \kappa'_l \kappa'_m}{\kappa^2 + 2\kappa_l \kappa'_l + \kappa'^2}\beta_{lmik}. \tag{5-152}$$

Equation (5-152) can be used to eliminate the quantities $\alpha_{ik}, \alpha_{ik}(-\kappa-\kappa', \kappa')$, and so forth from eq. (5-149). In order to solve eqs. (5-138), (5-149), and (5-152) simultaneously a relation between β_{iik} and φ_{iki} is required. Letting $r' = 0$ in eq. (5-142) and comparing the result with eq. (5-135) shows that

$$\varphi_{iki}(\boldsymbol{\kappa}) = \int_{-\infty}^{\infty} \beta_{iik}(\boldsymbol{\kappa}, \boldsymbol{\kappa}') d\boldsymbol{\kappa}'. \tag{5-153}$$

The set of eqs. (5-138), (5-149), (5-152), and (5-153) is still not determinate, because there are more unknowns than equations. It is proposed to obtain a solution applicable at times before the final period, as well as during the final period of decay, by neglecting the terms in eqs. (5-149) and (5-152) corresponding to quadruple correlation terms. Corresponding to the case of the final period, in which a solution was obtained by neglecting the triple correlation terms in the two-point equation, it should be possible to obtain a solution for times before the final period by neglecting the quadruple correlation terms in the three-point equations. If a solution applicable at still earlier times is desired it is necessary to consider four- or five-point equations. In each case the set of equations is made determinate by neglecting terms containing the highest-order correlations.

Equation (5-152) shows that if terms containing quadruple correlations are neglected, then the terms containing pressure correlations also must be neglected. Thus, neglecting all the terms on the right side of eq. (5-149), the equation can be integrated between t_0 and t to give

$$\kappa_k \beta_{iik} = \kappa_k (\beta_{iik})_0 \exp[-2\nu(t-t_0)(\kappa^2 + \kappa\kappa' \cos\theta + \kappa'^2)], \tag{5-154}$$

where $(\beta_{iik})_0$ is the value of β_{iik} at $t = t_0$ (an initial time or virtual origin), and θ is the angle between κ and κ'. Note that $(\beta_{ijk})_0$ also equals β_{iik} for small values of κ and κ' at various times, so that β_{ijk} has a stationary form at small values of κ and κ', at least for times at which the quadruple correlations are negligible. Substitution of eqs. (5-154) and (5-153) in eq. (5-138) results in

$$\frac{d\varphi_{ii}}{dt} + 2\nu\kappa^2\varphi_{ii} = \int_0^\infty 2\pi i\kappa_k[\beta_{iik} - \beta_{iik}(-\kappa, -\kappa')]_0 \kappa'^2$$
$$\times \left[\int_{-1}^1 \exp[-2\nu(t - t_0)(\kappa^2 + \kappa\kappa'\cos\theta + \kappa'^2)]\, d(\cos\theta)\right] d\kappa', \tag{5-155}$$

where $d\kappa' = d\kappa_{1'}d\kappa_{2'}d\kappa_{3'}$ is written in terms of κ' and θ as $-2\pi\kappa'^2 d(\cos\theta) d\kappa'$.

In order to make further calculations it is necessary to assume a relation that gives $[\beta_{iik} - \beta_{iik}(-\kappa, -\kappa')]_0$ as a function of κ and κ'. The theory itself, of course, does not supply this relation; it gives only the state of the turbulence at various times if the initial state is known. The relation assumed here is

$$i\kappa_k[\beta_{iik} - \beta_{iik}(-\kappa, -\kappa')]_0 = -\beta_0(\kappa^m \kappa'^n - \kappa^n \kappa'^m), \tag{5-156}$$

where β_0 is a constant determined by the initial conditions, so that its value in general depends on m and n. As is seen later, this expression gives a transfer term that satisfies eq. (5-98). Here we choose $m = 4$ and $n = 6$. Later we investigate the effect of the choices for m and n on some of the results. The negative sign is placed in front of β_0 in order to make the transfer of energy from small to large wavenumbers for positive values of β_0. By virtue of eqs. (5-155) and (5-156) one can write $\varphi_{ii} = \varphi_{ii}(\kappa)$.

Energy spectrum and spectral transfer. Substituting eq. (5-156) in (5-155), writing φ_{ii} in terms of the energy spectrum function [4],

$$E(\kappa) = 2\pi\kappa^2\varphi_{ii}(\kappa), \tag{5-157}$$

and carrying out the integration with respect to θ in eq. (5-155) result in

$$dE(\kappa)/dt + 2\nu\kappa^2 E(\kappa) = T(\kappa), \tag{5-158}$$

where $T(\kappa)$ is the energy transfer term and is given by

$$T(\kappa) = -\frac{\beta_0}{2\nu(t - t_0)} \int_0^\infty (\kappa^5\kappa'^7 - \kappa^7\kappa'^5)\{\exp[-2\nu(t - t_0)(\kappa^2 - \kappa\kappa' + \kappa'^2)]$$
$$- \exp[-2\nu(t - t_0)(\kappa^2 + \kappa\kappa' + \kappa'^2)]\}\, d\kappa'. \tag{5-159}$$

On multiplying each term in eq. (5-158) by $d\kappa$ it is seen that the first term represents the rate of change of energy in the wavenumber band $d\kappa$, the second term is energy dissipated within the band, and $T(\kappa)$ is the net energy transferred into the band.

Equation (5-159) also can be given an interesting physical interpretation. Letting the integrand in the equation be $P(\kappa, \kappa')$, we have

$$T(\kappa) = \int_0^\infty P(\kappa, \kappa')\, d\kappa'. \tag{5-160}$$

142 TURBULENT FLUID MOTION

Multiplying both sides of eq. (5-160) by $d\kappa$, we note that, as in eq. (5-158), $T(\kappa)d\kappa$ is the net energy flowing into the wavenumber band $d\kappa$ from all other wavenumbers. The quantity $P(\kappa, \kappa')d\kappa'd\kappa$ is the energy flowing from the wavenumber band $d\kappa'$ into the band $d\kappa$. It might be called the distribution function for contributions to $T(\kappa)$ from various wavenumbers or eddy sizes.

Carrying out the integration indicated in eq. (5-159), that is, summing up the contributions to $T(\kappa)$ from all wavenumber bands, gives

$$T(\kappa) = -\frac{(\pi/2)^{1/2}}{256}\beta_0 \exp[-(3/2)\kappa^2 \nu(t-t_0)]$$

$$\times \left[105\frac{\kappa^6}{[\nu(t-t_0)]^{9/2}} + 45\frac{\kappa^8}{[\nu(t-t_0)]^{7/2}} - 19\frac{\kappa^{10}}{[\nu(t-t_0)]^{5/2}} - 3\frac{\kappa^{12}}{[\nu(t-t_0)]^{3/2}}\right].$$

(5-161)

If we integrate eq. (5-161) over all wavenumbers we find that

$$\int_0^\infty T(\kappa)\, d\kappa = 0,$$

(5-162)

indicating that the expression for $T(\kappa)$ satisfies eq. (5-98). This is a consequence of the antisymmetry of eq. (5-156).

For obtaining the energy spectrum function E, eq. (5-158) can be written in integral form as

$$E = \exp[-2\nu\kappa^2(t-t_0)]\int \exp[2\kappa^2\nu(t-t_0)]T\, dt + C(\kappa)\exp[-2\nu\kappa^2(t-t_0)],$$

(5-163)

where T is given by eq. (5-161). We let the constant of integration be given by

$$C(\kappa) = J_0\kappa^4/3\pi.$$

(5-164)

There is some theoretical basis for eq. (5-164) [15], but $C(\kappa)$ for a particular flow may depend on how the turbulence is generated [29].

Carrying out the integration indicated in eq. (5-163) and substituting eq. (5-164) result in the following expression for the energy spectrum function:

$$E = \frac{J_0\kappa^4}{3\pi}\exp[-2\nu\kappa^2(t-t_0)] - \frac{\pi^{1/2}\beta_0}{256\nu}\exp[-(3/2)\kappa^2\nu(t-t_0)]$$

$$\times \left[-\frac{15\sqrt{2}}{\nu^{7/2}}\frac{\kappa^6}{(t-t_0)^{7/2}} - \frac{12\sqrt{2}}{\nu^{5/2}}\frac{\kappa^8}{(t-t_0)^{5/2}} + \frac{7\sqrt{2}}{3\nu^{3/2}}\frac{\kappa^{10}}{(t-t_0)^{3/2}} \right.$$

$$\left. + \frac{16\sqrt{2}}{3\nu^{1/2}}\frac{\kappa^{12}}{(t-t_0)^{1/2}} - \frac{32}{3}\kappa^{13}F\left(\kappa\left[\frac{\nu(t-t_0)}{2}\right]^{1/2}\right)\right]$$

(5-165)

where

$$F(\omega) = e^{-\omega^2}\int_0^\omega e^{x^2}dx,$$

$$\omega = \kappa\left[\frac{\nu(t-t_0)}{2}\right]^{1/2}.$$

Values of $F(\omega)$ are tabulated in ref. [30]. The first term on the right side of eq. (5-165) is the usual expression for E in the final period of decay. The last term is the contribution to E due to energy transfer.

The expression for the energy decay is obtained from eq. (5-134) by setting $r = 0$, $j = i$, $d\kappa = -2\pi\kappa^2 d(\cos\theta)d\kappa$, and $E = 2\pi\kappa^2\varphi_{ii}$. Thus

$$\frac{\overline{u_i u_i}}{2} = \int_0^\infty E\, d\kappa. \tag{5-166}$$

Substituting eq. (5-165) in (5-166) and performing the integration, that is, summing up the contributions to the energy from all wavenumbers results in the following expression for the energy decay law:

$$\frac{\overline{u_i u_i}}{2} = \frac{J_0}{32(2\pi)^{1/2}} \nu^{-5/2}(t-t_0)^{-5/2} + 0.2296\beta_0 \nu^{-8}(t-t_0)^{-7}. \tag{5-167}$$

The last term in eq. (5-165) was integrated by expanding it in a series. For large times the last term in eq. (5-167) becomes negligible, and the equation reduces to the well known $-5/2$-power decay law for the final period. If higher-order correlation equations are considered in the analysis—that is, if the quadruple correlations are not neglected—it appears that more terms in higher powers of $(t - t_0)$ would be added to eq. (5-167).[7]

Figure 5-3 shows our theoretical dimensionless energy-spectrum function $E^* = J_0^{1/3} E/\nu^{8/3}$ plotted against dimensionless wavenumber $\kappa^* = J_0^{1/3}\kappa/\nu^{2/3}$ for various dimensionless times $t^* = \nu^{7/3} t/J_0^{2/3}$. The curves are calculated from eq. (5-165) after converting it to dimensionless form.

The values of the parameters $\beta_0^* = \nu^5 \beta_0 / J_0$ and $t_0^* = \nu^{7/3} t_0 / J_0^{2/3}$, which depend on initial conditions, are determined by decay data for grid-generated turbulence reported in ref. [32] (see Fig. 5-4). The Reynolds number MU/ν is 950, where M is the mesh size and U is the mean velocity of the fluid flowing through the grid. The corresponding microscale Reynolds number $R_\lambda = \overline{u_i u_i}^{1/2}\lambda/(3\nu)$ is between 5 and 8, where λ is the microscale (dissipation length) [4]. The time t in the theory is taken to be $t = x/U$, where x is the distance downstream from the grid and t_0 is the virtual origin of the turbulence (the time at which the turbulence intensity $\overline{u_i u_i}/3$ would become infinite if eq. (5-167) applied for all times $\geq t_0$). Note that t_0 corresponds to x_0 (see Fig. 5-4).

The unusual shape of the curve for $t^* = 1$ may be due to the fact that the theory is not accurate for a time that early (see Fig. 5-4). The wavenumber κ has the dimension $1/\text{length}$ and, as mentioned earlier, can be considered as the reciprocal of an eddy size. Large wavenumbers correspond to small eddies, and small wavenumbers to large eddies. Equation (5-166) shows that E represents the distribution of contributions to the total energy from various wavenumbers or eddy sizes. As time increases, the bulk of the energy moves to smaller wavenumbers or to larger eddies. The high velocity gradients and, consequently, high shear stresses occurring in the smaller eddies cause them to dissipate more rapidly than the large ones. The viscous dissipation thus produces a sink for the energy at the higher wavenumbers.

---

[7] We have also carried out an analogous analysis of decaying turbulent temperature fluctuations. The analysis and results are reported in ref. [31].

144 TURBULENT FLUID MOTION

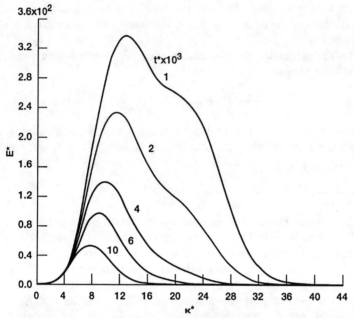

Figure 5-3 Variation of theoretical energy spectrum with time for approach to final period. Reynolds number based on mesh size and mean velocity, 950. $\beta_0^* = 1.55 \times 10^{-11}$, $t_0^* = -6.33 \times 10^{-3}$.

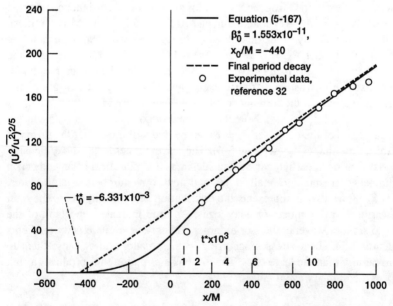

Figure 5-4 Turbulence-decay plot for determining parameters in Fig. 5-3. $J_0^* = J_0 U^{1/2}/(\nu M)^{5/2} = 1.929 \times 10^4$, $\beta_0^* = \nu^5 \beta_0 / J_0^4 = 1.553 \times 10^{-11}$, $x_0/M = -440$, $M = 0.159\,cm$, $R = UM/\nu = 950$, $t^* = (x/M)/(J_0^* R)^{2/3}$

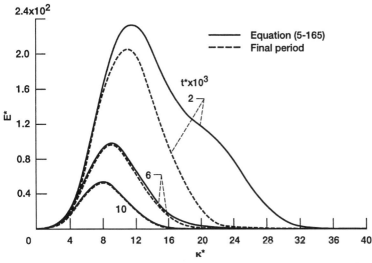

Figure 5-5 Comparison of theoretical spectra with those for final period. $R = 950$, $\beta_0^* = 1.55 \times 10^{-11}$, $t_0^* = -6.33 \times 10^{-3}$, $5 < R_\lambda < 8$.

Interaction between spectral energy transfer and dissipation. Figure 5-5 gives a comparison between the spectra obtained from eq. (5-165) and those for the final period obtained by retaining only the first term in eq. (5-165). The difference between the curves, of course, is caused by the transfer of energy from low wavenumbers to higher ones. The energy transfer tends to fill the sink produced by viscous dissipation at the higher wave numbers. As a result the slopes on the high-wavenumber sides of the spectra are more gradual than they would be in the absence of energy transfer. At later times, however, when inertia effects become small, the spectra assume a more or less symmetrical shape.

Thus the function of the inertia terms in the equations is to excite the higher-wavenumber or small-eddy regions of the spectrum by transferring energy into those regions. The high-wavenumber portion of the spectrum thus is determined primarily by the inertia effects, whereas the low-wavenumber portion is determined by the viscous terms in the equations. This may seem to be opposite of what one would expect, inasmuch as we usually consider the high-wavenumber or small-eddy region to be dominated by viscous effects. It is true that viscous dissipation is highest in the high-wavenumber region because of the high shear stresses between the small eddies. However, the small eddies owe their existence in the first place to the transfer of energy into that region, that is, to inertia effects.

It is this interaction between the energy transfer and the dissipation (see eq. [5-158]) that is responsible for the high rate of dissipation in turbulence; the dissipation is high because the high wavenumber region of the spectrum is excited by the transfer of energy into that region. The shear stresses are higher there than in a region that could be excited in the absence of energy transfer. Another way of saying all this is that the energy transfer generally has a stabilizing effect on the flow.

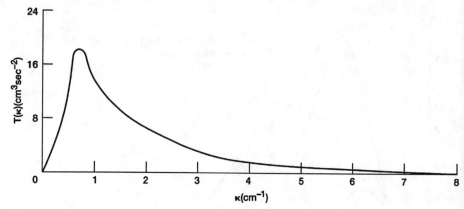

Figure 5-6 Energy spectrum for homogeneous turbulence obtained from a numerical solution of the instantaneous Navier-Stokes equations by Clark, Ferziger, and Reynolds [33]. $R_\lambda = 36.6$.

Inertia and dissipation tend to shift the energy in opposite directions on the wavenumber scale. Moreover, the mechanisms for the two effects appear to be different. Whereas inertia tends to transfer the energy to higher wavenumbers by a breakup of large eddies into smaller ones (or by a stretching of vortex filaments), dissipation tends to shift the energy to smaller wavenumbers by selective annihilation of eddies, the small eddies being the first to go. As the turbulence decays, the dissipation effects, of course, eventually must win out, because the inertia effects become negligible at the low Reynolds numbers occurring at large times.

A solution of the Navier-Stokes equations for decaying homogeneous turbulence has been obtained numerically by Clark, Ferziger, and Reynolds [33]. The resulting energy spectrum for a particular time is shown in Fig. 5-6. As in the case of our calculated spectra (Fig. 5-5) the energy transfer to higher wavenumbers causes the slopes on the high-wavenumber side of the spectrum to be more gradual than those on the low-wavenumber side. The larger effect in Fig. 5-6 apparently is caused by the higher Reynolds number there. That Reynolds number ($R_\lambda = 36.6$) evidently is too high for the theory used in obtaining Fig. 5-5 to be applicable.

Experimental energy spectra reported in refs. [34] through [38] are qualitatively similar to and show the same trends as the calculated spectra in Figs. 5-5 and 5-6. In particular, the effect of Reynolds number is shown clearly in Fig. 5-7, in which normalized experimental results from refs. [37] and [38] are compared. The shapes of the spectra again result from the energy transfer to higher wavenumbers. That energy transfer is of course greater for the higher-Reynolds-number curve, where inertial effects in comparison with viscous effects are greater, and the slopes on the high-wavenumber side of the spectrum are more gradual than they are in the curve for lower Reynolds numbers.

Spectrum for infinite Reynolds number. Also shown in Fig. 5-7 is the spectrum for infinite Reynolds number that results from the hypotheses of Kolmogorov and

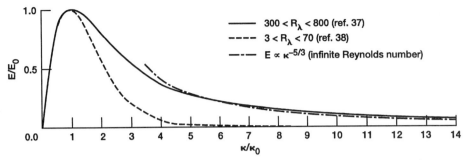

Figure 5-7 Experimental and $-5/3$-power energy spectra. Subscript zeroes refer to values at peak.

dimensional considerations [39–42]. Kolmogorov hypothesized that the physical quantity representative of the dynamics of high–Reynolds-number turbulence is ε, the average rate of transfer of kinetic energy (per unit mass) between large and small scales of motion. The dimensions of ε are (length/time)^2/time = length^2/time^3. Another quantity that is often a determining parameter for turbulence is the kinematic viscosity ν, with dimensions length^2/time. However, the region of wavenumber space that is affected by the action of viscous forces moves out from the origin toward a wavenumber of infinity as the Reynolds number of the turbulence increases. In the limit of infinite Reynolds number the energy sink produced by viscous dissipation is displaced to infinity, and the influence of viscous forces is negligible for wavenumbers κ of finite magnitude [4].

Thus, one can write for the energy-spectrum function E (with dimensions length^3/time^2),

$$E = E(\varepsilon, \kappa), \tag{5-168}$$

where the geometric quantity κ (with dimension length^{-1}) is included in the functional relationship because we want to end up with a relation between E and κ. The only dimensionally correct expression for E that can be formed from ε and κ is

$$E = \text{const.}\, \varepsilon^{2/3} \kappa^{-5/3}. \tag{5-169}$$

This is the celebrated $-5/3$-power Kolmogorov-Obukhov spectrum for infinite Reynolds number. It is plotted in Fig. 5-7, in which agreement with the experimental data for high Reynolds numbers is indicated for a range of wavenumbers.

Note that the Reynolds numbers we have considered include both very low and very high ones. However, the Navier-Stokes equations were not used for the latter, as they were for low Reynolds numbers where a solution of those equations was obtained. But the spectra for low and high Reynolds numbers are qualitatively similar; the energy is just spread out over a wider range of wavenumbers as the Reynolds number increases. There are no bifurcations in going from low to high Reynolds numbers (except possibly in the transition region).

For further consideration of the energy transfer we return to the analysis for moderately low Reynolds numbers.

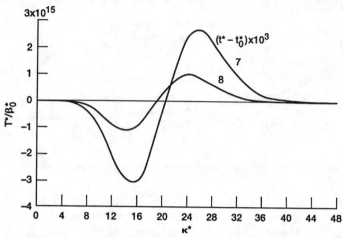

Figure 5-8 Variation of dimensionless energy transfer term with κ^* and $t^* - t_0^*$.

Further analysis of the spectral-energy transfer. Although we have considered in some detail the effect of spectral transfer on energy spectra, we have not looked at the transfer spectra themselves. Those have been calculated for the approach to the final period from eq. (5-161) and are plotted in dimensionless form in Fig. 5-8. The quantity T^* equals $J_0 T(\kappa)/\nu^5$, where $T(\kappa)$ is the energy-transfer term in eq. (5-158). The transfer term gives the net energy transfer into a wavenumber band from all other wavenumbers (eq. [5-159]). The curves indicate net energy loss from energy bands at low wavenumbers and an energy gain to those at higher wavenumbers. The total area under each curve is zero, in agreement with eq. (5-162), thus indicating that the total contribution of $T(\kappa)$ to $d(\overline{u_i u_i})/dt$ is zero (see eqs. [5-158], [5-162], and [5-166]). It should be emphasized that $T(\kappa)$ represents a difference between the energy flowing into and out of a wavenumber band. The actual energy transfer at a point where $T(\kappa)$ is low or zero may be quite high, as is shown subsequently.

As has been mentioned, the integrand in eq. (5-159) can be interpreted as giving the distribution of contributions to $T(\kappa)$ from various wavenumbers or eddy sizes. The integrand $P(\kappa, \kappa')$ (see eqs. [5-159] and [5-160]) can be written in dimensionless form as

$$\frac{P\nu^7(t-t_0)^7}{\beta_0} = -\frac{1}{2}\left[\left(\frac{\kappa'}{\kappa}\right)^7 - \left(\frac{\kappa'}{\kappa}\right)^5\right]\{\kappa[\nu(t-t_0)]^{1/2}\}^{12}$$

$$\times \left(\exp\left\{-2\{\kappa[\nu(t-t_0)]^{1/2}\}^2\left[1-\frac{\kappa'}{\kappa}+\left(\frac{\kappa'}{\kappa}\right)^2\right]\right\}\right.$$

$$\left. - \exp\left\{-2[\kappa[\nu(t-t_0)]^{1/2}]^2\left[1+\frac{\kappa'}{\kappa}+\left(\frac{\kappa'}{\kappa}\right)^2\right]\right\}\right). \quad (5\text{-}170)$$

Figures 5-9a, b, and c show $P\nu^7(t-t_0)^7/\beta_0$ plotted against κ'/κ for several values of $\kappa[\nu(t-t_0)]^{1/2}$. In agreement with eq. (5-105) (integrated over all directions in wavevector space), $P(\kappa/\kappa) = 0$. That is, energy transfer takes place only between wavenumber bands that, at least to some extent, are separated. The curves indicate that the energy entering a wavenumber band at κ comes from a range of wavenumbers κ' or eddy sizes rather than exclusively from neighboring wavenumbers. Similarly, the energy passes on to a range of wavenumbers. Thus, the energy in general is transported between wavenumber bands that are separated. This transport might occur by a breaking up of large eddies into small ones. The positive area under each curve corresponds to the total energy entering a wavenumber band at κ, the negative area to the total energy leaving. The curve corresponding to $T = 0$ indicates a considerable amount of energy entering and leaving at κ, although the net energy gain is of course zero. The asymmetrical curves indicate that if a small amount of energy is entering at κ and a large amount is leaving, the energy comes from wavenumbers close to κ and goes to more distant wavenumbers. The opposite is true if the energy entering at κ is comparatively large.

In order to obtain an idea of the average energy transfer for all values of κ, we can integrate eq. (5-170) over κ for constant κ'/κ and obtain a quantity that we call Q:

$$\frac{Q[(t-t_0)\nu]^{15/2}}{\beta_0} = -\frac{10\,395(\pi/2)^{1/2}}{16\,384}\left\{\left[\left(\frac{\kappa'}{\kappa}\right)^7 - \left(\frac{\kappa'}{\kappa}\right)^5\right]\right.$$
$$\left.\times\left[\frac{1}{[1-(\kappa'/\kappa)+(\kappa'/\kappa)^2]^{13/2}} - \frac{1}{[1+(\kappa'/\kappa)+(\kappa'/\kappa)^2]^{13/2}}\right]\right\}.$$
(5-171)

Equation (5-171) is plotted in Fig. 5-10. The curve indicates that on the average, energy enters and leaves wavenumber bands about as shown in Fig. 5-9b, in which the net energy transfer is zero.

All of the results for $T(\kappa)$, $P(\kappa, \kappa')$, and $Q(\kappa, \kappa')$ given so far are obtained by letting $m = 4$ and $n = 6$ in eq. (5-156) for the initial conditions. There was early theoretical support for those values, but later work indicates that in the real world they may be somewhat lower. Empirically, the transfer term $T(\kappa)$ seems, in fact, to start out at the origin with a power of κ close to one [38, 43]; that corresponds to $m = 0$ and $n = -1$. Then eq. (5-170) is replaced by

$$\frac{P[\nu(t-t_0)]^{3/2}}{\beta_0} = -\frac{1}{2}\left(\frac{\kappa'}{\kappa} - 1\right)\kappa[\nu(t-t_0)]^{1/2}$$
$$\times\left(\exp\left\{-2\{\kappa[\nu(t-t_0)]^{1/2}\}^2\left[1 - \frac{\kappa'}{\kappa} + \left(\frac{\kappa'}{\kappa}\right)^2\right]\right\}\right.$$
$$\left.- \exp\left\{-2\{\kappa[\nu(t-t_0)]^{1/2}\}^2\left[1 + \frac{\kappa'}{\kappa} + \left(\frac{\kappa'}{\kappa}\right)^2\right]\right\}\right),$$
(5-172)

and eq. (5-161) becomes

$$\frac{T(\kappa)[\nu(t-t_0)]^2}{\beta_0} = \frac{\sqrt{\pi/2}}{4}\kappa[\nu(t-t_0)]^{1/2}\left(2\,\text{erf}\left\{\frac{1}{\sqrt{2}}\kappa[\nu(t-t_0)]^{1/2}\right\} - 1\right)$$
$$\times \exp\left[-\frac{3}{2}\kappa^2\nu(t-t_0)\right],$$
(5-173)

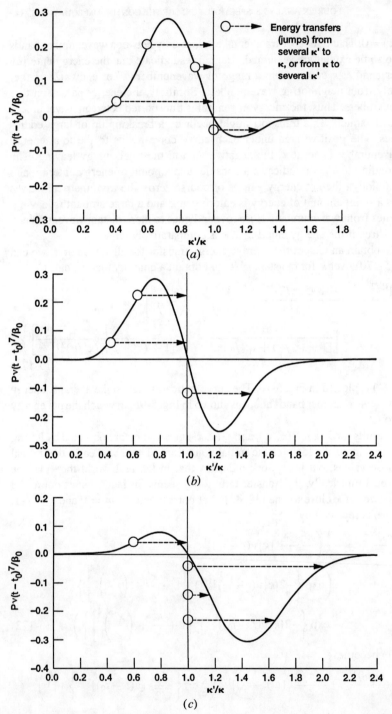

Figure 5-9 Plot showing contributions to energy transfer at κ from various wave numbers κ'. (a) $\kappa[\nu(t - t_0)]^{1/2} = 2.15$, T a positive maximum; (b) $\kappa[\nu(t - t_0)]^{1/2} = 1.71$, $T = 0$; (c) $\kappa[\nu(t - t_0)]^{1/2} = 1.29$, T a negative maximum.

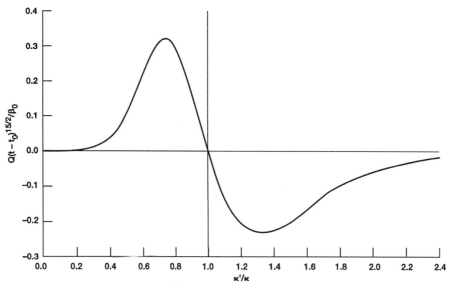

Figure 5-10 Plot showing contributions to energy transfer from various wave numbers κ' at average κ according to eq. (5-171) and Fig. 5-9.

where the symbol erf designates the error function. Finally, as an intermediate case, let $m = 0, n = 2$. Then

$$\frac{P[\nu(t - t_0)]^3}{\beta_0} = -\frac{1}{2}\left[\left(\frac{\kappa'}{\kappa}\right)^3 - \left(\frac{\kappa'}{\kappa}\right)\right]\{\kappa[\nu(t - t_0)]^{1/2}\}^4$$
$$\times \left(\exp\left\{-2\{\kappa[\nu(t - t_0)]^{1/2}\}^2\left[1 - \frac{\kappa'}{\kappa} + \left(\frac{\kappa'}{\kappa}\right)^2\right]\right\}\right.$$
$$\left. - \exp\left\{-2\{\kappa[\nu(t - t_0)]^{1/2}\}^2\left[1 + \frac{\kappa'}{\kappa} + \left(\frac{\kappa'}{\kappa}\right)^2\right]\right\}\right) \quad (5\text{-}174)$$

and

$$\frac{T(\kappa)[\nu(t - t_0)]^{7/2}}{\beta_0} = \frac{3\sqrt{\pi/2}}{16}(\{\kappa[\nu(t - t_0)]^{1/2}\}^4 - \{\kappa[\nu(t - t_0)]^{1/2}\}^2)$$
$$\times \exp\left[-\frac{3}{2}\kappa^2\nu(t - t_0)\right]. \quad (5\text{-}175)$$

We calculate spectra for dimensionless T and P from eqs. (5-172) through (5-175). In Figs. 5-11 and 5-12 results are compared with the previous ones for $m = 4, n = 6$. Note that there is a relation between the shapes of the curves for $T(\kappa)$ and those for $P(\kappa'/\kappa)$, and that the energy transfer is less local (the energy jumps are greater) if T starts out linearly at $\kappa = 0$, as does P at $\kappa' = 0$ (eqs. [5-172] and [5-173]). As mentioned earlier, that case, or a case close to it, seems to be indicated empirically.

It has been noted that the localness of the energy transfer is related to the shape of the $T(\kappa)$ spectrum, in particular to its shape near $\kappa = 0$. Here we develop an approximate method for obtaining the localness of the energy transfer from the T spectrum. To that end,

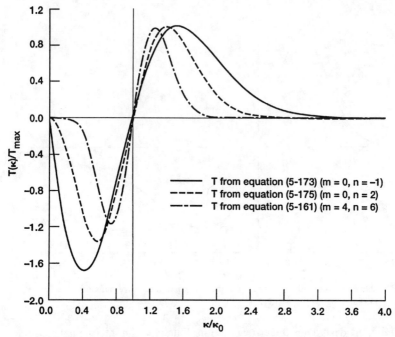

Figure 5-11 Effect of m and n in the initial contidion given by eq. (5-156) on the calculated net energy-transfer term $T(\kappa)$ (eq. [5-158]). κ_0 is the wavenumber where $T(\kappa) = 0$.

note that the T spectrum often can be represented by a truncated power-exponential series (or asymptotic expansion) in κ. See, for example, the theoretical spectra represented by eqs. (5-161) and (5-175), and the empirical equation for T in ref. [38]. Thus, one can write

$$T(\kappa) = \left(a_0 + a_1\kappa + a_2\kappa^2 + a_3\kappa^3 + \ldots\right)\left(e^{-\alpha_1\kappa} + e^{-\alpha_2\kappa^2} + \ldots\right), \qquad (5\text{-}176)$$

where dependencies on time are included in the a_i and α_i. Corresponding to eq. (5-176) we have for $P(\kappa, \kappa')$ the expansion

$$\begin{aligned}
P(\kappa, \kappa') = &[(c_{00}) + (c_{10}\kappa + c_{01}\kappa') + (c_{20}\kappa^2 + c_{11}\kappa\kappa' + c_{02}\kappa'^2) \\
&+ (c_{30}\kappa^3 + c_{21}\kappa^2\kappa' + c_{12}\kappa\kappa'^2 + c_{03}\kappa'^3) \\
&+ (c_{40}\kappa^4 + c_{31}\kappa^3\kappa' + c_{22}\kappa^2\kappa'^2 + c_{13}\kappa\kappa'^3 + c_{04}\kappa'^4) \\
&+ (c_{50}\kappa^5 + c_{41}\kappa^4\kappa' + c_{32}\kappa^3\kappa'^2 + c_{23}\kappa^2\kappa'^3 + c_{14}\kappa\kappa'^4 + c_{05}\kappa'^5) \\
&+ (c_{60}\kappa^6 + c_{51}\kappa^5\kappa' + c_{42}\kappa^4\kappa'^2 + c_{33}\kappa^3\kappa'^3 + c_{24}\kappa^2\kappa'^4 + c_{15}\kappa\kappa'^5 + c_{06}\kappa'^6) \\
&+ \cdots][e^{-\alpha_{10}\kappa - \alpha_{01}\kappa'} + e^{-\alpha_{20}\kappa^2 - \alpha_{02}\kappa'^2} + \cdots], \qquad (5\text{-}177)
\end{aligned}$$

where terms of the same degree in κ and κ' are grouped together in parentheses. By letting $j = i$ in eq. (5-104) and integrating over all directions in (κ, κ') space, we get

$$P(\kappa, \kappa') = -P(\kappa', \kappa), \qquad (5\text{-}178)$$

or that P is antisymmetric in κ and κ'.

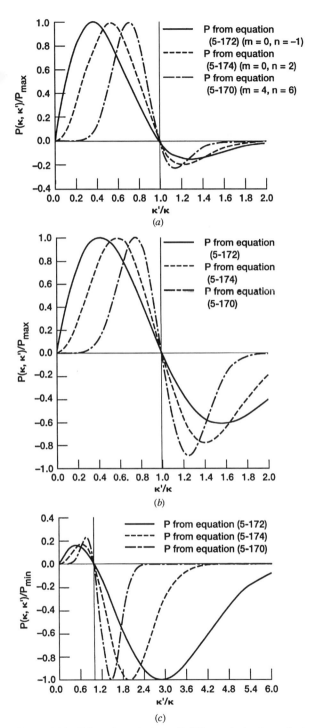

Figure 5-12 Effect of m and n in eq. (5-156) on calculated contributions $P(\kappa, \kappa')$ to net energy transfer $T(\kappa)$ (Fig. 5-11) at wavenumber κ from wavenumbers κ'. (a) $T(\kappa)$ a maximum (from Fig. 5-11). (b) $T(\kappa) = 0$ (from Fig. 5-11). (c) $T(\kappa)$ a minimum (from Fig. 5-11).

Then eq. (5-177) becomes
$$P(\kappa, \kappa') = \{c_{10}(\kappa - \kappa') + c_{20}(\kappa^2 - \kappa'^2) + [c_{30}(\kappa^3 - \kappa'^3) + c_{21}(\kappa^2\kappa' - \kappa\kappa'^2)]$$
$$+ [c_{40}(\kappa^4 - \kappa'^4) + c_{31}(\kappa^3\kappa' - \kappa\kappa'^3)] + [c_{50}(\kappa^5 - \kappa'^5) + c_{41}(\kappa^4\kappa' - \kappa\kappa'^4)$$
$$+ c_{32}(\kappa^3\kappa'^2 - \kappa^2\kappa'^3)] + [c_{60}(\kappa^6 - \kappa'^6) + c_{51}(\kappa^5\kappa' - \kappa\kappa'^5)$$
$$+ c_{42}(\kappa^4\kappa'^2 - \kappa^2\kappa'^4)] + \cdots\}[e^{-\alpha_{12}(\kappa+\kappa')} + e^{-\alpha_{20}(\kappa^2+\kappa'^2)} + \cdots]. \quad (5\text{-}179)$$

The terms retained in eq. (5-179) for a particular problem must satisfy the equation given for $T(\kappa)$ (say eq. [5-175]), where $T(\kappa)$ is related to $P(\kappa, \kappa')$ by

$$T(\kappa) = \int_0^\infty P(\kappa, \kappa') \, d\kappa'. \quad (5\text{-}180)$$

Equation (5-180) is obtained by contracting the indices i and j in eq. (5-99) and integrating over all directions in κ and κ' space. In addition, the following systematic procedure is used for truncating eq. (5-179): Those terms are retained that are of the lowest possible degree in κ, κ', such that the given equation for $T(\kappa)$ is satisfied.

If, for example, T is given by eq. (5-175), eq. (5-176) becomes

$$T(\kappa) = (a_2\kappa^2 + a_4\kappa^4)e^{-\alpha_2\kappa^2}, \quad (5\text{-}181)$$

and eq. (5-179) becomes

$$P(\kappa, \kappa') = c_{42}(\kappa^4\kappa'^2 - \kappa^2\kappa'^4) \, e^{-\alpha_{20}(\kappa^2+\kappa'^2)} \quad (5\text{-}182)$$

where, according to our procedure described in the previous paragraph, only the terms of the lowest possible degree in κ, κ' that satisfy eqs. (5-180) and (5-181) have been retained.

Substituting eqs. (5-175) and (5-182) into (5-180), we get

$$\frac{P[\nu(t - t_0)]^3}{\beta_0} = \frac{9\sqrt{3}}{16}\{\kappa[\nu(t - t_0)]^{1/2}\}^6\left[\left(\frac{\kappa'}{\kappa}\right)^2 - \left(\frac{\kappa'}{\kappa}\right)^4\right]$$
$$\times \exp\left\{-\frac{3}{2}\{\kappa[\nu(t - t_0)]^{1/2}\}^2\left[1 + \left(\frac{\kappa'}{\kappa}\right)^2\right]\right\}. \quad (5\text{-}183)$$

Figure 5-13 compares the approximate results calculated from eq. (5-183) with the theoretical results from eq. (5-174). The degree of localness calculated from the approximate equation is in good qualitative agreement with that obtained from the theoretical calculation. The agreement is particularly good in Fig. 5-13b, where the energy transfer into the wavenumber band at κ equals that leaving (if κ/κ_0 for the T spectrum equals 1).

The agreement in Fig. 5-13, in which our approximate method is used for calculating a known $P(\kappa, \kappa')$, gives us some confidence in that method. Thus, we use it to calculate the degree of localness of the energy transfer that corresponds to the experimental equation for $T(\kappa)$ obtained by Ling and Huang [38], where $P(\kappa, \kappa')$ is unknown. There, $T(\kappa)$ is given by

$$\frac{T(\kappa)[\nu(t - t_0)]^{5/2}}{A\nu^3} = \frac{10}{3}(2\sqrt{10}\{\kappa[\nu(t - t_0)]^{1/2}\}^4 - 3\{\kappa[\nu(t - t_0)]^{1/2}\}^3$$
$$- \sqrt{10}\{\kappa[\nu(t - t_0)]^{1/2}\}^2 - \kappa[\nu(t - t_0)]^{1/2})$$
$$\times \exp(-\sqrt{10}\{\kappa[\nu(t - t_0)]^{1/2}\}), \quad (5\text{-}184)$$

FOURIER ANALYSIS, SPECTRAL FORM OF THE CONTINUUM EQUATIONS 155

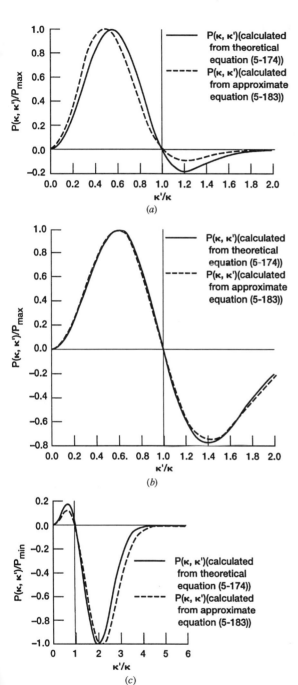

Figure 5-13 Comparison of contributions $P(\kappa, \kappa')$ to net energy transfer $T(\kappa)$ as calculated from theoretical equation and by approximate method. $T(\kappa)$ is obtained from eq. (5-175) (dashed curve in Fig. 5-11). (a) $T(\kappa)$ a maximum; (b) $T(\kappa) = 0$; (c) $T(\kappa)$ a minimum.

where A is defined by the evolution equation
$$\overline{u_1^2} = A(t - t_0)^2. \tag{5-185}$$
Then eq. (5-176) becomes
$$T(\kappa) = (a_1\kappa + a_2\kappa^2 + a_3\kappa^3 + a_4\kappa^4)e^{-\alpha_1\kappa} \tag{5-186}$$
and eq. (5-179) becomes
$$P(\kappa, \kappa') = [c_{21}(\kappa^2\kappa' - \kappa\kappa'^2) + c_{31}(\kappa^3\kappa' - \kappa\kappa'^3) + c_{41}(\kappa^4\kappa' - \kappa\kappa'^4)]e^{-\alpha_{12}(\kappa+\kappa')}, \tag{5-187}$$
where again only the terms of the lowest possible degree in κ, κ' that satisfy the determining equations (eqs. [5-180] and [5-186]) have been retained. Substituting eqs. (5-184) and (5-187) into (5-180) we get

$$\frac{P[\nu(t-t_0)]^2}{A\nu^3} = \left(-\frac{10^{5/2}}{3}\left[\frac{\kappa'}{\kappa} - \left(\frac{\kappa'}{\kappa}\right)^2\right]\{\kappa[\nu(t-t_0)]^{1/2}\}^3 - 100\left[\frac{\kappa'}{\kappa} - \left(\frac{\kappa'}{\kappa}\right)^3\right]\right.$$
$$\left. \times \{\kappa[\nu(t-t_0)]^{1/2}\}^4 + \frac{2(10)^{5/2}}{3}\left[\frac{\kappa'}{\kappa} - \left(\frac{\kappa'}{\kappa}\right)^4\right]\{\kappa[\nu(t-t_0)]^{1/2}\}^5\right)$$
$$\times \exp\left\{-\sqrt{10}\left(1 + \frac{\kappa'}{\kappa}\right)\kappa[\nu(t-t_0)]^{1/2}\right\}. \tag{5-188}$$

Results calculated from eq. (5-188) are plotted in Fig. 5-14. These plots of $P(\kappa, \kappa')$, which correspond to the experiment for $T(\kappa)$ from ref. [38], are similar to the theoretical results in Fig. 5-12 (a different case), but the degree of nonlocalness tends to be somewhat greater. It seems better not to attempt to characterize the energy transfer as local or nonlocal, because the dividing line between the two is necessarily arbitrary. Strictly speaking, however, as mentioned earlier, all spectral energy transfer is nonlocal because of the condition $P(\kappa, \kappa) = 0$, which follows from eq. (5-105) (integrated over all directions in wavevector space with $j = i$). Thus, energy transfer can take place only between wavenumber bands that, at least to some extent, are separated.

The tendency of the energy to jump between wavenumber regions that are separated appears to be in accord with the idea that turbulence tends to form concentrated regions of large velocity gradients [4]. Thus, if a low wavenumber eddy becomes unstable and forms a region of large velocity gradients, there is a transfer of energy from low to significantly higher wavenumbers. The concept is developed further in section 4-3-2-3, in which it is shown that in a turbulent flow there must be regions of relative quiescence interspersed with regions in which the instantaneous mixing is intense and localized.

A poem by Betchov [44] emphasizes the energy jumps occurring in spectral energy transfer more than does the one by Richardson (see section 5-1-2-1). Betchov's poem is

Big whirls lack smaller whirls,
To feed on their velocity.
They crash and form the finest curls
Permitted by viscosity.

It may seem that the equilibrium (cascade) theory [4] can apply in the presence of a rather high degree of nonlocalness of the spectral energy transfer if the Reynolds number of the turbulence is huge. In that case the energy spectrum extends over many decades

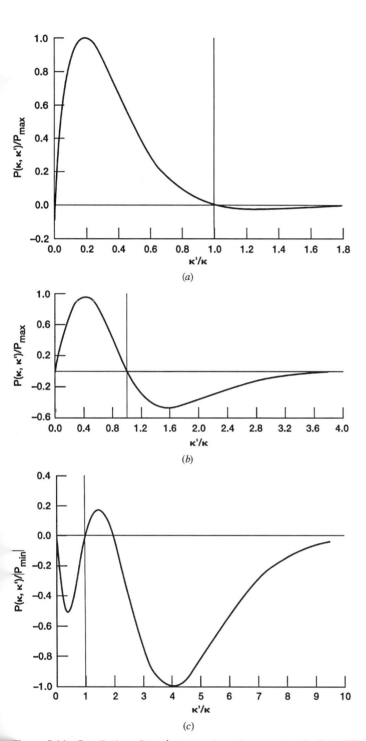

Figure 5-14 Contributions $P(\kappa, \kappa')$ to experimental energy transfer $T(\kappa)$ [38] from wavenumbers κ'. $P(\kappa, \kappa')$ calculated from approximate eq. (5-188). (a) $T(\kappa)$ a maximum (eq. [5-184]); (b) $T(\kappa) = 0$ (eq. [5-184]); (c) $T(\kappa)$ a minimum (eq. [5-184]).

of wavenumbers. Thus, there could be a cascade in which the energy is passed from low to high wavenumbers by successive moderately large jumps. However, the turbulence Reynolds number required to make the small eddies independent of the large ones is larger than if the energy transfer is more local.

Note, however, that Kraichnan [45], for example, has given arguments against the independence of the motion of the very small and the very large eddies, even at large Reynolds numbers. The crux of his argument appears to be that even though the active high-wavenumber components are much smaller than the low-wavenumber ones if the turbulence Reynolds number is high, both of those often consist of long vortex filaments that are tangled together. Because the small-scale filaments can extend axially over considerable distances, it is hard to see how their motions can be independent of the motions of the large-scale components, particularly because the former are likely to be stretched or compressed by the latter (see also refs. [44] and [46–48]).

As a further development of the discussion in the previous paragraph, we return to spectral-energy transfer as related to the interaction in triads of Fourier components (see eq. [5-80]). The spectral-transfer term in eq. (5-80) is

$$T_{ij}(\kappa) = \int_{-\infty}^{\infty} P_{ij}(\kappa, \kappa') d\kappa' \qquad (5\text{-}189)$$

where

$$P_{ij}(\kappa, \kappa') = i[\kappa_k \varphi_i^*(\kappa - \kappa')\varphi_j(\kappa)\varphi_k^*(\kappa') - \kappa_k \varphi_i^*(\kappa)\varphi_j(\kappa - \kappa')\varphi_k(\kappa')]. \qquad (5\text{-}190)$$

The asterisks designate complex conjugates. As mentioned earlier, the presence of the integral in eq. (5-189) means that contributions from a range of wavevectors κ' between $-\infty$ and $+\infty$ make up the total net transfer at κ. Moreover, according to eq. (5-190), the transfer takes place by the interaction in triads of Fourier components at the wavevectors κ, κ' and $\kappa - \kappa'$. Note that if a triad is composed of Fourier components at the wavevectors κ, p and q, then $\kappa + p + q = 0$ if $p = -\kappa'$ and $q = -(\kappa - \kappa')$.

An important observation is that the triads in eq. (5-190) consist of products, as opposed to, say, sums, of Fourier components at the three wavevectors. So all three components have an influence on $P_{ij}(\kappa, \kappa')$ and thus on $T_{ij}(\kappa)$, even if one of the components is much smaller than the other two, and even if one is at a much lower wavenumber than the other two.

Thus, if it turns out that the triads having the most influence on $T_{ij}(\kappa)$ (or on $T(\kappa)$) generally consist of Fourier components at three wavevectors of approximately equal magnitude, one might conclude that the interactions between large and small eddies are relatively unimportant. If, on the other hand, the magnitude of the wavevector of one of the Fourier components in the more influential triads differs greatly from the others, then interactions between large and small eddies have an important effect on the energy transfer. The latter is found to be the case in the direct numerical simulations of Domaradzki and Rogallo [49]. They found that their energy transfer was due mainly to wavevector triads that have one leg much shorter than the other two.[8] Yeung and Brasseur [48], using a somewhat different approach, arrived at the same conclusion.

---

[8]Ref. [49] also obtained results for the localness of energy transfer. It was found, in agreement with

These results reinforce the physical argument considered in ref. [45] and in this section and suggest that the motion of the small eddies, even at high Reynolds numbers, may not be independent of the motion of the large ones. An important part of this argument is the observation made here that Fourier components of the triads in eq. (5-190) occur as products. Thus, although the low wavenumber component of a triad may be weak, it affects the energy transfer as much as does the other components, because they are multiplied by one another. This might be related to the tangledness of the small and large vortex filaments in the physical argument; noting that the small- and large-scale components are multiplied together, and that many of the triads have one leg much shorter than the other two, may be another way of looking at the fact that they are twisted together over long distances as vortex filaments. At any rate, both arguments tend to indicate the lack of independence of the large and the small eddies. So the independence hypothesis [39] seems to be open to some question, based on the results to date. Of course, these comments are not meant to detract from the agreement with experiment of the Kolmogorov-Obukov $-5/3$-power spectrum. That spectrum, however, may require a different foundation.

We return now to the solution of the Navier-Stokes equations for decaying homogeneous turbulence obtained numerically by Clark, Ferziger, and Reynolds [33]. That solution yielded energy-transfer spectra in addition to energy spectra. The energy-transfer spectrum corresponding to the energy spectrum in Fig. 5-6 is plotted in Fig. 5-15. The shape of the spectrum is similar to the shapes of our theoretical spectra in Fig. 5-11, but as in the case of the energy (see Fig. 5-6), the energy transfer (at least the positive portion) is spread out over a wider range of wavenumbers. Because of the condition that the energy transfer integrated over all wavenumbers must be zero, this spreading causes the negative trough of the spectrum to become narrow and deep. The difference between the shapes of the spectra in Figs. 5-11 and 5-15 is apparently due to the higher Reynolds number for the latter. The effect of Reynolds number on the shape of transfer spectra is shown clearly by a comparison of experimental results from ref. [37] ($300 < R_\lambda < 800$) with those from ref. [38] ($3 < R_\lambda < 30$) (Fig. 5-16).

5-3-2-3 Correlation-term–discard closure for short times of turbulence decay and comparison with experiment. Thus far we have refrained from comparing our analytical results with experiment. In general, the significance of such comparisons is uncertain

our results from eq. (5-188) and Fig. 5-14, that contributions to the energy transfer are spread over a range of wavenumbers. But the results from ref. [49] showed somewhat more localness. However, comparison of Fig. 1a and fig. 4 of ref. [49] shows that the calculations for $P(\kappa, \kappa')$ are outside the range where the numerical simulations represent the experiment of ref. [38]. Thus the simulations are not strictly comparable with our results, which *are* for the experiment of ref. [38]. Also, the comments in ref. [49] concerning the arbitrariness in our methodology are not well-founded, considering the restrictions we placed on $P(\kappa, \kappa')$. Note, however, that the logic used here in obtaining eq. (5-188) is much improved over that in ref. [50]. At any rate it should be mentioned that the common practice (in ref. [49] and elsewhere) of designating energy transfers within a ratio of two as local is arbitrary. By virtue of eq. (5-105) all spectral transfer is nonlocal.

160 TURBULENT FLUID MOTION

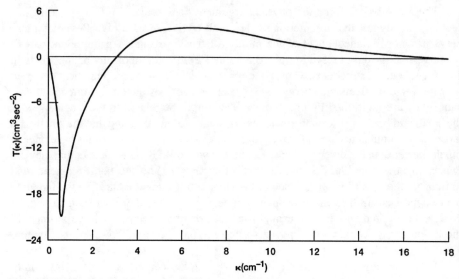

Figure 5-15 Energy-transfer spectrum for homogeneous turbulence obtained from a numerical solution of the instantaneous Navier-Stokes equations by Clark, Ferziger, and Reynolds [33]. $R = 36.6$.

because of lack of knowledge of the initial conditions. But it may be possible to make meaningful comparisons for at least one short–decay-time case.

Here we return to the correlation-term–discard closure (or expansion in powers of Reynolds number or time) in order to use it to calculate the decay of homogeneous turbulence for initial conditions obtained from experiment [51]. (It should be noted that the previous calculations in section 5-3-2-2 were based on initial conditions that allowed the solution to approach the final period of decay at large times, rather than on experimental initial conditions.) The two-point eq. (5-158), which was obtained by neglecting quadruple-correlation terms in the three-point correlation equation, can be written as

$$dE/dt + 2\nu\kappa^2 E = T(\kappa) \tag{5-158}$$

where

$$T(\kappa) = \int_0^\infty P(\kappa, \kappa') \, d\kappa' \tag{5-191}$$

and where, from eqs. (5-191) and (5-159), one can write

$$P(\kappa, \kappa') = \frac{f_1(\kappa, \kappa')}{t - t_1} \{\exp[-2\nu(t - t_1)(\kappa^2 + \kappa\kappa' + \kappa'^2)] \\ - \exp[-2\nu(t - t_1)(\kappa^2 - \kappa\kappa' + \kappa'^2)]\}. \tag{5-192}$$

Note that we have replaced the particular initial condition $\beta_0(\kappa^5\kappa'^7 - \kappa^7\kappa'^5)/(2\nu)$ at t_0 in eq. (5-159) with the general condition $f_1(\kappa, \kappa')$ at t_1 in eq. (5-192). As usual, E is the energy spectrum function, related to the total turbulent energy $\overline{u_i u_i}/2$ by

$$\frac{1}{2}\overline{u_i u_i} = \int_0^\infty E(\kappa) \, d\kappa; \tag{5-166}$$

FOURIER ANALYSIS, SPECTRAL FORM OF THE CONTINUUM EQUATIONS **161**

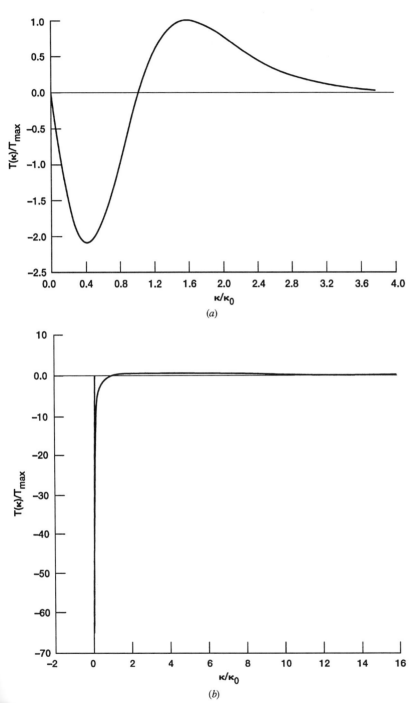

Figure 5-16 Effect of Reynolds number on experimental energy-transfer spectra. κ_0 is the wavenumber where $T(\kappa) = 0$. (a) $3 < R_\lambda < 30$ [38], (b) $300 < R_\lambda < 800$ [37].

162 TURBULENT FLUID MOTION

$T(\kappa)$ is the energy transfer function, which gives the net energy transfer into a wavenumber region at wavenumber κ from all other wavenumbers κ'; $P(\kappa, \kappa')$ gives the contribution from κ' to the energy transfer at κ, t is the time; the subscript 1 designates an initial value; ν is the kinematic viscosity; and u_i is a velocity component. A repeated subscript indicates a summation, and an overbar indicates an averaged value.

The function $f_1(\kappa, \kappa')$ in eq. (5-192) is evaluated by setting P equal to the initial condition $P_1(\kappa, \kappa')$ if $t = t_1$ and using the fact that

$$\lim_{t \to t_1}(t - t_1)^{-1}\{\exp[-2\nu(t - t_1)(\kappa^2 + \kappa\kappa' + \kappa'^2)]$$
$$- \exp[-2\nu(t - t_1)(\kappa^2 - \kappa\kappa' + \kappa'^2)]\} = -4\nu\kappa\kappa'. \quad (5\text{-}193)$$

This gives

$$f_1(\kappa, \kappa') = -P_1(\kappa, \kappa')/(4\nu\kappa\kappa'). \quad (5\text{-}194)$$

One of the difficulties in comparing closure schemes with experiment has been the unavailability of initial values of $P(\kappa, \kappa')$. Reasonable estimates of that quantity (P_1 in eq. [5-194]) however, can be obtained from Ling and Huang's experiment [38] and our approximate eq. (5-188). The initial condition for $E(\kappa)$ in eq. (5-158) is obtained from ref. [38].

Results calculated according to the present correlation-term–discard closure are compared with the experiment of Ling and Huang in Figs. 5-17 through 5-19. Figure 5-17 also shows results for $T = 0$ (pure viscous decay). The quantity A is the proportionality constant in the power decay law for $\overline{u_1^2}$ and has the dimensions of length^2 (eq. [5-185]).

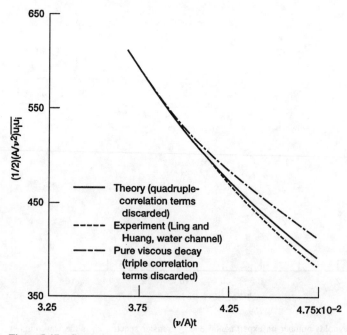

Figure 5-17 Comparison of theory with the experiment of ref. [38] for decay of turbulent energy.

FOURIER ANALYSIS, SPECTRAL FORM OF THE CONTINUUM EQUATIONS **163**

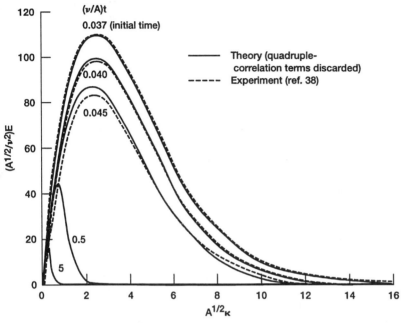

Figure 5-18 Comparison of theory and experiment for decay of three-dimensional turbulent-energy spectra.

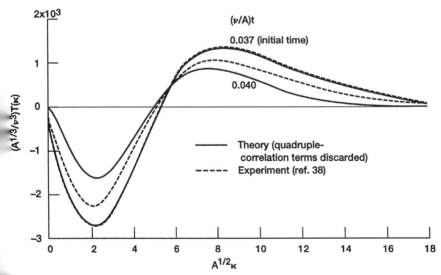

Figure 5-19 Comparison of theory and experiment for decay of energy-transfer spectra.

164 TURBULENT FLUID MOTION

The initial conditions for the analysis are specified at $(\nu/A)t_1 = 0.037$, where the turbulence Reynolds number $R_\lambda \equiv \overline{u_1^2}^{1/2} \lambda/\nu = 12$ (λ is the Taylor microscale).

As one might have expected, the lower-order results $\overline{u_i u_i}$ and $E(\kappa)$ compare better with experiment than the higher-order one, $T(\kappa)$. This is because a larger error is produced in quantities that are calculated directly from equations in which quadruple-correlation terms are neglected. Agreement with experiment is indicated only for moderately small times, and none of the results are as good as those obtained for closure by specification of initial conditions (to be considered next) or for a modified Kovasznay-type closure [52] (the latter is more an intuitive than a deductive procedure). For those closures good results for all initially specified quantities are obtained even for reductions in $\overline{u_i u_i}$ of over 80%. However, the present solution is well-behaved and shows no perceptible negative spectral energies even at large times (see curves for $[\nu/A]t_1$ of 0.5 and 5 in Fig. 5-18). (Some negligibly small negative values that occurred at high wavenumbers and large times are judged to be of no importance.) Improved results (for large times) evidently could be obtained by using a higher-order closure [22], but higher-order initial conditions then would have to be specified.

Calculations also were made for high Reynolds numbers and compared with high–Reynolds-number data [37]. Not surprisingly, the agreement is much poorer than for the low–Reynolds-number data shown here. If reasonable results are to be obtained for the high Reynolds numbers, the equations have to be closed by neglecting correlations of a much higher order, and the amount of calculation probably becomes prohibitive. Thus, this conceptually simplest deductive closure scheme works best for short times at moderately low Reynolds numbers, or for the approach to the final period of decay. For the latter case, the initial conditions must be chosen so that the solution approaches the final period at large times, as in refs. [21] and [22]. Because this is the simplest scheme, it is the most convenient one for illustrating the spectral-transfer process in turbulence. Further discussion of the correlation-term–discard closure is given in ref. [53].

5-3-2-4 Closure by specification of sufficient random initial conditions. It is mentioned in the previous section that although a low-order correlation-term–discard closure is the simplest and thus the most convenient closure for studying the processes in turbulence, serious difficulties are encountered if its extension to high Reynolds numbers or to large turbulence-decay times is attempted. There is, however, another way of looking at the problem of homogeneous turbulence. In order not to lose sight of our goal, we first give a statement of that problem. The statement given by Batchelor [4] is essentially the following: Given the statistical state of a homogeneous turbulent field at an initial instant, the problem is to predict the evolution of the turbulence (in probability) as a function of time. Note that the initial development of turbulence from a nonturbulent state produced by, say, flow through a grid is not considered here. Rather we are concerned with the evolution of turbulence after a time at which the flow is already turbulent. In order to specify completely a turbulent field at an initial time, it is necessary to give all of the multipoint velocity correlations or their spectral equivalents at that time [4]. It is not hard to show that, given these multipoint correlations and the correlation equations, all the time derivatives of the turbulent energy tensor and of other pertinent turbulence quantities can be calculated. These time derivatives can then be used in a series, for instance a Taylor series, to calculate the evolution of the turbulent energy tensor (or of the equivalent energy spectrum tensor) and of other turbulence quantities.

It is noted that if the turbulence is treated in this way, we no longer have the problem of closing the infinite set of correlation or spectral equations. The correlation equations are used only to relate the correlations at an initial time to their time derivatives, and those correlations must be given in order to have a complete specification of turbulence at that time. Of course, in practice only a small number of the correlations, and thus of their time derivatives, ordinarily are available, but a sufficient number may be known to give a reasonably good representation. It might be pointed out that even in those analyses that require a closure assumption, the turbulence should be specified initially by its correlations or spectra because the correlation equations require initial conditions.

Kraichnan [54] has studied the convergence properties of series such as those considered here. As mentioned in another article by that author [55], it is not necessary that an expansion be convergent in order to be useful, because divergent series can provide excellent asymptotic approximations [56].

Although the present method circumvents the closure problem in the usual sense, there is still the question of the legitimate truncation of the time series to obtain explicit results. Here we are not concerned primarily with convergence questions but use as a test the agreement of the results with experiment. Although a Taylor series might give good results if sufficient statistical information is available at the initial time, it is seen that an exponential series that arises in a study of the nonlinear decay of a disturbance in a fluid [57] is much more satisfactory. This is not surprising, because the exponential series is an iterative solution of the Navier-Stokes equations and thus contains information that is not contained in the Taylor series. The resulting formulation gives results that are in quite good agreement with the available experimental data [58, 59].

Initial time derivatives and simple expansions. In the preceding introductory section it is noted that if the multipoint correlations are known at an initial instant, as they must be for a complete specification of the turbulence at that instant, then the time derivatives of the correlations can be calculated from the correlation equations. For illustrative purposes we consider the derivatives of the turbulent energy tensor $\overline{u_i u'_j}$, where u_i and u'_j are respectively velocity components at the points P and P' separated by the vector r, and the overbar indicates an averaged value. Then the first time derivative of $\overline{u_i u'_j}$ at $t = t_1$ is given directly by the two-point correlation eqs. (5-70) through (5-72) evaluated at $t = t_1$:

$$\left(\frac{\partial \overline{u_i u'_j}}{\partial t}\right)_{t=t_1} = -\frac{\partial}{\partial r_k}\left[\left(\overline{u_i u'_j u'_k}\right)_{t=t_1} - \left(\overline{u_i u'_j u_k}\right)_{t=t_1}\right]$$

$$-\left[\frac{1}{\rho}\frac{\partial}{\partial r_j}(\overline{u_i \sigma'})_{t=t_1} - \frac{\partial}{\partial r_i}(\overline{\sigma u'_j})_{t=t_1}\right] + 2\nu \frac{\partial^2 \left(\overline{u_i u'_j}\right)_{t=t_1}}{\partial r_k \partial r_k}, \quad (5\text{-}195)$$

where the correlation between the (mechanical) pressure σ (eq. [3-14]) and the velocity u'_j is given by

$$\frac{1}{\rho}\frac{\partial^2 \left(\overline{\sigma u'_j}\right)_{t=t_1}}{\partial r_k \partial r_k} = -\frac{\partial^2 \left(\overline{u_l u_k u'_j}\right)_{t=t_1}}{\partial r_k \partial r_l} \quad (5\text{-}196)$$

166 TURBULENT FLUID MOTION

and a similar equation for $(\overline{u_i \sigma'})_{t=t_1}$. The pertinent solution of eq. (5-196) is [4]

$$\frac{1}{\rho}(\overline{\sigma u'_j})_{t=t_1} = \frac{1}{4\pi} \int \frac{1}{|r-s|} \frac{\partial^2 (\overline{u''_i u''_k u''_j})_{t=t_1}}{\partial s_i \partial s_k} ds,$$

where u''_i is the velocity at the point $x'' = x' - s$, and the integration is over all s space. This solution is for an infinite fluid, for which case the boundary conditions are that $\overline{\sigma u'_j}$ is bounded for $r = 0$ and zero for $r = \infty$. The quantity ρ is the density, ν is the kinematic viscosity, and σ is the pressure. A repeated subscript in a term indicates a summation, with the subscript successively taking on the values 1, 2, and 3. The correlation equations, of course, are derived from the Navier-Stokes equations. The quantity $\partial \overline{u_i u'_j}/\partial t$ at $t = t_1$ can be calculated from eqs. (5-195) and (5-196) if $\overline{u_i u'_j}$ and the two-point triple correlations are known at $t = t_1$.

The second time derivative of $\overline{u_i u'_j}$ is obtained by differentiating the two-point correlation equations and evaluating the result at t_1. This gives

$$\left(\frac{\partial^2 \overline{u_i u'_j}}{\partial t^2}\right)_{t=t_1} = -\frac{\partial}{\partial r_k}\left[\left(\frac{\partial}{\partial t}\overline{u_i u'_j u'_k}\right)_{t=t_1} - \left(\frac{\partial}{\partial t}\overline{u_i u'_j u_k}\right)_{t=t_1}\right]$$

$$-\frac{1}{\rho}\left[\frac{\partial}{\partial r_j}\left(\frac{\partial \overline{u_i \sigma'}}{\partial t}\right)_{t=t_1} - \frac{\partial}{\partial r_i}\left(\frac{\partial \overline{\sigma u'_j}}{\partial t}\right)_{t=t_1}\right] + 2\nu \frac{\partial^2}{\partial r_k \partial r_k}\left(\frac{\partial \overline{u_i u'_j}}{\partial t}\right)_{t=t_1}$$

(5-197)

and

$$\frac{1}{\rho}\frac{\partial^2}{\partial r_k \partial r_k}\left(\frac{\partial \overline{\sigma u'_j}}{\partial t}\right)_{t=t_1} = -\frac{\partial^2}{\partial r_k \partial r_l}\left(\frac{\partial \overline{u_l u_k u'_j}}{\partial t}\right)_{t=t_1}.$$

(5-198)

The quantity $[(\partial/\partial t)(\overline{u_i u'_j u'_k})]_{t=t_1}$ in eq. (5-197) is obtained from the three-point correlation equations (section 5-3-2-2) written for $t = t_1$ and $r' = r$. (The vector r' separates the points P and P''.) Thus,

$$\left(\frac{\partial}{\partial t}\overline{u_i u'_j u'_k}\right)_{t=t_1} = \left\{\frac{\partial}{\partial r_l}(\overline{u_i u'_j u''_k u_l})_{t=t_1} + \frac{\partial}{\partial r'_l}(\overline{u_i u'_j u''_k u_l})_{t=t_1} - \frac{\partial}{\partial r_l}(\overline{u_i u'_j u''_k u'_l})_{t=t_1}\right.$$

$$-\frac{\partial}{\partial r'_l}(\overline{u_i u'_j u''_k u''_l})_{t=t_1} - \frac{1}{\rho}\left[-\frac{\partial}{\partial r_i}(\overline{\sigma u'_j u''_k})_{t=t_1} - \frac{\partial}{\partial r'_i}(\overline{\sigma u'_j u''_k})_{t=t_1}\right.$$

$$+\frac{\partial}{\partial r_j}(\overline{\sigma' u_i u''_k})_{t=t_1} + \frac{\partial}{\partial r'_k}(\overline{\sigma'' u_i u'_j})_{t=t_1}\right] + 2\nu\left[\frac{\partial^2 (\overline{u_i u'_j u''_k})_{t=t_1}}{\partial r_l \partial r_l}\right.$$

$$\left.\left.+\frac{\partial^2 (\overline{u_i u'_j u''_k})_{t=t_1}}{\partial r_l \partial r'_l} + \frac{\partial^2 (\overline{u_i u'_j u''_k})_{t=t_1}}{\partial r'_l \partial r'_l}\right]\right\}_{r'=r},$$

(5-199)

where $(\overline{\sigma u'_j u''_k})_{t=t_1}$ is given by

$$\frac{1}{\rho}\left[\frac{\partial^2 (\overline{\sigma u'_j u''_k})_{t=t_1}}{\partial r_l \partial r_l} + 2\frac{\partial^2 (\overline{\sigma u'_j u''_k})_{t=t_1}}{\partial r_l \partial r'_l} + \frac{\partial^2 (\overline{\sigma u'_j u''_k})_{t=t_1}}{\partial r'_l \partial r'_l}\right] = -\frac{\partial^2 (\overline{u_l u_m u'_j u''_k})_{t=t_1}}{\partial r_m \partial r_l}$$

$$-\frac{\partial^2 (\overline{u_l u_m u'_j u''_k})_{t=t_1}}{\partial r_m \partial r'_l} - \frac{\partial^2 (\overline{u_l u_m u'_j u''_k})_{t=t_1}}{\partial r'_m \partial r_l} - \frac{\partial^2 (\overline{u_l u_m u'_j u''_k})_{t=t_1}}{\partial r'_m \partial r'_l}.$$

(5-200)

Similar equations are obtained for the other pressure–velocity correlations. The boundary conditions for eq. (5-200) are similar to those for eq. (5-196); that is, $\overline{\sigma u'_j u''_k}$ is bounded for r or $r' = 0$ and zero for r or $r' = \infty$. Also, an expression for $[(\partial/\partial t)(\overline{u_i u'_j u_k})]_{t=t_1}$ in eq. (5-197) is obtained by letting $r' = 0$ instead of r in eq. (5-199). Thus, if the turbulence is specified sufficiently well at $t = t_1$ that the double, triple, and quadruple velocity correlations are known, $(\partial^2 (\overline{u_i u'_j})/\partial t^2)_{t=t_1}$ can be calculated. Similarly, higher-order derivatives are obtained by considering four or more point correlations in the turbulent field [22]. With the time derivatives of $\overline{u_i u'_j}$ known at $t = t_1$, a Taylor series gives $\overline{u_i u'_j}$ as a function of time as

$$\overline{u_i u'_j} = (\overline{u_i u'_j})_{t=t_1} + \left(\frac{\partial \overline{u_i u'_j}}{\partial t}\right)_{t=t_1}(t-t_1) + \frac{1}{2!}\left(\frac{\partial^2 \overline{u_i u'_j}}{\partial t^2}\right)_{t=t_1}(t-t_1)^2 + \cdots. \quad (5\text{-}201)$$

A similar analysis can be carried out in wavenumber space. For instance, the energy spectrum function $E(\kappa)$, which shows the contributions at various wavenumbers to $\overline{u_i u'_j}/2$, can be written as

$$E(\kappa) = E(\kappa)_{t=t_1} + \left(\frac{\partial E(\kappa)}{\partial t}\right)_{t=t_1}(t-t_1) + \frac{1}{2!}\left(\frac{\partial^2 E(\kappa)}{\partial t^2}\right)_{t=t_1}(t-t_1)^2 + \cdots, \quad (5\text{-}202)$$

where it is understood that E is a function of t as well as of κ, and where $\partial E(\kappa)/\partial t$ is obtained from the Fourier transform of the two-point correlation equation (eq. [5-138]) as

$$\left(\frac{\partial E(\kappa)}{\partial t}\right) = \int_A \frac{1}{2}\{-2\nu\kappa^2 \varphi_{ii}(\boldsymbol{\kappa}) + i\kappa_k[\varphi_{iki}(\boldsymbol{\kappa}) - \varphi_{iki}(-\boldsymbol{\kappa})]\}\, dA(\boldsymbol{\kappa}), \quad (5\text{-}203)$$

where $dA(\boldsymbol{\kappa})$ is an element of surface area of a sphere of radius κ, $\boldsymbol{\kappa}$ is the wavevector corresponding to the spatial vector $\mathbf{r}$, and φ_{ii} and φ_{iki} are respectively the Fourier transforms of $\overline{u_i u'_i}$ and $\overline{u_i u_k u'_i}$. Extracting from the integral that portion that can be written in terms of $E(\kappa)$ and setting the rest of the integral equal to $T(\kappa)$ give

$$\frac{\partial E(\kappa)}{\partial t} = T(\kappa) - 2\nu\kappa^2 E(\kappa). \quad (5\text{-}204)$$

Equation (5-204) is, of course, the scalar form of the two-point spectral equation. The transfer term $T(\kappa)$ produces energy transfer between wavenumbers and arises from the triple-correlation term in eq. (5-195) (with $i = j$). (Note that the pressure–velocity correlation terms in eq. [5-195] drop out for $i = j$.) The second time derivative of $E(\kappa)$ is

$$\left(\frac{\partial^2 E(\kappa)}{\partial t^2}\right)_{t=t_1} = \left(\frac{\partial T(\kappa)}{\partial t}\right)_{t=t_1} - 2\nu\kappa^2 \left(\frac{\partial E(\kappa)}{\partial t}\right)_{t=t_1}$$

$$= \left(\frac{\partial T(\kappa)}{\partial t}\right)_{t=t_1} - 2\nu\kappa^2 T(\kappa)_{t=t_1} + (2\nu\kappa^2)^2 E(\kappa)_{t=t_1}.$$

The quantity $(\partial T(\kappa)/\partial t)_{t=t_1}$ can be calculated from the two- and three-point spectral equations if the two- and three-point spectral quantities in those equations are known at $t = t_1$. From eqs. (5-149), (5-152), and (5-153) one obtains

$$\frac{\partial T(\kappa)}{\partial t} = \int_A \int_{-\infty}^{\infty} \frac{1}{2}(-2\nu\kappa^2\{i\kappa_k[\beta_{iik}(\boldsymbol{\kappa}) - \beta_{iik}(-\boldsymbol{\kappa})]\} + f(\beta_{ijk}, \beta_{ijkl}))\, d\boldsymbol{\kappa}'\, dA(\boldsymbol{\kappa}),$$

168 TURBULENT FLUID MOTION

where κ' is the wavevector corresponding to r', $d\kappa = d\kappa_1 d\kappa_2 d\kappa_3$, and β_{ijk} and β_{ijkl} are respectively the Fourier transforms of $\overline{u_i u'_j u''_k}$ and $\overline{u_i u_j u'_k u''_l}$. If by analogy with the procedure used for obtaining eq. (5-204) we extract from the integral that portion which can be written in terms of spectral quantities already defined ($E[\kappa]$ and $T[\kappa]$), we have

$$\left(\frac{\partial T(\kappa)}{\partial t}\right) 2\nu\kappa^2 T(\kappa) = \int_A \int_{-\infty}^{\infty} \frac{1}{2} f(\beta_{ijk}, \beta_{ijkl}) \, d\kappa' \, dA(\kappa) = V\left[\beta_{ijk}(\kappa'), \beta_{ijkl}(\kappa')\right], \quad (5\text{-}205)$$

where V is a quantity related to the three-point spectral tensors β_{ijk} and β_{ijkl}. More precisely we can say that V is a functional of β_{ijk} and β_{ijkl}, because each value of V depends on values of β_{ijk} and β_{ijkl} at all points of κ' space. With eq. (5-205), the expression for $(\partial^2 E/\partial t^2)_{t=t_1}$ becomes

$$\left(\frac{\partial^2 E}{\partial t^2}\right)_{t=t_1} = V_{t=t_1} - 4\nu\kappa^2 T_{t=t_1} + (2\nu\kappa^2)^2 E_{t=t_1}. \quad (5\text{-}206)$$

The Taylor series for E then becomes

$$E(\kappa) = E_{t=t_1} + (T_{t=t_1} - 2\nu\kappa^2 E_{t=t_1})(t - t_1)$$
$$+ \frac{1}{2!}\left[V_{t=t_1} - 4\nu\kappa^2 T_{t=t_1} + (2\nu\kappa^2)^2 E_{t=t_1}\right](t - t_1)^2 + \cdots. \quad (5\text{-}207)$$

Equation (5-207) is used in conjunction with available experimental data at an initial time [35] in an attempt to calculate the variation with time of $E(\kappa)$ and thus of $\overline{u_i u_i}$. However, with the available initial data ($E_{t=t_1}$ (Fig. 5-20), $T_{t=t_1}$, and $V_{t=t_1}$), reasonable results are not obtained except at small times (Fig. 5-21). It thus appears that, in order to obtain good results by using a simple Taylor series, initial statistical information of much higher order than that which is available has to be given. Thus, an alternative approach that makes more efficient use of the initial statistical information and also incorporates additional information from the equations of motion is considered.

Workable formulation for the development of turbulence from a given initial state. In order to obtain a more efficient means for calculating the evolution of turbulence than by a Taylor series in time, we consider an iterative solution of the Navier-Stokes equations similar to that in ref. [57]. In addition to the initial statistical information and calculated time derivatives, we then have information about the form of the decay law from the equations of motion.

Although attention is confined to determinate initial conditions in ref. [57], for the present purposes we can just as well assume the initial velocity fluctuations to be random or turbulent. Thus, we consider a field of homogeneous turbulence to be made up of a very high density of eddies or harmonic disturbances in wavenumber space. For all practical purposes then, because the density of disturbances is very high, the spectrum of the turbulence can be considered continuous. The velocity and pressure at any point in the field are given by

$$\frac{\partial u_i}{\partial t} - \nu \frac{\partial^2 u_i}{\partial x_k \partial x_k} = -\frac{1}{\rho}\frac{\partial \sigma}{\partial x_i} - \frac{\partial(u_i u_k)}{\partial x_k} \quad (5\text{-}208)$$

FOURIER ANALYSIS, SPECTRAL FORM OF THE CONTINUUM EQUATIONS **169**

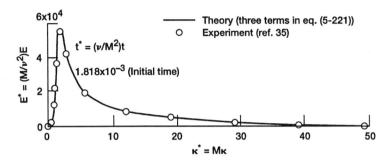
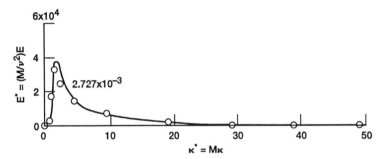
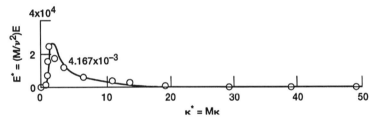

Figure 5-20 Comparison of theory with experiment of ref. [35] for decay of turbulent-energy spectra.

and

$$\frac{1}{\rho}\frac{\partial^2 \sigma}{\partial x_k \partial x_k} = -\frac{\partial^2 (u_k u_l)}{\partial x_k \partial x_l}. \tag{5-209}$$

The latter equation is obtained by taking the divergence of eq. (5-208) and applying the continuity equation.

From the spectrum of harmonic disturbances we arbitrarily select two cosine terms with wavevectors $\boldsymbol{q}$ and $\boldsymbol{r}$. Then, the velocity associated with those disturbances is

$$u_i^{cc} = a_i \cos \boldsymbol{q} \cdot \boldsymbol{x} + b_i \cos \boldsymbol{r} \cdot \boldsymbol{x}, \tag{5-210}$$

where the superscript cc on the velocity indicates that it depends on two cosine terms. The results that follow are the same if two sine terms or a sine and a cosine term are considered. If u_i^{cc} is substituted for u_i in the right sides of eqs. (5-208) and (5-209), the time variations of a_i and b_i plus additional harmonic terms are obtained. If we then substitute that new expression into eqs. (5-208) and (5-209), another expression containing

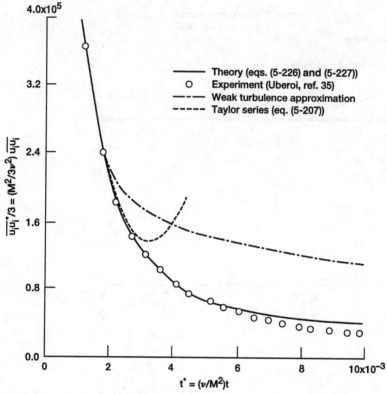

Figure 5-21 Comparison of theory with experiment of ref. [35] for decay of average component of velocity variance.

still more harmonic terms is obtained. In each approximation, the linear terms of the Navier-Stokes equations are considered as unknown and the nonlinear terms as known from the preceding approximation. As shown in ref. [57], continuation of this process leads to

$$u_i^{cc} = \sum_\kappa \left(A_{i,\kappa}^{c'} \cos \kappa \cdot x + A_{i,\kappa}^{s'} \sin \kappa \cdot x \right), \tag{5-211}$$

where

$$A_{i,\kappa}^{c'} = \sum_q a_{i,\kappa,q}^c \exp\left[-b_{\kappa,q}^c (t - t_1)\right] \tag{5-212}$$

and

$$A_{i,\kappa}^{s'} = \sum_r a_{i,\kappa,r}^s \exp\left[-b_{\kappa,r}^s (t - t_1)\right]. \tag{5-213}$$

Comparison of eqs. (5-211) through (5-213) with the first and second approximations in ref. [57] shows that $b_{\kappa,1}^c = b_{\kappa,1}^s = \nu \kappa^2$. Also, we note that because the two harmonic components in eq. (5-210) were selected arbitrarily, expressions similar to eqs. (5-211) through (5-213) are obtained for any other two components. But the nonlinear interaction of any number of harmonic components can be expressed as the sum of the interactions of pairs of components (eqs. [37] and [38] of ref. [57]). Thus u_i, the velocity resulting from all the harmonic components, is of the form of eqs. (5-211) through (5-213) and can be written as

$$u_i = \sum_\kappa \left(A^c_{i,\kappa} \cos \kappa \cdot x + A^s_{i,\kappa} \sin \kappa \cdot x \right), \tag{5-214}$$

where

$$A^{()}_{i,\kappa} = a^{()}_{i,\kappa,1} \exp\left[-\nu\kappa^2(t-t_1)\right] + \sum_{\substack{q \\ q \neq 1}} a^{()}_{i,\kappa,q} \exp\left[-b^{()}_{\kappa,q}(t-t_1)\right]. \tag{5-215}$$

The summations in eqs. (5-214) and (5-215), of course, contain more terms by many orders of magnitude than those in eqs. (5-211) through (5-213). Because the initial conditions are random, the quantities $A^{()}_{i,\kappa}$, $a^{()}_{i,\kappa,q}$, and $b^{()}_{\kappa,q}$ are assumed to be random variables. The space-averaged value of u_i^2 (no sum on i) is obtained from eq. (5-214) by squaring, integrating over a cycle, and using the orthogonality property of sines and cosines. This gives

$$\overline{u_i^2} = \sum_\kappa \frac{1}{2}\left[\left(A^c_{i,\kappa}\right)^2 + \left(A^s_{i,\kappa}\right)^2\right], \tag{5-216}$$

where

$$\left[A^{()}_{i,\kappa}\right]^2 = \left[a^{()}_{i,\kappa,1}\right]^2 \exp[-2\nu\kappa^2(t-t_1)] + \sum_{\substack{q \\ q \neq 1}} \left[a^{()}_{i,\kappa,q}\right]^2 \exp[-2b^{()}_{\kappa,q}(t-t_1)]$$

$$+ \sum_{\substack{q,r \\ q \neq r}} a^{()}_{i,\kappa,q} a^{()}_{i,\kappa,r} \exp\{-2[b^{()}_{\kappa,q}+b^{()}_{\kappa,r}](t-t_1)\}. \tag{5-217}$$

According to eq. (5-217), $(A^c_{i,\kappa})^2$ and $(A^s_{i,\kappa})^2$ in eq. (5-216) have the same form, so that we need not carry along the superscripts c and s.

We want to obtain an averaged form of eq. (5-217) that is a smoothed function of the magnitude of the vector κ (but not of its direction). In order to do that, we divide the interval of $\kappa = (\kappa_i \kappa_i)^{1/2}$ over which disturbances occur into a large number of small increments $\Delta\kappa$. The terms in $\sum_\kappa$ in eqs. (5-216) and (5-217) are divided into groups, each of which corresponds to a particular $\Delta\kappa$. Note that, although the magnitudes of the various vectors lying in a particular $\Delta\kappa$ are approximately equal, their directions, of course, can vary. The group of terms corresponding to each $\Delta\kappa$ is then subdivided into groups in each of which the values of the $b_{i,\kappa,q}$ in $\sum_{\substack{q \\ q \neq 1}}$ do not every appreciably from a value of $b_s(\kappa)$. The index s designates a particular increment in the values of the $b_{i,\kappa,q}$. Also, for each s, $a^2_{i,\kappa,q}$ has an average value that we designate by $\langle a^2_{i,\kappa}\rangle_s$. The summation $\sum_{\substack{q \\ q \neq 1}}$ in eq. (5-217), which applies to a particular κ, then is replaced by

$$\sum_s n_{s,(i)} \langle a^2_{i,\kappa}\rangle_s(\kappa) \exp[-2b_s(\kappa)(t-t_1)]$$

which applies to a particular $\Delta\kappa$, and where $n_{s(i)}$ is the number of terms in $\sum_{q\neq 1}^{q}$ that are assigned to the group s for the component i. The parentheses around i indicate that there is no summation on that subscript. A similar regrouping can be carried out for the terms in $\sum_{q\neq 1}^{q,r}$. However, that summation turns out to be zero, if we assume that the random $a_{i,\kappa}$ are uncorrelated, because $\langle a_{i,\kappa,q}a_{i,\kappa,r}\rangle_s$ will be zero for $q \neq r$. Then the average value of $A_{i,\kappa}^2$ in the increment $\Delta\kappa$ becomes (see eq. [5-217])

$$\langle A_{i,\kappa}^2\rangle(\kappa) = \langle a_{i,\kappa,1}^2\rangle(\kappa)\exp[-2\nu\kappa^2(t-t_1)]$$
$$+ \sum_s \left(\frac{n_s}{n_\kappa}\right)_{(i)} \langle a_{i,\kappa}^2\rangle_s(\kappa)\exp[-2b_s(\kappa)(t-t_1)], \quad (5\text{-}218)$$

where n_κ is the number of terms in $\sum_{q\neq 1}^{q}$ that lie in $\Delta\kappa$. The expression for $\overline{u_i^2}$ (eq. [5-216]) then becomes

$$\overline{u_i^2} = \sum_\kappa \Big\{ \langle a_{i,\kappa,1}^2\rangle(\kappa)\exp[-2\nu\kappa^2(t-t_1)]$$
$$+ \sum_s \left(\frac{n_s}{n_\kappa}\right)_{(i)} \langle a_{i,\kappa}^2\rangle_s(\kappa)\exp[-2b_s(\kappa)(t-t_1)] \Big\} \quad (5\text{-}219)$$

To obtain an expression for the energy spectrum function E, we note that (eq. [5-166])

$$\frac{1}{2}\overline{u_i u_i} = \int_0^\infty E\, d\kappa, \quad (5\text{-}166)$$

where $\overline{u_i u_i} = \overline{u_1^2} + \overline{u_2^2} + \overline{u_3^2}$. Equations (5-219) and (5-166) then give

$$\int_0^\infty E\, d\kappa = \sum_\kappa \frac{1}{2} \Big\{ \frac{\langle a_{i,\kappa,1}^2 a_{i,\kappa,1}\rangle}{\Delta\kappa} \exp[-2\nu\kappa^2(t-t_1)]$$
$$+ \sum_s \left(\frac{n_s}{n_\kappa}\right)_i \frac{\langle a_{i,\kappa}^2\rangle_s}{\Delta\kappa} \exp[-2b_s(t-t_1)] \Big\} \Delta\kappa, \quad (5\text{-}220)$$

where there is now a summation on i. If $\Delta\kappa$ is very small, we can write, to a very good approximation,

$$E(\kappa) = B^2(\kappa)\exp[-2\nu\kappa^2(t-t_1)] + \sum_s B_s^2(\kappa)\exp[-2b_s(\kappa)(t-t_1)]. \quad (5\text{-}221)$$

Equation (5-221) gives the evolution in time of the energy spectrum function from an initial state that is specified by B, B_s, and b_s in the equation.

As shown in the previous section, if the turbulence is specified at an initial instant, the time derivatives of E can be calculated at that instant by using the Fourier-transformed correlation equations. Thus, it is desirable to write B, B_s, and b_s in eq. (5-221) in terms of E and its derivatives at the initial time. That can be done by evaluating eq. (5-221) and its time derivatives at $t = t_1$ and solving the resulting system of equations for the B, B_s, and b_s.

In what follows, we first retain only two terms of eq. (5-221). Equation (5-221) then can be written conveniently as

$$E = E_{t=t_1}\{C(\kappa)\exp[-2\nu\kappa^2(t-t_1)] + (1-C)\exp[-2b(\kappa)(t-t_1)]\}, \quad (5\text{-}222)$$

where $0 \le C \le 1$.

For $C = 1$ eq. (5-222) reduces to the well-known expression for the final period of decay (see eqs. [5-121] and [5-157] for $\varphi_{ii}[\kappa] = \varphi_{ii}[\kappa]$). For the general case ($C \ne 1$) we can determine C and b in terms of the first and second derivatives of eq. (5-222) for $t = t_1$ and then evaluate those derivatives by using the two-point spectral equations (see eqs. [5-204] through [5-206]). The following procedure turns out to be simpler, however. By substituting eq. (5-222) into the spectral equation (5-204) we get for the energy-transfer term

$$T = 2(1-C)(\nu\kappa^2 - b)E_{t=t_1}\exp[-2b(t-t_1)] = T_{t=t_1}\exp[-2b(t-t_1)]. \quad (5\text{-}223)$$

Then,

$$\frac{\partial T}{\partial t} = -2bT_{t=t_1}\exp[-2b(t-t_1)] = \left(\frac{\partial T}{\partial t}\right)_{t=t_1}\exp[-2b(t-t_1)]. \quad (5\text{-}224)$$

Comparing the last two members of eq. (5-224) and using eq. (5-205) gives

$$b = \nu\kappa^2 - \frac{V_{t=t_1}}{2T_{t=t_1}}. \quad (5\text{-}225)$$

From eqs. (5-223) and (5-225) we have

$$C = 1 - \frac{T_{t=t_1}^2}{V_{t=t_1}E_{t=t_1}}. \quad (5\text{-}226)$$

Equations (5-222) and (5-223) then become

$$E = E_{t=t_1}\left\{C\exp[-2\nu\kappa^2(t-t_1)] + (1-C)\exp\left[-2\left(\nu\kappa^2 - \frac{V_{t=t_1}}{2T_{t=t_1}}\right)(t-t_1)\right]\right\} \quad (5\text{-}227)$$

and

$$T = T_{t=t_1}\exp\left[-2\left(\nu\kappa^2 - \frac{V_{t=t_1}}{2T_{t=t_1}}\right)(t-t_1)\right]. \quad (5\text{-}228)$$

From eq. (5-205)

$$V = V_{t=t_1}\exp\left[-2\left(\nu\kappa^2 - \frac{V_{t=t_1}}{2T_{t=t_1}}\right)(t-t_1)\right], \quad (5\text{-}229)$$

where C is given by eq. (5-226).

Equations (5-227) and (5-228) are obtained by retaining two terms on the right side of eq. (5-221). We consider next a higher-order approximation in which three terms are retained in that equation. If eq. (5-221), with three terms retained, is substituted into eq. (5-204), we get for T

$$T = 2B_1^2(\kappa^2 - b_1)\exp[-2b_1(t-t_1)] + 2B_2^2(\kappa^2 - b_2)\exp[-2b_2(t-t_1)]. \quad (5\text{-}230)$$

174 TURBULENT FLUID MOTION

Equation (5-230) contains four unknown functions, which are to be determined by the initial conditions. For that purpose we use eq. (5-230) and its first three derivatives evaluated at $t = t_1$. Thus, we obtain

$$b_1 = -\frac{T_1 T_2 - T_{t=t_1} T_3}{4(T_1^2 - T_{t=t_1} T_2)} + \left\{ \left[\frac{T_1 T_2 - T_{t=t_1} T_3}{4(T_1^2 - T_{t=t_1} T_2)} \right]^2 - \frac{T_2^2 - T_1 T_3}{4(T_1^2 - T_{t=t_1} T_2)} \right\}^{1/2}, \quad (5\text{-}231)$$

$$b_2 = -\frac{T_1 T_2 - T_{t=t_1} T_3}{4(T_1^2 - T_{t=t_1} T_2)} - \left\{ \left[\frac{T_1 T_2 - T_{t=t_1} T_3}{4(T_1^2 - T_{t=t_1} T_2)} \right]^2 - \frac{T_2^2 - T_1 T_3}{4(T_1^2 - T_{t=t_1} T_2)} \right\}^{1/2}, \quad (5\text{-}232)$$

$$B_1^2 = \frac{2b_2 T_2 + T_3}{16b_1^2(\kappa^2 - b_1)(b_2 - b_1)}, \quad (5\text{-}233)$$

and

$$B_2^2 = \frac{2b_1 T_2 + T_3}{16b_2^2(\kappa^2 - b_2)(b_1 - b_2)}, \quad (5\text{-}234)$$

where T_1, T_2, and T_3 are the first, second, and third time derivatives of $T(\kappa)$ at $t = t_1$. The first derivative T_1 can be written in terms of the functional $V_{t=t_1}$, which gives a representation of three-point spectral quantities (see eq. [5-205]). Equations for higher-order functionals can be obtained by the procedure used for obtaining eq. (5-205) for V. Thus, by using the four-point spectral equations (10), (12), and (18) of ref. [22] we get

$$\frac{\partial V}{\partial t} = -2\nu\kappa^2 V + R \quad (5\text{-}235)$$

where R is a functional of three- and four-point spectral quantities. Similarly,

$$\frac{\partial R}{\partial t} = -2\nu\kappa^2 R + S \quad (5\text{-}236)$$

where S is a functional of three-, four-, and five-point spectral quantities. By using eqs. (5-205), (5-235), and (5-236), the first, second, and third time derivatives of $T(\kappa)$ at $t = t_1$ in eqs. (5-231) through (5-234) can be written in terms of higher-order spectral quantities as

$$T_1 = -2\nu\kappa^2 T_{t=t_1} + V_{t=t_1}, \quad (5\text{-}237)$$

$$T_2 = (2\nu\kappa^2)^2 T_{t=t_1} - 4\nu\kappa^2 V_{t=t_1} + R_{t=t_1}, \quad (5\text{-}238)$$

and

$$T_3 = -(2\nu\kappa^2)^3 T_{t=t_1} + 3(2\nu\kappa^2)^2 V_{t=t_1} - 6\nu\kappa^2 R_{t=t_1} + S_{t=t_1}. \quad (5\text{-}239)$$

Results and discussion. A comparison between the experimental data of Uberoi [35] and the present theory (eqs. [5-227] through [5-229]) is given in Figs. 5-20 through 5-23. (Another pertinent experimental investigation is that of Van Atta and Chen [36] They measured directly the individual terms in the two-point spectral equation; however, their data are for only one time.) The comparison in Figs. 5-20 through 5-23 i

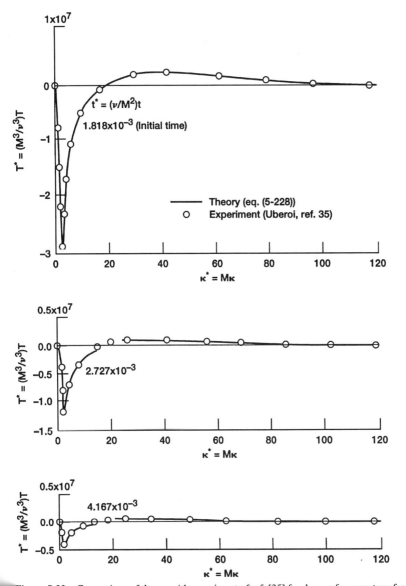

Figure 5-22 Comparison of theory with experiment of ref. [35] for decay of energy-transfer spectra.

made for an initial time corresponding to $X/M = 48$ in the experiment of ref. [35] ($t^* = (\nu/M^2)t = 0.001818$, X is the distance downstream from the grid, and M is the mesh size of the grid). For the initial specification of the turbulence, values of $E(\kappa)$ and $T(\kappa)$ are obtained from Figs. 5, 9, and 10 of ref. [35]. Initial values of V are not given directly in ref. [35] but are estimated from the decay data for $T(\kappa)$ and eq. (5-205). Except for experimental error those values are the same as those that might have been measured directly.

176 TURBULENT FLUID MOTION

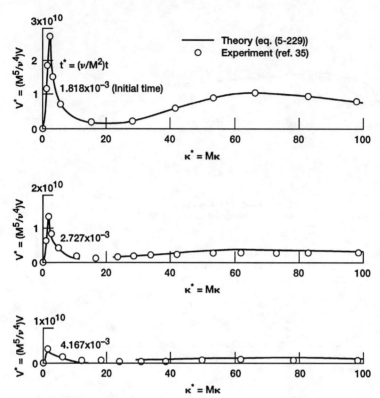

Figure 5-23 Comparison of theory with experiment of ref. [35] for decay of higher-order spectral quantity V (eq. [5-205]).

The agreement between the predicted and experimental energy spectra for the same initial conditions (see Fig. 5-20) appears to be quite good considering the difficulty of the measurements. The calculation of the experimental values of E required the differentiation of measured one-dimensional spectra and an assumption of isotropy.

Predicted and experimental values for the decay of $\overline{u_i u_i}$ are plotted in Fig. 5-21. The agreement between theory and experiment is excellent for values of t^* up to about 0.006. (Note that spectra are measured only for values of t^* between 0.00182 and 0.00417.) Elimination of the moderate deviation for $t^* > 0.006$ might require a higher-order theory (more terms in eq. [5-221]), together with additional initial statistical information. Alternatively, the deviation might be due to the amplification at large times of slight inaccuracies in the measured initial spectra. The theoretical values for t^* less than 0.00182 are calculated by working backwards from the measured initial spectra. Also included in Fig. 5-21 is a Taylor series solution that uses the same initial information as the exponential series and the curve for the weak turbulence approximation. It might be pointed out that the curve for the weak turbulence approximation is not the $-5/2$-power decay law usually given for the final period [4] but is the curve obtained by using the measured initial energy spectrum and eq. (5-222) with $C = 1$.

Spectra for the energy-transfer term $T(\kappa)$ are plotted in Fig. 5-22. The experimental and theoretical curves are in good agreement except near the value of κ, where $T_{t=t_1}$ changes sign. The deviation there results from a mathematical singularity in eq. (5-228) if $T_{t=t_1} = 0$. However, that deviation does not seem to be serious, because the real physical curve in that region can be estimated easily graphically or by using an interpolation formula. This is particularly true because it is known that the total area enclosed by the $T(\kappa)$ spectrum should be zero (eq. [5-162]). It appears likely that the difficulty can be eliminated if another term is retained in eq. (5-221). (More is said about that possibility in the next paragraph.) The deviation also carries over to some extent into the results for $E(\kappa)$ and $\overline{u_i u_i}$. However, if one does not use values of κ close to the point at which $T_{t=t_1}$ changes sign for calculating $E(\kappa)$ and $\overline{u_i u_i}$, the inaccuracies in those quantities are small. It appears that the overall agreement between theory and experiment obtained by using eqs. (5-226) through (5-228) should be considered encouraging.

For the sake of completeness, spectra of the functional V (eqs. [5-205] and [5-229]), the third initial condition specified for the turbulence, are plotted in Fig. 5-23. The agreement between theory and experiment is probably within the uncertainty in estimating V from the decay data in ref. [35], except in the vicinity of the point at which $T_{t=t_1}$ changes sign. Thus, the theory predicts the evolution in time of $E(\kappa)$, $T(\kappa)$, and $V(\kappa)$, if those quantities are specified at an initial time.

We have not been able to apply a higher-order theory to Uberoi's data, that is, to evaluate three instead of two terms in eq. (5-221) by using the initial data given in his article. However, we can apply a higher-order theory to the analysis in section 5-3-2-2, because for that analysis we can, in effect, calculate as much initial information as is desired. That analysis neglects quadruple-correlation terms in the three-point correlation equations and should apply, for a particular set of initial conditions, at times somewhat before the final period of decay. The initial conditions, as well as values at later times, are given by closed-form equations in that analysis and thus are better defined than may be possible in an experiment. For the present purposes, the analytical results from section 5-3-2-2 in fact might be thought of as experimental results in which the initial conditions are specified exactly (an analytical experiment).

The case considered here corresponds to Fig. 5-5. Values of dimensionless $E(\kappa)$, $T(\kappa)$, and time derivatives of $T(\kappa)$ for the initial specification of the turbulence ($t_1^* = 0.002$) are obtained from eqs. (5-165) and (5-161). We can eliminate the time derivatives of $T(\kappa)$ by introducing V (eq. [5-205]) and the higher-order functionals R and S (eqs. [5-235] and [5-236]). In the present case, those quantities all are representations of correlations of order no higher than the third, because terms involving correlations of higher order than the third are assumed negligible in the analysis of section 5-3-2-2.

Figure 5-24 gives a comparison between results for $T(\kappa)$ calculated from the present theory and those from section 5-3-2-2. The quantity J_0 is a constant related to conditions at $t_0^* = -0.00633$ in the equations of section 5-3-2-2. The starred quantities in Figs. 5-24 through 5-26 are the same as those in Figs. 5-20 through 5-23 if we let $J_0 = M^3 \nu^2$. As expected, if $T(\kappa)$ is calculated from eq. (5-228), the agreement with the results of section 5-3-2-2 is good except in the region in which $T_{t=t_1}$ changes sign. However, if a higher-order theory is used by retaining three terms in the expression for E (two terms

178 TURBULENT FLUID MOTION

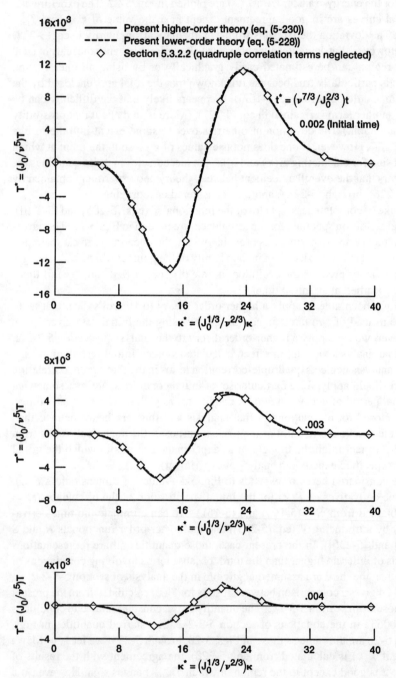

Figure 5-24 Comparison of present theory with that of section 5-3-2-2 for decay of energy-transfer spectrum

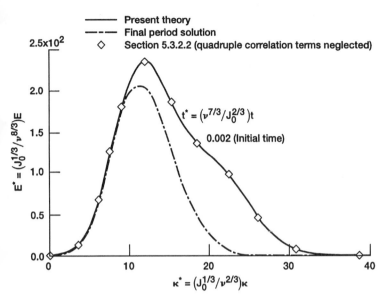
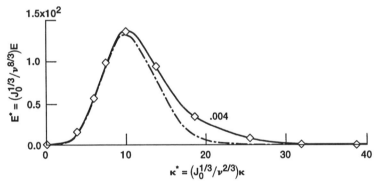
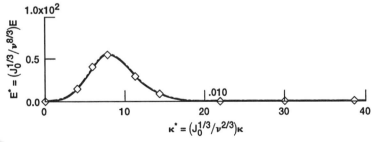

Figure 5-25 Comparison of present (higher-order) theory with that of section 5-3-2-2 for decay of turbulent-energy spectrum.

180 TURBULENT FLUID MOTION

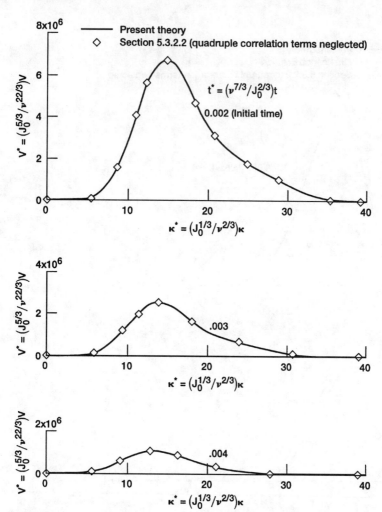

Figure 5-26 Comparison of present (higher-order) theory with that of section 5-3-2-2 for decay of higher-order spectral quantity V (eq. [5-205]).

in the expression for $T[\kappa]$—see eq. [5-230]), the agreement is excellent at essentially all values of κ. It might be expected that a similar improvement will be obtained in Fig. 5-22 if a higher-order theory can be used for comparison with the experimental data of Uberoi.

Because of the good agreement obtained for T in Fig. 5-24, one would expect the calculated energy spectra E also to be in good agreement with those from section 5-3-2-2. Figure 5-25 shows that this is indeed the case. The energy spectrum, in this case, decays in a highly nonsimilar fashion. In order to show the effects of spectral energy transfer on the energy spectrum, curves for the final period of decay (first term of eq. [5-165]) also are included in Fig. 5-25.

Figures 5-26 through 5-28 show plots for the decay of the higher-order spectral quantities V, R, and S. The agreement between the present higher-order theory and

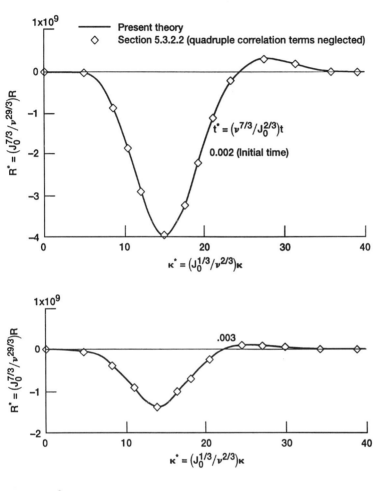

Figure 5-27 Comparison of present (higher-order) theory with that of section 5-3-2-2 for decay of higher-order spectral quantity R (eq. [5-235]).

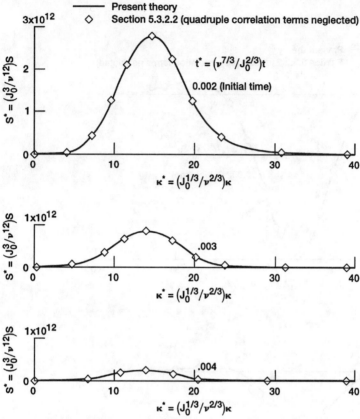

Figure 5-28 Comparison of present (higher-order) theory with that of section 5-3-2-2 for decay of higher-order spectral quantity S (eq. [5-236]).

the results of section 5-3-2-2 is very good. Although the effects of the singularity at $\kappa = 15.33$ are greater for these higher-order quantities than for the lower-order ones, they still are not apparent unless points very close to the singularity are used in plotting the curves. For points close to the singularity, an interpolation formula can be used. Thus, by specifying the initial conditions for E, T, V, R, and S, we can predict the evolution in time of those quantities by using the present higher-order theory. That is, the required number of initial conditions is no greater than the number of quantities whose decay we can predict.

The higher-order theory (three exponential terms retained in eq. [5-221] also can be compared with grid-turbulence data obtained in a water channel by Ling and Huang [38]. For that comparison, the experimental input can be conveniently obtained from an empirical equation for E (eq. [5-184]). The higher-order spectra were not measured directly in their experiment but can be calculated from their equation for E by using eqs. (5-204), (5-205), (5-235), and (5-236). Except for possible experimental error those values will be the same as those that might have been measured directly. The

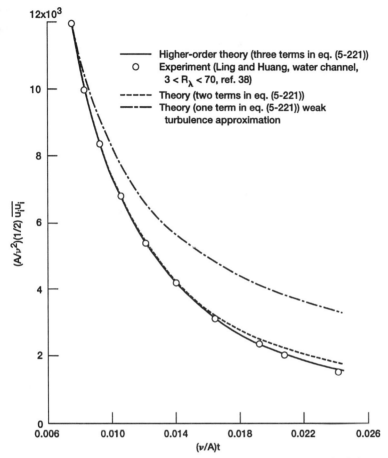

Figure 5-29 Comparison of theory with experiment of ref. [38] for decay of turbulent energy.

comparison is shown in dimensionless form in Figs. 5-29 through 5-34. The quantity A is an experimental constant (the proportionality constant in the power decay law for $\overline{u_i^2}$). It has the dimensions of length^2 and is related to conditions at time t_0 (eq. [5-185]). As in the preceding comparisons, unphysical singularities occur in the theoretical spectra at certain values of κ, particularly in the higher-order spectra. Thus, in the vicinity of those points, four-point interpolation formulas are used.

Figure 5-29 compares theory and experiment for the decay of turbulent energy if the initial state is specified at $(\nu/A)t_1 = 0.0075$. Theoretical curves are shown for one, two, and three exponential terms retained in eq. (5-221). The curve for three terms is in good agreement with the experiment for the whole decay period. The curve for two terms is in almost as good agreement. That is not the case for the spectra, for which only the curves for three terms agree closely (see the curves for E in Fig. 5-30). Comparison of the curve in Fig. 5-29 for one term retained (weak turbulence approximation) with the experimental curve shows the effect of inertia on the decay process. As in Fig. 5-21, the curve for the

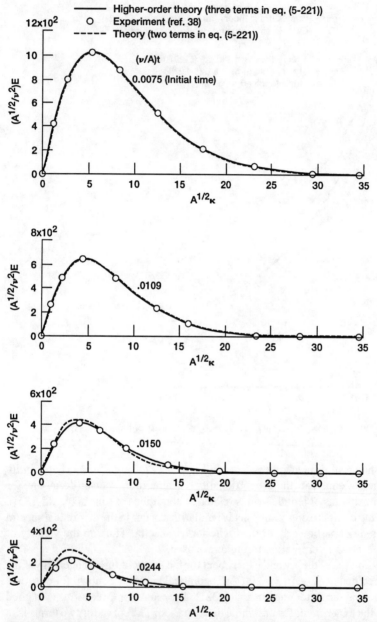

Figure 5-30 Comparison of theory with experiment of ref. [38] for decay of three-dimensional turbulent-energy spectra.

weak turbulence approximation in Fig. 5-29 is not the $-5/2$-power decay law usually given for the final period, because measured initial energy spectra are used here.

Figures 5-30 through 5-34 give a comparison of theory and experiment for the decay of the spectra used to specify the initial state of the turbulence at t_1. The curves indicate good agreement with the higher-order theory. That is, the theory is able to predict the decay of all of the spectra used to specify the initial turbulence, if three exponential terms are retained in eq. (5-221).

High–Reynolds-number turbulence. Thus far, we have given the basic theory for closure by specification of initial conditions and calculated results for low- and moderate-turbulence Reynolds numbers ($3 < R_\lambda < 70$). Here, we compare calculated results with the higher–Reynolds-number experimental data of Ling and Saad [37]. The Reynolds numbers for those data are high enough to obtain a $-5/3$-power region in the energy spectrum ($300 < R_\lambda < 800$). The exponential-series expression for the energy spectrum function E, an iterative solution of the Navier-Stokes equation, is given by eq. (5-221):

$$E(\kappa) = B^2(\kappa)\exp[-2\nu\kappa^2(t-t_1)] + \sum_s B_s^2(\kappa)\exp[-2b_s(\kappa)(t-t_1)].$$

Equation (5-221) gives the evolution in time of the energy spectrum from an initial state at time t_1, which is specified by the B, B_s, and b_s in the equation. The first term is the well-known expression for the decay of E in the final period (weak turbulence approximation). The rest of the terms in eq. (5-221) therefore give the contribution of inertia to E. In the present discussion we retain a maximum of four exponential terms in eq. (5-221). This is one more term than it is necessary to retain for the low– and moderate–Reynolds-number data considered in the previous section.

With four terms retained in eq. (5-221), we have to specify seven spectra at t_1 in order to evaluate the functions B, B_1, B_2, B_3, and b_1, b_2, and b_3. Evidently, we need more spectra to describe the initial turbulence at higher Reynolds numbers because a wider range of eddy sizes is excited, and the turbulence structure is more complicated than at lower Reynolds numbers.

In addition to $E(\kappa, t_1)$, we use spectrum functions designated by $T(\kappa)$, V, R, S, L, and M, all of which are specified at the initial time t_1. The quantity $T(\kappa)$ is the well-known energy-transfer function given by eq. (5-204):

$$T(\kappa, t) = \frac{\partial E(\kappa, t)}{\partial t} + 2\nu\kappa^2 E$$

The quantities V, R, S, L, and M are two-point functionals of three- to seven-point spectral quantities (see preceding sections). Each functional depends on a field of values of the multipoint spectral quantities. For instance,

$$V(\kappa) = \int_A \int_{-\infty}^{\infty} f[\beta_{ijk}(\kappa, \kappa'), \beta_{ijkl}(\kappa, \kappa')]\, d\kappa'\, dA(\kappa) = V[\beta_{ijk}(\kappa, \kappa'), \beta_{ijkl}(\kappa, \kappa')],$$
(5-240)

where κ and κ' are vectors in wavenumber space corresponding, respectively, to r and r' in physical space. The β_{ijk} and β_{ijkl} are Fourier transforms of $\overline{u_i u_j' u_k''}$ and $\overline{u_i u_j u_k' u_l''}$, where the u_i, u_j', and so forth are velocity components at various points, and the overbars

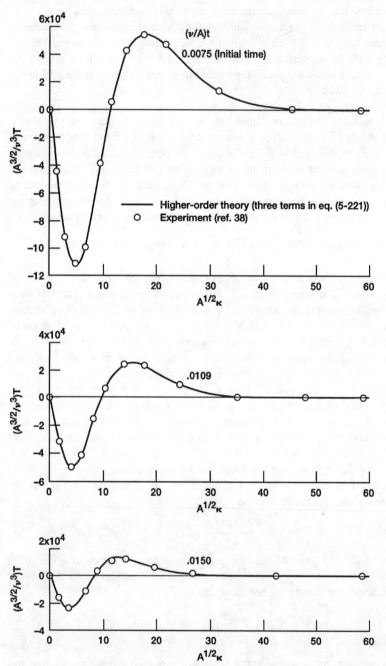

Figure 5-31 Comparison of theory with experiment of ref. [38] for decay of energy-transfer spectra.

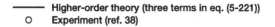
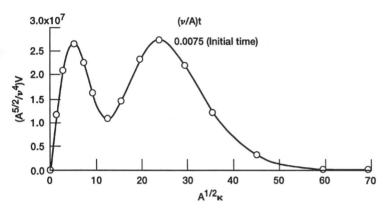
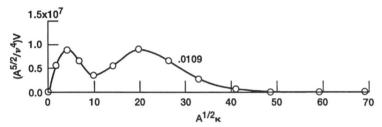
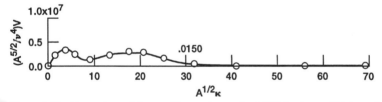

Figure 5-32 Comparison of theory with experiment of ref. [38] for decay of higher-order spectral quantity V (eq. [5-205]).

indicate averaged values. The element $d\kappa = d\kappa_1 d\kappa_2 d\kappa_3$, and dA is an element of surface area of radius κ. Equations similar to (5-240), but which involve a larger number of wavevectors, can be written for the other higher-order spectra.

By integrating a three-point spectral equation over κ' and $A(\kappa)$ (see preceding sections), we get, using eq. (5-240), eq. (5-205):

$$V(\kappa, t) = \partial T(\kappa, t)/\partial t + 2\nu\kappa^2 T.$$

Similarly, by performing integrations over wavenumber space of higher-order multipoint spectral equations, and using relations similar to (5-240), we get

$$R(\kappa, t) = \partial V(\kappa, t)/\partial t + 2\nu\kappa^2 V, \tag{5-235}$$

$$S(\kappa, t) = \partial R(\kappa, t)/\partial t + 2\nu\kappa^2 R, \tag{5-236}$$

$$L(\kappa, t) = \partial S(\kappa, t)/\partial t + 2\nu\kappa^2 S, \tag{5-241}$$

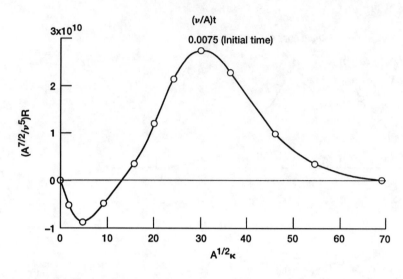
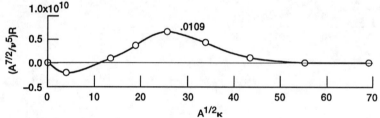
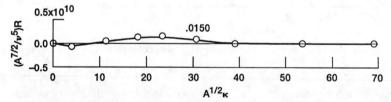

Figure 5-33 Comparison of theory with experiment of ref. [38] for decay of higher-order spectral quantity R (eq. [5-235]).

and

$$M(\kappa, t) = \partial L(\kappa, t)/\partial t + 2\nu\kappa^2 L. \qquad (5\text{-}242)$$

Equations (5-205) and (5-235) through (5-242) can be thought of as two-point alternatives to the multipoint spectral equations. They are much easier to work with because, although conceptually the spectra contained in them are functionals of multipoint spectral quantities, they are, for purposes of computation, simply functions of κ (and t). For

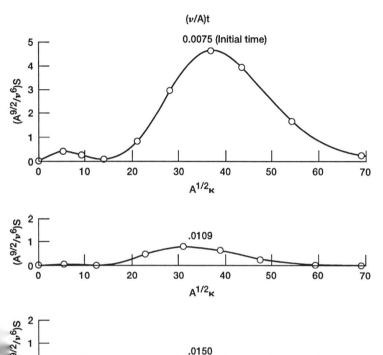

Figure 5-34 Comparison of theory with experiment of ref. [38] for decay of higher-order spectral quantity S (eq. [5-236]).

instance (see eq. [5-240]),

$$V[\beta_{ijk}(\kappa, \kappa'), \beta_{ijkl}(\kappa, \kappa')] = V(\kappa).$$

It is easy to show that V, R, and so forth can be obtained from eqs. (5-204), (5-240), (5-205), (5-235) through (5-242), and the time derivatives of E. We thus have a simple way of calculating initial functionals for a given set of data.

The higher-order spectral quantities are somewhat similar to $T(\kappa)$, inasmuch as they contain the effects of transfer between wavenumbers. However, they differ in that they also contain other effects, so that the areas under those spectra are not necessarily zero, as in the case of $T(\kappa)$.

For comparing the theory with the experiment of Ling and Saad [37] the experimental input can be conveniently obtained from an empirical equation of E, eq. (8) in their paper. The higher-order spectra were not measured directly in their experiment, but could be calculated from their equation for E and eqs. (5-204), (5-240), (5-205), and (5-235) through (5-242). Except for experimental error those values are the same as those that might have been measured directly. The B, B_s, and b_s in eq. (5-221) can be related to the initial spectra by successive differentiations of that equation and Ling and Saad's eq. (8) with respect to time, setting $t = t_1$, and using eqs. (5-204), (5-240), (5-205), and (5-235) through (5-242).

In might be pointed out that although we use the initial time variation of E in obtaining the initial conditions, we do so only because of the method of measurement of those conditions. Regardless of how they are measured, the initially specified spectra are, conceptually, functionals of multipoint spectral quantities that in principle could be measured directly at one initial time. However, the method used here is much simpler. Moreover, it does not require us to know E for the whole decay period but only at enough early times to calculate the required initial conditions, the latter being a much smaller amount of information.

The amount of required initial information still may seem large, and the theory therefore somewhat uninteresting. Indeed it would be so if the specification of many initial spectra were necessary to calculate the evolution of one or two lower-order quantities. The evolution of all of the initially specified spectra, however, can be calculated, as is seen subsequently. If the objective of a turbulence theory is to calculate the time evolution of as much statistical information as possible, then the structure of the present theory should be no disadvantage. Although it may not be particularly convenient from a practical standpoint, the initial specification of a number of interacting quantities appears necessary for the problem posed.

Before giving results obtained from eq. (5-221), we consider a Taylor series with a maximum of seven initial spectra (the same maximum number that are used with eq. [5-221]), and a modified Kovasznay-type closure (modified to include an effect of initial T) [52]. Results for those calculations are given in Fig. 5-35, in which the initial state is specified at $t_1^* = 0.0048$. The quantity u_i is a velocity component, the overbar indicates an averaged value, and an asterisk on a quantity indicates that it has been nondimensionalized by the kinematic viscosity and an experimental constant A, the proportionality constant in the power decay law for $\overline{u_1^2}$ (eq. [1] of ref. [37]). In contrast to the A for lower Reynolds numbers in Figs. 5-29 through 5-34 (see eq. [5-185]), the A in Figs. 5-35 through 5-42 has the dimensions ($[\text{length}]^2 [\text{time}]^{1.3}$). The turbulent energy $(1/2)\overline{u_i u_i}$ is obtained by integrating E over all wavenumbers. Figure 5-35 indicates that the Taylor-series results agree with experiment only for small times. Evidently many more terms (and initial spectra) are required in order to obtain accurate results for $\overline{u_i u_i}$ at large times. Note that even if enough terms are retained in the Taylor series to give accurate results for $\overline{u_i u_i}$, the decay of the higher-order spectra that then has to be specified initially cannot be calculated accurately. Thus, the use of a Taylor series in the present problem does not give a satisfactory solution, regardless of the number of terms retained.

The modified Kovasznay-type closure is in somewhat better agreement with experiment than the Taylor series, but at large times the agreement is still not good. This is in contrast to its good agreement at moderate and low Reynolds numbers [52]. It was introduced in ref. [52] in an effort to reduce the required number of initial spectra. Evidently that effort is successful only for moderate and low Reynolds numbers. It is possible, of course, that a more sophisticated method might be more successful (e.g., see ref. [60])

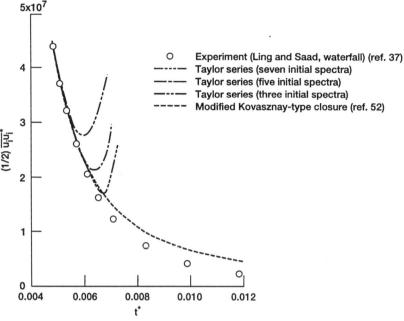

Figure 5-35 Comparison of theories with the high–Reynolds-number experiment of ref. [37] for decay of turbulent energy ($300 < R_\lambda < 800$). Turbulence equations closed by specification of initial conditions using a Taylor series and by a modified Kovasznay-type closure.

A comparison between theory and experiment using the exponential series (eq. [5-221]) is given in Figs. 5-36 through 5-42, where the initial state again is specified at $t_1^* = 0.0048$. As is the case for the lower Reynolds numbers considered in section 5-3-2-4, unphysical singularities occasionally occur in the present theoretical spectra. Inasmuch as the unphysical values are localized, particularly in the higher-order approximations, a smooth curve can be obtained without taking them into account.

Figure 5-36 gives a comparison between theory and experiment for the decay of turbulent energy. Theoretical curves are shown for one, two, three, and four exponential terms retained in eq. (5-221). The curve for four terms is in good agreement with experiment for the whole decay period.

Comparison of the curve in Fig. 5-36 for one term retained (weak turbulence approximation) with the curve for four terms shows the effect of inertia on the decay process. In contrast to the results for moderate Reynolds number (see Fig. 5-29), in which inertia and viscous effects were of the same order of magnitude, the inertia effects for the present high–Reynolds-number results are at least an order of magnitude greater than the purely viscous effects. Figure 5-37 compares results for the two ranges of Reynolds number. Most of the decay at high Reynolds numbers is due to inertial self-interaction of the turbulence, rather than to purely viscous effects. Another, perhaps better way of saying this is that at large Reynolds numbers most of the eddies making up the turbulence are inertial. That is, they would be absent if they were not excited (at higher wavenumbers) by the inertial or transfer term in the spectral equation. The ultimate dissipation of turbulent

192 TURBULENT FLUID MOTION

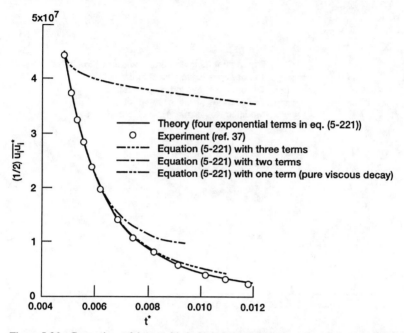

Figure 5-36 Comparison of theory with the high–Reynolds-number experiment of ref. [37] for decay of turbulent energy ($300 < R_\lambda < 800$). Turbulence equations closed by specification of sufficient initial conditions using an exponential series (eq. [5-221]).

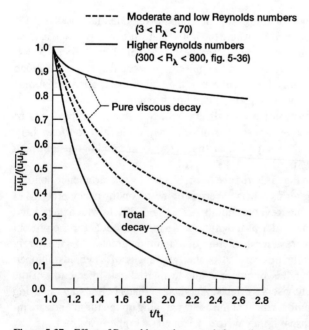

Figure 5-37 Effect of Reynolds number on turbulence-decay processes.

FOURIER ANALYSIS, SPECTRAL FORM OF THE CONTINUUM EQUATIONS 193

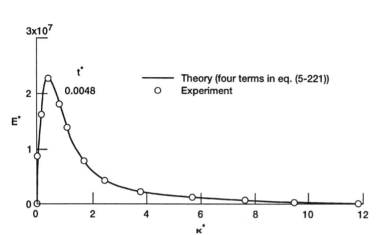
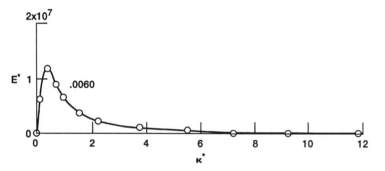
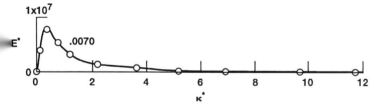
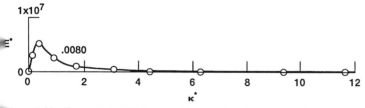

Figure 5-38 Comparison of higher-order theory with the high–Reynolds-number experiment [37] for decay of three-dimensional energy spectra.

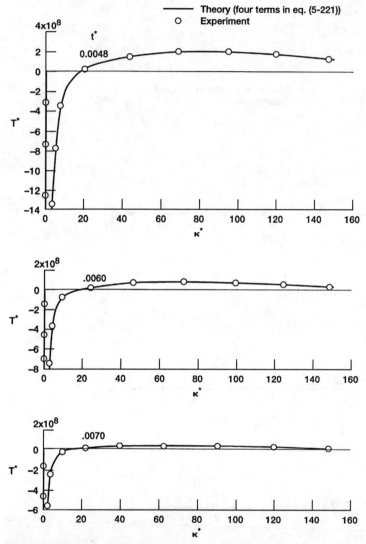

Figure 5-39 Comparison of higher-order theory with the high–Reynolds-number experiment [37] for decay of energy transfer spectra.

energy into heat still, of course, is produced by viscous action. Figures 5-38 and 5-39 show how the energy and the transfer spectra decay with time.

Figures 5-40 and 5-41 compare theory and experiment at a late time for all of the spectra that are initially specified to describe the initial turbulence. The prediction of the decay of all seven of the spectra that are specified initially is rather good. Thus, the present theory appears to solve an initial-value problem for high–Reynolds-number turbulence in which the decay of seven initially specified spectra is predicted. Note that the higher-order spectra occur at higher wavenumbers.

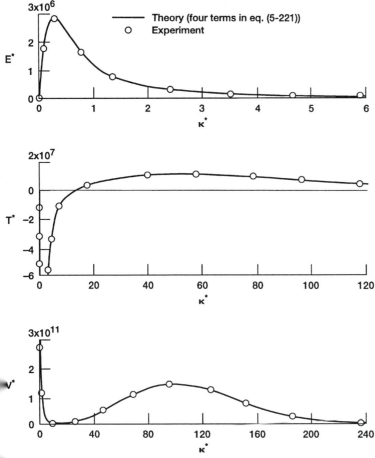

Figure 5-40 Comparison of higher-order theory with the high–Reynolds-number experiment [37] at a late time ($t^* = 0.01$) for the lower-order spectral quantities used to specify the initial turbulence.

Although the initial dissipation spectrum $\kappa^2 E$ is not specified independently, because of its importance in turbulence theory it is compared with experiment and with the energy spectrum at a late time in Fig. 5-42. Again, good agreement is indicated. The separation of the energy and dissipation spectra is good evidence that we are dealing with high Reynolds number turbulence.

Concluding remarks—the gap problem. If a homogeneous turbulent field is specified at an initial instant by its multipoint-velocity correlations (or their spectral equivalents), the initial time derivatives of those quantities can be calculated from the correlation or spectral equations. The development of the turbulence in time then can be obtained by using those derivatives in a series such as a Taylor power series. If the problem is formulated in this way, an assumption for closing the system of correlation equations is not required, because those equations are closed by the initially specified correlations

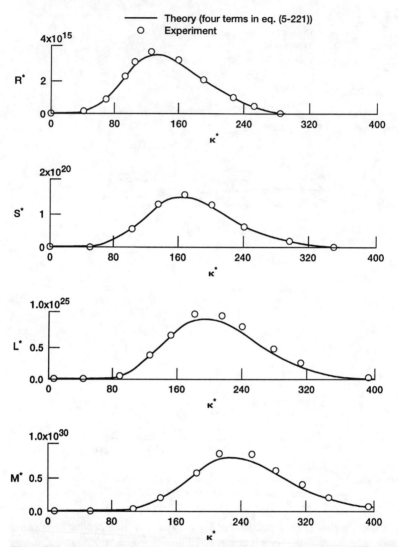

Figure 5-41 Comparison of higher-order theory with high–Reynolds-number experiment [37] at a late time ($t^* = 0.01$) for the higher-order spectral quantities used to specify the initial turbulence.

or spectral quantities. A Taylor-series expansion, however, does not give realistic results (except for small times) if the limited initial experimental data are used. An exponential series (eq. [5-221]) that is an iterative solution of the Navier-Stokes equations works much better.

By specifying n spectra at an initial time, where n is an odd integer greater than or equal to 3, we have been able to predict the evolution in time of those n spectra. We have not been able to obtain determinate results for $n < 3$, except for weak turbulence. From a practical standpoint it would be advantageous to calculate the evolution of the turbulence

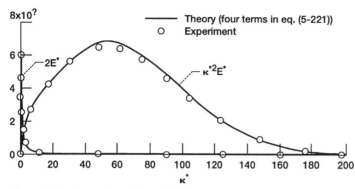

Figure 5-42 Comparison of theoretical energy dissipation spectrum at a late time ($t^* = 0.01$) with high-Reynolds-number experiment [37] and with energy spectra.

energy spectrum by specifying only that quantity initially. Unfortunately, because of the coupling between the members of the infinite hierarchy of multipoint correlation or spectral equations, it appears that we are not able to do that without making a closure assumption for the energy-transfer function, so that a satisfactory theory seems to require the initial specification of a number of interacting quantities.

The prediction of the evolution of the energy spectrum, in fact, requires the specification of an infinite number of initial multipoint correlations or spectra (or functionals of those quantities) [4]. If we claim that we should be able to predict the decay of the energy spectrum by specifying at an initial instant only that spectrum, we are in effect saying that the Fourier components of the energy spectrum decay independently, as in the final period.

Of course, in practice, one can specify only a finite number of the lower-order quantities. This has been called the "gap problem" [61]. It is the problem of bridging the gap between the infinite number of correlations that theoretically are necessary to calculate the evolution of the turbulence and a finite specifiable number of correlations. This is a difficulty that appears to be inherent in the problem of homogeneous turbulence and its initial specification. Most workers have attempted to bridge the gap by assuming that the initial distribution of turbulent fluctuations is exactly Gaussian (zero odd-order correlations). However, that is an artificial initial condition, probably never realized for real turbulence. The importance of accurate initial conditions is shown, for instance, by the data of Ling and Saad [37], measurements of which were made downstream from a turbulence-producing waterfall. The turbulence-decay law for the initial conditions produced by the waterfall is considerably different from that for initial conditions produced by a grid.

Here we bridge the gap simply by specifying a sufficient number of initial correlations or their spectral equivalents to calculate the evolution accurately. Fortunately, we do not have to specify the multipoint correlations or their spectral equivalents themselves (those are extremely difficult to measure), but only two-point functionals of the multipoint spectral quantities. It is seen that the evolution of all the quantities that are specified initially can be calculated. In that sense the solution may be considered complete. If, on the other hand, a large number of initial conditions are specified in order to predict the evolution of, say, one quantity, it might be objected that the initial conditions have been chosen to make the theory agree with experiment for that one quantity. However, that objection cannot be made if the evolution of all of the quantities that are specified initially can be calculated, as in the present theory. From a fundamental standpoint, the calculation of the evolution of those quantities is all that might be expected from a theory of evolving turbulence. Note that the present (higher-order) theory provides the only deductive procedure we have considered that works for high–Reynolds-number turbulence (see Figs. 5-35 through 5-42). Of course, with the continued improvement of computers and numerical methods, direct numerical simulation may offer an alternative.

5-3-2-5 Modified Kovasznay-type closure. In the concluding remarks of the previous section we mentioned the "gap problem"—the problem of bridging the gap between the infinite number of correlations or spectra that theoretically are necessary to calculate the evolution of turbulence and a finite specifiable number of correlations (or spectra). A theory for the decay of homogeneous turbulence is given that does not require a closure assumption in the usual sense, because the spectral equations are closed by the initial specification of the turbulence. By specifying n spectra at an initial time, where n is an odd integer greater than or equal to 3, the evolution in time of those n spectra is predictable. Good agreement with experiment and previous analytical results is obtained for $n = 3, 5$, or 7, depending on the Reynolds number, higher Reynolds numbers requiring more specified spectra.

It may be that the nature of the problem is such that three or more spectra have to be specified initially in order to calculate the evolution of any of them (except for weak turbulence). However, particularly in an applied problem, three or more initial spectra often are not available. In that case possible courses of action are first that the required initial spectra might be estimated from previous experimental or analytical results, or second that the introduction of physical or mathematical hypotheses into the theory might be allowed. In the remainder of the section the latter course of action is considered.

The analysis is limited to the two-point spectral equation, because a similar analysis that also uses the three-point equation would require the specification of at least three initial spectra. The two-point spectral equation is

$$\partial E(\kappa)/\partial t = -(\partial S(\kappa)/\partial \kappa) - 2\nu\kappa^2 E(\kappa), \qquad (5\text{-}243)$$

where E is the energy spectrum function, t is the time, S is the energy transfer at wavenumber κ, and ν is the kinematic viscosity. Equation (5-243) is the same as eq. (5-158) if

$$\frac{\partial S(\kappa)}{\partial \kappa} = -T(\kappa).$$

This equation ensures that eq. (5-162) is satisfied if $S(0) = S(\infty) = 0$.

In order to close eq. (5-243), a modification of Kovasznay's hypothesis [17] is used. this hypothesis is chosen mainly on the basis of simplicity, in order to illustrate how the initial specification of the turbulence might be simplified. Kovasznay's original hypothesis assumed that S is a function only of the energy at κ. Here, an effect of initial conditions is included. A sufficiently general functional relation for our purpose is

$$S = S(\kappa, E, I'), \qquad (5\text{-}244)$$

where $I'(\kappa)$ is a dimensionless function of initial conditions. From dimensional considerations

$$S/(E^{3/2}\kappa^{5/2}) = f[I'(\kappa)] = I(\kappa). \tag{5-245}$$

Then, the net rate of transfer of energy into a wavenumber band $d\kappa$ is

$$T\,d\kappa = -(\partial S/\partial \kappa)\,d\kappa \tag{5-246}$$

or, from eqs. (5-245) and (5-246)

$$T = -\frac{\partial}{\partial \kappa}(IE^{3/2}\kappa^{5/2}). \tag{5-247}$$

It is desired to determine $I(\kappa)$ as a function of initial conditions at $t = t_1$. Thus at t_1, eq. (5-247) becomes

$$T_1 = -\frac{\partial}{\partial \kappa}\left(IE_1^{3/2}\kappa^{5/2}\right)$$

or

$$I(\kappa) = -\int_0^\kappa T_1(\kappa')\,d\kappa'\left[\kappa^{5/2}E_1^{3/2}(\kappa)\right]^{-1} \tag{5-248}$$

where the subscript 1 refers to values at the initial time t_1. Equation (5-247) for the energy transfer function then becomes

$$T(\kappa) = \frac{\partial}{\partial \kappa}\left\{E^{3/2}(\kappa)\int_0^\kappa T_1(\kappa')\,d\kappa'\left[E_1^{3/2}(\kappa)\right]^{-1}\right\}. \tag{5-249}$$

Note that this expression for $T(\kappa)$ contains no adjustable constants or functions. Equation (5-243) becomes, by using eqs. (5-246) and (5-249),

$$\frac{\partial E}{\partial t} - \frac{\partial}{\partial \kappa}\left\{E^{3/2}(\kappa)\int_0^\kappa T_1(\kappa')\,d\kappa'\left[E_1^{3/2}(\kappa)\right]^{-1}\right\} = -2\nu\kappa^2 E. \tag{5-250}$$

Equation (5-250) is solved numerically for the initial conditions of the decay data in ref. [38], and the results for various times are compared with the data. The comparison is shown in dimensionless form in Figs. 5-43 through 5-45. The quantity A is a constant with the dimensions of length^2 [38] and is related to conditions at a time t_0, u_i is a velocity component, and the overbar indicates an averaged value. The curve for $S = 0$ (weak turbulence approximation) is included in Fig. 5-43 in order to show the effect of inertia on the decay process. Inertial and viscous effects appear to be of the same order of magnitude.

The plots for the decay of turbulent energy, the energy spectrum, and the energy-transfer spectrum show agreement between theory and experiment that is probably within the accuracy of the data. Thus, as in the case of the theory of section 5-3-2-4 we are able to predict the evolution of the spectra that are used for specifying the initial turbulence. By introducing a simple physical assumption, it evidently is possible to reduce the required number of initial spectra from three or more (preferably more) to two, E_1 and T_1. Moreover, as in our previous theory, the resulting equations contain no adjustable constants or functions. However, unlike the previous theory, the present equations do not predict the evolution of the higher–Reynolds-number data of ref. [37], except at short decay times (Fig. 5-35).

200 TURBULENT FLUID MOTION

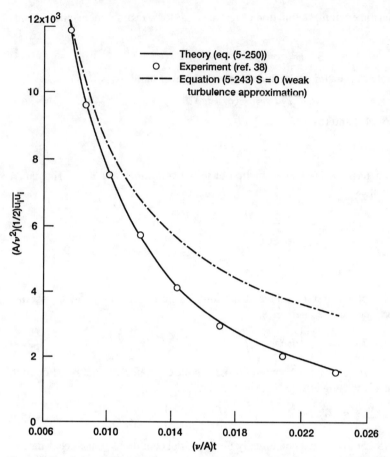

Figure 5-43 Comparison of theory (eq. [5-250]) and experiment [38] for decay of turbulent energy.

5-3-2-6 Further discussion of homogeneous turbulence without mean gradients—numerical solutions of the instantaneous equations.

Additional insights into the physics of homogeneous turbulence can be obtained by studying turbulent or turbulent-like numerical solutions of the unaveraged Navier-Stokes equations. We already have looked briefly at several numerical solutions for turbulence (ref. [33], and refs. [17], [49], and [50] of Chapter 4); it was pointed out that numerical solutions of the unaveraged equations are, in general, deductive, because no external information (modeling) is required. Here we consider several numerical solutions in somewhat greater detail [140].

The equations to be solved numerically, the incompressible Navier-Stokes equations, are

$$\frac{\partial u_i}{\partial t} + \frac{\partial (u_i u_k)}{\partial x_k} = -\frac{1}{\rho}\frac{\partial \sigma}{\partial x_i} + \nu \frac{\partial^2 u_i}{\partial x_k \partial x_k}, \quad (5\text{-}130)$$

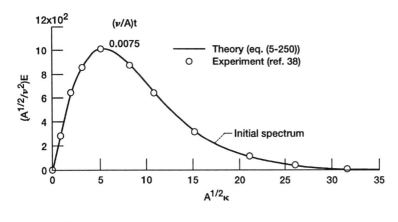
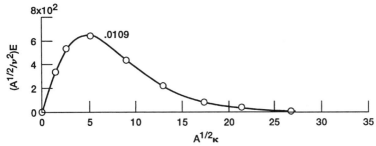
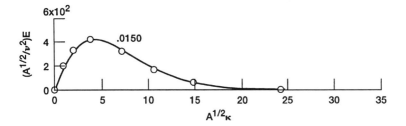
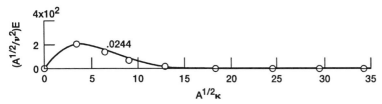

Figure 5-44 Comparison of theory (eq. [5-250]) and experiment [38] for decay of three-dimensional turbulent-energy spectra.

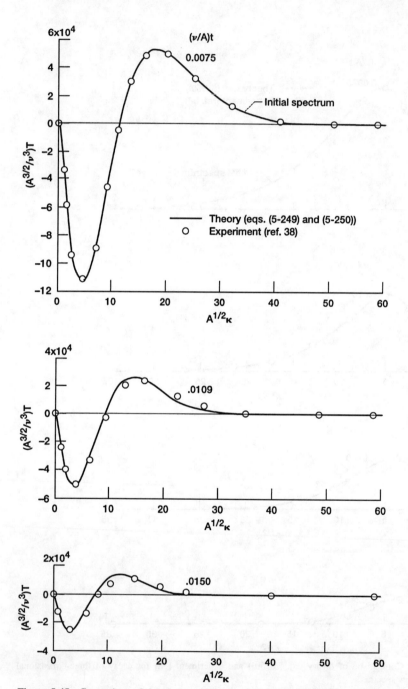

Figure 5-45 Comparison of theory (eqs. [5-249] and [5-250]) and experiment [38] for decay of energy-transfer spectra.

where the mechanical pressure (see eq. [3-14]) is given by the Poisson equation

$$\frac{1}{\rho}\frac{\partial^2 \sigma}{\partial x_l \partial x_l} = -\frac{\partial^2 (u_l u_m)}{\partial x_l \partial x_m} \quad (5\text{-}150)$$

and where, as usual, the subscripts can take on the values 1, 2, or 3, and a repeated subscript in a term indicates a summation. The quantities u_i and u'_j are instantaneous velocity components, x_i is a space coordinate, t is the time, ρ is the density, ν is the kinematic viscosity, and σ is the instantaneous (mechanical) pressure. Equation (5-150) is obtained by taking the divergence of eq. (5-130) and using continuity (eq. [3-4]).

The initial u_i in eqs. (5-130) and (5-150) are given at $t = 0$ by

$$u_i = \sum_{n=1}^{3} a_i^n \cos q^n \cdot x. \quad (5\text{-}251)$$

Equation (5-251), unlike the initial condition used for most numerical studies of turbulence, is nonrandom. But that condition might be considered comparable to the initial condition for turbulence generated experimentally by a regular grid in a wind tunnel. The quantity a_i^n is an initial velocity amplitude or Fourier coefficient of the velocity fluctuation, and q^n is an initial wavevector. In order to satisfy the continuity condition, we set (with a sum on i)

$$a_i^n q_i^n = 0. \quad (5\text{-}252)$$

For the present work let

$$a_i^1 = k(2, 1, 1), \quad a_i^2 = k(1, 2, 1), \quad a_i^3 = k(1, 1, 2), \quad (5\text{-}253)$$

and

$$q_i^1 = (-1, 1, 1)/x_0, \quad q_i^2 = (1, -1, 1)/x_0, \quad q_i^3 = (1, 1, -1)/x_0, \quad (5\text{-}254)$$

where k has the dimensions of a velocity and determines the intensity of the initial velocity fluctuation. The quantity x_0 is the single length scale of the initial velocity fluctuation (one over the magnitude of an initial wavenumber component). The quantities k and x_0, together with the kinematic viscosity ν and eqs. (5-253) and (5-254), then determine the initial Reynolds number $(\overline{u_0^2})^{1/2} x_0/\nu$, because the square of eq. (5-251), averaged over a spatial period, gives $\overline{u_0^2}$. In addition to satisfying the continuity equation (5-252), eqs. (5-251) through (5-254) give

$$\overline{u_1^2} = \overline{u_2^2} = \overline{u_3^2} = \overline{u_0^2} \quad (5\text{-}255)$$

at the initial time. Thus eqs. (5-251) through (5-254) give a particularly simple initial condition, in that we need specify only one component of the mean-square velocity fluctuation. Moreover, for no mean shear, they give an isotropic flow at later times, at least in the sense that $\overline{u_1^2} = \overline{u_2^2} = \overline{u_3^2}$, as is seen subsequently. In this way the present initial conditions differ from those of Taylor and Green [62]. Those initial conditions do not give isotropic results even at large times [24]. Note that it is necessary to have at least three terms in the summation in eq. (5-251) to satisfy eq. (5-255). We do not specify an initial condition for the pressure because it is determined by eq. (5-150) and the initial velocities.

In order to carry out numerical solutions subject to the initial condition given by eqs. (5-251) through (5-254), we use a stationary cubical grid with a maximum of 32^3 points and with faces at $x_i^* = x_i/x_0 = 0$ and 2π. For boundary conditions we assume periodicity for the fluctuating quantities; we consider turbulence (or a turbulent-like flow) in a box with periodic walls. That is, let

$$(u_i)_{x_j^*=2\pi+b_j^*} = (u_i)_{x_j^*=b_j^*} \tag{5-256}$$

and

$$\sigma_{x_j^*=2\pi+b_j^*} = \sigma_{x_j^*=b_j^*}, \tag{5-257}$$

where $b_j^* = b_j/x_0$, $x_j^* = x_j/x_0$, and b_j is a variable length. These equations are used to calculate numerical derivatives at the boundaries of the computational grid. For most of the results the spatial- and time-differencing schemes (which numerically conserve momentum and energy) are essentially those used by Clark et al. [33]. For the spatial derivatives in eqs. (5-130) and (5-150) we use centered fourth-order difference expressions (see, for example, ref. [63]). For instance, the fourth-order difference expression used for $\partial u_i/\partial x_k$ is

$$\left(\frac{\partial u_i}{\partial x_k}\right)_n = \frac{1}{12\Delta x_k}[(u_i)_{n-2} - 8(u_i)_{n-1} + 8(u_i)_{n+1} - (u_i)_{n+2}],$$

where Δx_k is the grid-point spacing, and the subscripts n, $n+1$, and so forth refer to grid points in the x_k direction. Fourth-order difference expressions often are considered more efficient than the usual second-order ones [64]. (Spectral methods devised by Orszag and associates are often still more efficient [64] but may be somewhat trickier to use.) Centered expressions (same number of points on both sides of n, see previous expression) can be used both at interior grid points and at the boundaries of the grid; if n refers to a point on a boundary, values for u_i outside of the grid that are required for calculating the numerical derivatives at the boundary are obtained from the boundary condition (eq. [5-256]).

For time-differencing we use a predictor–corrector method with a second-order (leapfrog) predictor and a third-order (Adams-Moulton) corrector [65]. If m represents a time step, and $(R_i)_m$ the right side of eq. (5-130), then at each grid point in space, the second-order leapfrog predictor for u_i at time step $m+1$ is

$$(u_i)_{m+1}^{(1)} = (u_i)_{m-1} + 2\Delta t(R_i)_m,$$

and the third-order Adams-Moulton corrector is

$$(u_i)_{m+1}^{(2)} = (u_i)_m + \frac{\Delta t}{12}[5(R_i)_{m+1}^{(1)} + 8(R_i)_m - (R_i)_{m-1}],$$

where Δt is the time increment. The quantity $(R_i)_{m+1}^{(1)}$ in the above corrector is calculated by using $(u_i)_{m+1}^{(1)}$ in the right side of eq. (5-130), where $(u_i)_{m+1}^{(1)}$ is calculated from the leapfrog predictor. Note that the leapfrog method (so called because it leaps over the time step m), although unstable for all Δt if used by itself for Navier-Stokes–type equations, is stable if used as a predictor.

The Poisson equation for the pressure (eq. [5-150]) is solved directly by a fast Fourier-transform method. This method of solution is found to preserve continuity quite well ($\nabla \cdot u \approx 0$) except near the ends of some of the runs, where the solutions begin to deteriorate. (Another indication of incipient solution deterioration near the ends of some of the runs is that the first three terms of eq. (5-255) are no longer accurately satisfied.)

Two known types of numerical instabilities can occur in the present solutions: a viscous instability connected with the first and last terms in eq. (5-130), which occurs if $\nu \Delta t/(\Delta x_k)^2$ is too large; and a convective instability connected with the first and second terms (or the first and third terms through eq. [5-150]), which occurs if $u_i \Delta t / \Delta x_k$ is too large. In these criteria Δt, Δx_k, and u_i are, respectively, a time step, distance step, and velocity. Thus, a particular solution should be numerically stable if, for a given Δx_k, the time step is sufficiently small. Numerical stability is typically obtained if the solution varies smoothly from time step to time step with no significant breaks in the slope from one step to the next. This is the case for all of the results given here. For the present solution very good temporal resolution is obtained automatically if Δt is sufficiently small to give numerical stability.

Temporal resolution is much easier to obtain than three-dimensional spatial resolution, which is more severely limited by the storage and power of computers. However, as is seen subsequently, good spatial resolution is obtained for Reynolds numbers and times not excessively large. Some of the averaged results are extrapolated to zero spatial mesh size in an effort to obtain greater accuracy. The fourth-order method of extrapolation (consistent with the fourth-order differencing used here) is given in ref. [66]. For the results given here the corrections are negligibly small.

Development of turbulent-like fluctuations. Figure 5-46 shows the calculated evolution of velocity-fluctuation components (normalized by the initial root-mean-square velocity) at two fixed points in space for the initial Reynolds number shown. Also plotted is the evolution of space-averaged root-mean-square velocity fluctuations. Because there is no input of energy, the fluctuating and space-averaged motions eventually decay to a state of rest. In spite of the nonrandom initial condition (eq. [5-251]), the velocity fluctuations have the appearance of those for a random turbulence. It is important to point out that the fluctuations are not due to numerical instability, because a large number of time steps (typically about 20) lies between changes of sign of du_i/dt.

We note that the spatially averaged values in Fig. 5-46 follow approximately the decay law $\overline{u^2} \sim t^{-n}$, where $n \sim 3$. This lies closer to the value for n of 3.3 observed for turbulence downstream of a waterfall [37] than to the value 1.2 generally observed for turbulence generated by flow through a grid in a wind tunnel [35]. The decay law is evidently very much dependent on the initial conditions for the turbulence.

Instantaneous velocity profiles on an off-center plane through the computational grid are plotted in Fig. 5.47 for various times. At the initial time $t = 0$ the profile is regular and has the shape given by eq. (5-251). However, it rapidly develops a turbulent-like (random) appearance as a result of the production of harmonics by the nonlinear self-interaction terms in eqs. (5-130) and (5-150). The profile shape is strongly time-dependent at early times. (Note that the vertical scale changes drastically as time increases.) The symbols are located at grid points, at which the instantaneous velocities are calculated. The eddies (fluctuations) are in general well-resolved numerically, particularly because a high-order differencing scheme is used in the numerical solution. (Fewer grid points

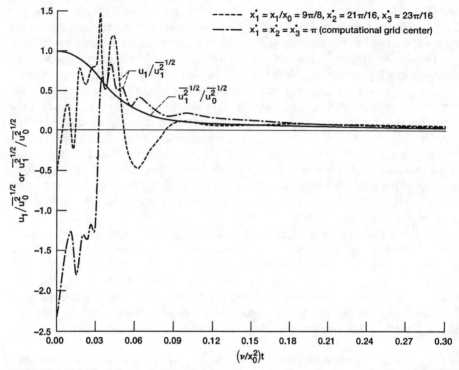

Figure 5-46 Calculated time evolution of turbulent-like velocity fluctuations (normalized by initial condition) for an initial Reynolds number $R_0 = \overline{u_0^2}^{1/2} x_0/\nu = 138.6$. Root-mean-square fluctuations are spatially averaged. 32^3 grid points.

may be required with a high-order scheme.) At any rate the curves appear to be well defined by the calculated values at the grid points. Further evidence that our calculated results are realistic is given in Fig. 5-48.

Note that the smaller eddies die out faster than the larger ones because of the higher shear stresses between the smaller eddies (see also the subsection on the interaction between spectral energy transfer and dissipation in section 5-3-2-2). For very large times essentially all of the higher harmonics have died out and the motion becomes linear. Then the profile assumes a regular shape not unlike that of the initial profile.

We have tried perturbing the initial condition to see if the flow in Figs. 5-46 and 5-47 is sensitively dependent on initial conditions, but the results are inconclusive. It appears, however, that the turbulent-like appearance of the flow is not due to sensitive dependence on initial conditions. That is not to say that sensitivity to small changes in initial conditions is not present in the flow, but the turbulent-like fluctuations evidently decay before such effects can be detected (before a perturbed flow has had a chance to diverge from an unperturbed one). This leads to the question of whether an initially nonturbulent low–Reynolds-number decaying flow, whether experimental or numerical ever can show the effects of sensitive dependence on initial conditions. The turbulent-like appearance of the curve in Figs. 5-46 and 5-47 is more likely caused by a proliferation of

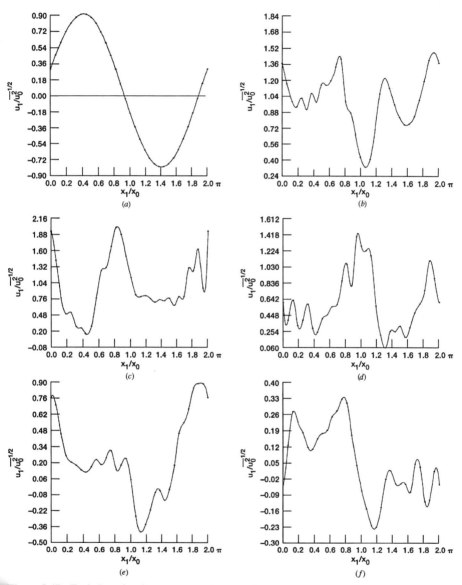

Figure 5-47 Evolution of an instantaneous velocity profile on an off-center plane through the computational grid. $x_2^* = x_2/x_0 = 21\pi/16$, $x_3^* = 23\pi/16$, $R_0 = \overline{u_0^2}^{1/2} x_0/\nu = 138.6$. Symbols are at grid points. (a) $(\nu/x_0^2)t = 0$; (b) $(\nu/x_0^2)t = 0.01443$; (c) $(\nu/x_0^2)t = 0.02887$; (d) $(\nu/x_0^2)t = 0.04330$; (e) $(\nu/x_0^2)t = 0.05773$; (f) $(\nu/x_0^2)t = 0.07217$.

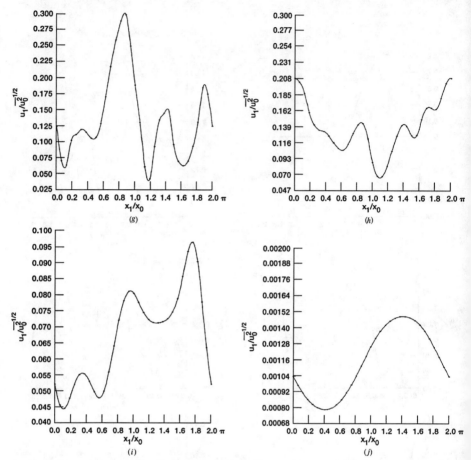

Figure 5-47(*Continued*) (g) $(\nu/x_0^2)t = 0.08660$; (h) $(\nu/x_0^2)t = 0.1155$; (i) $(\nu/x_0^2)t = 0.2454$; (j) $(\nu/x_0^2)t = 1.097$.

harmonic components by the nonlinear terms in eqs. (5-130) and (5-150). But if the decay is prevented by a forcing term, the effects of sensitive dependence on initial conditions can become important, even at the low Reynolds number in Figs. 5-46 and 5-47 (see Chapter 6). Moreover, we have shown that a higher–Reynolds-number decaying flow can be affected by sensitive dependence on initial conditions before the turbulent fluctuations have decayed [*J. Comp. Phys.*, June, 1992].

Evolution of mean-square velocity fluctuations. The effect of computational mesh size on space-averaged values of u_1^2, u_2^2, or u_3^2 (all three space-averaged values are equal) is shown in Fig. 5-48. Surprisingly good results for the decay are obtained, even with coarse grids.

Figure 5-49 shows the calculated evolution of mean-square velocity fluctuations (spatially averaged) for a series of initial Reynolds numbers. As the Reynolds number increases (ν and initial length scale x_0 held constant), the rate of decay of $\overline{u^2}$ increases

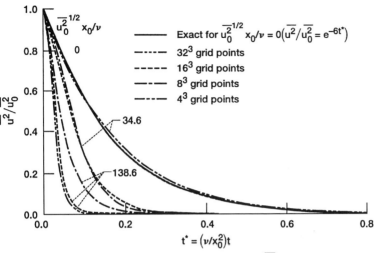

Figure 5-48 Effect of numerical mesh size on evolution of $\overline{u^2}$ at low and moderate Reynolds numbers. $\overline{u^2} \equiv \overline{u_1^2} = \overline{u_2^2} = \overline{u_3^2}$.

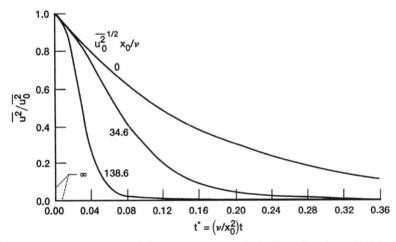

Figure 5-49 Calculated evolution of mean-square velocity fluctuations (normalized by initial value) for various initial Reynolds numbers. No mean shear. $\overline{u^2} \equiv \overline{u_1^2} = \overline{u_2^2} = \overline{u_3^2}$.

sharply, as in experimental turbulent flows (see Fig. 5-21). This can be attributed to the nonlinear excitation of small-scale turbulent-like fluctuations at the higher Reynolds numbers. The high shear stresses between the small eddies cause a rapid decay.

Microscales and nonlinear transfer of energy to smaller eddies. The development of the small-scale eddies is seen more clearly in Fig. 5-50, in which the microscale λ, normalized by its initial value, is plotted against dimensionless time. The microscale is

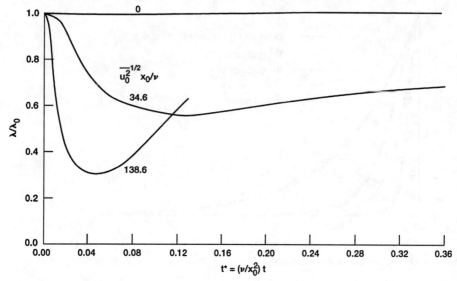

Figure 5-50 Calculated evolution of microscale of velocity fluctuations (normalized by initial value) for various initial Reynolds numbers.

defined by

$$\overline{\frac{\partial u_i}{\partial x_l}\frac{\partial u_i}{\partial x_l}} \propto \frac{\overline{u_i u_i}}{\lambda^2}. \tag{5-258}$$

For homogeneous turbulence with no mean gradients, λ can be calculated from eqs. (4-142) and (5-258):

$$\lambda^2 \propto -\nu \frac{\overline{u_i u_i}}{d\overline{u_i u_i}/dt}. \tag{5-259}$$

As the Reynolds number increases, the small-scale structure becomes finer. The microscale decreases until the fluctuation level (inertial effect) is low enough so that viscous forces prevent a further decrease. After λ decreases to a minimum, it begins to grow. (Results for coarser grids were not qualitatively different from these, but the minima were somewhat higher.) The increase of λ at later times is due to the selective annihilation of eddies by viscosity, the small eddies being the first to go. Thus, at large times, only the big eddies remain. It is this period of increasing λ that generally is observed experimentally in grid-generated turbulence (turbulence observed downstream of a grid of wires or bars whose plane is normal to the flow in a wind tunnel). The increases of λ with time observed experimentally (Fig. 7.2 in ref. [4]) are generally of the same order as those in Fig. 5-50 (doubling the time increases λ by a factor of about 1.5). The early period, in which λ decreases with time, is of interest as being illustrative of interwavenumber energy transfer. In order to generate the small-scale structure, turbulent energy must be transferred from big eddies to small ones.

For homogeneous turbulence without mean gradients the equation for the rate of change of turbulent kinetic energy, eq. (4-142), reduces to

$$\frac{\partial}{\partial t}\left(\frac{\overline{u_i u_i}}{2}\right) = -\nu \overline{\frac{\partial u_i}{\partial x_l}\frac{\partial u_i}{\partial x_l}}. \tag{5-260}$$

That is, only viscous dissipation contributes to the rate of change of kinetic energy, there being no indication that nonlinear transfer of energy between scales of motion is taking place. There may seem to be a paradox here in view of the large transfer of energy to small eddies indicated in Fig. 5-50. This is as it should be, however, because energy transfer between wavenumbers or scales of motion should not contribute to the rate of change of total energy. In order to consider interwavenumber energy transfer, we must use two-point equations. Thus it is shown in section 5-3-1-2 that the self-interaction transfer term $T_{ij}(\kappa)$ in the two-point spectral equation has the property that

$$\int_{-\infty}^{\infty} T_{ij}(\kappa)\, d\kappa = 0, \tag{5-98}$$

as a spectral transfer term should. The quantity κ is the wavevector. It is the spectral transfer term $T_{ij}(\kappa)$, or its Fourier transform $-\partial(\overline{u_i u'_j u'_k} - \overline{u_i u_k u'_j})/\partial r_k$ in eq. (5-133), that is responsible for the generation of the small-scale structure in Fig. 5-50. Those terms come from the nonlinear term $-\partial(u_i u_k)/\partial x_k$ in the unaveraged eq. (5-130).

Although eq. (5-98) shows that T_{ij} can transfer energy between wavenumbers without contributing to the rate of change of total energy $\partial \overline{u_i u_j}/\partial t$, it says nothing about the direction of the transfer or how important it is. For that we need calculations such as those in Fig. 5-50, which show that significant energy is transferred to smaller eddies (see also Figs. 5-15 and 5-16). The energy transfer can be thought of as being due to a breakup of big eddies into smaller ones, or as a stretching of vortex filaments to smaller diameters. In spite of this transfer to smaller eddies, experimental results generally show a growth of scale (see Fig. 7.2 in ref. [4]). This is because those results are usually for the later period shown in Fig. 5-50, in which, although energy is transferred to smaller eddies, the annihilation of small eddies by viscous action eventually wins out. The early period shown in Fig. 5-50 and in Fig. 2 of ref. [62] is of particular interest, in that the nonlinear transfer effects are truly dominant there; a sharp decrease in scale actually occurs as energy is transferred to smaller eddies.

Vorticity and dissipation. For homogeneous turbulence without mean gradients one can obtain a relation between the viscous dissipation term in eq. (4-142) and the vorticity or swirl in the turbulence. The vorticity ω is defined as the curl of u:

$$\omega(x) = \nabla \times u(x) \tag{5-261}$$

or

$$\omega_i = \varepsilon_{ijk} \frac{\partial u_k}{\partial x_j}, \tag{5-262}$$

where ε_{ijk} is the alternating tensor (see section 2-5-1). Then

$$\overline{\omega_k \omega_k} = \varepsilon_{ijk} \varepsilon_{lmk} \overline{\frac{\partial u_i}{\partial x_j} \frac{\partial u_l}{\partial x_m}} = (\delta_{il}\delta_{jm} - \delta_{im}\delta_{jl}) \overline{\frac{\partial u_i}{\partial x_j} \frac{\partial u_l}{\partial x_m}}$$

$$= \overline{\frac{\partial u_i}{\partial x_j} \frac{\partial u_i}{\partial x_j}} - \overline{\frac{\partial u_i}{\partial x_j} \frac{\partial u_j}{\partial x_i}}. \tag{5-263}$$

However, because

$$\frac{\partial u_j}{\partial x_j} = 0, \qquad (5\text{-}264)$$

$$\overline{\frac{\partial u_i}{\partial x_j}\frac{\partial u_j}{\partial x_i}} = \frac{\partial^2 \overline{u_i u_j}}{\partial x_i \partial x_j} = 0 \qquad (5\text{-}265)$$

for homogeneous turbulence. Equation (5-263) then becomes

$$\overline{\omega_k \omega_k} \equiv \overline{\omega^2} = \overline{\frac{\partial u_i}{\partial x_j}\frac{\partial u_i}{\partial x_j}} \equiv \frac{\varepsilon}{\nu}. \qquad (5\text{-}266)$$

Thus, for homogeneous turbulence or turbulent-like fluctuations, the mean-square vorticity is just the rate of viscous dissipation ε of turbulent energy divided by the kinematic viscosity (eq. [4-142]). So the more intense the swirl in the turbulence, the faster it dissipates.

Dissipation, vorticity generation, and pressure fluctuations. The energy dissipation term, the only term contributing to the rate of change of kinetic energy for homogeneous turbulence without mean gradients (eq. [5-260]) is plotted in Fig. 5-51. That is also the mean-square vorticity (see eq. [5-266]), but the two are distinct physical entitles. Although the curve for zero Reynolds number, where nonlinear effects are absent, decreases monotonically to zero, the curves for higher Reynolds numbers increase sharply for a while and then decrease. Thus the nonlinear terms in the Navier-Stokes equations are very effective vorticity generators and greatly enhance the dissipation at small and moderate times. For large times they appear to have the opposite effect, evidently because the turbulence itself decays rapidly to zero. Nonlinear effects, although they do not appear explicitly in the evolution equation for $\overline{u_i u_i}$ (eq. [5-260]), thus alter greatly the evolution by altering the dissipation term.

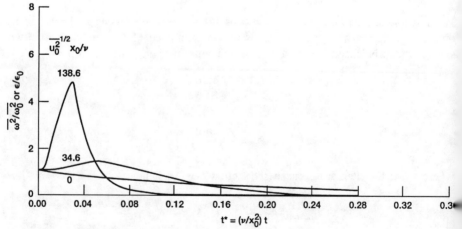

Figure 5-51 Calculated development of mean-square vorticity fluctuations $\overline{\omega^2}$ or dissipation ε (normalized by initial value) for various initial Reynolds numbers.

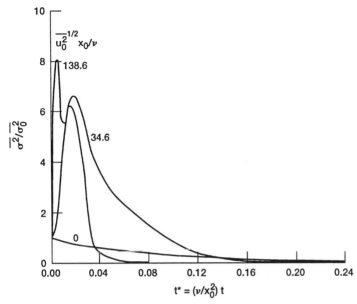

Figure 5-52 Calculated evolution of mean-square pressure fluctuation (normalized by initial value) for various initial Reynolds numbers.

Figure 5-52 shows mean-square pressure fluctuations plotted against dimensionless time. The enhancement of the pressure fluctuations, although not as great as that of the vorticity or dissipation, again is due to nonlinear effects: In this case the nonlinear terms on the right side of the Poisson equation for the pressure cause the effect.

Further discussion and summary of the processes in isotropic turbulence. Nonlinear velocity and pressure terms do not appear in the evolution equation for $\overline{u_i u_i}$ (eq. [5-260]). We can calculate, however, root-mean-square values of the nonlinear terms, as well as of the linear term, in the instantaneous evolution eq. (5-130). Three measures of the relative importance of inertial (nonlinear) and viscous effects are shown for a moderate Reynolds number in Fig. 5-53. The ratio of the nonlinear velocity term to the viscous term and the ratio of the pressure to the viscous term in eq. (5-130), together with the microscale Reynolds number, are plotted against dimensionless time. The terms are space-averaged root-mean-square values. All of those measures show a variation from a rather inertial to a weak fluctuating flow. For instance, R_λ varies from about 90 to 0.7. This is a much greater variation than has been obtained experimentally for a single run. The curves for the term ratios lie somewhat below that for R_λ. They indicate that except at early times the nonlinear inertial effects associated with velocity and with pressure do not differ greatly.

The appearance of both nonlinear velocity and pressure effects in eq. (5-130) and Fig. 5-53 may seem somewhat paradoxical in view of eq. (5-260), which says that neither contributes directly to $\partial \overline{u_i u_i}/\partial t$. The nonlinear velocity effects already have been discussed in this section; it is pointed out that such effects should not appear in eq. (5-260),

214 TURBULENT FLUID MOTION

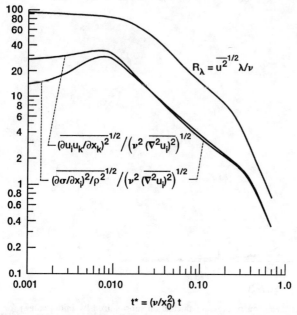

Figure 5-53 Three measures of relative importance of inertial and viscous effects versus dimensionless time. $\overline{u_0^2}^{1/2} x_0/\nu = 69.3$. $i = 1, 2,$ or 3.

because they only distribute energy in wavenumber space and so do not directly alter the total energy. Although there is no nonlinear velocity term in eq. (5-260), such a term appears in the two-point equation for $\partial \overline{u_i u_i'}/\partial t$. That equation, for the present case, is obtained from eq. (5-133) as

$$\frac{\partial}{\partial t}\overline{u_i u_i'} = 2\nu \frac{\partial^2 \overline{u_i u_i'}}{\partial r_k \partial r_k} - \frac{\partial}{\partial r_k}\left(\overline{u_i u_i' u_k'} - \overline{u_i u_k u_i'}\right) = 3\frac{\partial}{\partial t}\overline{u_i u_{(i)}'},$$

where r is again the vector extending from the unprimed to the primed point, and the pressure terms drop out because of continuity. The last term, in which the parentheses indicate no sum on i, is a consequence of the isotropy of the turbulent-like fluctuations. The equation for the rate of change of each component of $\overline{u_i u_i'}$ is contributed to by the nonlinear velocity term $-(\partial/\partial r_k)(\overline{u_i u_i' u_k'} - \overline{u_i u_k u_i'})$, but there is no contribution from the pressure. The strong effect of pressure shown in eq. (5-130) and Fig. 5-53 must be contained in higher-order equations in the hierarchy of averaged equations (moment equations) (see section 5-3-2-2 and ref. [22]). Thus, although two-point averaged equations contain a nonlinear effect of velocity, we must consider higher-order multipoint equations to obtain an effect of pressure. Terms in the unaveraged equations shown in Fig. 5-53 (averaged over space after the solution has been obtained) include effects of all orders. (Effects contained in the numerical results, however, may be limited by the fineness of the numerical grid.)

Although pressure effects appear in Fig. 5-53 and eq. (5-130), the physical significance of those effects is somewhat elusive, in contrast to the effects of viscous dissipation and spectral-energy transfer. If the turbulence or turbulent like fluctuations are anisotropic, a clear effect of pressure fluctuations is that they transfer net energy among directional components (see eqs. [4-140] and [4-142] and the discussion in section 4-3-3-1). That is discussed further when mean gradients are considered. If, in addition, the turbulence is inhomogeneous, pressure can produce a net spatial diffusion of energy (eq. [4-142]). Those are evidently the only physical effects of pressure fluctuations (at least that we know about). Thus, if the turbulence or turbulent-like fluctuations are homogeneous and isotropic, as they are here, it seems reasonable to attribute the observed pressure effects in the unaveraged equations (eq. [5-130]) to those processes. Even though there is no net interdirectional transfer or spatial diffusion of turbulence if the turbulence is isotropic, those processes still can be instantaneously or locally operative. They could cause, for instance, a diffusion of tagged particles. According to Fig. 5-53, they have an important indirect effect on the evolution of the turbulence.

From the findings of the present section we conclude that the following processes occur in isotropic turbulence: nonlinear radomization by proliferation of harmonic components or by sensitive dependence on initial conditions (discussed further in Chapter 6); nonlinear spectral transfer of turbulence among wavenumbers or eddy sizes (mainly to smaller eddies); spatial diffusion and transfer of turbulence among directional components by pressure forces, with zero net diffusion and transfer into each component, because the turbulence is isotropic; generation of vorticity or swirl; and dissipation of turbulence into heat by viscous action. From this description, as well as those in section 5-3-2-2 and Chapter 6, isotropic turbulence appears interesting and many-faceted. (This is in contrast to the characterization sometimes given that isotropic turbulence is tired or fossil turbulence.)

5-3-2-7 Turbulent diffusion and multitime–multipoint correlations One of the basic characteristics of turbulence is its great diffusive power. For example, the rate of diffusion of a glob of cream into a cup of unstirred coffee by molecular action is extremely slow. However, that rate is increased by orders of magnitude by a slight stirring (turbulent) action.

The theory of turbulent diffusion was originated by Taylor in 1922 [67] and since has been studied by a number of authors (e.g., refs. [68] and [69]). To illustrate turbulent diffusion, we calculate approximately the diffusion of particles originally concentrated at a source into a decaying turbulent field [70]. In order to simplify the problem, assume that the velocity fluctuations are small.

The distance in the x_2 direction that a fluid particle orginally at $x_2 = 0$ travels during the time interval $t' - t_a$ is

$$Y(t') = \int_{t_a}^{t'} u_2(t)\, dt. \tag{5-267}$$

Multiplication of this equation by $u_2(t') = dY/dt'$ gives

$$u_2(t')Y(t') = \frac{1}{2}\frac{dY^2}{dt'} = \int_{t_a}^{t'} u_2(t)u_2(t')\, dt. \tag{5-268}$$

Taking the particle average over all the "marked" particles that originally are concentrated at a source at $x_2 = 0$ and integrating with respect to t' result in

$$\overline{Y_b^2} = 2 \int_{t_a}^{t_b} \int_{t_a}^{t'} \left[\overline{u_2(t)u_2(t')}\right]_L dt \, dt', \tag{5-269}$$

where the subscript L designates a Lagrangian correlation that is based on the velocity of a moving particle at different times, rather than on the velocity at a fixed point at different times. The latter is generally called the *Eulerian time correlation*. Because it is easier to measure or calculate than the Lagrangian correlation, we would like to be able to relate the two types of correlations. For small velocity fluctuations it has been suggested by Burgers [71] that they should not differ greatly. This can be shown as follows:

Consider first the Eulerian time correlation $\overline{u_2(t)u_2(t')}$, where u_2 is the component of the velocity in the x_2 direction; similar results can be obtained for the other velocity components. The Eulerian correlation can be expanded in a Taylor series as

$$\overline{u_2(t)u_2(t')} = \left[\overline{u_2(t)u_2(t')}\right]_{t'=t} + \left[\frac{\partial}{\partial t'}\overline{u_2(t)u_2(t')}\right]_{t'=t}(t'-t)$$

$$+ \frac{1}{2}\left[\frac{\partial^2}{\partial t'^2}\overline{u_2(t)u_2(t')}\right]_{t'=t}(t'-t)^2 + \ldots = \overline{u_2^2(t)} + \overline{u_2(t)\left[\frac{\partial u_2(t')}{\partial t'}\right]_{t'=t}}$$

$$\times (t'-t) + \frac{1}{2}\overline{u_2(t)\left[\frac{\partial^2 u_2(t')}{\partial t'^2}\right]_{t'=t}}(t'-t)^2 + \ldots. \tag{5-270}$$

Similarly, the Lagrangian correlation is expanded as

$$\left[\overline{u_2(t)u_2(t')}\right]_L = \overline{u_2^2(t)} + \overline{u_2(t)\left[\frac{du_2(t')}{dt'}\right]_{t'=t}}(t'-t) + \frac{1}{2}\overline{u_2(t)\left[\frac{d^2u_2(t')}{dt'^2}\right]_{t'=t}}(t'-t)^2 + \ldots, \tag{5-271}$$

where the substantial or particle derivative is given as

$$\frac{du_2(t')}{dt'} = \frac{\partial u_2(t')}{\partial t'} + u_k \frac{\partial u_2(t')}{\partial x_k}.$$

For small velocity fluctuations,

$$\frac{du_2(t')}{dt'} \cong \frac{\partial u_2(t')}{\partial t'}. \tag{5-272}$$

Also

$$\frac{d^2 u_2(t')}{dt'^2} \cong \frac{\partial}{\partial t'}\frac{du_2(t')}{dt'} \cong \frac{\partial^2 u_2(t')}{\partial t'^2}, \tag{5-273}$$

and so on for higher-order derivatives. From eqs. (5-270) through (5-273),

$$\overline{u_2(t)u_2(t')} \cong \left[\overline{u_2(t)u_2(t')}\right]_L \tag{5-274}$$

is obtained, which is the relation to be proved. It should be noted that relation (5-274) is most accurate for small values of $t'-t$ as well as for small velocity fluctuations, inasmuch as the approximate relation (5-273) has to be applied a greater number of times to the higher-order derivatives in eq. (5-271) than to the lower-order ones (see eq. [5-273]).

FOURIER ANALYSIS, SPECTRAL FORM OF THE CONTINUUM EQUATIONS 217

It should also be emphasized that eq. (5-274) is obtained for the case of no mean motion. Thus, Eulerian time correlations measured with a stationary instrument in a moving stream may differ considerably from the Lagrangian correlations. However, if the instrument is moving with the stream, the two correlations are approximately equal if the turbulence level is not too high [72].

If the approximate relation (5-274) is introduced into eq. (5-269) and it is noted that, for isotropic turbulence, $\overline{u_2(t)u_2(t')} = \overline{u_i(t)u_i(t')}/3$, we get

$$\overline{Y_b^2} = \frac{2}{3}\int_{t_1}^{t_2}\int_{t_1}^{t'} \overline{u_i(t)u_i(t')}\, dt\, dt'. \tag{5-275}$$

To obtain the time correlation $\overline{u_i(t)u_i(t')}$ in eq. (5-275), first write the Navier-Stokes equations for the points P and P' separated by the distance vector r and the time increment Δt:

$$\frac{\partial u_i}{\partial t} + \frac{\partial (u_i u_k)}{\partial x_k} = -\frac{1}{\rho}\frac{\partial \sigma}{\partial x_i} + \nu\frac{\partial^2 u_i}{\partial x_k \partial x_k}, \tag{5-276}$$

$$\frac{\partial u'_j}{\partial t'} + \frac{\partial (u'_j u'_k)}{\partial x'_k} = -\frac{1}{\rho}\frac{\partial \sigma'}{\partial x'_j} + \nu\frac{\partial^2 u'_j}{\partial x'_k \partial x'_k}, \tag{5-277}$$

where, as usual, the subscripts can take on the values 1, 2, 3 and a repeated subscript in a term indicates a summation. The quantities u_i and u'_j are instantaneous velocity components, x_i is a space coordinate, t is the time, ρ is the density, ν is the kinematic viscosity, and σ is the instantaneous mechanical pressure. Multiplying the first equation by u'_j and the second by u_i and taking space averages result in

$$\frac{\partial \overline{u_i u'_j}}{\partial t} + \frac{\partial \overline{(u_i u'_j u_k)}}{\partial x_k} = -\frac{1}{\rho}\frac{\partial \overline{\sigma u'_j}}{\partial x_i} + \nu\frac{\partial^2 \overline{u_i u'_j}}{\partial x_k \partial x_k} \tag{5-278}$$

and

$$\frac{\partial \overline{u_i u'_j}}{\partial t'} + \frac{\partial \overline{(u_i u'_j u'_k)}}{\partial x'_k} = -\frac{1}{\rho}\frac{\partial \overline{\sigma' u_i}}{\partial x'_j} + \nu\frac{\partial^2 \overline{u_i u'_j}}{\partial x'_k \partial x'_k}, \tag{5-279}$$

where the fact that quantities at x_i and t are independent of x'_i and t' is used. By introducing the transformations $\partial/\partial x_i = -\partial/\partial r_i$, $\partial/\partial x'_i = \partial/\partial r_i$, $(\partial/\partial t)t' = (\partial/\partial t)_{\Delta t} - \partial/\partial \Delta t$, and $\partial/\partial t' = \partial/\partial \Delta t$, which are obtained by writing a correlation as a function of r_i, t, and Δt and differentiating, the following equations are obtained from (5-278) and (5-279):

$$\frac{\partial \overline{u_i u'_j}}{\partial t} + \frac{\partial}{\partial r_k}\overline{u_j u_k u'_i(-r, -\Delta t, t+\Delta t)} - \frac{\partial}{\partial r_k}\overline{u_i u_k u'_j(r, \Delta t, t)}$$

$$= \frac{1}{\rho}\frac{\partial}{\partial r_i}\overline{\sigma u'_j} - \frac{1}{\rho}\frac{\partial}{\partial r_j}\overline{\sigma u'_i(-r, -\Delta t, t+\Delta t)} + 2\nu\frac{\partial^2 \overline{u_i u'_j}}{\partial r_k \partial r_k}, \tag{5-280}$$

$$\frac{\partial \overline{u_i u'_j}}{\partial \Delta t} + \frac{\partial}{\partial r_k}\overline{u_j u_k u'_i(-r, -\Delta t, t+\Delta t)} = -\frac{1}{\rho}\frac{\partial}{\partial r_j}\overline{\sigma u'_i(-r, -\Delta t, t+\Delta t)} + \nu\frac{\partial^2 \overline{u_i u'_j}}{\partial r_k \partial r_k}. \tag{5-281}$$

Equations (5-280) and (5-281) are the space-time equivalents of the two-point eq. (5-133). They are obtained in a slightly different form, for the case of isotropic turbulence, by Bass [73].

In order to convert eqs. (5-280) and (5-281) to spectral form, the following three-dimensional Fourier transforms are introduced:

$$\overline{u_i u'_j}(r, \Delta t, t) = \int_{-\infty}^{\infty} \varphi_{ij}(\kappa, \Delta t, t) e^{i\kappa \cdot r} d\kappa \qquad (5\text{-}282)$$

$$\overline{u_j u_k u'_i}(r, \Delta t, t) = \int_{-\infty}^{\infty} \varphi_{jki}(\kappa, \Delta t, t) e^{i\kappa \cdot r} d\kappa \qquad (5\text{-}283)$$

$$\overline{\sigma u'_j}(r, \Delta t, t) = \int_{-\infty}^{\infty} \lambda_j(\kappa, \Delta t, t) e^{i\kappa \cdot r} d\kappa, \qquad (5\text{-}284)$$

where κ is a wavevector and $d\kappa = d\kappa_1 d\kappa_2 d\kappa_3$. By introducing these transforms, eqs. (5-280) and (5-281) become

$$\frac{\partial \varphi_{ij}}{\partial t} + i\kappa_k \varphi_{jki}(-\kappa, -\Delta t, t + \Delta t) - i\kappa_k \varphi_{ikj}(\kappa, \Delta t, t)$$
$$= \frac{1}{\rho} i\kappa_i \lambda_j - \frac{1}{\rho} i\kappa_j \lambda_i(-\kappa, -\Delta t, t + \Delta t) - 2\nu\kappa^2 \varphi_{ij} \qquad (5\text{-}285)$$

and

$$\frac{\partial \varphi_{ij}}{\partial \Delta t} + i\kappa_k \varphi_{jki}(-\kappa, -\Delta t, t + \Delta t) = -\frac{1}{\rho} \kappa_j \lambda_i(-\kappa, -\Delta t, t + \Delta t) - \nu\kappa^2 \varphi_{ij}. \qquad (5\text{-}286)$$

In order to convert the tensor eqs. (5-285) and (5-286) to scalar equations, contract the indices i and j:

$$\frac{\partial \varphi_{ii}}{\partial t} + 2\nu\kappa^2 \varphi_{ii} = i\kappa_k \varphi_{iki}(\kappa, \Delta t, t) + i(-\kappa_k)\varphi_{iki}(-\kappa, -\Delta t, t + \Delta t) \qquad (5\text{-}287)$$

$$\frac{\partial \varphi_{ii}}{\partial \Delta t} + \nu\kappa^2 \varphi_{ii} = i(-\kappa_k)\varphi_{iki}(-\kappa, -\Delta t, t + \Delta t). \qquad (5\text{-}288)$$

The pressure terms drop out of these scalar equations because of the continuity relation $\partial u_i/\partial x_i = \partial u'_i/\partial x'_i = 0$ and the relation $\partial/\partial x_i = -\partial/\partial x'_i$ (see eqs. [5-278] and [5-279]).

Equations (5-287) and (5-288), as they stand, contain too many unknowns for solutions to be obtained. For the final period of decay, however, the triple-correlation or inertial terms should be negligible compared with the double-correlation terms. Thus, the terms on the right sides of eqs. (5-287) and (5-288) are neglected, and the following solutions are obtained:

$$\varphi_{ii} = f_1(\kappa, \Delta t) e^{-2\nu\kappa^2(t-t_0)} \qquad (5\text{-}289)$$

and

$$\varphi_{ii} = f_2(\kappa, t) e^{-\nu\kappa^2 \Delta t}. \qquad (5\text{-}290)$$

In order for these equations to be consistent,

$$E = f(\kappa) e^{-\nu\kappa^2 \Delta t} e^{-2\nu\kappa^2(t-t_0)}, \qquad (5\text{-}291)$$

where the energy spectrum function $E = 2\pi\kappa^2\varphi_{ii}$ has been introduced. Evaluate $f(\kappa)$ by letting $E = J_0\kappa^4/3\pi$ if κ is small [74]. This gives

$$E = \frac{J_0\kappa^4}{3\pi}e^{-2\nu\kappa^2(t-t_0+\frac{1}{2}\Delta t)}, \tag{5-292}$$

where J_0 is a constant that depends on initial conditions. For $\Delta t = 0$, eq. (5-292) reduces to the usual expression for the energy spectrum function in the final period, which involves only one time. By integrating eq. (5-292) with respect to κ, the time correlation is obtained as

$$\frac{\overline{u_i u_i'}}{2} = \frac{J_0}{32(2\pi)^{1/2}} \nu^{-5/2} \left(t - t_0 + \frac{1}{2}\Delta t\right)^{-5/2}. \tag{5-293}$$

Equations (5-275) and (5-293) give, for diffusion by isotropic turbulence in the final period of decay,

$$\overline{Y_b^2} = \frac{J_0 \nu^{-5/2}}{9\sqrt{\pi}} \left\{ \frac{1}{\sqrt{2}} \left[\frac{1}{(t_a - t_0 + \Delta t_b)^{1/2}} + \frac{1}{(t_a - t_0)^{1/2}} \right] - \frac{1}{(t_a - t_0 + \Delta t_b/2)^{1/2}} \right\}, \tag{5-294}$$

where t_a is again the time at which diffusion begins and $\Delta t_b = t_b - t_a$ is the time during which the root-mean-square turbulent diffusion distance goes from 0 to $(\overline{Y_b^2})^{1/2}$. For large diffusion times,

$$\overline{Y_b^2} = \frac{J_0 \nu^{-5/2}}{9\sqrt{2\pi}(t_a - t_0)}. \tag{5-295}$$

That is, the turbulent diffusion distance reaches a constant value and becomes independent of Δt_b for large diffusion times. This differs from the case of stationary turbulence, in which $\overline{Y_b^2}$ increases linearly with Δt_b for large diffusion times [67]. The reason it reaches a constant value for decaying turbulence is that for large times the turbulence goes to zero, so that no more turbulent diffusion can take place. For early times (small $t_a - t_0$), both eqs. (5-294) and (5-295) show that the diffusion distances are much larger then those for later times because of the higher turbulence level at early times.

In ref. [70] the analysis for turbulent diffusion is extended to somewhat earlier times than those for which eq. (5-294) is applicable, and the analytical results are compared with the experiment of Uberoi and Corrsin [75]. For the comparison, the diffusion time in the analysis is the distance downstream from the beginning of the line source (of heat) in the experiment divided by the speed of the mean stream. The agreement between theory and experiment is quite good for large decay times (low turbulence levels) and for small diffusion times; for other conditions, some deviation is indicated. The latter apparently is due to the assumed equality of Eulerian and Lagrangian correlations, that equality being the most accurate for small velocity fluctuations and short diffusion times (see discussion following eq. [5-274]).

5-4 HOMOGENEOUS TURBULENCE AND HEAT TRANSFER WITH UNIFORM MEAN-VELOCITY OR -TEMPERATURE GRADIENTS

In section 5-3 we consider the simplest type of turbulence—statistically homogeneous turbulence without mean gradients. The turbulence processes occurring there are viscous dissipation, nonlinear directional transfer of turbulent activity, and nonlinear spectral transfer between scales of motion. The absence of other, often overshadowing processes makes homogeneous turbulence without mean gradients an ideal vehicle for studying dissipation, nonlinear directional transfer, and nonlinear spectral transfer. Those processes, particularly nonlinear spectral transfer and its interaction with dissipation, are discussed in some detail in section 5-3.

However, real turbulence usually occurs in the presence of mean gradients, particularly in the presence of mean shear. So we consider here a slightly more complicated turbulence than that in section 5-3 by studying the effect of uniform mean gradients on homogeneous turbulence and heat transfer. In that way we introduce turbulence production and other processes that depend on mean gradients.

Following a plan similar to that in section 5-3, we first consider the basic equations for homogeneous turbulence with mean gradients. Then we give some illustrative solutions. As in section 5-3, the analytical solutions considered usually are of the simplest kind, in order to avoid excessive mathematical complexity. Somewhat more widely applicable numerical solutions are discussed if available and appropriate.

5-4-1 Basic Equations

The equations for the fluctuations u_i, τ, and σ are obtained in Chapter 4 as eqs. (4-22) through (4-24):

$$\frac{\partial u_i}{\partial t} = -\frac{\partial (u_i u_k)}{\partial x_k} - \frac{1}{\rho}\frac{\partial (\sigma - \sigma_e)}{\partial x_i} + \nu \frac{\partial^2 u_i}{\partial x_k \partial x_k} - \beta g_i \tau - u_k \frac{\partial U_i}{\partial x_k} - U_k \frac{\partial u_i}{\partial x_k} + \frac{\partial \overline{u_i u_k}}{\partial x_k}, \quad (4\text{-}22)$$

$$\frac{\partial \tau}{\partial t} = -\frac{\partial (\tau u_k)}{\partial x_k} + \alpha \frac{\partial^2 \tau}{\partial x_k \partial x_k} - u_k \frac{\partial T}{\partial x_k} - U_k \frac{\partial \tau}{\partial x_k} + \frac{\partial \overline{\tau u_k}}{\partial x_k}, \quad (4\text{-}23)$$

and

$$\frac{1}{\rho}\frac{\partial (\sigma - \sigma_e)}{\partial x_i \partial x_i} = -\frac{\partial^2 (u_i u_k)}{\partial x_i \partial x_k} - \beta g_i \frac{\partial \tau}{\partial x_i} - 2\frac{\partial u_i}{\partial x_k}\frac{\partial U_k}{\partial x_i} + \frac{\partial^2 \overline{u_i u_k}}{\partial x_i \partial x_k}, \quad (4\text{-}24)$$

where the subscripts (except e) can take on the values 1, 2, or 3, and a repeated subscript in a term signifies a summation. The instantaneous velocities, temperatures, and mechanical pressures have respectively been divided into mean and fluctuating components U_i and u_i, T and τ, and P and σ. The quantity x_i is a space coordinate, t is the time, ρ is the density, ν is the kinematic viscosity, g_i is a component of the body force, and $\beta \equiv -(1/\rho)(\partial \rho/\partial T)_\sigma$ is the thermal expansion coefficient of the fluid. The quantities T_e and σ_e are respectively the equilibrium temperature and pressure. In obtaining the buoyancy term in eq. (4-22), the density is assumed to depend effectively only on temperature and is not far removed from its equilibrium value (value for no heat transfer). Note that the equilibrium temperature is uniform, whereas the equilibrium pressure is not. We retain buoyancy effects in eqs. (4-22) through (4-24) because those effects are considered in some of the cases to follow.

Equations (4-22) through (4-24) apply at a point P in the turbulent fluid. Similar equations at another point P' can be obtained simply by priming the variables and changing the subscript i to, say, j. Equations involving correlations between fluctuating quantities at points P and P' then can be constructed by methods similar to those used for obtaining eq. (5-133), or eqs. (4-147) through (4-150) in the previous chapter. The resulting equations for homogeneous turbulence with uniform velocity and temperature gradients are

$$\frac{\partial \overline{u_i u'_j}}{\partial t} + \overline{u_k u'_j}\frac{\partial U_i}{\partial x_k} + \overline{u_i u'_k}\frac{\partial U_j}{\partial x_k} + \frac{\partial U_k}{\partial x_l} r_l \frac{\partial \overline{u_i u'_j}}{\partial r_k} + \frac{\partial}{\partial r_k}\left(\overline{u_i u'_j u'_k} - \overline{u_i u_k u'_j}\right)$$

$$= -\frac{1}{\rho}\left(\frac{\partial \overline{u_i \sigma'}}{\partial r_j} - \frac{\partial \overline{\sigma u'_j}}{\partial r_i}\right) + 2\nu \frac{\partial^2 \overline{u_i u'_j}}{\partial r_k \partial r_k} - \beta g_i \overline{\tau u'_j} - \beta g_i \overline{u_j \tau'}, \quad (5\text{-}296)$$

$$\frac{\partial \overline{\tau \tau'}}{\partial t} + \overline{u_k \tau'}\frac{\partial T}{\partial x_k} + \overline{\tau u'_k}\frac{\partial T}{\partial x_k} + \frac{\partial U_k}{\partial x_l} r_l \frac{\partial \overline{\tau \tau'}}{\partial r_k} + \frac{\partial}{\partial r_k}\left(\overline{\tau \tau' u'_k} - \overline{\tau u_k \tau'}\right) = 2\alpha \frac{\partial^2 \overline{\tau \tau'}}{\partial r_k \partial r_k}, \quad (5\text{-}297)$$

$$\frac{\partial \overline{\tau u'_j}}{\partial t} + \overline{u_k u'_j}\frac{\partial T}{\partial x_k} + \overline{\tau u'_k}\frac{\partial U_j}{\partial x_k} + \frac{\partial U_k}{\partial x_l} r_l \frac{\partial \overline{\tau u'_j}}{\partial r_k} + \frac{\partial}{\partial r_k}\left(\overline{\tau u'_j u'_k} - \overline{\tau u_k u'_j}\right)$$

$$= -\frac{1}{\rho}\frac{\partial \overline{\tau \sigma'}}{\partial r_j} + (\alpha + \nu)\frac{\partial^2 \overline{\tau u'_j}}{\partial r_k \partial r_k} - \beta g_j \overline{\tau \tau'}, \quad (5\text{-}298)$$

$$\frac{\partial \overline{u_i \tau'}}{\partial t} + \overline{u_i u'_k}\frac{\partial T}{\partial x_k} + \overline{u_k \tau'}\frac{\partial U_i}{\partial x_k} + \frac{\partial U_k}{\partial x_l} r_l \frac{\partial \overline{u_i \tau'}}{\partial r_k} + \frac{\partial}{\partial r_k}\left(\overline{u_i \tau' u'_k} - \overline{u_i u_k \tau'}\right)$$

$$= -\frac{1}{\rho}\frac{\partial \overline{\sigma \tau'}}{\partial r_i} + (\alpha + \nu)\frac{\partial^2 \overline{u_i \tau'}}{\partial r_k \partial r_k} - \beta g_i \overline{\tau \tau'}, \quad (5\text{-}299)$$

$$\frac{1}{\rho}\frac{\partial^2 \overline{u_i \sigma'}}{\partial r_j \partial r_j} = -2\frac{\partial \overline{u_i u'_k}}{\partial r_j}\frac{\partial U_j}{\partial x_k} - \frac{\partial^2 \overline{u_i u'_j u'_k}}{\partial r_j \partial r_k} - \beta g_j \frac{\partial \overline{u_i \tau'}}{\partial r_j}, \quad (5\text{-}300)$$

$$\frac{1}{\rho}\frac{\partial^2 \overline{\sigma u'_j}}{\partial r_i \partial r_i} = 2\frac{\partial \overline{u_i u'_j}}{\partial r_k}\frac{\partial U_k}{\partial x_i} - \frac{\partial^2 \overline{u_i u_k u'_j}}{\partial r_i \partial r_k} + \beta g_i \frac{\partial \overline{\tau u'_j}}{\partial r_i}, \quad (5\text{-}301)$$

$$\frac{1}{\rho}\frac{\partial^2 \overline{\sigma \tau'}}{\partial r_i \partial r_i} = 2\frac{\partial \overline{u_i \tau'}}{\partial r_k}\frac{\partial U_k}{\partial x_i} - \frac{\partial^2 \overline{u_i u_k \tau'}}{\partial r_i \partial r_k} + \beta g_i \frac{\partial \overline{\tau \tau'}}{\partial r_i}, \quad (5\text{-}302)$$

and

$$\frac{1}{\rho}\frac{\partial^2 \overline{\tau \sigma'}}{\partial r_j \partial r_j} = -2\frac{\partial \overline{\tau u'_k}}{\partial r_j}\frac{\partial U_j}{\partial x_k} - \frac{\partial^2 \overline{\tau u'_j u'_k}}{\partial r_j \partial r_k} - \beta g_j \frac{\partial \overline{\tau \tau'}}{\partial r_j}, \quad (5\text{-}303)$$

where by virtue of the uniformity of the mean velocity gradients we have set $\partial U'_j/\partial x'_k = \partial U_j/\partial x_k$ and $U'_k - U_k = r_l \partial U_k/\partial x_l$.

5-4-2 Cases for Which Mean Gradients Are Large or the Turbulence Is Weak

Equations (5-296) through (5-303) form a determinate set if we neglect terms containing triple correlations. As in section 5-3-2-1, in which mean gradients are absent, those terms can be neglected if the turbulence is weak enough. However, it is important to notice that the turbulence in a flow with mean velocity or temperature gradients may not have to be as weak as that in a flow without mean gradients. If some of those gradients are large, the terms containing them can be large compared with the triple-correlation terms, even if the turbulence is moderately strong or strong. Thus, the applicability of the solutions to be obtained here is much wider than that of the solutions in section 5-3-2-1. The mean-gradient term, however, should be large enough to give a sensible solution.

For converting eqs. (5-296) through (5-303) to spectral form, we introduce three-dimensional Fourier transforms defined as follows:

$$\overline{u_i u'_j} = \int_{-\infty}^{\infty} \varphi_{ij} e^{i\kappa \cdot r} \, d\kappa, \tag{5-304}$$

$$\overline{\sigma u'_j} = \int_{-\infty}^{\infty} \lambda_j e^{i\kappa \cdot r} \, d\kappa, \tag{5-305}$$

$$\overline{u_i \sigma'} = \int_{-\infty}^{\infty} \lambda'_i e^{i\kappa \cdot r} \, d\kappa, \tag{5-306}$$

$$\overline{\sigma \tau'} = \int_{-\infty}^{\infty} \zeta e^{i\kappa \cdot r} \, d\kappa, \tag{5-307}$$

$$\overline{\tau \sigma'} = \int_{-\infty}^{\infty} \zeta' e^{i\kappa \cdot r} \, d\kappa, \tag{5-308}$$

$$\overline{\tau u'_j} = \int_{-\infty}^{\infty} \gamma_j e^{i\kappa \cdot r} \, d\kappa, \tag{5-309}$$

$$\overline{u_i \tau'} = \int_{-\infty}^{\infty} \gamma'_i e^{i\kappa \cdot r} \, d\kappa, \tag{5-310}$$

and

$$\overline{\tau \tau'} = \int_{-\infty}^{\infty} \delta e^{i\kappa \cdot r} \, d\kappa, \tag{5-311}$$

where κ is the wavevector and $d\kappa = d\kappa_1 d\kappa_2 d\kappa_3$. The magnitude of κ has the dimension 1/length and can be considered to be the reciprocal of a wavelength or eddy size. then, from eq. (5-304),

$$r_l \frac{\partial \overline{u_i u'_j}}{\partial r_k} = \int_{-\infty}^{\infty} -\left(\kappa_k \frac{\partial \varphi_{ij}}{\partial x_l} + \delta_{lk} \varphi_{ij}\right) e^{i\kappa \cdot r} \, d\kappa, \tag{5-312}$$

where, as usual, δ_{lk} is the Kronecker delta. Equation (5-312) can be obtained by differentiating eq. (5-304) with respect to r_k, writing the inverse transform, and then differentiating with respect to κ_l. Taking the Fourier transforms of eqs. (5-296) through (5-303) results in

$$\frac{\partial \varphi_{ij}}{\partial t} + \varphi_{kj}\frac{\partial U_i}{\partial x_k} + \varphi_{ik}\frac{\partial U_j}{\partial x_k} - \frac{\partial U_k}{\partial x_l}\left(\kappa_k \frac{\partial \varphi_{ij}}{\partial x_l} + \delta_{lk}\varphi_{ij}\right)$$
$$= -\frac{1}{\rho}(i\kappa_j \lambda_i' - i\kappa_i \lambda_j) - 2\nu\kappa^2 \varphi_{ij} - \beta g_i \gamma_j - \beta g_j \gamma_i', \quad (5\text{-}313)$$

$$\frac{\partial \delta}{\partial t} + (\gamma_k' + \gamma_k)\frac{\partial T}{\partial x_k} - \frac{\partial U_k}{\partial x_l}\left(\kappa_k \frac{\partial \delta}{\partial \kappa_l} + \delta_{lk}\delta\right) = -2\alpha\kappa^2 \delta, \quad (5\text{-}314)$$

$$\frac{\partial \gamma_j}{\partial t} + \varphi_{kj}\frac{\partial T}{\partial x_k} + \gamma_k \frac{\partial U_j}{\partial x_k} - \frac{\partial U_k}{\partial x_l}\left(\kappa_k \frac{\partial \gamma_j}{\partial x_l} + \delta_{lk}\gamma_j\right) = -\frac{1}{\rho}i\kappa_j \zeta' - (\alpha+\nu)\kappa^2 \gamma_j - \beta g_j \delta,$$
$$(5\text{-}315)$$

$$\frac{\partial \gamma_i'}{\partial t} + \varphi_{ik}\frac{\partial T}{\partial x_k} + \gamma_k' \frac{\partial U_i}{\partial x_k} - \frac{\partial U_k}{\partial x_l}\left(\kappa_k \frac{\partial \gamma_i'}{\partial x_l} + \delta_{lk}\gamma_i'\right) = -(\alpha+\nu)\kappa^2 \gamma_i' + \frac{1}{\rho}i\kappa_i \zeta - \beta g_i \delta,$$
$$(5\text{-}316)$$

$$-\frac{1}{\rho}\kappa^2 \lambda_i' = -2i\kappa_j \varphi_{ik}\frac{\partial U_j}{\partial x_k} - \beta g_j i\kappa_j \gamma_i', \quad (5\text{-}317)$$

$$-\frac{1}{\rho}\kappa^2 \lambda_j = -2i\kappa_k \varphi_{ij}\frac{\partial U_k}{\partial x_i} + \beta g_i i\kappa_i \gamma_j, \quad (5\text{-}318)$$

$$-\frac{1}{\rho}\kappa^2 \zeta = 2i\kappa_k \gamma_i'\frac{\partial U_k}{\partial x_i} + \beta g_i i\kappa_i \delta, \quad (5\text{-}319)$$

and

$$-\frac{1}{\rho}\kappa^2 \zeta' = -2i\kappa_j \gamma_k \frac{\partial U_j}{\partial x_k} - \beta g_j i\kappa_j \delta. \quad (5\text{-}320)$$

Equations (5-313) through (5-320) can be used to study a number of cases in which homogeneous turbulence is acted on by large mean gradients or for which the turbulence is weak. Some of those cases are considered in the following sections.

5-4-2-1 Uniformly and steadily sheared homogeneous turbulence.
The effect of a uniform transverse velocity gradient on a homogeneous turbulent field has been considered by a number of authors (e.g., refs. [7], [8], and [76–78]). The treatment here parallels that in ref. [7]. That reference carried the problem to the point of calculating spectra and of studying the processes associated with the turbulence.

Equations (5-313), (5-317), and (5-318) become, for a uniform transverse velocity gradient dU_1/dx_2 and no buoyancy ($g_i = 0$),

$$\frac{\partial}{\partial t}\varphi_{ij} + \delta_{i1}\varphi_{2j}\frac{dU_1}{dx_2} + \delta_{j1}\varphi_{i2}\frac{dU_1}{dx_2} - \kappa_1 \frac{\partial \varphi_{ij}}{\partial \kappa_2}\frac{dU_1}{dx_2} = -\frac{1}{\rho}(-i\kappa_i \lambda_j + i\kappa_j \lambda_i') - 2\nu\kappa^2 \varphi_{ij},$$
$$(5\text{-}321)$$

$$-\frac{1}{\rho}i\kappa_j \lambda_i' = 2\frac{\kappa_1 \kappa_j}{\kappa^2}\varphi_{i2}\frac{dU_1}{dx_2}, \quad (5\text{-}322)$$

and
$$\frac{1}{\rho}i\kappa_i\lambda_j = 2\frac{\kappa_1\kappa_i}{\kappa^2}\varphi_{2j}\frac{dU_1}{dx_2}. \tag{5-323}$$

Substituting eqs. (5-322) and (5-323) into (5-321) gives

$$\frac{\partial}{\partial t}\varphi_{ij} = -(\delta_{i1}\varphi_{2j}+\delta_{j1}\varphi_{i2})\frac{dU_1}{dx_2}+\kappa_1\frac{\partial\varphi_{ij}}{\partial\kappa_2}\frac{dU_1}{dx_2}+\left(2\frac{\kappa_1\kappa_i}{\kappa^2}\varphi_{2j}+2\frac{\kappa_1\kappa_j}{\kappa^2}\varphi_{i2}\right)\frac{dU_1}{dx_2}-2\nu\kappa^2\varphi_{ij}. \tag{5-324}$$

Equation (5-324) indicates that φ_{ij} is a function of the components of κ as well as of its magnitude. One can obtain a quantity that is a function only of κ by writing φ_{ij} in terms of spherical coordinates and integrating over the surface of a sphere of radius κ, as suggested by Batchelor [4]. This gives

$$\psi_{ij}(\kappa) = \int_A \varphi_{ij}(\kappa)\,dA(\kappa). \tag{5-325}$$

The quantity ψ_{ij} is the value of φ_{ij} averaged over all directions and multiplied by A. Similarly, each term of eq. (5-324) can be averaged. If we denote the average of the second term (multiplied by A) by $P_{ij}(\kappa)dU_1/dx_2$, the average of the third term by $T''_{ij}(\kappa)dU_1/dx_2$, and the average of the fourth term by $Q_{ij}(\kappa)dU_1/dx_2$, then the averaged equation becomes

$$\frac{\partial}{\partial t}\psi_{ij}(\kappa) = P_{ij}(\kappa)\frac{dU_1}{dx_2}+T''_{ij}(\kappa)\frac{dU_1}{dx_2}+Q_{ij}(\kappa)\frac{dU_1}{dx_2}-2\nu\kappa^2\psi_{ij}. \tag{5-326}$$

Contraction of the indices i and j in eq. (5-326) gives

$$\frac{\partial}{\partial t}\psi_{ii} = P_{ii}\frac{dU_1}{dx_2}+T'''_{ii}\frac{dU_1}{dx_2}-2\nu\kappa^2\psi_{ii}, \tag{5-327}$$

where $\frac{1}{2}\psi_{ii} = E$, the energy-spectrum function.

Interpretation of terms in spectral equations. The quantity $Q_{ij}(dU_1/dx_2)$ in eq. (5-326), which corresponds to the pressure-force term in eq. (5-296), does not appear in the contracted eq. (5-327), as pointed out in ref. [76]. That this is the case can be seen by substituting $\partial/\partial r_i = \partial/\partial x'_i$ and $\partial/\partial r_j = -\partial/\partial x_j$ in the pressure-force term in eq. (5-296) and applying continuity to the contracted equation. Thus, as in the case of homogeneous trubulence without a mean-velocity gradient, the pressure-force term exchanges energy between the directional components of the energy but makes no contribution to $\partial\psi_{ii}/\partial t$.

In order to interpret the quantity T''_{ii} or T''_{ij}, it is noted that $r_2\overline{\partial u_i u'_j}/\partial r_1$ in eq. (5-296) and $\kappa_1(\partial\varphi_{ij}/\partial\kappa_2)$ in eq. (5-324) are related by

$$r_2\frac{\overline{\partial u_i u'_j}}{\partial r_1} = -\int_{-\infty}^{\infty}\kappa_1\frac{\partial\varphi_{ij}}{\partial\kappa_2}e^{i\kappa\cdot r}\,d\kappa. \tag{5-328}$$

Evaluating this equation for $r=0$ gives

$$\int_{-\infty}^{\infty}\kappa_1\frac{\partial\varphi_{ij}}{\partial\kappa_2}d\kappa = 0, \tag{5-329}$$

and thus

$$\int_0^{\infty}T''_{ij}(\kappa)\,d\kappa = 0. \tag{5-330}$$

Thus, T'''_{ij} gives zero contribution to $\partial \overline{u_i u_j}/\partial t$. However, it can alter the distribution in wavenumber space of contributions to $\partial \overline{u_i u_j}/\partial t$ and thus can be interpreted as a transfer function. The quantity T'''_{ij} evidently is first interpreted as a transfer function in ref. [7]. More is said about it when spectra of T'''_{ii} are computed. The quantity T'''_{ii}, which arises because of the velocity gradient, has a function similar to that of the transfer term arising from triple correlations (neglected here) but should not be confused with the latter. The transfer term $T'''_{ij}(dU_1/dx_2)$ can be important even in the final period, whereas that arising from the triple correlations is absent in that case. Note that in the general case, in which the velocity gradient is not uniform, the transfer term is associated with an average velocity gradient $(U'_1 - U_1)/r_2$ (see the fourth term in eq. [4-147]).

In order to complete the interpretation of terms in eq. (5-327), we note that $P_{ii}(dU_1/dx_2)$, which corresponds to the second and third terms in eq. (5-296), represents the production of turbulent energy at wave number κ by work done on the turbulence by the velocity gradient. Finally, the term $-2\nu\kappa^2\psi_{ii}$ is the usual dissipation term.

Solutions of spectral equations. Next, solutions are given for some of the components of the tensor φ_{ij} in eq. (5-324). The nine equations represented by eq. (5-234) are simulatneous first-order partial-differential equations in the independent variables t and κ_2 and can be solved by methods given, for instance, in ref. [79].

The component of φ_{ij} most easily obtained is φ_{22}, inasmuch as it is independent of the other components. Thus, the expression for φ_{22} is obtained by solution of eq. (5-324) [79] as

$$\varphi_{22} = \{f[\kappa_1, a_{12}\kappa_1(t-t_0) + \kappa_2, \kappa_3]/\kappa^4\}$$
$$\times \exp\left\{-2\nu(t-t_0)\left[\kappa^2 + \frac{1}{3}\kappa_1^2 a_{12}^2(t-t_0)^2 + a_{12}\kappa_1\kappa_2(t-t_0)\right]\right\}, \quad (5\text{-}331)$$

where

$$a_{12} \equiv \partial U_1/\partial x_2, \quad (5\text{-}332)$$

and f is a function of integration that depends on initial conditions. In order to evaluate f, it is assumed that the turbulence is isotropic at $t = t_0$ (but not at other times). This is a possible assumption because the effect of the velocity gradient on the turbulent quantities becomes negligible as $t \to t_0$ (see eq. [5-331]). Thus, $(\varphi_{ij})_0$ is given by eq. (3-4-12) in ref. [4] in which, as in eq. (5-164), $E_0 = C(\kappa) = J_0\kappa^4/3\pi$. So

$$(\varphi_{ij})_0 = \frac{J_0}{12\pi^2}(\kappa^2\delta_{ij} - \kappa_i\kappa_j). \quad (5\text{-}333)$$

According to eq. (5-333), rotations and reflections (in wave number space) of the vector κ do not affect $(\varphi_{ij})_0$, provided that the velocity components in the turbulence field are measured relative to the same coordinates as are the components of κ. Thus, the field of turbulence is isotropic, according to the usual definition of isotropic turbulence (e.g., see ref. [4]). Moreover, eq. (5-333) satisfies continuity, because $\kappa_i(\varphi_{ij})_0 = \kappa_j(\varphi_{ij})_0 = 0$ (see equation [5-304].[9] Evaluation of f in eq. (5-331) by substituting eq. (5-333) at

---

[9] Note, however, that $(\varphi_{ij})_0$ is not an isotropic or numerical tensor according to the usual definition (see

$t = t_0$ gives

$$f(\kappa_1, \kappa_2, \kappa_3) = \frac{J_0 \kappa^4}{12\pi^2}(\kappa_1^2 + \kappa_3^2),$$

or

$$f[\kappa_1, \kappa_2 + a_{12}\kappa_1(t - t_0), \kappa_3] = \frac{J_0}{12\pi^2}\{\kappa_1^2 + [\kappa_2 + a_{12}\kappa_1(t - t_0)]^2 + \kappa_3^2\}^2 (\kappa_1^2 + \kappa_3^2).$$

Then

$$\varphi_{22} = \frac{J_0\{\kappa_1^2 + [\kappa_2 + a_{12}\kappa_1(t - t_0)]^2 + \kappa_3^2\}^2 (\kappa_1^2 + \kappa_3^2)}{12\pi^2 \kappa^4}$$

$$\times \exp\left\{-2\nu(t - t_0)\left[\kappa^2 + \frac{1}{3}\kappa_1^2 a^2(t - t_0)^2 + a_{12}\kappa_1\kappa_2(t - t_0)\right]\right\}. \quad (5\text{-}334)$$

Similarly the component φ_{12}, which is associated with the turbulent shear stress, is

$$\varphi_{12} = \frac{J_0\{\kappa_1^2 + [\kappa_2 + a_{12}\kappa_1(t - t_0)]^2 + \kappa_3^2\}^2}{12\pi^2 \kappa^2}$$

$$\times \exp\left\{-2\nu(t - t_0)\left[\kappa^2 + a_{12}\kappa_1\kappa_2(t - t_0) + \frac{1}{3}\kappa_1^2 a_{12}^2(t - t_0)^2\right]\right\}$$

$$\times \left\{\frac{\kappa_3^2}{\kappa_1(\kappa_1^2 + \kappa_3^2)^{\frac{1}{2}}}\left[\tan^{-1}\frac{\kappa_2}{(\kappa_1^2 + \kappa_3^2)^{\frac{1}{2}}} - \tan^{-1}\frac{\kappa_2 + a_{12}\kappa_1(t - t_0)}{(\kappa_1^2 + \kappa_3^2)^{\frac{1}{2}}}\right] - \frac{\kappa_1 \kappa_2}{\kappa^2}\right\},$$

(5-335)

where the function of integration was evaluated from eq. (5-333), as in the case of eq. (5-334). The same expression is obtained for φ_{21}; that is, $\varphi_{12} = \varphi_{21}$.

The contracted component φ_{ii}, which is associated with the turbulent energy, is obtained from eq. (5-324) as

$$\varphi_{ii} = \frac{J_0\{\kappa_1^2 + [\kappa_2 + a_{12}\kappa_1(t - t_0)]^2 + \kappa_3^2\}^2}{12\pi^2 \kappa^2}$$

$$\times \exp\left\{-2\nu(t - t_0)\left[\kappa^2 + a_{12}\kappa_1\kappa_2(t - t_0) + \frac{1}{3}a_{12}\kappa_1^2(t - t_0)^2\right]\right\}$$

$$\times \left\{\frac{\kappa^2}{\kappa_1^2 + [\kappa_2 + a_{12}\kappa_1(t - t_0)]^2 + \kappa_3^2} + 1 + \frac{\kappa_3^2 \kappa^2}{\kappa_1^2(\kappa_1^2 + \kappa_3^2)}\right.$$

$$\left.\times \left[\tan^{-1}\frac{\kappa_2}{(\kappa_1^2 + \kappa_3^2)^{\frac{1}{2}}} - \tan^{-1}\frac{\kappa_2 + a_{12}\kappa_1(t - t_0)}{(\kappa_1^2 + \kappa_3^2)^{\frac{1}{2}}}\right]^2\right\}, \quad (5\text{-}336)$$

section 2-4-3), except for $\kappa = 0$. That definition would require that the components of $(\varphi_{ij})_0$ have the same numerical values for arbitrary rotations of the coordinate axes, even if κ does not rotate. Thus, a field of isotropic turbulence is described by a nonisotropic tensor (eq. [5-333]), because the only second-order isotropic tensor (if defined in the usual way) is the product of a scalar and δ_{ij} (section 2-4-3). On the other hand, a tensor such as that given by eq. (5-333) is often called a *two-point* isotropic tensor [4].

where the function of integration again is evaluated by using eq. (5-334), and a_{12} is given by (5-332). Other components of φ_{ij} can be obtained in a similar manner. For example, it is shown in ref. [80] that φ_{11} is given by

$$\varphi_{11} = \frac{J_0\{\kappa_1^2 + [\kappa_2 + a_{12}\kappa_1(t-t_0)]^2 + \kappa_3^2\}^2}{12\pi^2}$$

$$\times \exp\left\{-2\nu(t-t_0)\left[\kappa^2 + \frac{1}{3}\kappa_1^2 a_{12}^2(t-t_0)^2 + a_{12}\kappa_1\kappa_2(t-t_0)\right]\right\}$$

$$\times \left\{\frac{\kappa_1^2}{(\kappa_1^2 + \kappa_3^2)}\left(\frac{\kappa_2^2}{\kappa^4} - \left\{\frac{\kappa_2 + a_{12}\kappa_1(t-t_0)}{\kappa_1^2 + [\kappa_2 + a_{12}\kappa_1(t-t_0)]^2 + \kappa_3^2}\right\}^2\right)\right.$$

$$+ \frac{[\kappa_2 + a_{12}\kappa_1(t-t_0)]^2 + \kappa_3^2}{\{\kappa_1^2 + [\kappa_2 + a_{12}\kappa_1(t-t_0)]^2 + \kappa_3^2\}^2} - \frac{2\kappa_2\kappa_3^2}{(\kappa_1^2 + \kappa_3^2)^{\frac{3}{2}}\kappa^2}$$

$$\times \left[\tan^{-1}\frac{\kappa_2}{(\kappa_1^2 + \kappa_3^2)^{\frac{1}{2}}} - \tan^{-1}\frac{\kappa_2 + a_{12}\kappa_1(t-t_0)}{(\kappa_1^2 + \kappa_3^2)^{\frac{1}{2}}}\right]$$

$$\left. + \frac{\kappa_3^4}{(\kappa_1^2 + \kappa_3^2)^2 \kappa_1^2}\left[\tan^{-1}\frac{\kappa_2}{(\kappa_1^2 + \kappa_3^2)^{\frac{1}{2}}} - \tan^{-1}\frac{\kappa_2 + a_{12}\kappa_1(t-t_0)}{(\kappa_1^2 + \kappa_3^2)^{\frac{1}{2}}}\right]^2\right\}.$$

(5-337)

Note the similarities (and differences) among eqs. (5-334) through (5-337).

The quantities φ_{22}, φ_{12}, φ_{ii}, and φ_{11} are of interest in themselves; however, it is somewhat easier to interpret quantities that have been averaged over all directions in κ space, as if done in eqs. (5-325) through (5-327). In order to do this, we write eqs. (5-334) through (5-337) in terms of spherical coordinates by setting $\kappa_1 = \kappa \cos\varphi \sin\theta$, $\kappa_2 = \kappa \sin\varphi \sin\theta$, and $\kappa_3 = \kappa \cos\theta$. Then eq. (5-325) becomes

$$\psi_{ij}(\kappa) = \int_0^\pi \int_0^{2\pi} \varphi_{ij}(\kappa, \varphi, \theta)\kappa^2 \sin\theta \, d\varphi \, d\theta. \tag{5-338}$$

Each of the terms in eqs. (5-326) and (5-327) can be obtained in a similar manner. For instance, from eqs. (5-324) and (5-326),

$$T_{ij}(\kappa) = \int_0^\pi \int_0^{2\pi} \kappa \cos\varphi \sin\theta \frac{\partial \varphi_{ij}}{\partial \kappa_2}(\kappa, \varphi, \theta)\kappa^2 \sin\theta \, d\varphi \, d\theta. \tag{5-339}$$

Discussion of computed spectra. Dimensionless spectra of $\frac{1}{2}\overline{u_2^2}$, $\frac{1}{2}\overline{u_1 u_2}$, and $\frac{1}{2}\overline{u_i u_i}$ for various values of dimensionless velocity gradient are plotted in Figs. 5-54 through 5-56. If plotted in the form shown, the spectrum curve for zero velocity gradient does not change with time, so that the various curves indicate how the velocity gradient influences the spectrum. The increase in the heights of the spectra with $(t-t_0)dU_1/dx_2$ appears to be associated with the production of turbulence by the mean-velocity gradient. The spectra move toward the left as velocity gradient increases, because most of the production takes place in the low wavenumber region (see Fig. 5-62). Actually, the spectral equation corresponding to $\overline{u_2^2}$ does not contain a production term (see eq. [5-324]). However, energy produced in the $\overline{u_1^2}$ component can be fed into or out of $\overline{u_2^2}$ by the pressure-force

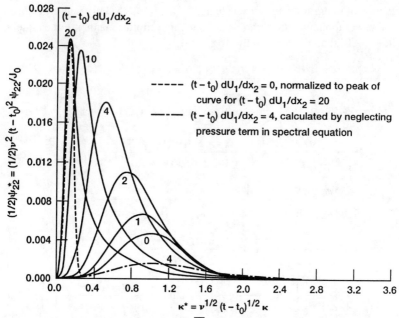

Figure 5-54 Dimensionless spectra of $(1/2)\overline{u_2^2}$ for uniform transverse velocity gradient.

terms that transfer energy between directional components. The magnitude of the effect of pressure forces is illustrated by the dot-dashed curve in Fig. 5-54, in which the pressure-force term in the spectral equation (fourth term in eq. [5-324]) is neglected. If the pressure-force term is neglected, the portion of eq. (5-334) in front of the exponential becomes $J_0(\kappa_1^2 + \kappa_3^2)/(12\pi^2)$, but the exponential is unchanged. Comparison of the dot-dashed curve with the solid curve for the same velocity gradient shows the considerable effects of pressure forces. That curve differs from the curve for zero velocity gradient because of the effect of the transfer term (third term in eq. [5-324]).

The spectrum of $\overline{u_2^2}$ moves to the left much more rapidly than does that of $\overline{u_i u_i}$, because of the action of pressure forces. The pressure-force term Q_{22} from the spectral eq. (5-326) is plotted in Fig. 5-57. For all values of dimensionless velocity gradient, Q_{22} is negative at the higher wavenumbers. This indicates that energy is transferred out of the spectrum of $\overline{u_2^2}$ at the higher wavenumbers and into the sum of the spectra of $\overline{u_1^2}$ and $\overline{u_3^2}$. Thus, the spectrum of $\overline{u_2^2}$ moves to the left faster with increasing velocity gradient than does that of $\overline{u_i u_i}$, the latter representing the average of the three components of the energy. This means that a typical eddy is elongated in the transverse direction x_2 by the action of the pressure forces.[10]

---

[10] In contrast to the behavior of Q_{22}, note that Q_{11} is always negative and Q_{33} is always positive, regardless of wavenumber [80]. Thus, pressure forces transfer energy out of the spectrum of $\overline{u_1^2}$ and into the spectrum of $\overline{u_3^2}$ at all wave numbers. Most of the directional energy transfer at high velocity gradients is from $\overline{u_1^2}$ to $\overline{u_3^2}$; smaller transfer is effected by Q_{22}. At smaller velocity gradients the energy transfers in the three directions are of the same order [80].

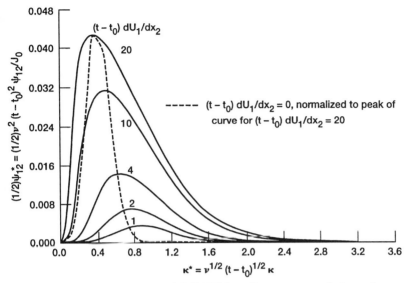

Figure 5-55 Dimensionless spectra of $(1/2)\overline{u_1 u_2}$ for uniform transverse velocity gradient.

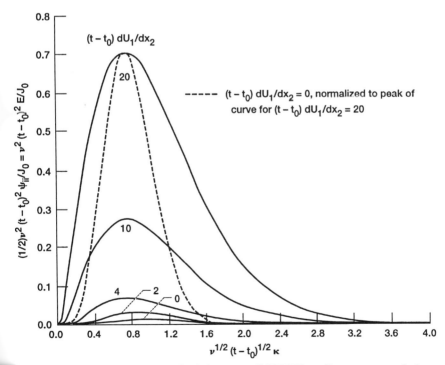

Figure 5-56 Dimensionless spectra of turbulent energy $(1/2)\overline{u_i u_i}$ for uniform transverse velocity gradient.

230 TURBULENT FLUID MOTION

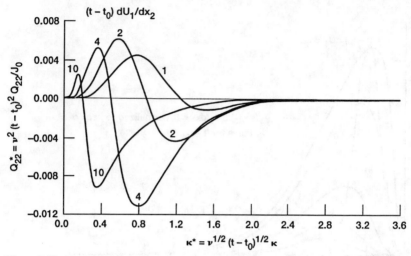

Figure 5-57 Plot of dimensionless pressure term in spectral equation (eq. [5-326]) for $i = j = 2$.

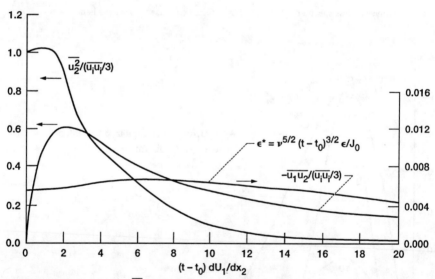

Figure 5-58 Plot showing $\overline{u_2^2}/((1/3)\overline{u_i u_i})$, $-\overline{u_1 u_2}/((1/3)\overline{u_i u_i})$, and ϵ^* against dimensionless velocity gradient.

The net areas under the curves of Q_{22} are positive for small velocity gradients and negative for large ones. Thus, the pressure forces transfer energy into the $\overline{u_2^2}$ component at low velocity gradients, whereas at high velocity gradients they transfer energy out of the $\overline{u_2^2}$ component into the sum of the $\overline{u_1^2}$ and $\overline{u_3^2}$ components. The resulting effect of this transfer on the components is shown in Fig. 5-58, in which $\overline{u_2^2}/(\frac{1}{3}\overline{u_i u_i})$ is plotted against dimensionless velocity gradient. The values of $\overline{u_2^2}$ and $\frac{1}{3}\overline{u_i u_i}$ are obtained by integrating

FOURIER ANALYSIS, SPECTRAL FORM OF THE CONTINUUM EQUATIONS   231

under the spectrum curves in Figs. 5-54 and 5-56. For isotropic turbulence (zero velocity gradient in the present case), $\overline{u_2^2}/(\frac{1}{3}\overline{u_iu_i})$ has the value 1. As the velocity gradient increses, $\overline{u_2^2}/(\frac{1}{3}\overline{u_iu_i})$ first increases very slightly and then decreases considerably because of the transfer of energy between directional components by the pressure forces. The quantities $\overline{u_1^2}/(\overline{u_iu_i}/3)$ and $\overline{u_3^2}/(\overline{u_iu_i}/3)$ are calculated in ref. [80] and are plotted in Fig. 4 of that reference. The calculated ordering of the three components of $\overline{u_iu_i}$ is $\overline{u_1^2} > \overline{u_3^2} > \overline{u_2^2}$, in agreement with experiment (see, e.g., ref. [11] of Chapter 4 and ref. [85] of this chapter).

Also plotted in Fig. 5-58 is the ratio $\overline{u_1u_2}/(\frac{1}{3}\overline{u_iu_i})$, where $\overline{u_1u_2}$ is obtained from the shear-stress spectra in Fig. 5-55. As expected, the shear stress is zero for zero velocity gradient. In other respects the curve is similar to that for $\overline{u_2^2}/(\frac{1}{3}\overline{u_iu_i})$, with the exception that it decreases more gradually as velocity gradient increases.

A quantity closely related to the shear stress is the eddy diffusivity ε, defined by

$$\varepsilon \equiv -\frac{\overline{u_1u_2}}{dU_1/dx_2}. \tag{5-340}$$

A plot of dimensionless eddy diffusivity against dimensionless velocity gradient is given in Fig. 5-58. Of some interest is the observation that the eddy diffusivity does not go to zero for zero velocity gradient. In this respect the eddy diffusivity (or eddy viscosity) is like the molecular viscosity. The result is reasonable, because one would expect the turbulence to be diffusive even in the absence of a velocity gradient. It is, however, in disagreement with the usual mixing-length theories, which predict zero eddy diffusivity for zero velocity gradient (see, e.g., the relation for ε given by eqs. [4-125] and [5-340]). This does not imply that the usual mixing-length theories are not useful for predicting mean velocity distributions in boundary layers and so forth, inasmuch as the velocity profile is insensitive to the value of eddy diffusivity in the region in which dU_1/dx_2 is small.[11] Figure 5-58 indicates that the value of dimensionless eddy diffusivity does not vary more than $\pm 10\%$ from a mean value, except at very high velocity gradients.

Local isotropy. The discussion in the preceding section concerning the transfer of energy between directional components by pressure forces is pertinent to the theory of local isotropy [42]. According to that theory the high-wavenumber (small-scale) components of the turbulence should be isotropic regardless of the directional orientation of the large-scale components. This tendency to isotropy of the small eddies is generally ascribed to the action of pressure forces. However, the present calculations indicate that in the presence of a mean velocity gradient the pressure forces can act to increase rather than decrease the anisotropy of the turbulence (see Fig. 5-58). The effect of this orientation on the small-scale components is clearly shown in Fig. 5-59, where $\psi_{22}/(\frac{1}{3}\psi_{ii})$ is plotted against dimensionless wavenumber. For isotropic turbulence that ratio would be 1, but the curves indicate that it is far from 1 in the high-wavenumber region, especially for high velocity gradients. These findings concerning the lack of local isotropy are in qualitative agreement with experimental results in ref. [81], in which turbulent spectra are measured in a low-speed boundary layer.

---

[11] Moreover, according to eq. (4-38), which is a more general mixing-length expression than that obtained from eqs. (4-125) and (5-340), ε does not necessarily go to zero for a velocity gradient of zero.

232 TURBULENT FLUID MOTION

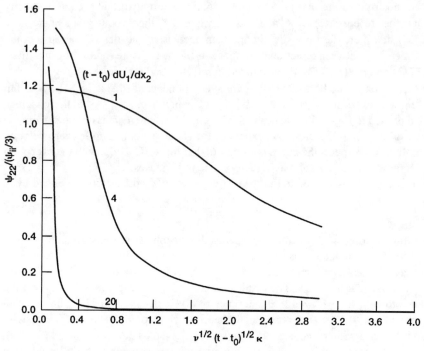

Figure 5-59 Plot showing $\psi_{22}/((1/3)\psi_{ii})$ against dimensionless wavenumber.

It should be pointed out that the action of the pressure forces in working against isotropy in the present analysis is due to the presence of the velocity gradient (eq. [5-326]). Triple correlations, which are neglected here, also can affect the pressure forces at high-turbulence Reynolds numbers (see eq. [4-149]) and may tend to increase the isotropy. However, it appears that one should be extremely cautious in assuming local isotropy in a boundary layer or pipe flow, inasmuch as high-turbulence Reynolds numbers in those cases generally correspond to high velocity gradients, except possibly at a large distance from the wall. Reference [82], in which the effects of anisotropy due to pressure forces apparently are not considered, indicates that local isotropy in a channel might conceivably occur, but only at extremely high Reynolds numbers. Local isotropy may be a better assumption for the turbulent wake of a cylinder [83].

Another quantity of interest in connection with local isotropy is the spectrum of the shear stress ψ_{12} (see Fig. 5-55). For local isotropy to exist, that quantity should go to zero faster with increasing wavenumber than does the average intensity component $\frac{1}{3}\psi_{ii}$. However, as pointed out in ref. [81], that is a necessary but not a sufficient condition for local isotropy.

Values of $\psi_{12}/(\frac{1}{3}\psi_{ii})$ are plotted against dimensionless wavenumber in Fig. 5-60. The curves indicate that this function, in general, decreases with increasing wavenumber, in agreement with the experimental findings of ref. [81], [84], and [85].

Energy transfer between wavenumbers. As discussed previously in connection with eq. (5-327), the term $T''_{ii}dU_1/dx_2$ makes no contribution to the change of total energy,

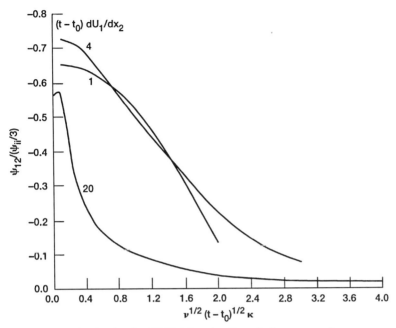

Figure 5-60 Plot showing $\psi_{12}/((1/3)\psi_{ii})$ against dimensionless wavenumber.

but it can transfer energy between wavenumbers or eddies of various sizes. Spectra of T''_{ii} for various values of dimensionless velocity gradient are plotted in Fig. 5-61. The curves are predominantly negative for small wavenumbers and positive for large ones, so that in general energy is transferred from small wavenumbers to large ones. Thus, the effect here is similar to that of the transfer term due to triple correlations. The transfer apparently affects the shape of the spectra in Figs. 5-54 through 5-56 by exciting the higher wavenumber regions of those spectra, as in the case of the transfer due to triple correlations (see, e.g., section 5-3-2-2). This is shown by the dashed curves for zero velocity gradient normalized to the peaks of the curves for $(t - t_0)dU_1/dx_2 = 20$ in Figs. 5-54 through 5-56.

A natural explanation of the transfer of energy to the high wavenumber regions by the mean-velocity gradient would be that the velocity gradient stretches the vortex lines associated with the turbulence. Some related problems in this connection are considered in ref. [86]. This picture also might explain the small amount of reverse transfer shown in Fig. 5-61 for low wavenumbers at small velocity gradients, because the velocity gradient should be able to compress, as well as stretch, the vortex lines if they are properly oriented. This reverse transfer is found to be more pronounced in the transfer term associated with $\overline{u_2^2}$ (not given here).

Production, energy-containing, and dissipation regions. Production, energy, and dissipation spectra, normalized to the same ordinate for comparison, are plotted in Fig. 5-62. The production and dissipation spectra correspond to the production and dissipation terms in eq. (5-327). Curves are shown for a negligibly small and for a comparatively large dimensionless velocity gradient. For the small velocity gradient the production,

234 TURBULENT FLUID MOTION

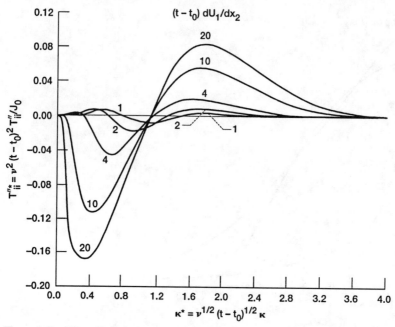

Figure 5-61 Dimensionless spectra of transfer term due to mean-velocity gradient in spectral equation for ψ_{ii} (eq. [5-327]).

energy-containing, and dissipation regions are only slightly separated, whereas at the higher velocity gradient they are more widely separated. The turbulent production by the mean velocity gradient occurs mostly in the low-wavenumber or large-eddy region; the dissipation occurs in the higher-wavenumber region. Although the three regions separate as velocity gradient increases, there is still considerable overlap at a value of $(t - t_0)dU_1/dx_2$ of 50. Energy from the mean velocity gradient feeds into the turbulence over a considerable range of wavenumbers. The energy goes into the turbulence through the $\overline{u_1^2}$ component, inasmuch as the production terms are absent from the spectral equations for $\overline{u_2^2}$ and $\overline{u_3^2}$ (see eq. [5-324]). However, the energy can be transferred between the various directional components by the pressure forces as discussed previously.

The separation of the energy-containing and dissipation regions at high velocity gradients is similar to the separation of those regions at high-turbulence Reynolds numbers without a velocity gradient (section 5-3-2-2). In both cases the separation appears to be a consequence of the change in shape of the energy spectrum produced by the transfer of energy to high wavenumbers (see Figs. 5-54 through 5-56).

Summary of turbulent energy processes. The sequence of turbulent energy processes in a flow with strong shear or weak turbulence might be summarized as follows. The turbulent energy is produced by the mean-velocity gradient. This production occurs in the $\overline{u_1^2}$ component of the energy and predominately in the large eddy region. The pressure

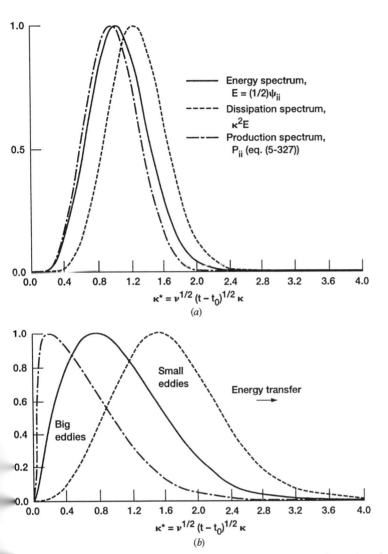

Figure 5-62 Comparison of production, energy, and dissipation spectra. Curves in each plot are normalized to same height. (a) $(t - t_0)dU_1/dx_2 \to 0$; (b) $(t - t_0)dU_1/dx_2 = 50$.

forces, which depend here on the velocity gradient, transfer the energy between various directional components. In doing this, they may increase the anisotropy of the turbulence, and, in particular, they oppose local isotropy in the high-wavenumber region. In cases in which the effect of the triple correlations on the pressure forces is not small, the turbulence might be somewhat more locally isotropic. The mean-velocity gradient, like the triple correlations, transfers energy from the large eddies to the small ones. This transfer can be interpreted as a stretching of the vortex lines by the mean-velocity gradient. Finally, the energy is dissipated by viscous action in the small-eddy region.

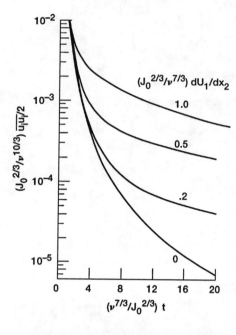

Figure 5-63 Effect of uniform shear on decay of turbulent energy. In all cases the turbulence ultimately decays.

Decay of the total turbulent energy. A dimensionless plot of the decay of turbulent energy for various velocity gradients is presented in Fig. 5-63. As velocity gradient increases, the rate of decrease of the turbulent energy with time decreases because of energy fed into the turbulent field by the mean-velocity gradient. Although the changes produced by the velocity gradient are considerable (note that the vertical scale is logarithmic), the turbulence at all times decays.

The curves in Fig. (5-63) are plotted with three dimensionless groups in order to show the effects of time and mean velocity gradient separately. The curves are, however, similar and can be compressed into one by using the proper similarity parameters. The result is shown in Fig. 5-64, in which $\nu^{5/2}(t-t_0)^{5/2}\overline{u_i u_i}/2J_0$ is plotted against $(t-t_0)dU_1/dx_2$.

The ultimate decay of turbulent energy in all parts of wavenumber space can be seen from the structure of eq. (5-336). At large times the argument of the exponential in that equation is negative for all values of κ_i and a_{12}.

Further understanding of the dynamics of shear-flow turbulence can be obtained by studying the random vorticity in sheared turbulence. One aspect of the interaction of turbulent vorticity with a mean shear is considered in the next section.

Direction of maximum turbulent vorticity in a shear flow. The vorticity in a turbulent shear flow has been mentioned briefly in connection with the energy transfer between wavenumbers. Here we further discuss turbulent vorticity, in particular the alignment of vorticity by a mean-velocity gradient and the direction of maximum vorticity.[12]

---

[12] The direction maximum turbulent stress or intensity is, of course, also of interest (see, e.g., ref. [87]), but the discussion here is confined to vorticity.

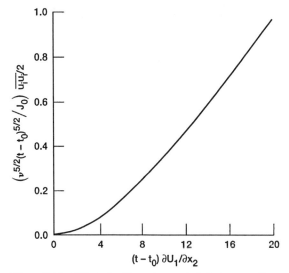

Figure 5-64 Effect of uniform shear on turbulent energy. Curves from Fig. 5-63 collapsed into one.

Intuitively, one might expect that the random vortices tend to become aligned in the direction of maximum mean strain rate, that is, at 45° to the flow direction. Taylor [88] was evidently the first one to emphasize the stretching of turbulent vortex filaments as a mechanism for the production of turbulence. Theodorsen [89], using a possibly overmechanized model, has attributed the maintenance of turbulence in a shear flow to the stretching of "horseshoe vortices" that are inclined to the direction of mean flow at an average of 45°. Weske and Plantholt [90] have discussed that concept further and have been able to generate such vortices artificially in a pipe flow.

Although it seems reasonable that the maximum turbulent vorticity should occur at 45° to the flow direction, further analysis indicates that it can occur at that angle only if the transverse component of vorticity equals the longitudinal component; that, of course, is not the case in an anisotropic shear flow. However, there may be a range of velocity gradients at which the maximum vorticity can occur at angles reasonably close to 45°.

Because the turbulent vorticity $\overline{\omega_i \omega_j}$ is a second-order tensor, its components relative to a rotated coordinate system (see eq. [2-12]) are given by

$$\overline{\omega_i^* \omega_j^*} = b_{ik} b_{jl} \overline{\omega_k \omega_l}, \qquad (5\text{-}341)$$

where ω_i is an instantaneous vorticity component, and the overbars indicate averaged values. Quantities without an asterisk give components relative to coordinates x_i, and quantities with an asterisk are relative to coordinates x_i^*.[13] The quantity b_{ik} is the cosine of the angle between x_i^* and x_k. The summation convention is operative in eq. (5-341).

---

[13] The use of asterisks here on rotated coordinates and on quantities measured relative to rotated coordinates is consistent with the usage in chapter 2. Note, however, that asterisks are sometimes also used to designate dimensionless quantities and complex conjugates.

If we consider a counterclockwise rotation about x_3 so that x_1^* makes an angle α with x_1, eq. (5-341) reduces (by setting $i = j = 1$) to

$$\overline{\omega_1^{*2}} = \frac{1}{2}\overline{\omega_1^2}(1 + \cos 2\alpha) + \overline{\omega_1\omega_2}\sin 2\alpha + \frac{1}{2}\overline{\omega_2^2}(1 - \cos 2\alpha). \tag{5-342}$$

The component $\overline{\omega_2^{*2}}$ then is obtained by substituting $\alpha + 90°$ for α in eq. (5-342). The angle for which $\overline{\omega_1^{*2}}$ is a maximum is obtained from eq. (5-342) by setting $\partial\overline{\omega_1^{*2}}/\partial\alpha = 0$. This gives

$$\alpha_{\omega_{max}} = \frac{1}{2}\tan^{-1}\frac{2\overline{\omega_1\omega_2}}{\overline{\omega_1^2} - \overline{\omega_2^2}}. \tag{5-343}$$

Equation (5-343) shows (as mentioned previously) that for a finite $\overline{\omega_1\omega_2}$, $\alpha_{\omega_{max}}$ can be equal to 45° only for $\overline{\omega_1^2} = \overline{\omega_2^2}$, or for isotropic turbulence. (A similar result is obtained for velocity fluctuations by replacing ω in eqs. [5-341] through [5-343] by u, because both $\overline{\omega_i\omega_j}$ and $\overline{u_iu_j}$ are second-order tensors.)

In order to calculate the angle $\alpha_{\omega_{max}}$ from eq. (5-343), components of $\overline{\omega_i\omega_j}$ in the unrotated coordinate system must be known. To that end, the two-point tensor $\overline{\omega_i\omega_j'}(r)$ can be related to $\overline{u_iu_j'}(r)$ by writing eq. (2-24) at points P and P'. Then

$$\overline{\omega_i\omega_j'} = \varepsilon_{ilm}\varepsilon_{jpq}\overline{\frac{\partial u_m}{\partial x_l}\frac{\partial u_q'}{\partial x_p'}} = -\varepsilon_{ilm}\varepsilon_{jpq}\overline{\frac{\partial^2 u_m u_q'}{\partial r_l \partial r_p}}, \tag{5-344}$$

where the relations $\partial/\partial x_i' = \partial/\partial r_i$ and $\partial/\partial x_i = \partial/\partial r_i$ are used and $\mathbf{r}$ is the vector extending from point P to point P'. Then defining the spectral tensor of $\overline{\omega_i\omega_j'}$ by

$$\overline{\omega_i\omega_j'}(r) = \int_{-\infty}^{\infty}\Omega_{ij}(\kappa)e^{i\kappa\cdot r}d\kappa, \tag{5-345}$$

and using

$$\overline{u_iu_j'}(r) = \int_{-\infty}^{\infty}\varphi_{ij}(\kappa)e^{i\kappa\cdot r}d\kappa, \tag{5-304}$$

we get

$$\Omega_{ij} = \varepsilon_{ilm}\varepsilon_{jpq}\kappa_l\kappa_p\varphi_{mq}. \tag{5-346}$$

However,

$$\varepsilon_{ilm}\varepsilon_{jpq} = \delta_{ij}\delta_{lp}\delta_{mq} + \delta_{ip}\delta_{lq}\delta_{mj} + \delta_{iq}\delta_{lj}\delta_{mp} - \delta_{ij}\delta_{lq}\delta_{mp} - \delta_{ip}\delta_{lj}\delta_{mq} - \delta_{iq}\delta_{lp}\delta_{mj}, \tag{5-347}$$

as can be seen by substituting numerical values for the subscripts and using eqs. (2-23) or (2-23a) and (2-3). Then eq. (5-346) becomes

$$\Omega_{ij} = (\delta_{ij}\kappa^2 - \kappa_i\kappa_j)\varphi_{ll} - \kappa^2\varphi_{ji}, \tag{5-348}$$

which is the same as eq. (3-2-3) of ref. [4]. Finally $\overline{\omega_i\omega_j'}(0) = \overline{\omega_i\omega_j}$ is obtained from eq. (5-345).

Figure 5-65 shows the variation of $\alpha_{\omega_{max}}$ with dimensionless velocity gradient. The quantity t in the dimensionless velocity gradient is the time of decay, and t_0 is an initial

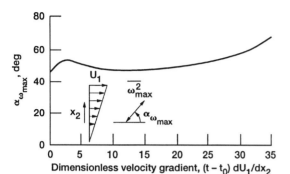

Figure 5-65 Variation of angle of maximum turbulent vorticity with dimensionless velocity gradient.

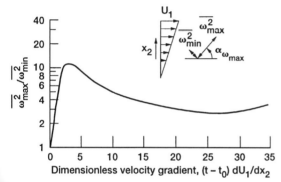

Figure 5-66 Variation of ratio of maximum to minimum mean-square turbulent vorticity with dimensionless velocity gradient.

time when the turbulence is isotropic. U_1 is the mean velocity, and x_2 is a transverse coordinate. As expected from the form of eq. (5-343), the angle for which the turbulent vorticity is a maximum is 45° only if the dimensionless velocity gradient is zero (isotropic turbulence). However, it remains at values slightly greater than 45° (between 48° and 53°) for a considerable range of dimensionless velocity gradients. It is of interest that the experimental vortices of ref. [90] appear to be inclined to the mean flow at similar angles.

Figure 5-66 shows the degree of alignment of the turbulent vortices in the direction of maximum vorticity. For no alignment of the vortices, the ratio $\overline{\omega_{max}^2}/\overline{\omega_{min}^2}$ is, of course 1. On the other hand, if the vortices all are aligned at the angle $\alpha_{\omega max}$, the ratio $\overline{\omega_{max}^2}/\overline{\omega_{min}^2}$ is infinite. The figure shows that the ratio actually varies between 1 and about 12 for values of dimensionless velocity gradient shown. The decrease in degree of vortex alignment as dimensionless velocity gradient increases beyond 3 may be an effect of pressure fluctuations.

Effect of initial condition. For the results given in section 5-4-2-1 thus far, the turbulent energy always decays with time (see Fig. 5-63). Although energy is fed into the turbulence by the mean-velocity gradient, so that the turbulence decays at a slower rate than it would have with no shear, the turbulent energy produced by the shear is less than that dissipated by viscous action. For the initial condition on the spectrum of the turbulence it is assumed that eq. (5-333) holds at $t = t_0$. Equation (5-333) is for an initially isotropic turbulence and gives infinite initial total turbulent energy (per unit mass), although the energy at any finite wavenumber is finite. (At any time greater than t_0 the total turbulent energy is also finite.)

Recently, Hasen [91] used an expression for $(\varphi_{ij})_0$ equivalent to that in eq. (5-333) multiplied by a negative exponential in κ^2. In that case the total initial energy was finite. Her results, which were limited to two-dimensional initial disturbances, showed that the turbulent energy can increase for a finite range of times. It is of interest to determine whether the same effect occurs for an initial three-dimensional isotropic turbulence [92]. In order to do that we modiy eq. (5-333) to give

$$(\varphi_{ij})_0 = \frac{J_0}{12\pi^2}(\delta_{ij}\kappa^2 - \kappa_i\kappa_j)e^{-\kappa^2/\kappa_0^2}, \tag{5-349}$$

where $1/\kappa_0$ can be considered as an initial scale for the turbulence.

The equations for φ_{ij} given thus far in section 5-4-2-1 are derived for the initial condition given in eq. (5-333). Those equations (eqs. [5-334] through [5-337]) can be modified so that the initial condition for φ_{ij} is given by eq. (5-349) by multiplying their right sides by the exponential

$$\exp\left\{-\frac{1}{\kappa_0^2}\left[\kappa_1^2 + \left(\kappa_2 + \frac{dU_1}{dx_2}\kappa_1(t - t_0)\right)^2 + \kappa_3^2\right]\right\}.$$

Integration of φ_{ij} over all wavenumber space then gives the turbulent energy tensor $\overline{u_i u_j}$, where u_i and u_j are velocity components and the overbar indicates an averaged value.

Figure 5-67 shows a dimensionless plot of turbulent energy against time for various values of a Reynolds number R. The quantity $\frac{1}{2}\overline{u_i u_i} = \frac{1}{2}(\overline{u_1^2} + \overline{u_2^2} + \overline{u_3^2})$ is the turbulent energy, and $(\overline{u_i u_i})_0$ is the value of $\overline{u_i u_i}$ at the initial time t_0. The Reynolds number is based on mean velocity gradient dU_1/dx_2 and initial turbulence scale $1/\kappa_0$. The quantity U_1 is the mean velocity, x_2 is distance in the direction of the mean-velocity gradient, and ν is the kinematic viscosity. The curves for the plot are obtained by using the modified equation (5-336) and eq. (5-304) (with $r = 0$).

For small values of R the energy decreases monotonically with time, as in Fig. 5-63, in which eq. (5-333) rather than (5-349) is used for the initial condition. (Note that eq. [5-333] is obtained from eq. [5-349] by letting $R = 0$, or $\kappa_0 = \infty$.) For larger values of R, however, the energy increases during finite time intervals. Thus, for the larger Reynolds numbers there are time regions for which the energy fed into the turbulent field by the mean shear exceeds that dissipated by viscous action. For large times the energy again decays, and it appears that a steady-state turbulence is not attained.

The ultimate decay of turbulent energy in all parts of wavenumber space, according to the analysis in the present section, can be seen from the structure of eq. (5-336) and of the above exponential (by which eq. [5-336] is multiplied here). At large times the

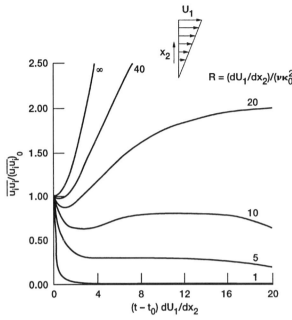

Figure 5-67 Variation of total turbulent energy (per unit mass) with dimensionless time and Reynolds number R.

argument of the exponential in eq. (5-336), as well as that in the above exponential, is negative for all values of κ_i and dU_1/dx_2.

Some comments on the maintenance of turbulence. Because the turbulent energy can increase for finite time intervals, one might ask what happens physically at large times to cause the energy to again decrease. In order to answer that question, we look at the $\overline{u_2^2}$ component of the energy. Figure 5-68 shows a dimensionless plot of $\overline{u_2^2}$, the component of turbulent energy in the direction of mean-velocity gradient. It is seen that regardless of the value of R, there are no time regions during which $\overline{u_2^2}$ increases, and for large times $\overline{u_2^2}$ becomes a small fraction of the total energy. This is evidently because there is no turbulence production term in the spectral equation for $\overline{u_2^2}$; so that the only way energy can be fed into the $\overline{u_2^2}$ component is by means of the pressure-velocity correlation terms, which can transfer energy between directional components. However, as shown in Fig. 5-57, the pressure-velocity terms tend to extract energy from the $\overline{u_2^2}$ component, rather than to deposit it there (except at small velocity gradients or times). Thus, the reason that all of the components of the turbulence decay for large times is evidently that the energy is drained out of the u_2 component of the velocity, with the result that the turbulent shear $\overline{u_1 u_2}$ goes to zero. There is then no mechanism for maintaining the turbulence, because the turbulent energy is produced by the work done on the Reyonlds strss $\overline{u_1 u_2}$ by the velocity gradient.

On the other hand, consider a somewhat stronger turbulence or small mean shear, where the triple-correlation terms in the equations for the pressure-velocity correlations

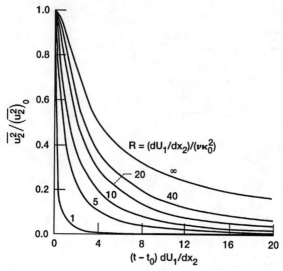

Figure 5-68 Variation of component of turbulent energy in direction of mean velocity gradient with dimensionless time and Reynolds number R.

are not negligible. Equation (5-301) becomes (neglecting buoyancy),

$$\frac{1}{\rho} \frac{\partial^2 \overline{\sigma u'_j}}{\partial r_l \partial r_l} = 2 \frac{\partial U_l}{\partial x_2} \frac{\overline{\partial u_2 u'_j}}{\partial r_l} - \frac{\partial^2 \overline{u_l u_k u'_j}}{\partial r_l \partial r_k}, \qquad (5\text{-}350)$$

where unprimed and primed quantities refer, respectively, to values at points P and P' separated by the vector r, σ is the pressure, and ρ is the density. A repeated subscript in a term indicates a summation. In this case it seems likely that the pressure-force terms transfer energy into the $\overline{u_2^2}$ components, because their probable effect, if triple correlations are present, is to make the turbulence more isotropic [4]. The result may be that a nondecaying solution can be obtained with a uniform velocity gradient in a homogeneous turbulence, as seems to be the case in the experiments of Rose [93]. This in no way conflicts with the results of section 4-3-4, because there inhomogeneities in the turbulence may have an additional sustaining effect on the turbulent energy.

Thus, if the above speculation concerning the directional transfer of energy is correct, it may be that the triple correlations play a crucial, although indirect, role in maintaining the turbulence by transferring energy between directional components through the pressure–velocity correlations. This effect of the triple correlations may be more important in turbulent shear flow than the transfer of energy between wavenumbers by those correlations. The latter seems to be simulated reasonably well by the energy transfer between wavenumbers that is produced by the mean-velocity gradient (see Fig. 5-61). We shall return to the problem of the maintenance (or growth) of shear-flow turbulence in subsequent sections. First, however, we want to get an idea as to whether predictions from our linearized theory are realistic.

5-4-2-2 Comparison of theory with experiment for uniformly sheared turbulence.

Approximately uniformly sheared turbulence has been investigated experimentally in refs. [93] through [95]. Comparison of those results with the somewhat idealized theory of section 5-4-2-1 indicates that the latter shows qualitative features that are very much like those observed.

In order to obtain a quantitative comparison between theory and experiment, more realistic initial conditions than those given by eq. (5-333) or (5-349) should be used. Initially isotropic turbulence is assumed in the theory, whereas the initial turbulence in the experiments was not isotropic. Also, the shape of the assumed initial energy spectrum is probably not realistic. In the present section we use an initial anisotropic spectral tensor that appears to be general enough to represent the initial experimental turbulence realistically. The theoretical results for the evolution of the turbulence are compared with those obtained experimentally to see whether a reasonable quantitative correspondence exists.

The evolution of the spectrum tensor corresponding to $\overline{u_i u'_j}(r)$ is given by eq. (5-324):

$$\frac{\partial \varphi_{ij}}{\partial t} = -(\delta_{i1}\varphi_{2j} + \delta_{j1}\varphi_{i2})\frac{dU_1}{dx_2} + \kappa_1 \frac{\partial \varphi_{ij}}{\partial \kappa_2}\frac{dU_1}{dx_2} + \left(2\frac{\kappa_1 \kappa_i}{\kappa^2}\varphi_{2j} + 2\frac{\kappa_1 \kappa_j}{\kappa^2}\varphi_{i2}\right)\frac{dU_1}{dx_2} - 2\nu\kappa^2 \varphi_{ij}. \tag{5-324}$$

where the spectrum tensor φ_{ij} is defined by

$$\overline{u_i u'_j}(r) = \int_{-\infty}^{\infty} \varphi_{ij}(\kappa) e^{i\kappa \cdot r} d\kappa, \tag{5-304}$$

and κ is the wavenumber vector. The quantity φ_{ij} is the spectral component of $\overline{u_i u_j}$ at κ. In order to interpret the terms in eq. (5-324), we first multiply the equation through by $d\kappa$. Then, the first term on the right side gives the rate of production of φ_{ij} in $d\kappa$ by work done on φ_{ij} by the mean-velocity gradient. The second term gives the net rate of transfer of φ_{ij} into $d\kappa$ from other wavenumber regions by $\partial U_1/\partial x_2$. If this term is integrated over all κ, the result is zero. The third term gives the rate of transfer of φ_{ij} between its directional components by pressure forces associated with $\partial U_1/\partial x_2$. This term drops out for $i = j$. Finally, the last term gives the rate of dissipation of φ_{ij} in $d\kappa$.

In section 5-4-2-1 it is assumed that the initial turbulence is isotropic and is given by eq. (5-333) or (5-349). The former equation is a special case of the latter, which is

$$(\varphi_{ij})_0 = \frac{J_0}{12\pi^2}(\delta_{ij}\kappa^2 - \kappa_i \kappa_j)e^{-l^2\kappa^2}, \tag{5-351}$$

J_0 and l are constants of the initial conditions, and the subscript 0 designates values at the initial time.

However, eq. (5-349) is not general enough to represent the initial experimental turbulence in refs. [93] through [95]. That turbulence is anisotropic and may have a spectrum whose shape is not given by eq. (5-349). An expression for $(\varphi_{ij})_0$ that may be sufficiently general to represent the initial experimental turbulence is

$$(\varphi_{ij})_0 = \frac{c_{lm}\kappa^2}{4\pi}\left(\delta_{il} - \frac{\kappa_i \kappa_l}{\kappa^2}\right)\left(\delta_{jm} - \frac{\kappa_j \kappa_m}{\kappa^2}\right)\exp\left[-(\kappa l_{(lm)})^{n_{(lm)}}\right], \tag{5-352}$$

where, for φ_{11}, φ_{22}, φ_{33}, and φ_{12} (the only components of φ_{ij} of interest here),

$$c_{lm} = \frac{3}{14} \frac{n_{(lm)} l_{(lm)}^5}{\Gamma(5/n_{(lm)})} [10\overline{(u_l u_m)}_0 - \delta_{lm} \overline{(u_k u_k)}_0], \tag{5-353}$$

and where the l_{lm} and n_{lm} are constants of the initial conditions, and Γ is the gamma function. Parentheses are placed on some of the subscripts to indicate that there is no sum on those subscripts. Equations (5-352) and (5-353) are consistent with eq. (5-304) for $r = 0$. As an aid to obtaining the initial constants l_{lm}, they can be related to the initial derivatives $(\partial \overline{u_i u_j}/\partial t)_0$ by substituting eqs. (5-352) and (5-353) in (5-324), multiplying (5-324) by $d\kappa$, integrating, and using eq. (5-304) (with $r = 0$). This gives

$$l_{lm} = \left(\frac{\Gamma(7/n_{(lm)})}{\Gamma(5/n_{(lm)})} \frac{[10\overline{(u_l u_m)}_0 - \delta_{lm} \overline{(u_k u_k)}_0]}{10 I_{lm} - \delta_{lm} I_{kk}} \right)^{1/2}, \tag{5-354}$$

where

$$I_{ij} = -\frac{1}{2\nu} \left(\frac{\partial \overline{u_i u_j}}{\partial t} \right)_0 - B_{ijkl} \frac{\overline{(u_k u_l)}_0}{\nu} \frac{dU_1}{dx_2}, \tag{5-355}$$

and where the nonzero components of B_{ijkl} are $B_{1112} = 23/49$, $B_{2212} = 16/49$, $B_{3312} = 10/49$, $B_{1211} = 29/490$, $B_{1222} = 68/490$, and $B_{1233} = -1/490$.

The values of the remaining constants of the initial conditions, the n_{lm}, depend on the shapes of the initial spectra. Because the initial spectra were not measured in the experiments, we determined the n_{lm} from the evolution of the $\overline{u_i u_j}$. For the data of references 93 through 95, values for the nonzero n_{lm} were obtained as $n_{11} = 3/4$, $n_{22} = 1/2$, $n_{33} = 3/4$, and $n_{12} = 4$.

Equation (5-324) is solved in section 5-4-2-1 for the special initial condition given by eq. (5-351). For general initial conditions we obtain, for φ_{11}, φ_{22}, φ_{33}, and φ_{12}.

$$\varphi_{ij} = H_{ij} \exp\left[-2\nu \kappa^2 (t - t_0) \left(1 + a_{12}^*(\kappa_1 \kappa_2/\kappa^2) + \frac{1}{3} a_{12}^{*2} \kappa_1^2/\kappa^2 \right) \right], \tag{5-356}$$

where

$$a_{12}^* = (dU_1/dx_2)(t - t_0), \tag{5-357}$$

$$H_{22} = (\varphi_{22})_0 D^2, \tag{5-358}$$

$$H_{12} = D[(\varphi_{12})_0 + (\varphi_{22})_0 R], \tag{5-359}$$

$$H_{11} = (\varphi_{11})_0 + 2(\varphi_{12})_0 R + (\varphi_{22})_0 R^2, \tag{5-360}$$

$$H_{ii} = (\varphi_{ii})_0 + \frac{2}{\kappa_1/\kappa}[(\varphi_{12})_0 E + (\varphi_{22})_0 I], \tag{5-361}$$

$$H_{33} = H_{ii} - H_{11} - H_{22}, \tag{5-362}$$

$$D = 1 + 2a_{12}^*(\kappa_1 \kappa_2/\kappa^2) + a_{12}^{*2} \kappa_1^2/\kappa^2, \tag{5-363}$$

$$R = \frac{E}{\kappa_1/\kappa} - \frac{2\kappa_1/\kappa}{D} F, \tag{5-364}$$

$$I = \frac{1}{2}\left[(D^2-1)\frac{\kappa_1}{\kappa} + \frac{E^2}{\kappa_1/\kappa}\right] - \frac{\kappa_1}{\kappa}\frac{F}{D}\left[E - (D+1)\frac{\kappa_2}{\kappa} - a_{12}^*\frac{\kappa_1}{\kappa}\right], \tag{5-365}$$

$$E = -\frac{D}{(1-\kappa_2^2/\kappa^2)^{1/2}} \tan^{-1}\left(\frac{a_{12}^*(\kappa_1/\kappa)(1-\kappa_2^2/\kappa^2)^{1/2}}{1 + a_{12}^* \kappa_1 \kappa_2/\kappa^2} \right), \tag{5-366}$$

and

$$F = \frac{D}{2(1 - \kappa_2^2/\kappa^2)} \left[\frac{\kappa_2}{\kappa}(D-1) - a_{12}^* \frac{\kappa_1}{\kappa} + E \right]. \tag{5-367}$$

The quantity $(\varphi_{ij})_0$ is the value of φ_{ij} at the initial time t_0. In the present study it is obtained from eq. (5-352). The correlations $\overline{u_i u_j}$ and $\overline{u_i u'_j}$ are calculated from eq. (5-304).

Plots for the evolution of components of $\overline{u_i u_j}$ are shown in Fig. 5-69. Those evolutions are used to determine the constants in the initial-condition eq. (5-352). Values of t for the experiments are calculated as the longitudinal distance divided by the centerline velocity. The agreement of the calculated curves with experiment indicates that the initial condition given by eq. (5-352) is evidently general enough for our purposes. Note that the total strain in the experiments of ref. [95] is appreciably greater than that in the other experiments. The results near the end of the curve for $\overline{u_2^2}^{1/2}$ in Fig. 5-69c indicate that those results may be partially outside the range of applicability of the analysis, although the accuracy of the experimental results in that region is uncertain.

Once the initial conditions have been determined from eq. (5-352) and Fig. 5-69, one can make a comparison of theory with experiment for several turbulence quantities (Figs. 5-70 through 5-72). Evolutions of turbulence macroscales L_i and Taylor microscales λ_i are plotted in Fig. 5-70. The scales are defined in the usual way as

$$L_i = \int_0^\infty \overline{u_1 u'_1}(r_i, 0, 0) dr_i / \overline{u_1^2} \tag{5-368}$$

and

$$\lambda_1^2 = -\frac{2\overline{u_1^2}}{\left(\partial^2 \overline{u_1 u'_1}/\partial r_i^2\right)_{r=0}}. \tag{5-369}$$

In Fig. 5-70 both theory and experiment indicate that the microscale λ_1 increases with time. Similar results are obtained for the other scales. The experimental scale ratios L_2/L_1 and λ_2/λ_1 are close to the values for isotropic turbulence (1/2 and $1/\sqrt{2}$). The theoretical values of L_2/L_1 and λ_2/λ_1 are close to the isotropic values at early times and tend to be, respectively, somewhat higher and lower than those values at later times.

This growth of the turbulence scales with time appears to be a characteristic of unbounded fields of turbulence with or without mean velocity gradients. An exception occurs at short times, if the spectral transfer of energy to large wavenumbers causes a decrease of scale (see Fig. 5-50). The growth of scales, which eventually wins out, is due mainly to the selective annihilation of eddies by viscous action, the smaller eddies being the first to go because of the larger shear stresses between them. If all eddy sizes are present at some time, as they are in eq. (5-352), growth continues indefinitely, so that a steady-state situation for uniform shear flow in which all turbulence quantities are constant with time appears unlikely.

Even if there is an upper limit on the initial eddy size, say the grid spacing, larger eddies may be generated by inertial transfer. Figure 5-73 shows that for the range of values of $(t - t_0)dU_1/dx_2$ of interest here (from 0 to about 2) there is considerable reverse energy transfer to lower wavenumbers (larger eddies) produced by the shear. This is indicated by the positive areas at low wavenumbers in Fig. 5-73 and may be a

246 TURBULENT FLUID MOTION

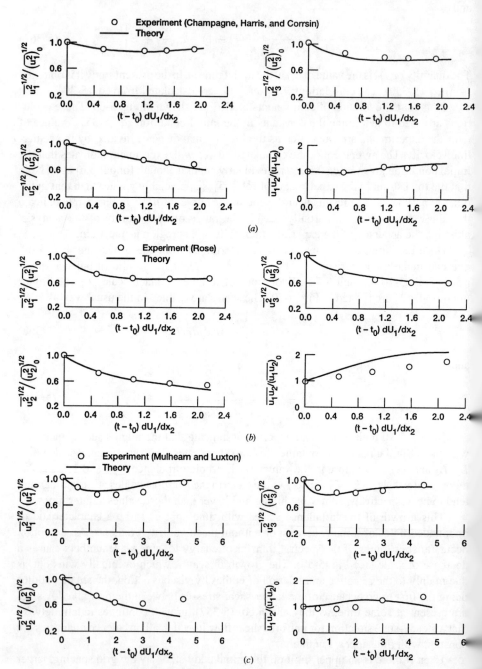

Figure 5-69 Evolution of one-point turbulence components; plots for determining constants (eq. [5-352]) for initial conditions. (*a*) Data from ref. [94]; (*b*) data from ref. [93]; (*c*) data from ref. [95].

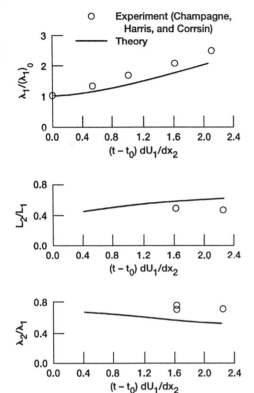

Figure 5-70 Comparison of theory and experiment for evolution of scale ratios. Data from ref. [94].

reason the experiment in ref. [93] sometimes shows scales larger than the turbulence-generator spacing at large distances downstream. Of course, eventually the scale size is limited by the size of the test section, but when that occurs, the turbulence is no longer nearly homogeneous.

Two-point correlations. Of particular interest are the two-point velocity correlations in Figs. 5-71 and 5-72. They indicate that the negative region observed experimentally in the $\overline{u_1 u_1'}(0, 0, r_3)$ correlation, and the absence of such a region in the other correlations, are predicted by the theory. In isotropic turbulence negative values occur, of course, in both the $\overline{u_1 u_1'}(0, r_2, 0)$ and $\overline{u_1 u_1'}(0, 0, r_3)$ correlations as a result of the continuity condition [4]. The confinement of such regions to the $\overline{u_1 u_1'}(0, 0, r_3)$ correlations in Figs. 5-71 and 5-72 seems to be associated with the turbulent shear flow. This is observed in the experiments both of Champagne, Harris, and Corrsin [94] and of Rose [93]. The successful prediction of the two-point velocity correlations, particularly the correct negative and positive regions, is somewhat of a triumph of the linearized theory. This, of course, does not mean that the turbulence is linear, but only that the nonlinear effects seem to be overshadowed in the present cases.

Comparison of the curves for $\overline{u_1 u_1'}(0, r_2, 0)$ in Figs. 5-71 and 5-72 indicates that the shape is nearly the same for the two times shown. Similarly, Fig. 5-74 shows that the

248 TURBULENT FLUID MOTION

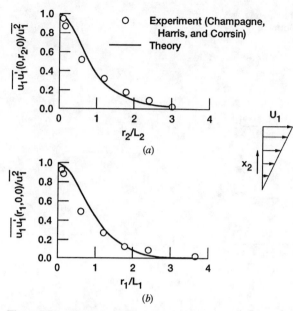

Figure 5-71 Comparison of theory and experiment for two-point velocity correlations. $(dU_1/dx_2)(t-t_0) = 1.641$. Data from ref. [94]. (a) Two points separated along direction of velocity gradient; (b) two points separated along flow direction.

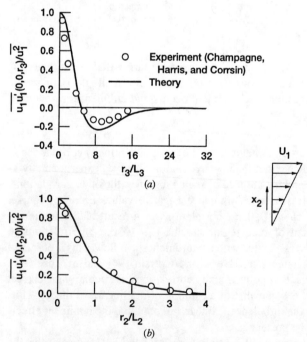

Figure 5-72 Comparison of theory and experiment for two-point velocity correlations. $(dU_1/dx_2)(t-t_0) = 2.27$. Data from ref. [94]. (a) Two points separated along direction normal to both flow direction and direction of velocity gradient; (b) two points separated along direction of velocity gradient.

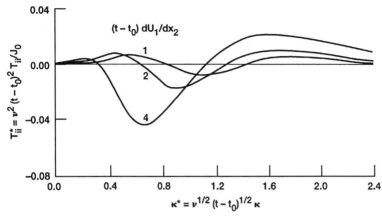

Figure 5-73 Dimensionless spectra of energy-transfer term due to mean-velocity gradient (eq. [5-324]) (second term on right side) integrated over all directions in wavenumber space. Equation (5-351) with $l = 0$ is used for initial condition.

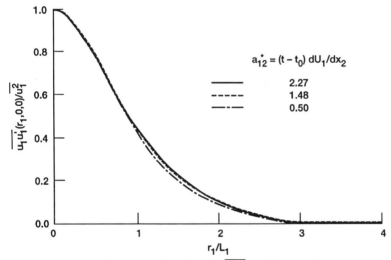

Figure 5-74 Curves showing the similarity of the $\overline{u_1 u_1'}(r_1, 0, 0)$ correlations at various times.

shape of the $\overline{u_1 u_1'}(r_1, 0, 0)$ curve is nearly preserved over a considerable time span. Thus, although the turbulence scales grow with time, the distribution of eddy sizes seems to remain similar as the scales grow. This is somewhat like the final period of decay without shear, in which the correlations and spectra remain similar as the turbulence decays and the scales grow [4].

In view of the complexity of the overall turbulence process, with contributions to the change in turbulence components being produced by turbulence production, by transfer between wavenumbers and directional components, and by dissipation (eq. [5-324]), it seems remarkable that the combined effect is to change the turbulence in such a way that the eddy-size distribution remains nearly similar. Of course, one cannot expect that this similarity is preserved indefinitely. The spectra in ref. [7] show that for large $(t - t_0)dU_1/dx_2$ the shapes of the spectra change. However, for the values of $(t - t_0)dU_1/dx_2$ in the experiments considered here, although the effects of the shear are great enough to produce a large influence on the turbulence levels (in the absence of shear all components of the turbulence would decrease monotonically in Fig. 5-69), they are not large enough to alter the shapes of the correlation curves appreciably.

Although the results indicate that in most cases a good correspondence exists between theory and experiment, a higher-order theory that retains turbulence self-interaction terms in the equations might give some improvement. However, the small inhomogeneities that occur to some extent in all of the experiments may have as important an effect as the self-interaction.

5-4-2-3 Heat transfer and temperature fluctuations in a uniformly sheared turbulence.
Turbulent heat transfer and flow in passages and boundary layers usually are analyzed using a phenomenological approach. That is, assumptions are introduced into the analysis to relate the turbulent shear stress and turbulent heat transfer to the mean flow. Examples of these analyses are given, for example, in sections 4-3-2-6 through 4-3-2-13 and in refs. [96] through [105]. This approach is very useful and makes it possible to generalize large quantities of experimental data. In fact, it appears to be the only feasible way, at present, of analyzing the complex high–Reynolds-number flows occurring in boundary layers and passages.

Although the phenomenological analyses are very useful, we can obtain a great deal more insight into the turbulent processes by using a statistical approach based on the equations of motion and energy. This, of course, is the approach generally followed in this chapter. These studies should help to put the phenomenological analyses on a sounder basis. Because of the complexity of turbulence it is necessary to limit one's self at least at the beginning to simple models when studying it from a fundamental standpoint. Thus, Corrsin [106] and Dunn and Reid [107] study heat transfer in isotropic turbulence with a uniform mean temperature gradient. (The term *isotropic* as usual indicates that the statistical properties of the turbulence are independent of direction.)

Here we extend the analysis of section 5-4-2-1 to include uniform heat transfer and temperature fluctuations [108]. The mean temperature gradient, as well as the shear, is uniform. Locally, the heat transfer and flow in this case are somewhat similar to those in passages and boundary layers if the scales of the turbulence in the flows are reasonably small compared with the scales of the inhomogeneities.

The fluid properties are assumed constant, so that the turbulent velocity field is independent of the temperature field. Thus the results for turbulence with a uniform velocity gradient from section 5-4-2-1 can be used for obtaining the turbulent heat transfer and temperature fluctuations. It is shown in section 5-4-2-1 that a homogeneous turbulent field with a uniform velocity gradient decays with time. Although energy is fed into the turbulence from the mean-velocity gradient, the production of turbulence is never great enough to offset the dissipation. The fluctuating temperature field and the turbulent heat transfer also change with time.

Because of the decay of the turbulence with time it is necessary to produce it initially by some means, for instance, by passing a sream through a grid. Then various distances downstream from the grid correspond to various times of decay. Approximately uniform transverse velocity and temperature gradients in the stream can be produced by passing the flow through parallel channels before passing it through the grid. The temperature and velocity of the fluid in each channel are adjusted to produce the desired velocity and temperature gradients across the stream emerging from the channels. Because of the higher velocities through some parts of the grid it might be necessary to vary the thickness of the wries in the grid to produce an approximately homogeneous turbulence. Heating of the grid is not necessary because, as is seen subsequently, temperature fluctuations can arise from the interaction of the turbulence and the mean temperature gradient.

As in section 5-4-2-1 the mean gradients are assumed to be large enough, or the turbulence weak enough, for the triple-correlation terms occurring in the analysis to be neglected or at least for those terms to be small enough compared with other terms to give a sensible solution. Thus eqs. (5-313) through (5-320), with $g_i = 0$ (buoyancy neglected), apply here. As in section 5-4-2-1, the mean velocity is in the x_1 direction, and the mean-velocity gradient is in the x_2 direction. The mean-temperature gradient also is taken to be in the x_2 direction. Then eqs. (5-313) through (5-320) become

$$\frac{\partial}{\partial t}\varphi_{ij} + \delta_{i1}\varphi_{2j}\frac{dU_1}{dx_2} + \delta_{j1}\varphi_{i2}\frac{dU_1}{dx_2} - \kappa_1\frac{\partial \varphi_{ij}}{\partial \kappa_2}\frac{dU_1}{dx_2} = -\frac{1}{\rho}(-i\kappa_i\lambda_j + i\kappa_j\lambda_i') - 2\nu\kappa^2\varphi_{ij},$$
(5-370)

$$\frac{\partial \delta}{\partial t} - \frac{dU_1}{dx_2}\kappa_1\frac{\partial \delta}{\partial \kappa_2} + \frac{dT}{dx_2}(\gamma_2 + \gamma_2') = -2\alpha\kappa^2\delta$$
(5-371)

$$\frac{\partial \gamma_j}{\partial t} - \frac{dU_1}{dx_2}\kappa_1\frac{\partial \gamma_j}{\partial \kappa_2} + \varphi_{2j}\frac{dT}{dx_2} + \delta_{1j}\gamma_2\frac{dU_1}{dx_2} = -\frac{1}{\rho}i\kappa_j\zeta' - (\alpha+\nu)\kappa^2\gamma_j$$
(5-372)

$$\frac{\partial \gamma_i'}{\partial t} - \frac{dU_1}{dx_2}\kappa_1\frac{\partial \gamma_i'}{\partial \kappa_2} + \delta_{i1}\gamma_2'\frac{dU_1}{dx_2} + \varphi_{i2}\frac{dT}{dx_2} = \frac{1}{\rho}i\kappa_i\zeta - (\alpha+\nu)\kappa^2\gamma_i'.$$
(5-373)

$$-\frac{1}{\rho}i\kappa_j\lambda_i' = 2\frac{\kappa_1\kappa_j}{\kappa^2}\varphi_{i2}\frac{dU_1}{dx_2},$$
(5-374)

$$\frac{1}{\rho}i\kappa_i\lambda_j = 2\frac{\kappa_1\kappa_i}{\kappa^2}\varphi_{2j}\frac{dU_1}{dx_2},$$
(5-375)

$$\frac{1}{\rho}i\kappa_i\zeta = 2\frac{\kappa_1\kappa_i}{\kappa^2}\gamma_2'\frac{dU_1}{dx_2},$$
(5-376)

and

$$-\frac{1}{\rho}i\kappa_j\zeta' = 2\frac{\kappa_1\kappa_j}{\kappa^2}\gamma_2\frac{dU_1}{dx_2}.$$
(5-377)

Substituting eqs. (5-376) and (5-377) into the right-hand sides of eqs. (5-373) and (5-372), letting $i = j = 2$, and comparing the resulting equations show that $\gamma_2 = \gamma_2'$ for all times if they are equal at an initial time. Here it is assumed that the temperature fluctuations are initially zero, so that the above relation will hold. If

$$\frac{dU_1}{dx_2} \equiv a_{12} \tag{5-378}$$

and

$$\frac{dT}{dx_2} \equiv b_2 \tag{5-379}$$

we finally obtain

$$\frac{\partial \gamma_2}{\partial t} - a_{12}\kappa_1 \frac{\partial \gamma_2}{\partial \kappa_2} = -b_2 \varphi_{22} + \left[2a_{12}\frac{\kappa_1 \kappa_2}{\kappa^2} - \left(\frac{1}{Pr}+1\right)\nu\kappa^2\right]\gamma_2 \tag{5-380}$$

and

$$\frac{\partial \delta}{\partial t} - a_{12}\kappa_1 \frac{\partial \delta}{\partial \kappa_2} = -2b_2 \gamma_2 - 2\alpha\kappa^2 \delta, \tag{5-381}$$

where the Prandtl number $Pr = \nu/\alpha$.

In order to obtain solutions of eqs. (5-380) and (5-381) it is assumed, as in section 5-4-2-1, that the turbulence is initially isotropic, so that we can use eq. (5-333) for the initial φ_{ij}. The turbulence is not, of course, isotropic at later times.

Note that φ_{ij}, according to eqs. (5-370) and (5-333), is not a function of temperature (those equations do not contain T, δ, γ_i, or γ_i'). Thus the solution already obtained in section 5-4-2-1 can be used to obtain φ_{22} in eq. (5-380):

$$\varphi_{22} = \frac{J_0\{\kappa_1^2 + [\kappa_2 + a_{12}\kappa_1(t-t_0)]^2 + \kappa_3^2\}^2 (\kappa_1^2 + \kappa_3^2)}{12\pi^2 \kappa^4}$$

$$\times \exp\left\{-2\nu(t-t_0)\left[\kappa^2 + \frac{1}{3}\kappa_1^2 a^2(t-t_0)^2 + a_{12}\kappa_1\kappa_2(t-t_0)\right]\right\}, \tag{5-334}$$

where J_0 and t_0 are constants that depend on initial conditions. For a Prandtl number Pr of 1 the solution of (5-380) is

$$\kappa^2 \exp\left\{2\nu(t-t_0)\left[\kappa^2 + a_{12}\kappa_1\kappa_2(t-t_0) + \frac{1}{3}a_{12}^2\kappa_1^2(t-t_0)^2\right]\right\}\gamma_2$$

$$= \frac{J_0\{\kappa_1^2 + [\kappa_2 + a_{12}(t-t_0)]^2 + \kappa_3^2\}^2(\kappa_1^2 + \kappa_3^2)^{1/2}}{12\pi^2 a_{12}\kappa_1}$$

$$\times b_2 \tan^{-1} \frac{\kappa_2}{(\kappa_1^2 + \kappa_3^2)^{1/2}} + f[\kappa_1, \kappa_2 + a_{12}\kappa_1(t-t_0), \kappa_3] \tag{5-382}$$

where f is a function of integration. The method of solution is given in ref. [79]. In order to evaluate f, it is assumed that the temperature fluctuations are zero for $t = t_0$. Thus substituting $\gamma_2 = 0$ for $t = t_0$ in eq. (5-382),

$$f(\kappa_1\kappa_2, \kappa_3) = -\frac{J_0(\kappa_1^2 + \kappa_2^2 + \kappa_3^2)^2(\kappa_1^2 + \kappa_3^2)^{1/2}}{12\pi^2 a_{12}\kappa_1} b_2 \tan^{-1} \frac{\kappa_2}{(\kappa_1^2 + \kappa_3^2)^{1/2}} \tag{5-383}$$

or

$$f[\kappa_1\kappa_2 + a_{12}\kappa_1(t-t_0), \kappa_3] = -\frac{J_0\{\kappa_1^2 + [\kappa_2 + a_{12}\kappa_1(t-t_0)]^2 + \kappa_3^2\}^2 (\kappa_1^2 + \kappa_3^2)^{1/2}}{12\pi^2 a_{12}\kappa_1}$$

$$\times b_2 \tan^{-1} \frac{\kappa_2 + a_{12}\kappa_1(t-t_0)}{(\kappa_1^2 + \kappa_3^2)^{1/2}}. \tag{5-384}$$

Substitution of eq. (5-384) in (5-382) gives, for the Fourier transform of $\overline{\tau u_2'}$ for a Prandtl number of 1,

$$\gamma_2 = \frac{J_0\{\kappa_1^2 + [\kappa_2 + a_{12}\kappa_1(t-t_0)]^2 + \kappa_3^2\}^2 (\kappa_1^2 + \kappa_3^2)^{1/2}}{12\pi^2 a_{12}\kappa_1\kappa^2}$$

$$\times b_2 \exp\left\{-2\nu(t-t_0)\left[\kappa^2 + a_{12}\kappa_1\kappa_2(t-t_0) + \frac{1}{3}a_{12}^2\kappa_1^2(t-t_0)^2\right]\right\}$$

$$\times \left[\tan^{-1}\frac{\kappa_2}{(\kappa_1^2 + \kappa_3^2)^{1/2}} - \tan^{-1}\frac{\kappa_2 + a_{12}\kappa_1(t-t_0)}{(\kappa_1^2 + \kappa_3^2)^{1/2}}\right] \tag{5-385}$$

For $Pr \neq 1$, the solution for γ_2 can be written as

$$\gamma_2 = \frac{b_2 J_0(\kappa_1^2 + \kappa_3^2)}{a_{12}\kappa_1\kappa^2 12\pi^2} \{\kappa_1^2 + [\kappa_2 + a_{12}\kappa_1(t-t_0)]^2 + \kappa_3^2\}^2 \exp\left\{\frac{[(1/Pr)-1]\nu\kappa_2}{a_{12}\kappa_1}\right.$$

$$\times \left(\kappa_1^2 + \frac{\kappa_2^2}{3} + \kappa_3^2\right) - 2\nu(t-t_0)\left[\kappa^2 + a_{12}\kappa_1\kappa_2(t-t_0) + \frac{1}{3}a_{12}^2\kappa_1^2(t-t_0)^2\right]\right\}$$

$$\times \int_{\kappa_2}^{\kappa_2 + a_{12}\kappa_1(t-t_0)} \frac{1}{\kappa_1^2 + \xi^2 + \kappa_3^2} \exp\left\{-\left[\frac{[(1/Pr)-1]\nu}{a_{12}\kappa_1}\xi\left(\kappa_1^2 + \frac{\xi^2}{3} + \kappa_3^2\right)\right]\right\}d\xi.$$

$$\tag{5-385a}$$

The expression for the Fourier transform of $\overline{\tau\tau'}$ for a Prandtl number of 1 is obtained by solution of eq. (5-381):

$$\delta = \frac{J_0\{\kappa_1^2 + [\kappa_2 + a_{12}\kappa_1(t-t_0)]^2 + \kappa_3^2\}^2 b_2^2}{12\pi^2 a_{12}^2\kappa_1^2} \exp\left\{-2\nu(t-t_0)\left[\kappa^2 + a_{12}\kappa_1\kappa_2(t-t_0)\right.\right.$$

$$\left.\left. + \frac{1}{3}a_{12}^2\kappa_1^2(t-t_0)^2\right]\right\}\left[\tan^{-1}\frac{\kappa_2}{(\kappa_1^2 + \kappa_3^2)^{1/2}} - \tan^{-1}\frac{\kappa_2 + a_{12}\kappa_1(t-t_0)}{(\kappa_1^2 + \kappa_3^2)^{1/2}}\right]^2$$

$$\tag{5-386}$$

where δ is set equal to zero for $t = t_0$.

The spectral quantities γ_2 and δ are functions of the components of the wavevector κ as well as of its magnitude. It is somewhat easier to interpret quantities that are functions only of the magnitude κ. We can obtain such quantities in the usual way by integrating γ_2 and δ over all directions in wavenumber space. Thus, define a quantity Γ_2 by the equation

$$\Gamma_2(\kappa) = \int_0^A \gamma_2(\kappa) dA(\kappa) \tag{5-387}$$

where A is the area of the surface of a sphere of radius κ. Then, because

$$\overline{\tau u_2} = \int_0^\infty \Gamma_2 \, d\kappa \tag{5-388}$$

(let $r = 0$ in equation [5-309]), $\Gamma_2 \, d\kappa$ gives the contribution from wavenumber band $d\kappa$ to $\overline{\tau u_2}$. Thus a plot of Γ_2 against κ shows how contributions to $\overline{\tau u_2}$ are distributed among the various wavenumbers or eddy sizes.

Equations (5-385) and (5-386) can be written in terms of spherical coordinates by setting

$$\kappa_1 = \kappa \cos\varphi \sin\theta, \quad \kappa_2 = \kappa \sin\varphi \sin\theta, \quad \kappa_3 = \kappa \cos\theta. \tag{5-389}$$

Then eq. (5-387) becomes

$$\Gamma_2(\kappa) = \int_0^\pi \int_0^{2\pi} \gamma_2(\kappa, \varphi, \theta) \kappa^2 \sin\theta \, d\varphi \, d\theta. \tag{5-390}$$

A similar equation for δ integrated over all directions in wavenumber space is

$$\Delta(\kappa) = \int_0^\pi \int_0^{2\pi} \delta(\kappa, \varphi, \theta) \kappa^2 \sin\theta \, d\varphi \, d\theta. \tag{5-391}$$

Letting $r = 0$ in eq. (5-311)

$$\overline{\tau^2} = \int_0^\infty \Delta \, d\kappa \tag{5-392}$$

so that, as in the case of Γ_2, $\Delta \, d\kappa$ gives contributions from the wavenumber band $d\kappa$ to $\overline{\tau^2}$.

Computed spectra. Spectra of $\overline{\tau u_2}$ and $\overline{\tau^2}$ for various values of dimensionless velocity gradient are plotted in Figs. 5-75 and 5-76. The integrations in eqs. (5-390), (5-391), (5-385), and (5-385a) are carried out numerically.

If plotted using the similarity variables shown, the curves for zero velocity gradient do not change with time, so that comparison of the various curves indicates how the velocity gradient alters the spectrum. Thus the curves in Figs. 5-75 and 5-76 that lie above those for $a_{12}^* = 0$ indicate that for those cases $\overline{\tau u_2}$ or $\overline{\tau^2}$ at a particular time is greater than it would be for no velocity gradient. The turbulence itself, of course, is decaying with time. Figure 5-75 shows the effect of Prandtl number on the spectrum of $\overline{\tau u_2}$. As Prandtl number increases, the peaks of the spectra move toward the higher wavenumber region, the change being greater at the lower values of a_{12}^*. High wavenumbers correspond to small eddies, inasmuch as the wavenumber represents the reciprocal of an eddy size (or wave length).

For zero velocity gradient the results are the same as those obtained by Dunn and Reid [107]. As the velocity gradient increases, the peaks of the spectra of $\overline{\tau u_2}$ move to lower wavenumbers because the spectrum of the production term $b_2\varphi_{22}$ in eq. (5-380) moves to the left (see Fig. 5-54). Because the production term in the equation for the spectrum of $\overline{\tau^2}$, eq. (5-381), is proportional to γ_2, the peaks of the spectra of $\overline{\tau^2}$ also move to lower wavenumbers.

FOURIER ANALYSIS, SPECTRAL FORM OF THE CONTINUUM EQUATIONS 255

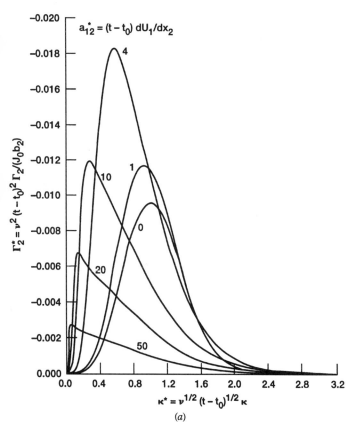

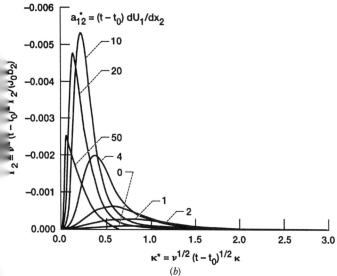

Figure 5-75 Dimensionless spectra of $\overline{\tau u_2}$ for uniform transverse velocity and temperature gradients. (a) Prandtl number, 1; (b) Prandtl number, 0.01.

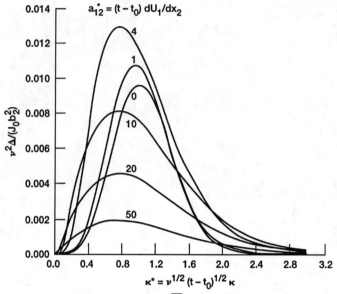

Figure 5-76 Dimensionless spectra of $\overline{\tau^2}$ for uniform tranverse velocity and temperature gradients. Prandtl number, 1.

The spectra change from approximately symmetric curves to curves having more gradual slopes on the high wavenumber sides as a_{12}^* increases. The changes in shape of the spectra apparently are caused by a transfer of activity from low wavenumbers to high wavenumbers or from big eddies to small ones. This transfer generally is associated with triple correlations [31], but in the present case, in which triple correlations are neglected, it is associated with the velocity gradient. Thus we can interpret the second terms in eqs. (5-380) and (5-381) as transfer terms. In order to do that, note that for our case $r_l \partial \overline{\tau u_j'} / \partial r_k$ in eq. (5-298) becomes $r_2 \partial \overline{\tau u_2'} / \partial r_1$, which is related to $\kappa_1 \partial \gamma_2 / \partial \kappa_2$ in eq. (5-380) by

$$r_2 \frac{\partial \overline{\tau u_2'}}{\partial r_1} = -\int_{-\infty}^{\infty} \kappa_1 \frac{\partial \gamma_2}{\partial \kappa_2} e^{i\kappa \cdot r} d\kappa. \tag{5-393}$$

For $r = 0$, this becomes

$$\int_{-\infty}^{\infty} \kappa_1 \frac{\partial \gamma_2}{\partial \kappa_2} d\kappa = 0. \tag{5-394}$$

Similarly, in eq. (5-381),

$$\int_{-\infty}^{\infty} \kappa_1 \frac{\partial \delta}{\partial \kappa_2} d\kappa = 0. \tag{5-395}$$

So these terms give zero total contribution to $\partial \overline{\tau u_2} / \partial t$ or to $\partial \overline{\tau^2} / \partial t$. However, they can alter the distribution in wavenumber space of contributions to $\partial \overline{\tau u_2} / \partial t$ or $\partial \overline{\tau^2} / \partial t$ and thus can be interpreted as transfer terms. A similar term in the equation for $\overline{u_i u_j'}$ is obtained in section 5-4-2-1.

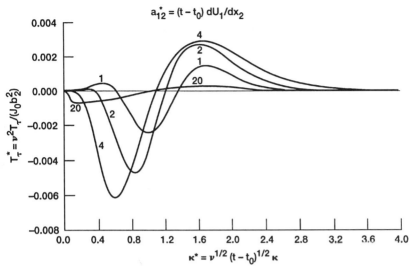

Figure 5-77 Dimensionless spectra of transfer term due to mean-velocity gradient in spectral equation for $\overline{\tau^2}$. Prandtl number, 1.

The expressions for the transfer terms in eqs. (5-380) and (5-381) can be integrated over all directions in wavenumber space by using equations similar to eqs. (5-390) and (5-391) in order to obtain quantities that are functions only of κ and dU_1/dx_2. A plot of the integrated transfer term corresponding to $\overline{\tau^2}$ is given in dimensionless form for a Prandtl number of 1 in Fig. 5-77. This term corresponds to the second term in eq. (5-381) with the exception that it has not been multiplied by a_{12}. The total area enclosed by each curve is zero, in agreement with eq. (5-395). The curves are predominately negative at low wavenumbers and positive at higher ones, so that, in general, contributions to $\overline{\tau^2}$ are transferred from low wavenumbers to high ones. In this way the higher wavenumber portions of the spectra of $\overline{\tau^2}$ in Fig. 5-76 are excited by the transfer of activity into those regions, so that the shapes of the spectra are altered. This effect is similar to that due to triple correlations [31]. In the present case a natural explanation of the effect is that the transfer to higher wavenumbers is due to the stretching of the vortex lines in the turbulence by the velocity gradient. The velocity gradient also should be able to compress some of the vortex lines, particularly at low velocity gradients at which the orientation of the vortex lines would tend to be random. This might explain the small amount of reverse transfer at low wavenumbers and low velocity gradients in Fig. 5-77.

Production, temperature-fluctuation, and conduction regions. By analogy with the interpretation of the equation for turbulent energy in section 5-4-2-1, one can say that the third term in eq. (5-381) produces temperature fluctuations by the action of the mean-temperature gradient on the turbulent heat transfer $\overline{\tau u_2}$. In the corresponding production term in the turbulent energy equation the mean-velocity gradient does work on the turbulent shear stress. The last term in eq. (5-381) is the conduction or dissipation term and tends to destroy the temperature fluctuations by conducting heat away from

258 TURBULENT FLUID MOTION

regions of high local temperature. This action is similar to the action of viscosity on the velocity fluctuations.

The production and conduction terms in eq. (5-381) can be integrated over all directions in wavenumber space by substituting Γ_{12} and Δ for γ_2 and δ respectively in those terms. These terms, together with the spectrum of $\overline{\tau^2}$ are plotted in normalized form in Figs. 5-78a and b for two values of a_{12}^* and a Prandtl number of 1. For the low dimensionless velocity gradient the production, temperature fluctuation, and conduction regions

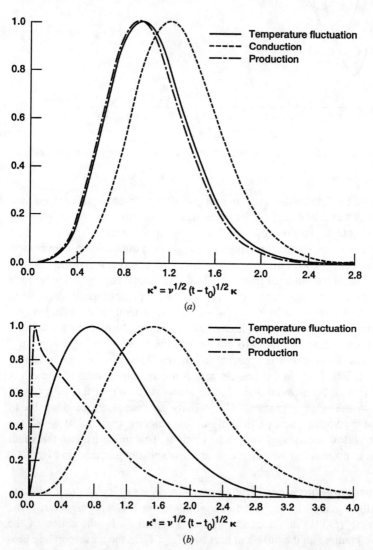

Figure 5-78 Comparison of production, temperature fluctuation, and conduction spectra from spectra equation for $\overline{\tau u_2}$. Prandtl number, 1. Curves are normalized to same height. (a) $a_{12}^* = (t - t_0) dU_1/dx_2 = 1$ (b) $a_{12}^* = (t - t_0) dU_1/dx_2 = 50$.

are but slightly separated. On the other hand, for high velocity gradient ($a_{12}^* = 50$), the production takes place mostly in the low-wavenumber or big-eddy region, and the conductive attenuation occurs in the high-wavenumber region. The conductive attenuation occurs mostly in the high-wavenumber region because conduction effects tend to "smear out" the small-scale temperature fluctuations more readily than the large ones. Note that the appearance of the curves in Fig. 5-78 is similar to that of the curves for the turbulent energy in Fig. 5-62.

One might summarize the history of the temperature fluctuations at high velocity gradients as follows: The temperature fluctuations are produced by the mean-temperature gradient mainly in the big-eddy region. This temperature-fluctuation activity or "energy" is transferred from the big temperature eddies to smaller ones by the action of the velocity gradient. Finally the temperature energy is dissipated by conduction effects in the small-eddy region. The separation at high velocity gradients of the three regions shown in Fig. 5-78b is analogous to the separation of the production, energy-containing, and dissipation regions associated with the turbulent energy $\overline{u_i u_i}/2$ (see Fig. 5-62b).

Temperature–velocity correlation coefficient. The temperature–velocity correlation coefficient as introduced by Corrsin [106] is defined as $\overline{\tau u_2}/(\overline{\tau^2}\,\overline{u_2^2})^{1/2}$. For perfect correlation between τ and u_2, this coefficient has a value of 1. The coefficient can be calculated by measuring the areas under the spectrum curves in Figs. 5-75 and 5-76 and in Fig. 5-54. A plot of the temperature–velocity correlation coefficient against dimensionless velocity gradient is given for a Prandtl number of 1 in Fig. 5-79. For zero-velocity gradient, perfect correlation between the temperature and velocity fluctuations is indicated. It should be mentioned that this result applies only to a Prandtl number of 1. The Prandtl-number dependence of the coefficient for zero-velocity gradient is given by eq. (78) of ref. [107]. As the velocity gradient increases, Fig. 5-79 indicates that the correlation between the

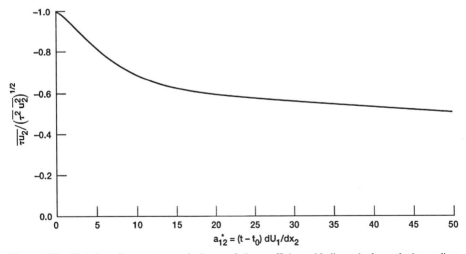

Figure 5-79 Variation of temperature–velocity correlation coefficient with dimensionless velocity gradient. Prandtl number, 1.

temperature and velocity is partially destroyed. At a value of a_{12}^* of 50 the correlation coefficient has decreased to about 0.5.

Ratio of eddy diffusivities for heat transfer to momentum transfer. The eddy diffusivities for heat transfer and for momentum transfer are defined as

$$\varepsilon_h = -\frac{\overline{\tau u_2}}{dT/dx_2} \tag{5-396}$$

and

$$\varepsilon = -\frac{\overline{u_1 u_2}}{dU_1/dx_2}. \tag{5-397}$$

The eddy diffusivity ratio $\varepsilon_h/\varepsilon$ is of considerable importance in the phenomenological theories of turbulent heat transfer and usually is assumed to be one. In fact that assumption gives the best agreement between analysis and experiment (see Fig. 4-4), except possibly at very low Prandtl or Peclet numbers [109, 110]. A dimensionless eddy diffusivity for heat transfer $v^{5/2}(t-t_0)^{3/2}\varepsilon_h/J_0$ can be obtained from the areas under the curves in Fig. 5-75. A similar dimensionless eddy diffusivity for momentum transfer is given in Fig. 5-58. The ratio $\varepsilon_h/\varepsilon$ plotted in Figs. 5-80 and 5-81. Figure 5-81 is included because the eddy diffusivity ratio for $a_{12}^* = 0$ is not given in Fig. 5-80. This case corresponds to isotropic turbulence and can be calculated from the results in Fig. 5-58 and ref. [107]. For small velocity gradients, $\varepsilon_h/\varepsilon$ is greater than 1 except for the low Prandtl number. However, as the velocity gradient increases, $\varepsilon_h/\varepsilon$ ultimately decreases and approaches 1 at large velocity gradients. This is shown on a spectral basis in Fig. 5-82, in which the dimensionless spectra of ε_h and ε for a Prandtl number of 1 are compared. As the velocity gradient increases, the spectrum curves of ε_h and ε approach each other rapidly

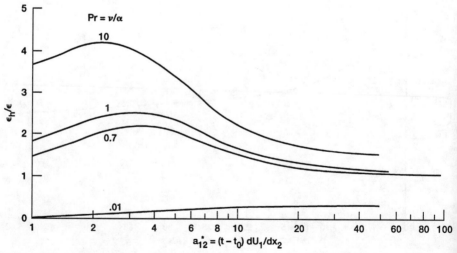

Figure 5-80 Variation of ratio of eddy diffusivity for heat transfer to that for momentum transfer with dimensionless velocity gradient.

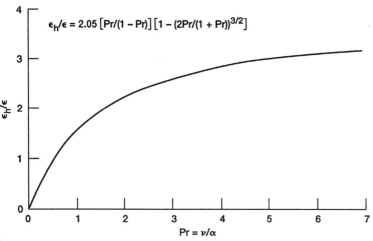

Figure 5-81 Variation of ϵ_h/ϵ with Prandtl number for isotropic turbulence with velocity gradient = 0.

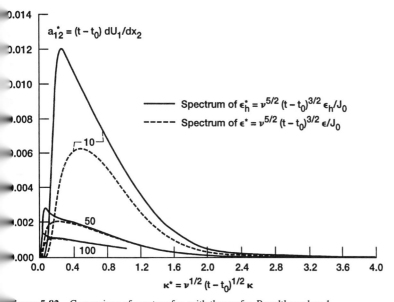

Figure 5-82 Comparison of spectra of ϵ_h with those of ϵ. Prandtl number, 1.

in the high-wavenumber or small-eddy region and somewhat more slowly in the low-wavenumber region.

The approach to 1 of $\varepsilon_h/\varepsilon$ as the velocity gradient increases occurs at all Prandtl numbers. This can be seen by inspection of eq. (5-385a), which indicates that for large values of the velocity gradient a_{12}, the effect of Prandtl number on γ_2 and thus on ε_h is negligible. However, the effect of Prandtl number is much greater at low values of Pr than at higher ones. This is because the terms in eq. (5-385a) that contain the Prandtl number vary much more rapidly with low values of that quantity than with high ones.

Figure 5-80 indicates that as the velocity gradient increases, the approach of $\varepsilon_h/\varepsilon$ to 1 is most rapid for Prandtl numbers on the order of one and least rapid for very low Prandtl numbers.

It is of interest to compare the various terms in the differential equations for γ_2/b_2 and φ_{12}/a_{12} at high values of a_{12}. The quantities γ_2/b_2 and φ_{12}/a_{12}, if integrated over wavenumber space, give ε_h and ε. Equation (5-380) can be written in terms of γ_2/b_2 as

$$\frac{\partial(\gamma_2/b_2)}{\partial t} - a_{12}\kappa_1\frac{\partial(\gamma_2/b_2)}{\partial \kappa_2} = -\varphi_{22} + 2a_{12}\frac{\kappa_1\kappa_2}{\kappa^2}\left(\frac{\gamma_2}{b_2}\right) - \left(\frac{1}{Pr}+1\right)\nu\kappa^2\left(\frac{\gamma_2}{b_2}\right). \tag{5-398}$$

From eq. (5-324),

$$\frac{\partial(\varphi_{12}/a_{12})}{\partial t} - a_{12}\kappa_1\frac{\partial(\varphi_{12}/a_{12})}{\partial \kappa_2} = -\varphi_{22} + 2\frac{a_{12}\kappa_1\kappa_2}{\kappa^2}\left(\frac{\varphi_{12}}{a_{12}}\right) + 2\frac{\kappa_1^2}{\kappa^2}\varphi_{22} - 2\nu\kappa^2\left(\frac{\varphi_{12}}{a_{12}}\right). \tag{5-399}$$

These equations for γ_2/b_2 and for φ_{12}/a_{12} are the same except for the last term in eq. (5-398) and the last two terms in eq. (5-399). It appears, however, from the forms of the equations that these terms should not be important for high values of a_{12}. The next-to-last term in eq. (5-399) arises from the pressure fluctuations.

Although eqs. (5-398) amd (5-399) are similar for large values of a_{12}, the initial conditions for γ_2/b_2 and φ_{12}/a_{12} are different, the initial form for φ_{12} being given by eq. (5-333), whereas γ_2 is initially zero. However, a numerical check indicates that γ_2/b_2 and φ_{12}/a_{12}, as well as the integrated values ε_h and ε, are essentially equal for large values of a_{12}^*. This suggests that the initial conditions have a negligible effect on the results for large times or velocity gradients.

The calculations in this section are for the case in which both the velocity and temperature gradients are in the x_2 direction. Fox [111] has considered a temperature gradient in an arbitrary direction on a plane normal to the flow direction. He showed that for large values of $(t - t_0)dU_1/dx_2$, the thermal eddy diffusivity corresponding to the temperature gradient normal to the velocity gradient can be much larger than that for the temperature gradient in the direction of the velocity gradient. That seems reasonable because $\overline{\tau u_2}$ should be smaller than $\overline{\tau u_3}$ because of the smaller velocity fluctuations in the x_2 direction (see section 5-4-2-1). The results do not support an assumption sometime made that the radial and circumferential thermal eddy diffusivities in a turbulent tube flow are equal.

It is hard to make comparisons between the present results and a steady-state pipe flow or boundary layer, inasmuch as a_{12}^* contains time. However, we can make a rough estimate of the order of magnitude of a_{12}^* for a steady-state case as follows. From the turbulent energy spectra in Fig. 5-56, $\kappa_{\text{average}}^* \sim 1$. Then an average length, $1/\kappa_{\text{average}} = L$ associated with the turbulence is $[\nu(t - t_0)]^{1/2}$. Let δ be the radius of the pipe or the thickness of the boundary layer and U be a characteristic mean velocity. Letting $t - t_0 \sim L^2/\nu$ (see previously), $dU_1/dx_2 \sim U/\delta$, and $L \sim 0.3\delta$, a_{12}^* is on the order of 0. $U\delta/\nu$. Thus for values of mean-flow Reynolds numbers usually obtained in the turbulent flows, $\varepsilon_h/\varepsilon$, according to Fig. 5-80, probably will be close to 1 for gases and liquids. For liquid metals $\varepsilon_h/\varepsilon$ may be less than 1, in qualitative agreement with those analyses that use a modified mixing-length theory to account for heat conduction to or from an eddy as it moves transversely in a mean-temperature gradient [109, 110]. In making the above comparisons, it should be remembered, of course, that the present calculations are for an idealized case, which has only a partial correspondence to a passage or boundary layer. A discussion of possible differences between the two cases is given in ref. [87].

Except for some qualitative discussion in Chapter 1 we have not yet considered the effect of buoyancy on turbulence. To do that we first study a simple (at least conceptually) case in which buoyancy effects are present, but mean velocity gradients are absent.

5-4-2-4 Turbulence in the presence of a vertical body force and temperature gradient.
The analysis described here is concerned with the effect of buoyancy forces on a homogeneous turbulent field [112]. The buoyancy effects are produced by a uniform vertical temperature gradient and body force. Equations (5-313) through (5-320) become, for $\partial U_i/\partial x_i = 0$,

$$\frac{\partial \varphi_{ij}}{\partial t} = -\frac{1}{\rho}\left(i\kappa_j \lambda'_i - i\kappa_i \lambda_j\right) - 2\nu\kappa^2 \varphi_{ij} - \beta g_i \gamma_j - \beta g_j \gamma'_i, \quad (5\text{-}400)$$

$$-\frac{1}{\rho}\kappa^2 \lambda_j = \beta g_k i\kappa_k \gamma_j, \quad (5\text{-}401)$$

$$-\frac{1}{\rho}\kappa^2 \lambda'_i = -\beta g_k i\kappa_k \gamma'_i, \quad (5\text{-}402)$$

$$\frac{\partial \gamma_j}{\partial t} = -\varphi_{kj}\frac{\partial T}{\partial x_k} - \frac{1}{\rho}i\kappa_j \zeta' - (\alpha+\nu)\kappa^2 \gamma_j - \beta g_j \delta, \quad (5\text{-}403)$$

$$\frac{\partial \gamma'_i}{\partial t} = -\varphi_{ik}\frac{\partial T}{\partial x_k} + \frac{1}{\rho}\kappa_i \zeta - (\alpha+\nu)\kappa^2 \gamma'_i - \beta g_i \delta, \quad (5\text{-}404)$$

$$-\frac{1}{\rho}\kappa^2 \zeta = \beta g_k i\kappa_k \delta, \quad (5\text{-}405)$$

$$-\frac{1}{\rho}\kappa^2 \zeta' = -\beta g_k i\kappa_k \delta, \quad (5\text{-}406)$$

and

$$\frac{\partial \delta}{\partial t} = -(\gamma'_k + \gamma_k)\frac{\partial T}{\partial x_k} - 2\alpha\kappa^2 \delta. \quad (5\text{-}407)$$

Equations (5-400) through (5-407) should apply if the terms responsible for buoyancy effects are large enough, or the turbulence weak enough, for the triple-correlation terms to be reasonably small compared with other terms in the correlation equations. Substitution of eqs. (5-401) and (5-402) into (5-400) and eqs. (5-405) and (5-406) into (5-404) and (5-403) shows that $\varphi_{ij} = \varphi_{ji}$ and $\gamma_i = \gamma'_i$ for all times if they are equal at an initial time. Here it is assumed that the turbulence is initially isotropic and that the temperature fluctuations are initially zero, so that the above relations hold. Thus the set of eqs. (5-400) through (5-407) becomes

$$\frac{\partial \varphi_{ij}}{\partial t} = \beta g_k \frac{\kappa_k \kappa_j}{\kappa^2} \gamma_i + \beta g_k \frac{\kappa_k \kappa_i}{\kappa^2} \gamma_j - 2\nu \kappa^2 \varphi_{ij} - \beta g_i \gamma_j - \beta g_j \gamma_i, \quad (5\text{-}408)$$

$$\frac{\partial \gamma_j}{\partial t} = -\varphi_{kj} \frac{\partial T}{\partial x_k} + \beta g_k \frac{\kappa_k \kappa_j}{\kappa^2} \delta - (\alpha + \nu)\kappa^2 \gamma_j - \beta g_j \delta, \quad (5\text{-}409)$$

and

$$\frac{\partial \delta}{\partial t} = -2\gamma_k \frac{\partial T}{\partial x_k} - 2\alpha \kappa^2 \delta. \quad (5\text{-}410)$$

Assume that the only nonzero component of g is in the negative vertical direction, and let

$$g \equiv -g_3. \quad (5\text{-}411)$$

Also, assume that the uniform temperature gradient is in the vertical direction, and let

$$b_3 \equiv \partial T/\partial x_3. \quad (5\text{-}412)$$

Letting $i = j = 3$ in eqs. (5-408) through (5-410),

$$\frac{d\varphi_{33}}{dt} = -2\beta g \frac{\kappa_3^2}{\kappa^2} \gamma_3 - 2\nu \kappa^2 \varphi_{33} + 2\beta g \gamma_3, \quad (5\text{-}413)$$

$$\frac{d\gamma_3}{dt} = -b_3 \varphi_{33} - \beta g \frac{\kappa_3^2}{\kappa^2} \delta - (\alpha + \nu)\kappa^2 \gamma_3 + \beta g \delta, \quad (5\text{-}414)$$

and

$$\frac{d\delta}{dt} = -2b_3 \gamma_3 - 2\alpha \kappa^2 \delta. \quad (5\text{-}415)$$

Contracting i and j in eq. (5-408) gives

$$\frac{d\varphi_{ii}}{dt} = -2\nu \kappa^2 \varphi_{ii} + 2\beta g \gamma_3. \quad (5\text{-}416)$$

The pressure term (second term in eq. [5-413]) drops out of eq. (5-416), as can be seen from eq. (5-296) and the relations $\partial/\partial r_j = -\partial/\partial x_j$ and $\partial/\partial r_i = \partial/\partial x_i'$. Thus, as in the case of homogeneous turbulence without buoyancy effects, the pressure term transfers energy between the directional components of the energy but gives no contribution to the change of energy at a particular wavenumber.

Solution of spectral equations. A general solution of the simultaneous eqs. (5-413) through (5-415) is

$$\varphi_{33} = C_1 \exp[-(\alpha + \nu)\kappa^2 (t - t_0)] + C_2 \exp\{-[(\alpha + \nu)\kappa^2 - s](t - t_0)\}$$
$$+ C_3 \exp\{-[(\alpha + \nu)\kappa^2 + s](t - t_0)\}, \quad (5\text{-}417)$$

$$\gamma_3 = -(C_1 (\alpha - \nu)\kappa^2 \exp[-(\alpha + \nu)\kappa^2 (t - t_0)] + C_2 [-(\alpha - \nu)\kappa^2 - s]$$
$$\times \exp\{-[(\alpha + \nu)\kappa^2 - s](t - t_0)\} + C_3 [(\alpha - \nu)\kappa^2 + s]$$
$$\times \exp\{-[(\alpha + \nu)\kappa^2 + s](t - t_0)\}) \bigg/ \left[2\beta g \left(1 - \frac{\kappa_3^2}{\kappa^2}\right) \right], \quad (5\text{-}418)$$

and

$$\delta = \left(2C_1 b_3 \beta g \left(1 - \frac{\kappa_3^2}{\kappa^2}\right) \exp[-(\alpha + \nu)\kappa^2 (t - t_0)] + C_2 \left[(\alpha - \nu)^2 \kappa^4 - (\alpha - \nu)\kappa^2 s \right. \right.$$
$$\left. - 2b_3 \beta g \left(1 - \frac{\kappa_3^2}{\kappa^2}\right)\right] \exp\{-[(\alpha + \nu)\kappa^2 - s](t - t_0)\} + C_3 \left[(\alpha - \nu)^2 \kappa^4 + (\alpha - \nu)\kappa^2 s \right.$$
$$\left. \left. - 2b_3 \beta g \left(1 - \frac{\kappa_3^2}{\kappa^2}\right)\right] \exp\{-[(\alpha + \nu)\kappa^2 + s](t - t_0)\}\right) \bigg/ \left[2\beta^2 g^2 \left(1 - \frac{\kappa_3^2}{\kappa^2}\right)^2\right],$$
(5-419)

where

$$s \equiv \sqrt{(\alpha - \nu)^2 \kappa^4 - 4b_3 \beta g (1 - \kappa_3^2/\kappa^2)} \tag{5-420}$$

and C_1, C_2, and C_3 are constants of integration.

For determining the constants of integration, we use the initial conditions that, for $t = t_0$, the turbulence is isotropic and $\gamma_3 = \delta = 0$. The last two conditions correspond to the assumption that the temperature fluctuations are zero at $t = t_0$. This is true, for instance, if the turbulence is produced by an unheated grid. The mean-temperature gradient then would cause temperature fluctuations to arise at subsequent times. The assumption that the turbulence is isotropic at $t = t_0$ implies that for our case

$$(\varphi_{ij})_0 = \frac{J_0}{12\pi^2}(\kappa^2 \delta_{ij} - \kappa_i \kappa_j), \tag{5-333}$$

as given in section 5-4-2-1. The turbulence is not, of course, isotropic at subsequent times, as is seen subsequently. By using these initial conditions, the constants of integration are found to be

$$C_1 = -\frac{J_0 \kappa^2 b_3 \beta g \left(1 - \kappa_3^2/\kappa^2\right)^2}{6\pi^2 s^2}, \tag{5-421}$$

$$C_2 = \left(J_0 \kappa^2 \left(1 - \frac{\kappa_3^2}{\kappa^2}\right)\left[(\alpha - \nu)^3 \kappa^6 + (\alpha - \nu)^2 \kappa^4 s - 4(\alpha - \nu)\kappa^2 b_3 \beta g \left(1 - \frac{\kappa_3^2}{\kappa^2}\right)\right.\right.$$
$$\left.\left. - 2b_3 \beta g \left(1 - \frac{\kappa_3^2}{\kappa^2}\right) s\right]\right) \bigg/ (24\pi^2 s^3), \tag{5-422}$$

and

$$C_3 = -\left(J_0 \kappa^2 \left(1 - \frac{\kappa_3^2}{\kappa^2}\right)\left[(\alpha - \nu)^3 \kappa^6 - (\alpha - \nu)^2 \kappa^4 s - 4(\alpha - \nu)\kappa^2 b_3 \beta g \left(1 - \frac{\kappa_3^2}{\kappa^2}\right)\right.\right.$$
$$\left.\left. + 2b_3 \beta g \left(1 - \frac{\kappa_3^2}{\kappa^2}\right) s\right]\right) \bigg/ (24\pi^2 s^3). \tag{5-423}$$

For small values of κ, the quantity s, as given by eq. (5-420), becomes imaginary. In that case the following solution can be used:

$$\varphi_{33} = \exp[-(\alpha + \nu)\kappa^2(t=t_0)]\{C_1' + C_2'\cos[s'(t=t_0)] + C_3'\sin[s'(t=t_0)]\}, \quad (5\text{-}424)$$

$$\gamma_3 = -\left(\exp[-(\alpha+\nu)\kappa^2(t-t_0)]\{C_1'(\alpha-\nu)\kappa^2 + [C_2'(\alpha-\nu)\kappa^2 - C_3's']\right.$$
$$\left. \times \cos[s'(t-t_0)] + [C_3'(\alpha-\nu)\kappa^2 + C_2's']\sin[s'(t-t_0)]\}\right) \bigg/ \left[2\beta g\left(1 - \frac{\kappa_3^2}{\kappa^2}\right)\right], \quad (5\text{-}425)$$

$$\delta = \left(\exp -[(\alpha+\nu)\kappa^2(t-t_0)]\left\{2C_1'b_3\beta g\left(1 - \frac{\kappa_3^2}{\kappa^2}\right) + \left[C_2'(\alpha-\nu)^2\kappa^4\right.\right.\right.$$
$$\left. - C_3's'(\alpha-\nu)\kappa^2 - 2C_2'b_3\beta g\left(1 - \frac{\kappa_3^2}{\kappa^2}\right)\right]\cos[s'(t-t_0)] + \left[C_3'(\alpha-\nu)^2\kappa^4\right.$$
$$\left.\left.+ C_2's'(\alpha-\nu)\kappa^2 - 2C_3'b_3\beta g\left(1 - \frac{\kappa_3^2}{\kappa^2}\right)\right]\sin[s'(t-t_0)]\right\}\right) \bigg/ \left[2\beta^2 g^2\left(1 - \frac{\kappa_3^2}{\kappa^2}\right)^2\right], \quad (5\text{-}426)$$

where

$$s' \equiv \sqrt{4b_3\beta g\left(1 - \frac{\kappa_3^2}{\kappa^2}\right) - (\alpha-\nu)^2\kappa^4}, \quad (5\text{-}427)$$

$$C_1' = \left[J_0\kappa^2 b_3\beta g\left(1 - \frac{\kappa_3^2}{\kappa^2}\right)^2\right] \bigg/ (6\pi^2 s'^2), \quad (5\text{-}428)$$

$$C_2' = J_0\kappa^2\left(1 - \frac{\kappa_3^2}{\kappa^2}\right)\left[2b_3\beta g\left(1 - \frac{\kappa_3^2}{\kappa^2}\right) - (\alpha-\nu)^2\kappa^4\right] \bigg/ (12\pi s'^2), \quad (5\text{-}429)$$

and

$$C_3' = \left[J_0\kappa^4\left(1 - \frac{\kappa_3^2}{\kappa^2}\right) - (\alpha-\nu)\right] \bigg/ (12\pi^2 s').\text{[14]} \quad (5\text{-}430)$$

Finally, solution of eq. (5-416) gives

$$\varphi_{ii} = \left[\frac{\varphi_{33}}{1 - \left(\frac{\kappa_3^2}{\kappa^2}\right)}\right] + \left(\frac{J_0\kappa^2}{12\pi^2}\right) e^{-2\nu\kappa^2(t-t_0)}. \quad (5\text{-}431)$$

---

[14] We also have analyzed magnetofluid dynamic turbulence with a uniform imposed magnetic field (*Phys. Fluids*, vol. 6, no. 9, 1963, pp. 1250–1259). The solutions there are the same as those obtained here if some of the variables represent different quantities.

Although the quantities φ_{ij}, γ_i, and δ are of interest in themselves, it is somewhat easier to interpret quantities that have been integrated over all directions in wavenumber space as suggested by Batchelor [4]. Thus, a quantity ψ_{ij} can be defined by the equation

$$\psi_{ij}(\kappa) = \int_0^A \varphi_{ij} dA, \tag{5-432}$$

where A is the area of a sphere of radius κ. Then, because

$$\overline{u_i u_j} = \int_0^\infty \psi_{ij} d\kappa, \tag{5-433}$$

(let $r = 0$ in eq. [5-304]), $\psi_{ij} d\kappa$ gives the contribution from the wavenumber band $d\kappa$ to $\overline{u_i u_j}$.

The equations for φ_{33}, φ_{ii}, γ_3, and δ can be written in spherical coordinates by using the transformations

$$\kappa_1 = \kappa \cos\varphi \sin\theta,$$
$$\kappa_2 = \kappa \sin\varphi \sin\theta,$$

and

$$\kappa_3 = \kappa \cos\theta.$$

Then, because φ_{33} (as well as φ_{ii}, γ_3, and δ) is not a function of the angle φ, the expression for ψ_{33} from eq. (5-432) can be written as

$$\psi_{33} = 4\pi\kappa^2 \int_0^1 \varphi_{33} d(\cos\theta). \tag{5-434}$$

We can write similar expressions for φ_{ii}, γ_3, and δ integrated over all directions in wavenumber space:

$$\psi_{ii} = 4\pi\kappa^2 \int_0^1 \varphi_{ii} d(\cos\theta), \tag{5-435}$$

$$\Gamma_3 = 4\pi\kappa^2 \int_0^1 \gamma_3 d(\cos\theta), \tag{5-436}$$

and

$$\Delta = 4\pi\kappa^2 \int_0^1 \delta d(\cos\theta). \tag{5-437}$$

Letting $r = 0$ in eqs. (5-309) and (5-311),

$$\overline{\tau u_j} = \int_0^\infty \Gamma_j d\kappa \tag{5-438}$$

and

$$\overline{\tau^2} = \int_0^\infty \Delta d\kappa, \tag{5-439}$$

so that, as in the case of ψ_{ij}, $\Gamma_j d\kappa$ and $\Delta d\kappa$ give, respectively, contributions from the wavenumber band $d\kappa$ to $\overline{\tau u_j}$ and $\overline{\tau^2}$. Computed spectra of the various turbulent quantities are considered in the next section.

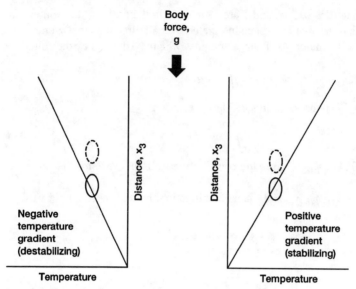

Figure 5-83 Expected effects of buoyancy forces on turbulent eddy.

Effect of buoyancy on the turbulence. Before we consider in detail the spectra computed from the foregoing analysis, it may be worthwhile to indicate physically how the buoyancy forces are expected to alter the turbulence. Figure 5-83 shows the effects of a negative and a positive vertical temperature gradient with the body force directed downward. For a ngeative temperature gradient, a turbulent eddy moving upward, for instance, usually is hotter than the surrounding fluid. If the fluid has a positive temperature-expansion coefficient, the eddy also is less dense than the surrounding fluid, so that buoyancy forces tend to accelerate it upward. Similarly, an eddy moving downward usually is accelerated downward. Thus, the negative temperature gradient tends to feed energy into the turbulent field, so that its effect is destabilizing. For a positive temperature gradient, it can be seen that the effect is opposite to that just described; that is, the buoyancy forces tend to stabilize the fluid.

Consider now whether it is possible, according to our analysis, for buoyancy to cause a growth of turbulence at large times. We recall from section 5-4-2-1 that a uniformly sheared turbulence ultimately decays. The energy fed into the turbulence by the mean gradient is less than that dissipated. However, the effect of buoyancy, as analyzed in the present section, is different. Consider eqs. (5-417) and (5-420). For $(\alpha - \nu)^2 \kappa^4 > 4b_3 \beta g (1 - \kappa_3^2/\kappa^2)$ and $s > (\alpha + \nu)\kappa^2$ the argument of the second exponential in eq. (5-417) is positive and increases without limit for large times, and so does φ_{33}. So at least for some regions of wavenumber space, more energy can be fed into the turbulence than is dissipated, and the turbulent energy ultimately grows.

A dimensionless plot of the evolution of the turbulence, as calculated by integrating φ_{ii} over wavenumber space, is presented in Fig. 5-84. The turbulent energy does indeed grow at large times, for a destabilizing temperature (density) gradient, as predicted in the previous paragraph.

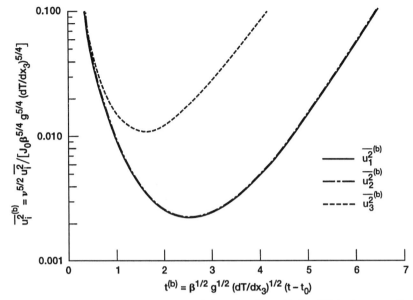

Figure 5-84 Evolution of homogeneous turbulence in a flow with destabilizing buoyancy. Prandtl number, 0.7.

Loeffler [112] has considered the effect of a gradient in electrical charge and an applied electric field on homogeneous turbulence. That problem is analogous to the present one for an infinite Prandtl number. It was determined that the turbulent energy increases without limit as time increases, if the electric field is in the direction of increasing charge density. For large times the turbulent energy is proportional to $(\exp t^{(b)})/(t^{(b)})^3$.[15]

Dimensionless energy spectra (spectra of $\overline{u_i u_i}$) are plotted in Fig. 5-85. For making the calculations, the indicated integration in eq. (5-435) is carried out numerically. If plotted using the similarity variables shown, the spectrum for no buoyancy forces ($g^* = 0$) does not change with time, so that comparison of the various curves indicates how buoyancy effects alter the spectrum. Thus, if a dimensionless spectrum curve lies above the curve for $g^* = 0$, the turbulent energy for that case is greater than it would be for no buoyancy forces. Curves are shown for Prandtl numbers ν/α of 0.7, 10, and 0.01. These Prandtl numbers correspond, respectively, as far as order of magnitude is concerned, to a gas, a liquid like water, and a liquid metal.

Negative values of the buoyancy parameter g^*, defined as $b_3\beta(t-t_0)^2 g$, correspond to negative temperature gradients, and positive values correspond to positive temperature gradients. (The quantity b_3 in the definition of g^* is the temperature gradient.) In agreement with the discussion in connection with Fig. 5-83, the areas under the spectrum curves increase for negative temperature gradients and, in general, decrease for positive

---

[15]Loeffler also pointed out that according to the present analysis the turbulence need not decay at large times for destabilizing temperature gradients. The implication in our ref. [113] that according to our analysis the turbulence always decays at large times therefore should be disregarded.

270 TURBULENT FLUID MOTION

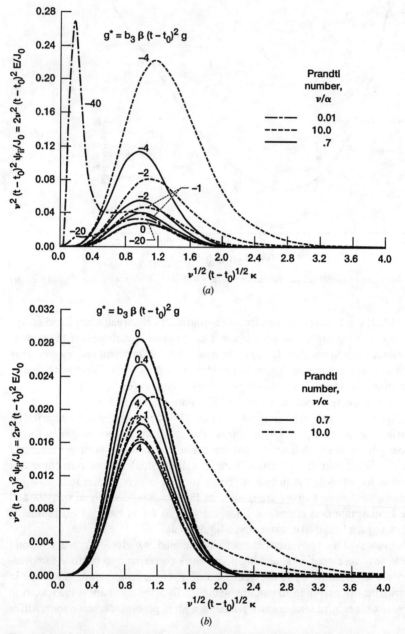

Figure 5-85 Dimensionless spectra of $\overline{u_i u_i}$ (energy spectra). (*a*) With buoyancy forces destabilizing; (*b*) with buoyancy forces stabilizing.

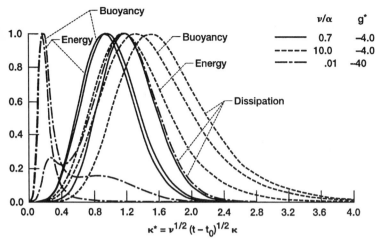

Figure 5-86 Comparison of normalized energy, dissipation, and buoyancy spectra with buoyancy forces destabilizing.

ones. A reversal of the expected trend is shown by the curve for a Prandtl number of 10 and a g^* of 4. The action of the buoyancy forces in producing turbulent energy is particularly evident for a Prandtl number of 0.01 and negative values of g^*. There, the buoyancy forces tend to produce an extra peak in the spectra in the low-wavenumber or large-eddy region.

Terms in the spectral energy equation, as well as energy spectra, are plotted in Fig. 5-86 for cases in which the buoyancy forces augment the turbulence. The curves are normalized to the same height for comparison. The terms for the energy equation are obtained by integrating the terms in eq. (5-416) over all directions in wavenumber space by using eqs. (5-435) and (5-436). The second term in eq. (5-416) gives the turbulent dissipation, and the last term gives the effect of buoyancy forces on the turbulence.

Consider first the curves in Fig. 5-86 for Prandtl numbers less than 1. Those curves indicate that the spectrum of the buoyancy term tends to coincide with the energy spectrum for Prandtl numbers less than 1. That is, the energy from the buoyancy forces feeds into most of the parts of the energy spectrum. On the other hand, the dissipation regions are considerably separated from the energy-containing regions, the separation being greater for the lower Prandtl number. The dissipation regions for the two Prandtl numbers are close together, thus indicating that buoyancy forces, which are influenced by Prandtl number, do not greatly influence the dissipation for Prandtl numbers less than 1. The dissipation occurs mostly at high wavenumbers, at which the effect of buoyancy forces is not important. The low-wavenumber parts of the energy spectrum, by contrast, are much more affected by buoyancy forces at low Prandtl numbers than at higher ones, because the eddies associated with the temperature–velocity correlations (see eq. [5-296]) are much larger at low Prandtl numbers. The spectra of the temperature–velocity correlations are considered subsequently (see Fig. 5-90).

The curves in Fig. 5-86 for a Prandtl number of 10 indicate that for high Prandtl numbers, in contrast to the case of Prandtl numbers less than 1, the buoyancy forces can

272 TURBULENT FLUID MOTION

act on the small eddies. As a result of this effect, the buoyancy forces alter the dissipation spectrum for high–Prandtl-number fluids.

Dimensionless spectra of $\overline{u_3^2}$, which is the component of the turbulent energy in the direction of the temperature gradient and body force, are presented in Fig. 5-87. The curves are somewhat similar to those for the spectra of $\overline{u_i u_i}$ and exhibit double peaks at the low Prandtl number. However, some of the spectra for $\overline{u_3^2}$ also have double peaks for a Prandtl number of 10. These are apparently caused by the action of the buoyancy forces

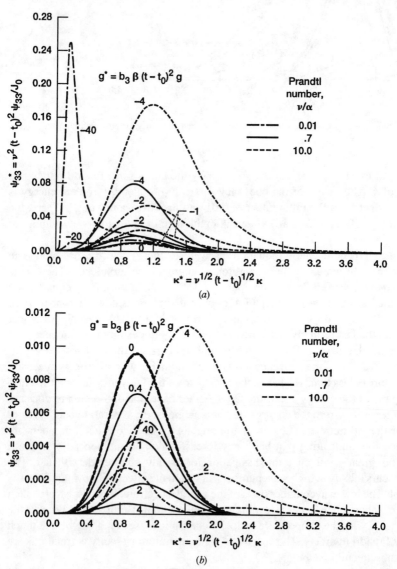

Figure 5-87 Dimensionless spectra of $\overline{u_3^2}$. (a) With buoyancy forces destabilizing; (b) with buoyancy forces stabilizing.

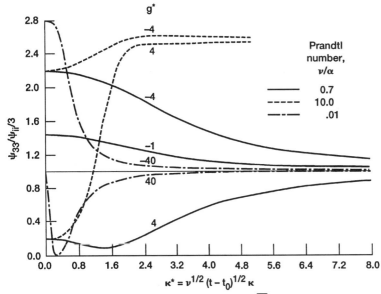

Figure 5-88 Curves showing ratio of spectrum curves for $\overline{u_3^2}$ to those for $\overline{u_i u_i}/3$.

on the small eddies. Another unexpected result is that the curve for a Prandtl number of 10 and a g^* of 4, although for a case in which the buoyancy forces would be expected to be stabilizing, lies above the curve for no buoyancy effects. The physical reason for this result is not clear. It may be that some of the eddies, in this case, oscillate several times before being damped out.

In general, the turbulence is anisotropic. The anisotropy of the turbulence is clearly seen in Fig. 5-88, in which the spectrum curves for u_3^2 divided by those for $\overline{u_i u_i}/3$ are plotted. For isotropic turbulence all values of $\psi_{33}/(\psi_{ii}/3)$ are 1, because $\psi_{ii}/3$ represents the average spectrum of the components of the energy. For destabilizing conditions $\overline{u_3^2}$ is higher than the average component, whereas for stabilizing conditions it is lower. This is physically reasonable, because the buoyancy forces are expected to act mainly on the vertical components of the velocities of the eddies. In fact, eq. (5-408) indicates that the buoyancy terms (the last two terms) occur only in the equation for φ_{33} for a vertical body force.

For Prandtl numbers less than 1 the anisotropy is most pronounced in the large-eddy region, so that apparently the buoyancy forces act mostly on the large eddies. In the small-eddy region the curves for Prandtl numbers less than 1 approach 1, so that the turbulence is isotropic in the smallest eddies. Thus, the theory of local isotropy seems to apply here. This observation is in opposition to that for turbulence, which is weak or for which a uniform mean-velocity gradient is large, for which local isotropy is absent (section 5-4-2-1). Also, the curves in Fig. 5-88 for a Prandtl number greater than 1 do not show local isotropy. Thus, local isotropy seems to be obtained only for Prandtl numbers less than 1 in the present analysis. The situation may be different for high-turbulence Reynolds numbers and small mean gradients.

It originally was thought that the difference between the results for Prandtl numbers less than and greater than 1 was caused by a difference in the effect of pressure forces in the two cases. A calculation with the pressure-force terms absent, however, indicated that those terms have but a minor effect on the results. It appears that the effect is due to the way the buoyancy forces act in the two cases and that the buoyancy forces can act even on the smaller eddies at high Prandtl numbers. This is in agreement with the curves in Fig. 5-86.

Spectra of the temperature variance $\overline{\tau^2}$ are plotted in Fig. 5-89. For $g^* = 0$, the results reduce to those of Dunn and Reid [107]. The trends with g^* are similar to those for the spectra of $\overline{u_i u_i}$; that is, the areas under the curves are larger for negative than for positive temperature gradients. However, the areas under the curves for low Prandtl numbers are much smaller than for the higher ones because, for the same viscosity, the high thermal conductivities associated with lower-Prandtl-number fluids tend to smear out the temperature fluctuations. As Prandtl number decreases, the spectra move into the lower wavenumber regions because the conduction effects tend to destroy the small temperature eddies more readily than larger ones.

The last spectra to be considered are those of the temperature–velocity correlations $\overline{\tau u_3}$. These are plotted in dimensionless form in Fig. 5-90. The quantity $\overline{\tau u_3}$ is proportional to the turbulent heat transfer. The total heat transfer q_3 is the sum of the laminar and turbulent heat transfer; it is given by

$$q_3 = -k \left(\frac{dT}{dx_3} \right) + \rho c_p \overline{\tau u_3},$$

where k is the thermal conductivity and c_p is the specific heat at constant pressure. Inasmuch as the temperature gradient b_3 occurs in the denominator of the dimensionless spectrum function in Fig. 5-90, those curves also can be considered as the spectra of the eddy diffusivity for heat transfer. The eddy diffusivity for heat transfer ε_h is defined by

$$\varepsilon_h \equiv -\frac{\overline{\tau u_3}}{dT/dx_3}.$$

The spectra indicate that, if the buoyancy forces are destabilizing, the turbulent heat transfer is greater than it would be without buoyancy effects. This is congruous with the effect of buoyancy forces on the turbulent intensity shown in Fig. 5-85. Similarly, for positive values of g^*, the turbulent heat transfer is less than it would be for no buoyancy forces. However, as g^* continues to increase, the turbulent heat transfer goes to zero and then changes sign. That is, the turbulence begins to transfer heat against the temperature gradient. This is shown somewhat more clearly in Fig. 5-91, in which the temperature–velocity correlation coefficient $\overline{\tau u_3}/[(\overline{\tau^2})^{1/2}(\overline{u_3^2})^{1/2}]$ is plotted against g^*. As g^* increases, the sign of the turbulent heat transfer oscillates. Although these are rather surprising results, turbulence on occasion has been observed to pump heat against a temperature gradient. This occurs, for instance, in a Ranque-Hilsch vortex tube, in which expansion and contraction of eddies in a pressure gradient can cause heat to flow against a temperature gradient. The effect observed here, however, appears to be caused by the action of the buoyancy forces on the eddies. In the stabilizing case, the buoyancy forces ordinarily act in the direction opposite to that in which an eddy starts to move (see Fig. 5-83), and so the sign of the velocity fluctuation might be changed without

FOURIER ANALYSIS, SPECTRAL FORM OF THE CONTINUUM EQUATIONS **275**

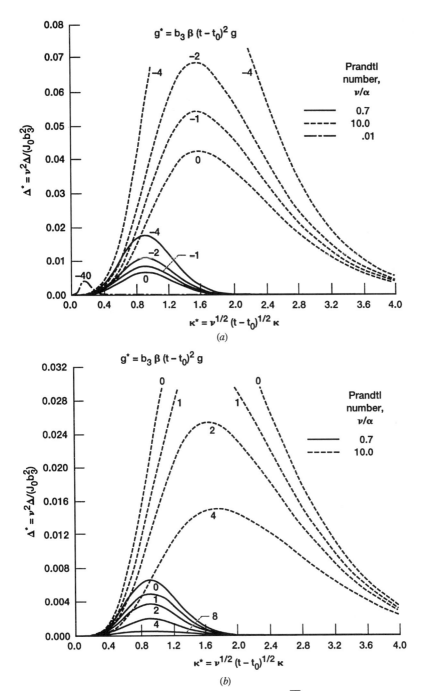

Figure 5-89 Dimensionless spectra of temperature variance $\overline{\tau^2}$. (*a*) With buoyancy forces destabilizing; (*b*) with buoyancy forces stabilizing.

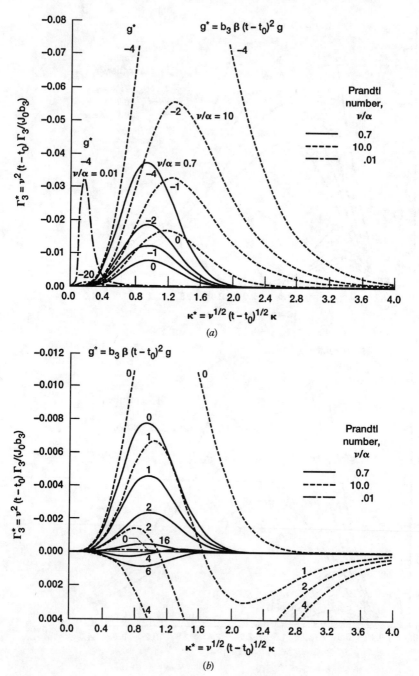

Figure 5-90 Dimensionless spectra of temperature–velocity correlations $\overline{\tau u_3}$. (a) With buoyancy forces destabilizing; (b) with buoyancy forces stabilizing.

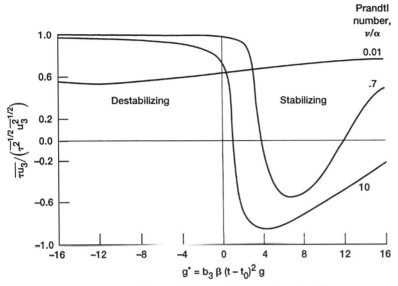

Figure 5-91 Temperature–velocity correlation coefficient as a function of buoyancy parameter.

necessarily changing the sign of the corresponding temperature fluctuation. Thus, it appears possible that the direction of the turbulent heat transfer can be reversed.

For negative values of g^*, Fig. 5-91 indicates that nearly perfect correlation between the temperature and velocity fluctuations is approached. This, again, can be explained by the action of the buoyancy forces. Thus, as was mentioned previously, an eddy moving upward in a negative temperature gradient usually is hotter than the surrounding fluid and so is pushed upward still more by the buoyancy forces. If an eddy moving upward happens to be cooler than the surrounding fluid, it is pushed downward. Therefore, positive contributions to $\overline{\tau u'_3}$ are amplified, whereas negative contributions are damped out by the buoyancy forces, so that the net effect is to increase the value of $\overline{\tau u'_3}$ toward 1.

It appears that by using the present method of analysis we can profitably study many of the turbulent processes. It is true that because we neglect triple-correlation terms we are not able to study the transfer of energy between eddies of various sizes, but that is only one of the important processes occurring in turbulence and can be studied separately. For instance, we could, as in section 5-3-2-2, consider three-point correlation equations and neglect fourth-order correlation terms. However, if that were done in the present case, in which buoyancy effects are considered, the problem might tend to get out of hand. Alternatively, if a mean-velocity gradient as well as a temperature gradient are included, we obtain a transfer of energy from large to small eddies as in section 5-4-2-1, even though triple correlations are neglected. That is done in the next section. It appears that the method of analysis followed here gives information about other turbulent processes such as the dissipation and the production or extraction of energy by buoyancy forces. Note, in particular, the solutions obtained for destabilizing buoyancy in which the turbulence does not die out at large times (Fig. 5-84).

5-4-2-5 Effects of combined buoyancy and shear on turbulence.

The effects of shear and of buoyancy are considered separately in sections 5-4-2-1 and 5-4-2-4. In real situations, for instance in the atmosphere, the two effects occur simultaneously. The speculative theory given in ref. [114] considers that case.

Herein, the methods used in sections 5-4-2-1 and 5-4-2-4 are extended to analyze the combined effects of buoyancy and shear on homogeneous turbulence ref. [115].[16] We consider the case in which the velocity and temperature gradients are in the x_3 direction (vertical) and the body force (gravity) is in the $-x_3$ direction. Then eqs. (5-313) through (5-320) become

$$\frac{\partial \varphi_{ij}}{\partial t} + \delta_{i1} a_{13} \varphi_{3j} + \delta_{j1} a_{13} \varphi_{i3} - a_{13} \kappa_1 \frac{\partial \varphi_{ij}}{\partial \kappa_3}$$
$$= -\frac{1}{\rho}(i\kappa_j \lambda'_i - i\kappa_i \lambda_j) - 2\nu\kappa^2 \varphi_{ij} + \beta\delta_{i3} g\gamma_j + \beta\delta_{j3} g\gamma'_i, \quad (5\text{-}440)$$

$$\frac{\partial \delta}{\partial t} + b_3 \gamma'_3 + b_3 \gamma_3 - a_{13} \kappa_1 \frac{\partial \delta}{\partial \kappa_3} = -2\alpha\kappa^2 \delta, \quad (5\text{-}441)$$

$$\frac{\partial \gamma_j}{\partial t} + b_3 \varphi_{3j} + a_{13}\delta_{1j}\gamma_3 - a_{13}\kappa_1 \frac{\partial \gamma_j}{\partial \kappa_3} = -\frac{1}{\rho} i\kappa_j \zeta' - (\alpha+\nu)\kappa^2 \gamma_j + \beta g \delta_{j3} \delta, \quad (5\text{-}442)$$

$$\frac{\partial \gamma'_i}{\partial t} + b_3 \varphi_{i3} + a_{13}\delta_{i1}\gamma'_3 - a_{13}\kappa_1 \frac{\partial \gamma'_i}{\partial \kappa_3} = -(\alpha+\nu)\kappa^2 \gamma'_i + \frac{1}{\rho} i\kappa_i \zeta + \beta g \delta_{i3} \delta, \quad (5\text{-}443)$$

$$-\frac{1}{\rho}\kappa^2 \lambda'_i = -2a_{13} i\kappa_1 \varphi_{i3} - \beta g i\kappa_3 \gamma'_i, \quad (5\text{-}444)$$

$$-\frac{1}{\rho}\kappa^2 \lambda_j = 2a_{13} i\kappa_1 \varphi_{3j} + \beta g i\kappa_3 \gamma_j, \quad (5\text{-}445)$$

$$-\frac{1}{\rho}\kappa^2 \zeta = 2a_{13} i\kappa_1 \gamma'_3 - \beta g i\kappa_3 \delta, \quad (5\text{-}446)$$

and

$$-\frac{1}{\rho}\kappa^2 \zeta' = -2a_{13} i\kappa_1 \gamma_3 + \beta g i\kappa_3 \delta, \quad (5\text{-}447)$$

where, as in earlier sections, $g = -g_3$, $a_{13} = dU_1/dx_3$, $b_3 = dT/dx_3$, and the spectral quantities are defined by eqs. (5-304) through (5-311). Substitution of eqs. (5-444) through (5-447) into eqs. (5-440) through (5-443) shows that $\varphi_{ij} = \varphi_{ji}$ and $\gamma_i = \gamma'_i$ for all times if they are equal at an initial time. It is assumed herein that the turbulence is initially isotropic and that the temperature fluctuations are initially zero, so that the above relations hold. Thus, the set of eqs. (5-440) through (5-447) becomes

$$\frac{\partial \varphi_{ij}}{\partial t} = a_{13}\kappa_1 \frac{\partial \varphi_{ij}}{\partial \kappa_3} - a_{13}(\delta_{i1}\varphi_{j3} + \delta_{1j}\varphi_{i3}) + 2a_{13}\left(\frac{\kappa_1 \kappa_j}{\kappa^2}\varphi_{i3} + \frac{\kappa_i \kappa_1}{\kappa^2}\varphi_{j3}\right)$$
$$+ \beta g \gamma_i \left(\delta_{j3} - \frac{\kappa_3 \kappa_j}{\kappa^2}\right) + \beta g \gamma_j \left(\delta_{i3} - \frac{\kappa_i \kappa_3}{\kappa^2}\right) - 2\nu\kappa^2 \varphi_{ij}, \quad (5\text{-}448)$$

$$\frac{\partial \gamma_i}{\partial t} = a_{13}\kappa_1 \frac{\partial \gamma_i}{\partial \kappa_3} - b_3 \varphi_{i3} + a_{13}\gamma_3 \left(2\frac{\kappa_i \kappa_1}{\kappa^2} - \delta_{i1}\right) + \beta g \delta \left(\delta_{i3} - \frac{\kappa_i \kappa_3}{\kappa^2}\right) - (\alpha+\nu)\kappa^2 \gamma_i, \quad (5\text{-}449)$$

---

[16] We have applied this analysis to the growth of random vortices to form tornadoes in an unstable atmosphere with vertical shear (see *J. Atmos. Sciences*, vol. 34, 1977, pp. 1502–1517).

and

$$\frac{\partial \delta}{\partial t} = a_{13}\kappa_1 \frac{\partial \delta}{\partial \kappa_3} - 2b_3\gamma_3 - 2\alpha\kappa^2\delta. \quad (5\text{-}450)$$

Equations (5-448) through (5-450) give contributions of various processes to the rates of change of spectral components of $\overline{u_i u_j}$, $\overline{\tau u_i}$, and $\overline{\tau^2}$, respectively. The second term in each equation is a transfer term that transfers activity into or out of a spectral component by the stretching or compressing of turbulent vortex filaments by the mean-velocity gradient, as discussed in sections 5-1-2-1, 5-4-2-1, and 5-4-2-3. The terms with κ^2 in the denominator are spectral components of pressure–velocity or pressure–temperature correlations and transfer activity between directional components (sections 5-4-2-1 and 5-4-2-3). The terms proportional to βg and δ_{i3} (or δ_{j3}) are buoyancy terms that augment or diminish the activity in a spectral component by buoyant action. The last terms in the equations are dissipation terms, which dissipate activity by viscous effects (eq. [5-448]) or by conduction effects (eq. [5-450]). The dissipation term in eq. (5-449) contains both viscous and conduction effects, inasmuch as it dissipates spectral components of velocity–temperature correlations. The remaining terms in the equations produce activity by velocity- or temperature-gradient effects. Although a buoyancy term does not appear in eq. (5-450), buoyancy affects δ (or $\overline{\tau^2}$) indirectly through the temperature gradient and γ_3 (or $\overline{\tau u_3}$).

For solving eqs. (5-448) through (5-450), the turbulence is assumed to be initially isotropic at $t = t_0$. That condition is again given by eq. (5-333):

$$(\varphi_{ij})_0 = \frac{J_0}{12\pi^2}(\kappa^2 \delta_{ij} - \kappa_i \kappa_j). \quad (5\text{-}333)$$

where J_0 is a constant that depends on initial conditions. For the initial conditions of δ and γ_i (at $t = t_0$), it is again assumed that

$$\delta_0 = (\gamma_i)_0 = 0. \quad (5\text{-}451)$$

That is, the turbulence-producing grid is assumed to be unheated, so that the temperature fluctuations are produced by the interactions of the mean-temperature gradient with the turbulence.

Solution of spectral equations. In order to reduce the set of partial differential equations, eqs. (5-448) through (5-450), to ordinary differential equations, the running variables ξ and η are considered, of which at κ_3 and t are particular values such that $\xi = \kappa_3$ if $\eta = t$. If ξ and η are introduced into the set of equations in place of κ_3 and t, the resulting equations, of course, automatically satisfy the original set. In addition,

$$\xi + a_{13}\kappa_1(\eta - \eta_0) = \text{constant} \quad (5\text{-}452)$$

during integration. Then

$$\left(\frac{\partial}{\partial \xi}\right)_\eta - \frac{1}{a_{13}\kappa_1}\left(\frac{\partial}{\partial \eta}\right)_\xi = \left(\frac{d}{d\xi}\right)_{\xi + a_{13}\kappa_1 \eta}, \quad (5\text{-}453)$$

where the subscripts outside the parentheses signify quantities that are held constant. Then eqs. (5-448) through (5-450) become

$$\frac{d\varphi_{33}(\xi)}{d\xi} = -4\frac{\xi\varphi_{33}}{\kappa_1^2 + \kappa_2^2 + \xi^2} - 2\frac{\beta g}{a_{13}\kappa_1}\left(1 - \frac{\xi^2}{\kappa_1^2 + \kappa_2^2 + \xi^2}\right)\gamma_3 + 2\nu\frac{\kappa_1^2 + \kappa_2^2 + \xi^2}{a_{13}\kappa_1}\varphi_{33}, \tag{5-454}$$

$$\frac{d\gamma_3(\xi)}{d\xi} = \frac{b_3}{a_{13}\kappa_1}\varphi_{33} - 2\frac{\xi\gamma_3}{\kappa_1^2 + \kappa_2^2 + \xi^2} - \frac{\beta g}{a_{13}\kappa_1}\left(1 - \frac{\xi^2}{\kappa_1^2 + \kappa_2^2 + \xi^2}\right)\delta$$

$$+ \left(\frac{\alpha}{\nu} + 1\right)\nu\frac{\kappa_1^2 + \kappa_2^2 + \xi^2}{a_{13}\kappa_1}\gamma_3, \tag{5-455}$$

$$\frac{d\delta(\xi)}{d\xi} = \frac{2b_3}{a_{13}\kappa_1}\gamma_3 + 2\alpha\frac{\kappa_1^2 + \kappa_2^2 + \xi^2}{a_{13}\kappa_1}\delta, \tag{5-456}$$

$$\frac{d\varphi_{13}(\xi)}{d\xi} = \frac{\varphi_{33}}{\kappa_1} - 2\frac{(\xi\varphi_{13} + \kappa_1\varphi_{33})}{\kappa_1^2 + \kappa_2^2 + \xi^2} - \frac{\beta g}{a_{13}\kappa_1}\left(1 - \frac{\xi^2}{\kappa_1^2 + \kappa_2^2 + \xi^2}\right)\gamma_1$$

$$+ \frac{\beta g}{a_{13}}\frac{\xi\gamma_3}{\kappa_1^2 + \kappa_2^2 + \xi^2} + 2\nu\frac{\kappa_1^2 + \kappa_2^2 + \xi^2}{a_{13}\kappa_1}\varphi_{13}, \tag{5-457}$$

$$\frac{d\varphi_{11}(\xi)}{d\xi} = \frac{2}{\kappa_1}\varphi_{13} - 4\frac{\kappa_1\varphi_{13}}{\kappa_1^2 + \kappa_2^2 + \xi^2} + \frac{2\beta g\xi\gamma_1}{a_{13}(\kappa_1^2 + \kappa_2^2 + \xi^2)} + 2\nu\frac{\kappa_1^2 + \kappa_2^2 + \xi^2}{a_{13}\kappa_1}\varphi_{11}, \tag{5-458}$$

$$\frac{d\varphi_{ii}(\xi)}{d\xi} = \frac{2}{\kappa_1}\varphi_{13} - \frac{2\beta g\gamma_3}{a_{13}\kappa_1} + 2\nu\frac{\kappa_1^2 + \kappa_2^2 + \xi^2}{a_{13}\kappa_1}\varphi_{ii}, \tag{5-459}$$

and

$$\frac{d\gamma_1(\xi)}{d\xi} = \frac{b_3\varphi_{13}}{a_{13}\kappa_1} - \left(2\frac{\kappa_1}{\kappa_1^2 + \kappa_2^2 + \xi^2} - \frac{1}{\kappa_1}\right)\gamma_3 + \frac{\beta g\xi\delta}{a_{13}(\kappa_1^2 + \kappa_2^2 + \xi^2)}$$

$$+ \left(\frac{\alpha}{\nu} + 1\right)\nu\frac{\kappa_1^2 + \kappa_2^2 + \xi^2}{a_{13}\kappa_1}\gamma_1. \tag{5-460}$$

Note that the first three of these equations are independent of the remaining ones.

The constant in eq. (5-452), subject to initial conditions given by eqs. (5-333) and (5-451), may be determined by letting $\xi = \xi_0$ if $\eta = \eta_0$, or $\xi_0 = \xi + a_{13}\kappa_1(\eta - \eta_0)$. This equation applies for any value of ξ, and thus

$$\xi_0 = \kappa_3 + a_{13}\kappa_1(t - t_0). \tag{5-461}$$

Equation (5-461) gives the value of ξ at which to start the integration for given values of κ_3, κ_1, a_{13}, and $t - t_0$. The initial conditions, eqs. (5-333) and (5-451), may be satisfied by letting

$$\begin{aligned}
\varphi_{33}(\xi) &= \frac{J_0}{12\pi^2}(\kappa_1^2 + \kappa_2^2), \\
\varphi_{13}(\xi) &= -\frac{J_0}{12\pi^2}\kappa_1\xi_0, \\
\varphi_{11}(\xi) &= \frac{J_0}{12\pi^2}(\kappa_2^2 + \xi_0^2), \\
\varphi_{ii} &= \frac{J_0}{6\pi^2}(\kappa_1^2 + \kappa_2^2 + \xi_0^2), \\
\gamma_i(\xi) &= \delta(\xi) = 0
\end{aligned} \quad \text{when } \xi = \xi_0.$$

and

The integration of eqs. (5-454) through (5-460) then goes from ξ_0 to $\xi = \kappa_3$. Final values of φ_{ij}, γ_i, and δ, for which $\xi = \kappa_3$ and $\eta = t$ are of most interest. The quantity ξ can be considered as a dummy variable of integration.

The following relations are introduced in order to rescale eqs. (5-454) through (5-460) and convert them to dimensionless form:

$$\nu^{1/2}(t - t_0)^{1/2}\kappa_i \to \kappa_i, \tag{5-462}$$

$$\nu^{1/2}(t - t_0)^{1/2}\xi \to \xi, \tag{5-463}$$

$$\frac{\nu(t - t_0)}{J_0}\varphi_{ij} \to \varphi_{ij}, \tag{5-464}$$

$$\frac{\nu}{J_0 b_3}\gamma_i \to \gamma_i, \tag{5-465}$$

$$\frac{\nu\delta}{J_0 b_3^2(t - t_0)} \to \delta, \tag{5-466}$$

$$(t - t_0)a_{13} \to a_{13}, \tag{5-467}$$

$$\frac{\beta g b_3}{a_{13}^2} \to Ri, \tag{5-468}$$

and

$$\frac{\nu}{\alpha} \to Pr, \tag{5-469}$$

where the arrow means "becomes." In addition, spherical coordinates are introduced into the equations by using the transformations

$$\begin{aligned}
\kappa_1 &= \kappa \cos\varphi \sin\theta, \\
\kappa_2 &= \kappa \sin\varphi \sin\theta, \\
\kappa_3 &= \kappa \cos\theta.
\end{aligned} \tag{5-470}$$

Equations (5-454) through (5-460) then become

$$\frac{d\varphi_{33}(\xi)}{d\xi} = -4\frac{\xi\varphi_{33}}{\kappa^2\sin^2\theta + \xi^2} - 2\frac{a_{13}Ri}{\kappa\cos\varphi\sin\theta}\left(1 - \frac{\xi^2}{\kappa^2\sin^2\theta + \xi^2}\right)\gamma_3$$

$$+ 2\frac{\kappa^2\sin^2\theta + \xi^2}{a_{13}\kappa\cos\varphi\sin\theta}\varphi_{33}, \tag{5-471}$$

282 TURBULENT FLUID MOTION

$$\frac{d\gamma_3(\xi)}{d\xi} = 4\frac{\varphi_{33}}{a_{13}\kappa\cos\varphi\sin\theta} - \frac{2\xi\gamma_3}{\kappa^2\sin^2\theta+\xi^2} - \frac{a_{13}\mathrm{Ri}\delta}{\kappa\cos\varphi\sin\theta}\left(1 - \frac{\xi^2}{\kappa^2\sin^2\theta+\xi^2}\right)$$
$$+ \left(\frac{1}{Pr}+1\right)\frac{\kappa^2\sin^2\theta+\xi^2}{a_{13}\kappa\cos\varphi\sin\theta}\gamma_3, \tag{5-472}$$

$$\frac{d\delta(\xi)}{d\xi} = \frac{2\gamma_3}{a_{13}\kappa\cos\varphi\sin\theta} + \frac{2}{Pr}\frac{\kappa^2\sin^2\theta+\xi^2}{a_{13}\kappa\cos\varphi\sin\theta}\delta,$$

$$\frac{d\varphi_{13}(\xi)}{d\xi} = \frac{\varphi_{33}}{\kappa\cos\varphi\sin\theta} - 2\frac{\xi\varphi_{13}+\kappa(\cos\varphi\sin\theta)\varphi_{33}}{\kappa^2\sin^2\theta+\xi^2} - \frac{a_{13}\mathrm{Ri}\gamma_1}{\kappa\cos\varphi\sin\theta} \tag{5-473}$$

$$\times\left(1 - \frac{\xi^2}{\kappa^2\sin^2\theta+\xi^2}\right) + \frac{a_{13}\mathrm{Ri}\xi\gamma_3}{\kappa^2\sin^2\theta+\xi^2} + 2\frac{\kappa^2\sin^2\theta+\xi^2}{a_{13}\kappa\cos\varphi\sin\theta}\varphi_{13}, \tag{5-474}$$

$$\frac{d\varphi_{ii}(\xi)}{d\xi} = \frac{2\varphi_{13}}{\kappa\cos\varphi\sin\theta} - \frac{4\kappa(\cos\varphi\sin\theta)\varphi_{13}}{\kappa^2\sin^2\theta+\xi^2} + \frac{2a_{13}\mathrm{Ri}\xi\gamma_1}{\kappa^2\sin^2\theta+\xi^2} + 2\frac{\kappa^2\sin^2\theta+\xi^2}{a_{13}\kappa\cos\varphi\sin\theta}\varphi_{11},$$
$$\tag{5-475}$$

$$\frac{d\varphi_{11}(\xi)}{d\xi} = \frac{2\varphi_{13}}{\kappa\cos\varphi\sin\theta} - \frac{2a_{13}\mathrm{Ri}\gamma_3}{\kappa\cos\varphi\sin\theta} + 2\frac{\kappa^2\sin^2\theta+\xi^2}{a_{13}\kappa\cos\varphi\sin\theta}\varphi_{ii}, \tag{5-476}$$

$$\frac{d\gamma_1(\xi)}{d\xi} = \frac{\varphi_{13}}{a_{13}\kappa\cos\varphi\sin\theta} + \frac{a_{13}\mathrm{Ri}\xi\delta}{\kappa^2\sin^2\theta+\xi^2} - \left(\frac{2\kappa\cos\varphi\sin\theta}{\kappa^2\sin^2\theta+\xi^2} - \frac{1}{\kappa\cos\varphi\sin\theta}\right)\gamma_3$$
$$+ \left(\frac{1}{Pr}+1\right)\frac{\kappa^2\sin^2\theta+\xi^2}{a_{13}\kappa\cos\varphi\sin\theta}\gamma_1, \tag{5-477}$$

For integrating these equations, ξ starts at

$$\xi_0 = \kappa\cos\theta + a_{13}\kappa\cos\varphi\sin\theta,$$

where

$$\varphi_{33}(\xi) = \left(\frac{1}{12\pi^2}\right)\kappa^2\sin^2\theta,$$

$$\varphi_{13}(\xi) = -\left(\frac{1}{12\pi^2}\right)\kappa(\cos\varphi\sin\theta)\xi_0,$$

$$\varphi_{11}(\xi) = \frac{1}{12\pi^2}(\kappa^2\sin^2\varphi\sin^2\theta + \xi_0^2),$$

$$\varphi_{ii}(\xi) = \frac{1}{6\pi^2}(\kappa^2\sin^2\theta + \xi_0^2),$$

and

$$\gamma_i(\xi) = \delta(\xi) = 0,$$

and goes to $\kappa\cos\theta$, where

$$\varphi_{ij}(\xi) = \varphi_{ij}(\kappa\cos\theta),$$

$$\gamma_i(\xi) = \gamma_i(\kappa\cos\theta),$$

and
$$\delta(\xi) = \delta(\kappa \cos \theta).$$

The integrations are carried out numerically for various fixed values of κ, θ, φ, a_{13}, Ri, and Pr. Directionally integrated spectrum functions can be obtained from

$$\begin{bmatrix} \psi_{ij} \\ \Gamma_i \\ \Delta \\ \Lambda_{ij} \\ \psi_{ij}^* \\ \Lambda_{ij}^* \end{bmatrix} = \int_0^\pi \int_0^{2\pi} \begin{bmatrix} \varphi_{ij} \\ \gamma_i \\ \delta \\ \Omega_{ij} \\ \varphi_{ij}^* \\ \Omega_{ij}^* \end{bmatrix} \kappa^2 \sin\theta \, d\varphi \, d\theta \tag{5-478}$$

In this equation, Ω_{ij} is the vorticity spectrum tensor given by eq. (5-345) and is related to φ_{ij} by eq. (5-348):

$$\Omega_{ij} = (\delta_{ij}\kappa^2 - \kappa_i\kappa_j)\varphi_{ll} - \kappa^2 \varphi_{ij}. \tag{5-348}$$

The starred quantities φ_{ij}^* and Ω_{ij}^* give, respectively, components of φ_{ij} and Ω_{ij} in a coordinate system rotated 45° about the x_2 axis from the x_1 axis toward the x_3 axis. Because φ_{ij} and Ω_{ij} are second-order tensors, components in the rotated system (see eqs. [2-12] and [5-342] and footnote[13] are

$$\begin{bmatrix} \varphi_{11}^* \\ \Omega_{11}^* \end{bmatrix} = \frac{1}{2}\begin{bmatrix} \varphi_{11} \\ \Omega_{11} \end{bmatrix} + \begin{bmatrix} \varphi_{13} \\ \Omega_{13} \end{bmatrix} + \frac{1}{2}\begin{bmatrix} \varphi_{33} \\ \Omega_{33} \end{bmatrix} \tag{5-479}$$

and

$$\begin{bmatrix} \varphi_{33}^* \\ \Omega_{33}^* \end{bmatrix} = \frac{1}{2}\begin{bmatrix} \varphi_{11} \\ \Omega_{11} \end{bmatrix} - \begin{bmatrix} \varphi_{13} \\ \Omega_{13} \end{bmatrix} + \frac{1}{2}\begin{bmatrix} \varphi_{33} \\ \Omega_{33} \end{bmatrix} \tag{5-480}$$

The spectrum functions given by eq. (5-478) can be integrated over all wavenumbers to give

$$\begin{bmatrix} \overline{u_i u_j} \\ \overline{\tau u_i} \\ \overline{\tau^2} \\ \overline{\omega_i \omega_j} \\ \overline{(u_i u_j)^*} \\ \overline{(\omega_i \omega_j)^*} \end{bmatrix} = \int_0^\infty \begin{bmatrix} \psi_{ij} \\ \Gamma_i \\ \Delta \\ \Lambda_{ij} \\ \psi_{ij}^* \\ \Lambda_{ij}^* \end{bmatrix} d\kappa, \tag{5-481}$$

where the asterisks again refer to components in a coordinate system rotated 45°.

All the calculated results given here are for a gas with a Prandtl number of 0.7. Dimensionless energy spectra (spectra of $\overline{u_i u_i}$) and spectra of $\overline{\tau^2}$ are plotted in Figs. 5-92 and 5-93. The spectra are plotted for several values of $(t-t_0)dU_1/dx_3$ and of $\beta g(t-t_0)^2 dT/dx_3$. The parameter $\beta g(t-t_0)^2 dT/dx_3$, rather than the Richardson number Ri $= \beta g(t-t_0)^2 dT/dx_3/(dU_1/dx_3)^2$, is used here because the use of $\beta g(t-t_0)^2 dT/dx_3$

284 TURBULENT FLUID MOTION

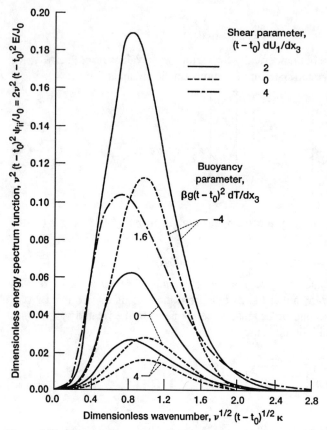

Figure 5-92 Dimensionless spectra of turbulent energy $\overline{u_i u_i}$. Prandtl number, 0.7.

and $(t - t_0)dU_1/dx_3$ enables us to consider buoyancy and shear effects separately. (The Richardson number contains both buoyancy and shear effects.) The quantity $\beta g(t - t_0)^2 dT/dx_3$ is related to Ri and the shear parameter by the equation

$$\beta g(t - t_0)^2 \frac{dT}{dx_3} = \left[(t - t_0) \frac{dU_1}{dx_3} \right]^2 \text{Ri}. \qquad (5\text{-}482)$$

If plotted by using the similarity variables shown in Figs. 5-92 and 5-93, the dimensionless spectra for buoyancy and shear parameters of zero do not change with time, and thus comparison of the various curves indicates how buoyancy and shear effects alter the spectra. Thus, if a dimensionless-spectrum curve lies above the curve for buoyancy and shear parameters of zero, the turbulent activity for that case is greater than it would be for no buoyancy and shear effects.

Positive values of the buoyancy parameter correspond to stabilizing conditions, and negative values correspond to destabilizing conditions. Figures 5-92 and 5-93 show that the trends with buoyancy parameter for a shear parameter of 2 are similar to those in Fig. 5-85 for no shear. That is, in the destabilizing case, buoyancy forces tend to feed

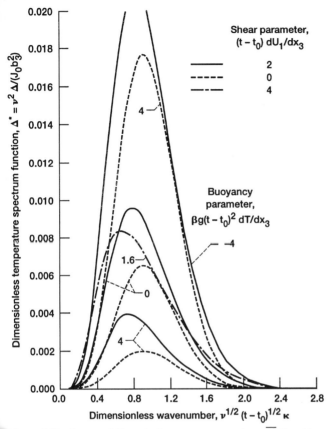

Figure 5-93 Spectra of dimensionless temperature variance $\overline{\tau^2}$. Prandtl number, 0.7.

energy or activity into the turbulent field, whereas in the stabilizing case they tend to extract it. Comparison of the curves with shear (solid curves) with those without shear (dashed curves) for values of buoyancy parameter of −4, 0, and 4 indicates that for all three cases the effect of the shear is to feed energy or activity into the turbulent field. Thus, for the destabilizing case, the buoyancy and shear have similar effects; but for the stabilizing case, they have opposite effects. Comparison of the curve in Fig. 5-92 for a dimensionless g of 4 and a shear parameter of 2 with that for those parameters equal to zero indicates that for the former curve the energy added by the shear effects approximately balances that extracted by buoyancy but that the wavenumber distributions for the two processes are slightly different.

As the shear parameter increases, the spectra become asymmetric, the slopes on the high-wavenumber sides of the curves becoming more gradual. The dot-dashed curves for a shear parameter of 4 and a buoyancy parameter of 1.6 are plotted to show this effect. As in Figs. 5-56 and 5-76 the effect is due to the transfer of energy or activity into the high-wavenumber regions by the transfer term associated with the mean-velocity gradient (see the discussion in the paragraph following eq. [5-450]).

286 TURBULENT FLUID MOTION

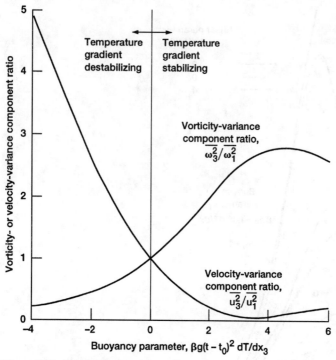

Figure 5-94 Velocity- and vorticity-variance component ratios plotted against buoyancy parameter for case of no shear. Prandtl number, 0.7.

The buoyancy forces might be expected to act more strongly on the turbulent-velocity components lying in the direction of those forces than on the other components. This expectation is confirmed in the plot of $\overline{u_3^2}/\overline{u_1^2}$ for no shear in Fig. 5-94. The ratio $\overline{u_2^2}/\overline{u_1^2}$ is greater than 1 in the destabilizing case and less than 1 in the stabilizing case. However, although $\overline{u_3^2}/\overline{u_1^2}$ becomes small, it does not approach zero for highly stable conditions. Apparently $\overline{u_1^2}$ begins to decrease as rapidly as or more rapidly than $\overline{u_3^2}$, as the buoyancy becomes large.

The vorticity component ratio $\overline{\omega_3^2}/\overline{\omega_1^2}$ also is plotted in Fig. 5-94. The trends for $\overline{\omega_3^2}/\overline{\omega_1^2}$ are opposite to those for $\overline{u_3^2}/\overline{u_1^2}$. For the stabilizing case, the vorticity tends to be aligned in the direction of the buoyancy forces. The turbulent velocities associated with that vorticity then tend to be normal to the buoyancy forces, in agreement with the curve for $\overline{u_3^2}/\overline{u_1^2}$. For the destabilizing case, the vorticity tends to lie in directions normal to the buoyancy forces. That tends to increase $\overline{u_3^2}/\overline{u_1^2}$ as shown in Fig. 5-94, although the ratio does not approach infinity, because even if all the vorticity lies in directions normal to the buoyancy forces, $\overline{u_1^2}$ or $\overline{u_2^2}$ is not zero.

In the preceding discussion, the effects of buoyancy forces on the turbulence components without mean shear were considered. Next, consider the case of shear with no buoyancy effects. In that case, the turbulent vorticity (or vortex filaments) is expected to tend to align in the direction of maximum strain, which is at 45° to the mean velocity

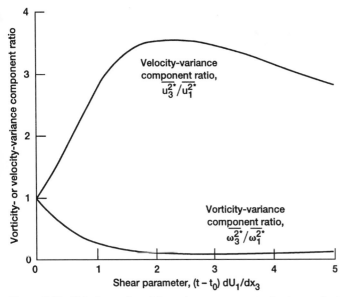

Figure 5-95 Velocity- and vorticity-variance component ratios in coordinate system rotated 45° plotted against shear parameter for case of no buoyancy effects.

Figure 5-95 shows turbulent vorticity and velocity components ratios in a coordinate system rotated 45° counterclockwise about the x_2 axis. If the vorticity all is aligned in the direction of maximum strain, $\overline{\omega_3^{2*}}/\overline{\omega_1^{2*}}$ is zero. The curve shows that there is a strong tendency for that alignment to occur at moderate values of shear parameter, but the degree of alignment does not continue to improve as the shear becomes large. The tendency for the vortex filaments to align in the direction of maximum strain is reflected in the trend for the turbulent velocity components to become maximum in a direction normal to the maximum strain, as shown in the curve for $\overline{u_3^{2*}}/\overline{u_1^{2*}}$. The degree of alignment, however, does not continue to improve as the shear becomes large.

Combined effects of buoyancy and shear on $\overline{u_3^2}/\overline{u_1^2}$ are shown in Fig. 5-96. The curves show that for no buoyancy effects the turbulence component $\overline{u_3^2}$, which is in the direction of the mean velocity gradient, is reduced in comparison with $\overline{u_1^2}$ by the shear. This trend also occurs for negative (destabilizing) values of dimensionless g and for small positive (stabilizing) values of dimensionless g. For more strongly stabilizing conditions, the trends become more complex, and the curves cross over one another.

The effects of buoyancy and shear are considered separately in Fig. 5-96, which utilizes dimensionless g and the shear parameter. Because the Richardson number contains both buoyancy and shear effects, one might suppose that its use would reduce or eliminate the need for another parameter. Figure 5-97 shows that is not the case, because $\overline{u_3^2}/\overline{u_1^2}$ is a strong function of both Richardson number and shear parameter.

The ratio of two turbulence components that are normal to the body forces and mean-velocity gradient plotted as a function of dimensionless g and the shear parameter are shown in Fig. 5-98. For no shear $\overline{u_2^2}/\overline{u_1^2}$ is 1 because the turbulence is axially symmetric.

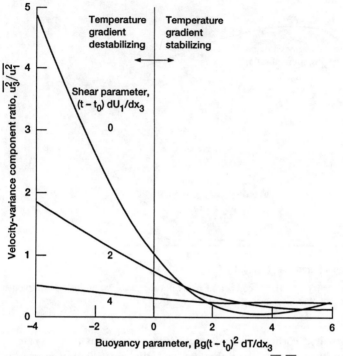

Figure 5-96 Plot showing velocity–variance component ratio $\overline{u_3^2}/\overline{u_1^2}$ as a function of buoyancy and shear parameters. Prandtl number, 0.7.

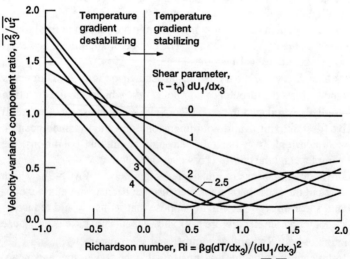

Figure 5-97 Plot showing velocity-variance component ratio $\overline{u_3^2}/\overline{u_1^2}$ as a function of Richardson number and shear parameters. Prandtl number, 0.7.

FOURIER ANALYSIS, SPECTRAL FORM OF THE CONTINUUM EQUATIONS **289**

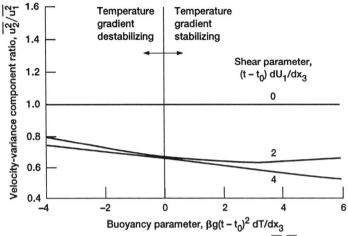

Figure 5-98 Plot showing velocity-variance component ratio $\overline{u_2^2}/\overline{u_1^2}$ as a function of buoyancy and shear parameters.

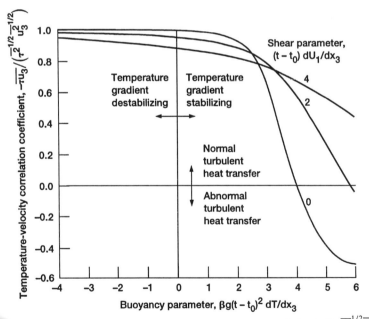

Figure 5-99 Plot showing temperature–velocity correlation coefficient $\overline{\tau u_3}/(\overline{\tau^2}^{1/2}\overline{u_3^2}^{1/2})$ as a function of buoyancy and shear parameters. Prandtl number, 0.7.

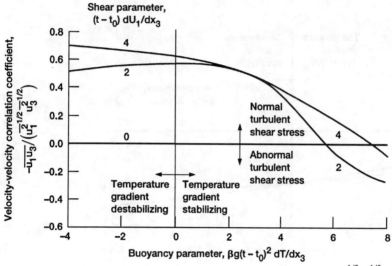

Figure 5-100 Plot showing velocity–velocity correlation coefficient $-\overline{u_1 u_3}/(\overline{u_1^2}^{1/2} \overline{u_3^2}^{1/2})$ as a function of buoyancy and shear parameters.

For a shear parameter $\neq 0$, the shear tends to destroy the axial symmetry and to reduce $\overline{u_2^2}/\overline{u_1^2}$ below 1.

Consider next the turbulent heat transfer and the turbulent shear stress. Temperature–velocity correlation coefficients $-\overline{\tau u_3}/[(\overline{\tau^2})^{1/2}(\overline{u_3^2})^{1/2}]$ are plotted in Fig. 5-99. The correlation $\overline{\tau u_3}$ is proportional to the turbulent heat transfer in the direction of the temperature gradient. The unusual feature of these results is that $\overline{\tau u_3}$ changes sign as dimensionless g becomes large. That is, for very stable conditions, the turbulence begins to pump heat against the temperature gradient. This phenomenon also is observed in the results in Fig. 5-91 for a shear parameter of zero. As the shear increases, the value of dimensionless g at which $\overline{\tau u_3}$ changes sign increases.

Velocity–velocity correlation coefficients for shear are plotted in Fig. 5-100. At small values of dimensionless g, the trends with shear parameter are opposite to those for Fig. 5-99; the values of $\overline{u_1 u_3}/[(\overline{u_1^2})^{1/2}(\overline{u_3^2})^{1/2}]$ are zero for a shear parameter of 0, and the values of $-\overline{\tau u_3}/[(\overline{\tau^2})^{1/2}(\overline{u_3^2})^{1/2}]$ are close to 1 for small shear parameter and dimensionless g. As was the case for $\overline{\tau u_3}$, $\overline{u_1 u_3}$ changes sign as g^* becomes large. As conditions become strongly stabilizing, the turbulence begins to pump the fluid in such a way as to tend to increase the velocity gradient. Thus, there occurs, for sufficiently large values of dimensionless g, a negative eddy viscosity as well as a negative eddy conductivity. Although a negative eddy conductivity can occur with only buoyancy effects present, the occurrence of a negative eddy viscosity requires combined buoyancy and shear.

A possible explanation for the theoretically observed negative eddy viscosity and conductivity can be given in terms of a modified mixing-length theory as illustrated in Fig. 5-101. Normal turbulent heat transfer or shear stress is shown in the left portion of the figure. An eddy originating at the mean velocity and temperature of the fluid at a point may move either upward or downward. By conduction and viscous effects, it tends to acquire the local mean temperature and velocity of the fluid as it moves, and thus it

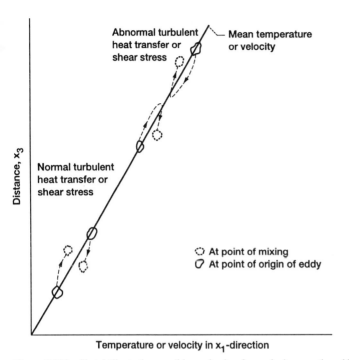

Figure 5-101 Sketch illustrating possible mechanism for producing negative eddy conductivity and viscosity.

path curves toward the mean velocity or temperature line. If the eddy mixes with the fluid, it tends to decrease the mean-temperature and -velocity gradients. The effective eddy conductivity and viscosity are positive, because they act in the same direction as the molecular conductivity and viscosity.

By contrast, for the abnormal case in which the buoyancy forces are strongly stabilizing, the original direction of motion of an eddy may be reversed. This reversal might happen because the buoyancy force, in the stabilizing case, acts in the direction opposite to that in which the eddy starts to move. Possible paths for the eddy on the distance–temperature or distance–velocity plane under these conditions are sketched on the right side of Fig. 5-101. As shown, the eddy path can cross the mean temperature or velocity line. As the eddy mixes with the fluid, it then tends to increase the mean temperature and velocity gradients, and thus the effective eddy conductivity and viscosity are negative. The actual mechanism may be more complicated than that considered here. The preceding explanation is given only to show that negative eddy conductivities and viscosities are physically reasonable. The turbulent heat transfer and shear stress do not necessarily change sign at the same value of dimensionless g, because the eddy paths on the distance–temperature plane and on the distance–velocity plane may be different because of differences between the conduction and viscous effects on the eddy as it moves. Comparison of Figs. 5-99 and 5-100 shows that the turbulent shear stress changes sign first as dimensionless g increases.

The ratio of eddy conductivity to eddy viscosity plotted against dimensionless g is shown in Fig. 5-102. The eddy conductivity and eddy viscosity are defined by the

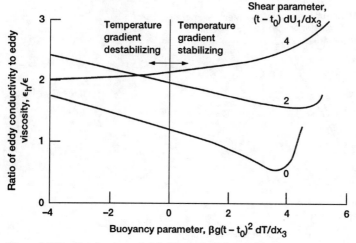

Figure 5-102 Plot showing ratio of eddy conductivity to eddy viscosity as a function of buoyancy and shear parameters. Prandtl number, 0.7.

relations

$$\varepsilon_h = -\frac{\overline{\tau u_3}}{dT/dx_3}$$

and

$$\varepsilon = -\frac{\overline{u_1 u_3}}{dU_1/dx_3}.$$

For small values of shear parameter, $\varepsilon_h/\varepsilon$ decreases with increasing buoyancy parameter except for large buoyancy. For a shear parameter of 4, $\varepsilon_h/\varepsilon$ increases with increasing dimensionless g. The sharp increases in $\varepsilon_h/\varepsilon$ near the ends of the curves occur because the eddy viscosity approaches zero and changes sign near those points.

Values of the correlation coefficient $\overline{\tau u_1}/[(\overline{\tau^2})^{1/2}(\overline{u_1^2})^{1/2}]$ are presented in Fig. 5-103. The correlation $\overline{\tau u_1}$ is proportional to turbulent heat transfer in the x_1 direction. The fact that there should be heat transfer in the x_1 direction is surprising, because there is a temperature gradient only in the x_3 direction. It appears, however, that $\overline{\tau u_1}$ can be nonzero because of the nonzero values of $\overline{\tau u_3}$ and $\overline{u_1 u_3}$. Because there is a correlation between τ and u_3 and between u_3 and u_1, the fact that a correlation should occur between τ and u_1 seems reasonable. It must be admitted, however, that heat transfer in a direction of zero temperature gradient runs contrary to normal intuition. It should be noted that the effect is not dependent on the presence of buoyancy forces (i.e., dimensionless g can be zero). The turbulent heat transfer in the x_1 direction is not necessarily small compared with that in the x_3 direction. Figure 5-104, which shows plotted values of $-\overline{\tau u_3}/\overline{\tau u_1}$, indicates that the turbulent heat transfer in the two directions can be of the same order of magnitude, even though the temperature gradient in the x_1 direction is zero.

Comparison of the present analytical results with available experimental data is of interest in order to see if there is a correspondence. Experimental data for grid-generated turbulence in the presence of combined buoyancy and shear are given in ref. [116] Unfortunately, it is hard to estimate the values of the shear parameter $(t - t_0)dU_1/dx_3$ in

FOURIER ANALYSIS, SPECTRAL FORM OF THE CONTINUUM EQUATIONS **293**

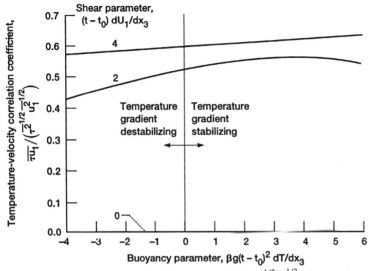

Figure 5-103 Plot showing correlation coefficient $\overline{\tau u_1}/(\overline{\tau^2}^{1/2}\,\overline{u_1^2}^{1/2})$ as a function of buoyancy and shear parameters. Prandtl number, 0.7.

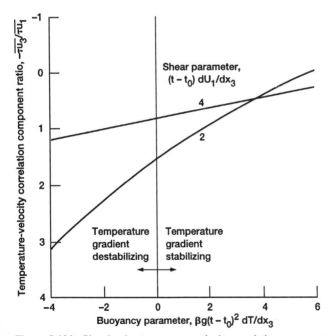

Figure 5-104 Plot showing temperature–velocity correlation component ratio $-\overline{\tau u_3}/\overline{\tau u_1}$ as a function of buoyancy and shear parameters.

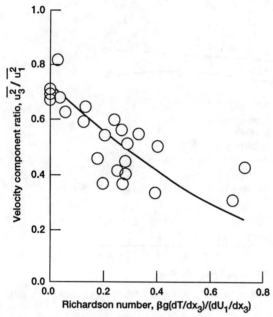

Figure 5-105 Some analytical and experimental results for station 5 in ref. [116] with a shear parameter $(t - t_0)dU_1/dx_3$ of 2.

the experiments because of uncertainties in the initial and other conditions. A comparison between analysis and data for $\overline{u_3^2}/\overline{u_1^2}$ at station 5 in ref. [116] is plotted in Fig. 5-105. For a shear parameter of 2 the trend with Richardson number for the data appears to correspond with that for the analysis.

Growth due to buoyancy of turbulence with shear. So far we have not considered the question of whether or not, according to our analysis, turbulence with combined buoyancy and shear ultimately decays with time; the dimensionless parameters plotted in Fig. 5-92 through 5-105 do not provide that information.

Studies of homogeneous turbulence with uniform mean shear for which mean gradients are large enough, or the turbulence weak enough, to neglect terms containing triple correlations are described in section 5-4-2-1 (see Figs. 5-54 through 5-68). In the results obtained there, the turbulent energy ultimately decayed with time. The energy produced by the mean velocity gradient was less than that dissipated. In Figs. 5-67 and 5-68, in which the initial condition is modified to give a finite initial turbulent energy, the energy sometimes increases for a while, but ultimately it still decays. This behavior is attributed to the fact that although the total energy is increasing, energy is being drained out of the component of the turbulence in the direction of the velocity gradient by the pressure–velocity correlaion terms associated with the mean shear. Because there is no turbulence production in that component, it quickly decays, with the result that the turbulent shear stress, and consequently the turbulence production in all the components, ultimately decreases.

Thus, the key to obtaining a nondecaying turbulent shear flow appears to lie in keeping the energy from being drained out of the turbulence component in the direction of the mean velocity gradient. In turbulence in which the nonlinear terms are important, the distribution of energy among the directional components is evidently accomplished by the pressure–velocity correlations, but in out case those correlations generally tend to make the turbulence more anisotropic. If, however, we superimpose on the shear flow destabilizing buoyancy forces in the direction of the mean-velocity gradient, it may be possible to obtain a nondecaying solution for our turbulence. Those buoyancy forces should tend to prevent the turbulence component in the direction of the mean velocity gradient from decaying, as in Fig. 5-84, in which mean shear was absent. It is shown that although our turbulence with mean shear ultimately decays if buoyancy is absent, the presence of destabilizing buoyancy counteracts the decay, and the turbulence grows at large times [117].

In this section we use for the initial φ_{ij} the relation for isotropic turbulence given by eq. (5-349):

$$(\varphi_{ij})_0 = \frac{J_0}{12\pi^2}(\delta_{ij}\kappa^2 - \kappa_i\kappa_j)e^{-\kappa^2/\kappa_0^2}, \qquad (5\text{-}349)$$

where J_0 is again a constant that depends on initial conditions, and κ_0 is an initial wavenumber that is characteristic of the turbulence. Except for the use of eq. (5-349) for $(\varphi_{ij})_0$ in place of eq. (5-333), the calculations in this section are done in the same way as those already given for combined shear and buoyancy. The two expressions for $(\varphi_{ij})_0$ differ only by an exponential and are identical for $\kappa_0 = \infty$. It again is assumed, of course, that δ_0 and $(\gamma_i)_0$ are zero. That is, as before, the turbulence producer (grid) is assumed to be unheated, so that the temperature fluctuations at later times are produced entirely by the interactions of the mean-temperature gradient with the turbulence.

The effect of destabilizing buoyancy forces (vertical-temperature gradient, negative) on weak homogeneous shear-flow turbulence in a gas is illustrated in Fig. 5-106. The superscript (a) on $\overline{u_i u_j}^{(a)}$ and $\kappa_0^{(a)}$ indicates that those parameters have been made dimensionless by using quantities related to the shear (in contrast to those related to the buoyancy, which are used later). Curves are shown for two values of Richardson number Ri and of the initial wavenumber parameter.

The curves indicate that for a Richardson number of 0 (no buoyancy effects) all components of the turbulent energy decrease with time. The turbulent shear stress $-\overline{u_1 u_3}$ also decreases with time, except near the initial time. (At $t^{(a)} = 0$ the turbulence is isotropic and $\overline{u_1 u_3}$ is 0.)

The decay of the components of turbulent energy for no buoyancy effects evidently occurs mainly because there is no production term in the equation for $\overline{u_3 u_3}$ (component in the direction of the mean-velocity gradient). This can be seen by letting $i = j = 3$ in eq. (5-448), in which case the production term (second term on the right side) drops out. In addition, the pressure–velocity correlation term in eq. (5-448) (third term on the right side) tends to drain energy out of the $\overline{u_3^2}$ component, as discussed in section 5-4-2-1. As a result the $\overline{u_3^2}$ component decays rapidly compared with the other components, which have energy fed into them by the mean-velocity gradient or by the pressure–velocity correlations (see Fig. 5-106). If $\overline{u_3^2}$ decays, the shear component $\overline{u_3 u_1}$ also must decay.

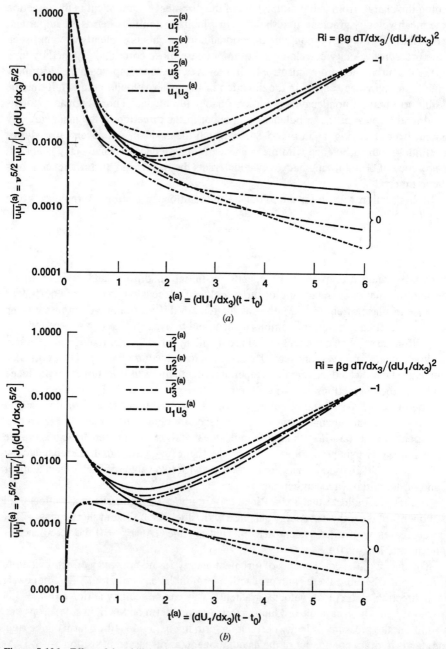

Figure 5-106 Effect of destabilizing buoyancy on the variation with time of turbulence in a uniform shear flow. Prandtl number, 0.7. (a) $\kappa_0^{(a)} = \nu^{1/2}\kappa_0/(dU_1/dx_3)^{1/2} = \infty$; (b) $\kappa_0^{(a)} = \nu^{1/2}\kappa_0/(dU_1/dx_3)^{1/2} = 1$.

There is then no mechanism for maintaining the turbulence, because that maintenance apparently takes place as a result of work done on the turbulent shear stress by the velocity gradient (see second term on the right side of eq. [5-448]). All of the turbulence components must then decay.

By contrast, for Ri $= -1$ (buoyancy forces destabilizing), all components of the turbulent energy decay for a while and then begin to increase without limit as time becomes large. This increase evidently occurs because the vertical buoyancy forces excite the $\overline{u_3^2}$ component of the turbulence and replenish the energy being drained out of it.

It might seem surprising that all components of the turbulence continue to increase with time rather than level off. There are no boundaries on the flow considered here, however, so the effective Reynolds number and Rayleigh number of the mean flow are infinite. As the scale or mixing length of the turbulence continues to grow, the eddies encounter larger and larger velocity and temperature differences, so that the effective driving forces on the turbulence continue to grow.

Comparison of Figs. 5-106a and b shows, as expected, that for $\kappa_0^{(a)} = \infty$ (all wavenumbers present), the components of the initial energy are infinite, whereas for $\kappa_0^{(a)} = 1$ they have a finite value. The turbulent shear stress $-\overline{u_3 u_1}$ starts at zero on both plots because the turbulent shear stress for isotropic turbulence is zero. For the case of $\kappa_0^{(a)} = \infty$, however, the value of $-\overline{u_3 u_1}$ jumps to infinity in an infinitely short time and then decreases. For $\kappa_0^{(a)} = 1$, $-\overline{u_3 u_1}$ first increases steadily and then either decreases (Ri $= 0$) or continues to increase (Ri $= -1$).

To give an idea of the distribution of the turbulent energy with wavenumber, energy spectra (spectra of $\overline{u_i u_i}$) are plotted in Fig. 5-107 for $\kappa_0^{(a)} = \infty$ and Ri $= 0$ and 1. For $t^{(a)} = 0$, ψ_{ii} is proportional to κ^4 (eq. [5-349]). As time increases, the spectra move to the small wavenumber regions; that is, the scale of the turbulence grows indefinitely large with time, because the fluid is unbounded.

Thus far we have been considering the effect of buoyancy on a shear-flow turbulence. Next we want to consider the related problem of the effect of imposing a mean shear on turbulence that is buoyancy-controlled. For doing this it is convenient to use the parameters $\overline{u_i^2}^{(b)}$, $t^{(b)}$, $\kappa_0^{(b)}$, $\overline{\tau^2}^{(b)}$, $\overline{\tau u_3}^{(b)}$, which have been made dimensionless by using quantities related to the buoyancy (Figs. 5-108, 5-109). The parameters used in Figs. 5-106 and 5-107, on the other hand, are nondimensionalized by using quantities related to the shear.

The effects of shear on buoyancy-controlled turbulence are illustrated in Figs. 5-108 and 5-109, in which $\overline{u_i^2}^{(b)}$, $\overline{\tau^2}^{(b)}$, and $\overline{\tau u_3}^{(b)}$ are plotted against $t^{(b)}$ for several values of Richardson number and $\kappa_0^{(b)}$. For the case of no shear (Ri $= -\infty$) the results are obtained from the integrated equations in section 5-4-2-4. All components of the turbulent energy, as well as the temperature fluctuations and the temperature–velocity correlations, increase as $t^{(b)}$ becomes large. This occurs even if shear is absent and the turbulence is completely controlled by the destabilizing buoyancy forces (Ri $= -\infty$). Although all turbulent energy components can increase with time if shear is absent, the component in the direction of the buoyancy forces is, in that case, at least an order of magnitude greater than the other components. On the other hand, if both buoyancy and shear are present, all components can be of the same order of magnitude.

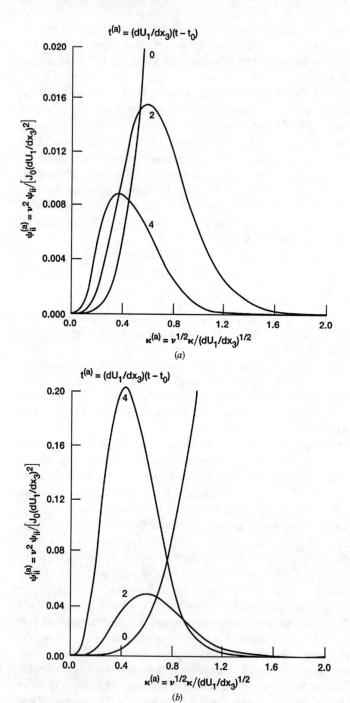

Figure 5-107 Variation with time of turbulent energy spectra (spectra of $\overline{u_i u_i}$). Prandtl number, 0.7. (a) Ri = 0; (b) Ri = −1.

FOURIER ANALYSIS, SPECTRAL FORM OF THE CONTINUUM EQUATIONS **299**

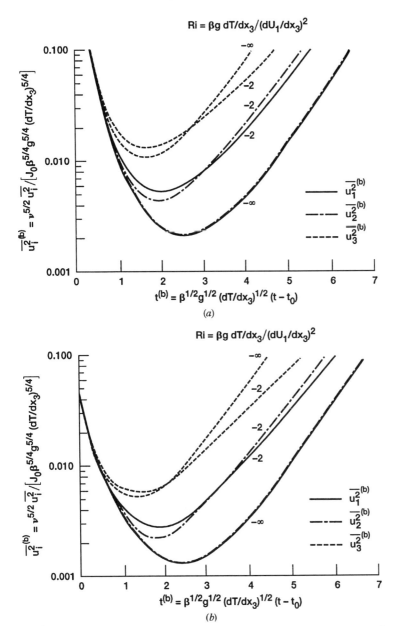

Figure 5-108 Effect of uniform shear on the variation with time of turbulence in a flow with destabilizing buoyancy. Prandtl number, 0.7. (a) $\kappa_0^{(b)} = \nu^{1/2}\kappa_0/[(-dT/dx_3)^{1/4}\beta^{1/4}g^{1/4}] = \infty$; (b) $\kappa_0^{(b)} = \nu^{1/2}\kappa_0/[(-dT/dx_3)^{1/4}\beta^{1/4}g^{1/4}] = 1$.

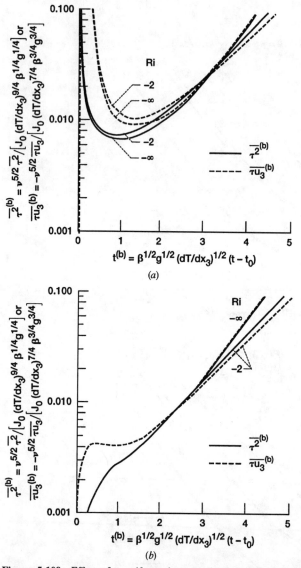

Figure 5-109 Effect of a uniform shear on the variation with time of temperature fluctuations and temperature–velocity correlations in a flow with destabilizing buoyancy. Prandtl number, 0.7. (a) $\kappa_0^{(b)} = \nu^{1/2}\kappa_0/[(-dT/dx_3)^{1/4}\beta^{1/4}g^{1/4}] = \infty$; (b) $\kappa_0^{(b)} = \nu^{1/2}\kappa_0/[(-dT/dx_3)^{1/4}\beta^{1/4}g^{1/4}] = 1$.

As the shear increases (as Ri goes from $-\infty$ to -2), the turbulent activity in general increases, at least at the earlier times. The shear does not seem to affect $\overline{\tau^2}^{(b)}$ or $\overline{\tau u_3}^{(b)}$ at the smaller times if $\kappa_0^{(b)} = 1$, however. At larger times, although $\overline{u_1^2}$ and $\overline{u_2^2}$ increase with increasing shear, $\overline{u_3^2}$, $\overline{\tau^2}$, and $\overline{\tau u_3}$ all decrease with increasing shear. These decreases appear to be related to the fact that at large times the presence of the shear causes energy to be drained out of the $\overline{u_3^2}$ component (as discussed previously), and thus out of $\overline{\tau u_3}$ and $\overline{\tau^2}$ (see eqs. [5-449] and [5-450]).

Summary of results for combined buoyancy and shear. The results for combined effects of vertical buoyancy forces and vertical velocity gradients indicate that, as in the case of no shear, destabilizing buoyancy forces can feed energy or activity into a turbulent field, whereas stabilizing buoyancy forces can extract it. The effect of the shear is to feed energy or activity into the turbulent field. Thus for the destabilizing case, the buoyancy and shear have similar effects; but for the stabilizing case, they work in opposite directions.

Energy or activity transfer between wavenumbers by the stretching of turbulent vortex filaments by the mean velocity gradient causes the spectra to become asymmetric; the slopes on the high-wavenumber sides of the spectra become more gradual.

For the destabilizing case, buoyancy forces tend to increase the vertical turbulence component in comparison to the horizontal component in the flow direction; the shear tends to decrease it. For weakly stabilizing conditions both the buoyancy and shear tend to decrease the ratio of vertical to horizontal turbulence components. For more strongly stabilizing conditions, the trends become less well defined.

The shear tends to align the turbulent vorticity in the direction of maximum mean strain, which is 45° from the flow direction. Destabilizing buoyancy forces tend to align the vorticity in horizontal directions, whereas stabilizing forces tend to align it vertically.

Some deductions from our analysis appear at first to be counterintuitive. If buoyancy forces are strongly stabilizing, the eddy conductivity and viscosity can be negative. This result appears reasonable if considered from the standpoint of a modified mixing-length theory. Also, turbulent heat transfer can occur in a horizontal as well as a vertical direction, even though the velocity and temperature gradients are both vertical. This latter effect is not dependent on the presence of buoyancy forces.

Uniformly sheared turbulence that is weak, or for which the mean-velocity gradient is large, ultimately decays with time (see section 5-4-2-1). However, the presence of destabilizing buoyancy forces in the direction of the mean-velocity gradient can prevent that decay. In that case the buoyancy forces replenish the energy being drained out of the component of the turbulence in the direction of the mean-velocity gradient by the shearing deformation, and the turbulent energy increases without limit as time increases. Apparently the energy can increase without limit because the effective Reyonlds and Rayleigh numbers are infinite in an unbounded fluid. As the scale or mixing length of the turbulence continues to grow, the eddies encounter larger and larger velocity and temperature differences, so that the effective driving forces acting on the turbulence continue to grow.

So far we have not considered the effects of normal strain on turbulence. That is done in the next sections.

5-4-2-6 Turbulence in an idealized flow through a cone. If fluid flows axially through a section of a cone, the fluid elements are distorted because of the changing cross-sectional area of the flow. If turbulence is initially present, it is modified by this distortion of the fluid. The interaction between turbulence and an idealized distorting mean flow in a cone is studied in this section. Flow in a cone is of particular interest because wind tunnels and rocket nozzles frequently contain conical sections.

The effect of an irrotational distortion (no shear) on turbulence has been studied in refs. [118] through [121]. In those studies the effects of viscosity are neglected and the distortion is assumed to occur so rapidly that the turbulent velocities have a negligible

302 TURBULENT FLUID MOTION

effect on the motion during distortion. The present work is more closely related to that of Pearson [78]. In his work the effects of viscosity are included and the requirement that the distortion be rapid is not imposed. If the distortion is not rapid, however, the turbulence must be weak enough to neglect terms containing triple correlations in comparison to other terms in the equations. Pearson's analysis assumes that the normal velocity gradients are uniform and that the turbulence is homogeneous.

The present analysis of turbulence in incompressible idealized flow through a cone [122] uses generalized two-point correlation equations that are based on the Navier-Stokes and continuity equations (see eqs. [4-145] to [4-150]). The normal velocity gradients $\partial U_1/\partial x_1$, $\partial U_2/\partial x_2$, and $\partial U_3/\partial x_3$ are allowed to vary axially but not transversely. The turbulence is assumed to be homogeneous in the transverse direction but only locally homogeneous in the axial direction. That is, the variation in intensity of the turbulence over a correlation (or mixing) length in the axial direction is assumed to be small. The mean axial velocity is assumed uniform over a cross-section, and mean shear stresses are assumed absent. The turbulence is initially isotropic but is allowed to become anisotropic under the distorting influence of the mean flow. Components of the turbulent velocity and vorticity variances are calculated, as well as components of the spectra of those quantities. A mean-strain energy-transfer term in the spectral equation that transfers energy components between wavenumbers also is considered. By using the momentum-heat-transfer analogy, the results are related to heat transfer between the fluid and the cone wall.

General two-point correlation equations for turbulence in an incompressible fluid with mean-velocity gradients are obtained in section 4-3-4 from the Navier-Stokes and continuity equations as follows:

$$\frac{\partial}{\partial t}\overline{u_i u_j'} + \overline{u_k u_j'}\frac{\partial U_i}{\partial x_k} + \overline{u_i u_k'}\frac{\partial U_j'}{\partial x_k'} + (U_k' - U_k)\frac{\partial}{\partial r_k}\overline{u_i u_j'} + \frac{1}{2}(U_k + U_k')\frac{\partial}{\partial (x_k)_m}\overline{u_i u_j'}$$

$$+ \frac{1}{2}\frac{\partial}{\partial (x_k)_m}\left(\overline{u_i u_j' u_k'} + \overline{u_i u_k u_j'}\right) + \frac{\partial}{\partial r_k}\left(\overline{u_i u_j' u_k'} - \overline{u_i u_k u_j'}\right)$$

$$= -\frac{1}{\rho}\left\{\frac{1}{2}\left[\frac{\partial}{\partial (x_i)_m}\overline{\sigma u_j'} + \frac{\partial}{\partial (x_j)_m}\overline{u_i \sigma'}\right] + \frac{\partial}{\partial r_j}\overline{u_i \sigma'} - \frac{\partial}{\partial r_i}\overline{\sigma u_j'}\right\}$$

$$+ \frac{1}{2}\nu\frac{\partial^2 \overline{u_i u_j'}}{\partial (x_k)_m \partial (x_k)_m} + 2\nu\frac{\partial^2 \overline{u_i u_j'}}{\partial r_k \partial r_k}, \tag{5-483}$$

$$\frac{1}{\rho}\left[\frac{1}{4}\frac{\partial^2 \overline{u_i \sigma'}}{\partial (x_j)_m \partial (x_j)_m} + \frac{\partial^2 \overline{u_i \sigma'}}{\partial (x_j)_m \partial r_j} + \frac{\partial^2 \overline{u_i \sigma'}}{\partial r_j \partial r_j}\right]$$

$$= -2\frac{\partial U_j'}{\partial x_k'}\left[\frac{1}{2}\frac{\partial \overline{u_i u_k'}}{\partial (x_j)_m} + \frac{\partial \overline{u_i u_k'}}{\partial r_j}\right] - \frac{1}{4}\frac{\partial^2 \overline{u_i u_j' u_k'}}{\partial (x_j)_m \partial (x_k)_m} - \frac{1}{2}\frac{\partial^2 \overline{u_i u_j' u_k'}}{\partial (x_j)_m \partial r_k}$$

$$- \frac{1}{2}\frac{\partial^2 \overline{u_i u_j' u_k'}}{\partial (x_k)_m \partial r_j} - \frac{\partial^2 \overline{u_i u_j' u_k'}}{\partial r_j \partial r_k}, \tag{5-484}$$

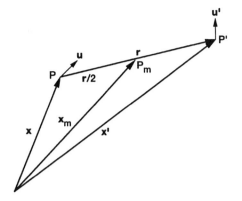

Figure 5-110 Vector configuration for two-point correlation equations.

and

$$\frac{1}{\rho}\left[\frac{1}{4}\frac{\partial^2 \overline{\sigma u'_j}}{\partial (x_i)_m \partial (x_i)_m} - \frac{\partial^2 \overline{\sigma u'_j}}{\partial (x_i)_m \partial r_i} + \frac{\partial^2 \overline{\sigma u'_j}}{\partial r_i \partial r_i}\right] = -2\frac{\partial U_i}{\partial x_k}\left[\frac{1}{2}\frac{\partial \overline{u_k u'_j}}{\partial (x_i)_m} - \frac{\partial \overline{u_k u'_j}}{\partial r_i}\right]$$

$$-\frac{1}{4}\frac{\partial^2 \overline{u_i u_k u'_j}}{\partial (x_i)_m \partial (x_k)_m} + \frac{1}{2}\frac{\partial^2 \overline{u_i u_k u'_j}}{\partial (x_i)_m \partial r_k} + \frac{1}{2}\frac{\partial^2 \overline{u_i u_k u'_j}}{\partial (x_k)_m \partial r_i} - \frac{\partial^2 \overline{u_i u_k u'_j}}{\partial r_i \partial r_k}. \quad (5\text{-}485)$$

The above equations are obtained from those in section 4-3-4, where $0 \leq n \leq 1$, by setting the *constant n* equal to $1/2$ and replacing the *subscript n* by m. Then the vector configuration in Fig. 4-17 becomes that in Fig. 5-110. The quantities u_i and u'_j are fluctuating velocity components at P and P', U_i and U'_j are mean velocity components, x_i is a space coordinate, t is the time, ρ is the density, ν is the kinematic viscosity, and σ is the instantaneous (mechanical) pressure. Bars over terms designate correlations or averaged quantities. The subscripts can take on the values 1, 2, or 3, and a repeated subscript in a term indicates a summation.

For locally homogeneous turbulence the turbulence is considered homogeneous over a correlation length, or the scale of the inhomogeneity is much greater than the scale of the turbulence. Thus, a quantity such as $\partial^2 \overline{u'_i u'_j}/\partial (x_k)_m \partial (x_k)_m$ in eq. (5-483) is negligible compared with $\partial^2 \overline{u_i u'_j}/\partial r_k \partial r_k$. (A calculation for axially decaying turbulence without mean velocity gradients (Fig. 3 in ref. [7]) shows that this is a good approximation except very close to the virtual origin of the turbulence.) In general, for locally homogenous turbulence,

$$\frac{\partial}{\partial (x_i)_m} \ll \frac{\partial}{\partial r_i}.$$

Also, for that type of turbulence the mean velocity varies linearly over distances for which correlations are appreciable so that

$$\frac{\partial U'_j}{\partial x'_k} = \frac{\partial U_j}{\partial x_k} = \frac{\partial (U_j)_m}{\partial (x_k)_m}, \quad U'_k - U_k = \frac{r_l \partial (U_k)_m}{\partial (x_l)_m}, \quad \frac{(U_k + U'_k)}{2} = (U_k)_m.$$

304 TURBULENT FLUID MOTION

Finally, in order to make the set of equations determinate, the turbulence is assumed weak enough to neglect terms containing triple correlations. It again should be noted that the turbulence in a flow with large velocity gradients may not have to be as weak as that in flow without velocity gradients. The terms containing those gradients may be large compared with triple-correlation terms, even if the turbulence is moderately strong. Equations (5-483) through (5-485) become, for the steady state at a fixed point,

$$\overline{u_k u'_j} \frac{\partial (U_i)_m}{\partial (x_k)_m} + \overline{u_i u'_k} \frac{\partial (U_j)_m}{\partial (x_k)_m} + \frac{\partial (U_k)_m}{\partial (x_l)_m} r_l \frac{\partial}{\partial r_k} \overline{u_i u'_j} + (U_k)_m \frac{\partial}{\partial (x_k)_m} \overline{u_i u'_j}$$

$$= -\frac{1}{\rho} \left(\frac{\partial}{\partial r_j} \overline{u_i \sigma'} - \frac{\partial}{\partial r_i} \overline{\sigma u'_j} \right) + 2\nu \frac{\partial^2 \overline{u_i u'_j}}{\partial r_k \partial r_k}, \tag{5-486}$$

$$\frac{1}{\rho} \frac{\partial^2 \overline{u_i \sigma'}}{\partial r_j \partial r_j} = -2 \frac{\partial (U_j)_m}{\partial (x_k)_m} \frac{\partial \overline{u_i u'_k}}{\partial r_j}, \tag{5-487}$$

and

$$\frac{1}{\rho} \frac{\partial^2 \overline{\sigma u'_j}}{\partial r_i \partial r_i} = 2 \frac{\partial (U_i)_m}{\partial (x_k)_m} \frac{\partial \overline{u_k u'_j}}{\partial r_i}. \tag{5-488}$$

Equations (5-486) through (5-488) are the correlation equations for steady-state locally homogeneous turbulence with mean-velocity gradients. The equations can be converted to spectral form by using the usual three-dimensional Fourier transforms already defined in eqs. (5-304) through (5-306):

$$\overline{u_i u'_j}(r) = \int_{-\infty}^{\infty} \varphi_{ij}(\kappa) e^{i\kappa \cdot r} d\kappa, \tag{5-304}$$

$$\overline{\sigma u'_j} = \int_{-\infty}^{\infty} \lambda_j e^{i\kappa \cdot r} d\kappa, \tag{5-305}$$

$$\overline{u_i \sigma'} = \int_{-\infty}^{\infty} \lambda'_i e^{i\kappa \cdot r} d\kappa, \tag{5-306}$$

where $d\kappa = d\kappa_1 d\kappa_2 d\kappa_3$. Then,

$$r_l \frac{\partial \overline{u_i u'_j}}{\partial r_k} = \int_{-\infty}^{\infty} -\left(\kappa_k \frac{\partial \varphi_{ij}}{\partial \kappa_l} + \delta_{lk} \varphi_{ij} \right) e^{i\kappa \cdot r} d\kappa, \tag{5-489}$$

where δ_{lk} is the Kronecker delta. Equation (5-489) can be obtained by differentiating eq. (5-304) with respect to r_k, writing the inverse transform, and then differentiating with respect to κ_l. Taking the Fourier transforms of eqs. (5-486) through (5-488) results in

$$\varphi_{kj} \frac{\partial (U_i)_m}{\partial (x_k)_m} + \varphi_{ik} \frac{\partial (U_j)_m}{\partial (x_k)_m} - \frac{\partial (U_k)_m}{\partial (x_l)_m} \left(\kappa_k \frac{\partial \varphi_{ij}}{\partial \kappa_l} + \delta_{lk} \varphi_{ij} \right) + (U_k)_m \frac{\partial}{\partial (x_k)_m} \varphi_{ij}$$

$$= -\frac{1}{\rho} (i\kappa_j \lambda'_i - i\kappa_i \lambda_j) - 2\nu \kappa^2 \varphi_{ij}, \tag{5-490}$$

$$-\frac{1}{\rho} i\kappa_j \lambda'_i = 2 \frac{\partial (U_l)_m}{\partial (x_k)_m} \frac{\kappa_l \kappa_j}{\kappa^2} \varphi_{ik}, \tag{5-491}$$

$$\frac{1}{\rho} i\kappa_i \lambda_j = 2 \frac{\partial (U_l)_m}{\partial (x_k)_m} \frac{\kappa_l \kappa_i}{\kappa^2} \varphi_{jk}. \tag{5-492}$$

Combining eqs. (5-490) through (5-492) and noting that $\delta_{lk}\partial U_l/\partial x_k = \partial U_l/\partial x_l = 0$ by continuity result in

$$(U_k)_m \frac{\partial}{\partial(x_k)_m}\varphi_{ij} = \frac{\partial(U_l)_m}{\partial(x_k)_m}\left[\left(2\frac{\kappa_l\kappa_j}{\kappa^2}-\delta_{jl}\right)\varphi_{ik}+\left(2\frac{\kappa_l\kappa_i}{\kappa^2}-\delta_{il}\right)\varphi_{kj}+\kappa_l\frac{\partial\varphi_{ij}}{\partial\kappa_k}\right]-2\nu\kappa^2\varphi_{ij}. \tag{5-493}$$

Equation (5-493) is the spectral equation for steady-state locally homogeneous turbulence that is weak or for which mean-velocity gradients are large.

Consider next the case in which the mean strain is irrotational, that is, in which the shearing components of the mean-velocity gradient are zero. If we let

$$a_{11} \equiv \frac{\partial(U_1)_m}{\partial(x_1)_m}, \qquad a_{22} \equiv \frac{\partial(U_2)_m}{\partial(x_2)_m}, \qquad a_{33} \equiv \frac{\partial(U_3)_m}{\partial(x_3)_m},$$

Equation (5-493) becomes, for irrotational strain,

$$(U_k)_m \frac{\partial}{\partial(x_k)_m}\varphi_{ij} = a_{(ll)}\left[\left(2\frac{\kappa_l\kappa_i}{\kappa^2}-\delta_{il}\right)\varphi_{jl}+\left(2\frac{\kappa_l\kappa_j}{\kappa^2}-\delta_{jl}\right)\varphi_{ki}+\kappa_l\frac{\partial\varphi_{ij}}{\partial\kappa_l}\right]-2\nu\kappa^2\varphi_{ij}, \tag{5-494}$$

where (ll) is not strictly a tensor subscript. For axisymmetric strain at each point in a cross-section, as occurs in uniform flow through a cone, $a_{22} = a_{33}$, and by continuity of the mean flow,

$$a_{22} = a_{33} = -\frac{1}{2}a_{11}. \tag{5-495}$$

The subscript 1 refers to the direction of an axis of symmetry for the turbulence. Because $a_{11} = f(x_1)$, integration of eq. (5-495) gives $U_2 = -(1/2)a_{11}x_2$ and $U_3 = -(1/2)a_{11}x_3$, or for circular cross-section the radial velocity $U_r = (1/2)a_{11}r$. In addition, it is assumed that the turbulence is homogeneous over a cross-section of the flow and that the turbulence changes only in the axial or x_1 direction, so that

$$(U_k)_m \frac{\partial \varphi_{ij}}{\partial(x_k)_m} = (U_1)_m \frac{\partial \varphi_{ij}}{\partial(x_1)_m}. \tag{5-496}$$

To simplify the notation, let $(U_1)_m \equiv U$, $(x_1)_m \equiv x$, and $a_{11} \equiv a$ in the remainder of this section. Then for φ_{11} eq. (5-494) becomes

$$\frac{U}{a}\frac{\partial\varphi_{11}}{\partial x} - \kappa_1\frac{\partial\varphi_{11}}{\partial\kappa_1} + \frac{1}{2}\kappa_2\frac{\partial\varphi_{11}}{\partial\kappa_2} + \frac{1}{2}\kappa_3\frac{\partial\varphi_{11}}{\partial\kappa_3} = \varphi_{11}\left(6\frac{\kappa_1^2}{\kappa^2}-2-\frac{2\nu}{a}\kappa^2\right), \tag{5-497}$$

where use was made of the continuity relation in the form $\kappa_2\varphi_{12} + \kappa_3\varphi_{13} = -\kappa_1\varphi_{11}$. Similarly, for φ_{ii},

$$\frac{U}{a}\frac{\partial\varphi_{ii}}{\partial x} - \kappa_1\frac{\partial\varphi_{ii}}{\partial\kappa_1} + \frac{1}{2}\kappa_2\frac{\partial\varphi_{ii}}{\partial\kappa_2} + \frac{1}{2}\kappa_3\frac{\partial\varphi_{ii}}{\partial\kappa_3} = -3\varphi_{11}+\varphi_{ii}-2\frac{\nu}{a}\kappa^2\varphi_{ii}, \tag{5-498}$$

Next we determine U and a as functions of x for flow through a cone of arbitrary cross-section. With the aid of the diagram in Fig. 5-111, these are obtained as

$$U = U_0\left(1+\frac{x-x_0}{r_0}\tan\alpha\right)^{-2} = \frac{U_0 r_0^2}{X^2 \tan^2\alpha}, \tag{5-499}$$

$$a = -\frac{2U_0\tan\alpha}{r_0}\left(1+\frac{x-x_0}{r_0}\tan\alpha\right)^{-3} = -\frac{2U_0 r_0^2}{X^3 \tan^2\alpha}, \tag{5-500}$$

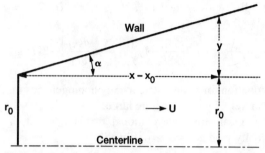

Figure 5-111 Section of cone considered in analysis. Flow cross section is arbitrary; α can be positive or negative.

where

$$X \equiv x - x_0 + \frac{r_0}{\tan\alpha} = -2\frac{U}{a}. \quad (5\text{-}501)$$

Then eqs. (5-497) and (5-498) become

$$X\frac{\partial\varphi_{11}}{\partial X} + 2\kappa_1\frac{\partial\varphi_{11}}{\partial\kappa_1} - \kappa_2\frac{\partial\varphi_{11}}{\partial\kappa_2} - \kappa_3\frac{\partial\varphi_{11}}{\partial\kappa_3} = \left(-12\frac{\kappa_1^2}{\kappa^2} + 4 - \frac{2\nu\tan^2\alpha}{U_0 r_0^2}X^3\kappa^2\right)\varphi_{11} \quad (5\text{-}502)$$

and

$$X\frac{\partial\varphi_{ii}}{\partial X} + 2\kappa_1\frac{\partial\varphi_{ii}}{\partial\kappa_1} - \kappa_2\frac{\partial\varphi_{ii}}{\partial\kappa_2} - \kappa_3\frac{\partial\varphi_{ii}}{\partial\kappa_3} = -2\left(1 + \frac{\nu\tan^2\alpha}{U_0 r_0^2}X^3\kappa^2\right)\varphi_{ii} + 6\varphi_{11}. \quad (5\text{-}503)$$

Equations (5-502) and (5-503) apply to either diverging or converging flow through a cone, depending on whether α is positive or negative.

Before eqs. (5-502) and (5-503) can be solved in an initial value problem, the turbulence must be specified, of course, at an initial position. It is assumed that the turbulence is isotropic at $x = x_0$, the virtual origin of the turbulence (or at $X = X_0 = r_0/\tan\alpha$ by eq. (5-501)), and that, as in previous sections, $(\varphi_{ij})_0$ is given by eq. (5-333):

$$(\varphi_{ij})_0 = \frac{J_0}{12\pi^2}(\kappa^2\delta_{ij} - \kappa_i\kappa_j), \quad (5\text{-}333)$$

where J_0 is a constant. Equation (5-333) gives results that, at all values of x, reduce to those for isotropic turbulence as the mean strain goes to zero. The use of that initial condition implies that Pearson's parameter $\tau = \nu\kappa_0^2/a$ approaches ∞, where κ_0 is a characteristic initial wavenumber of the turbulence [78]. Thus, the present results should be applicable for large kinematic viscosity, small initial turbulence scale, or small strain rate. The case $\tau \to \infty$ was not considered by Pearson.

Equations (5-502) and (5-503) are first-order partial-differential equations in four independent variables and can be solved by methods similar to those given in ref. [79]. Solutions of these equations in rescaled (dimensionless) from, subject to the initial condition given in eq. (5-333), are

$$\varphi_{11} = \frac{1}{12\pi^2}X^6(\kappa_2^2 + \kappa_3^2)\frac{1}{\kappa^4}(X^{-6}\kappa_1^2 + \kappa_2^2 + \kappa_3^2)^2 e^z \quad (5\text{-}504)$$

and

$$\varphi_{ii} = -\frac{\kappa^2}{\kappa_1^2}\varphi_{11} + \frac{1}{12\pi^2}\frac{X^6}{\kappa_1^2}(2X^{-6}\kappa_1^2 + \kappa_2^2 + \kappa_3^2)(X^{-6}\kappa_1^2 + \kappa_2^2 + \kappa_3^2)e^z, \quad (5\text{-}505)$$

where

$$-2X^2\left[\frac{1}{7}\kappa_1^2\frac{X^7-1}{X^6(X-1)} + \kappa_2^2 + \kappa_3^2\right] \to z, \quad (5\text{-}506)$$

$$\frac{X}{X_0} = 1 + (x - x_0)\frac{\tan\alpha}{r_0} \to X, \quad (5\text{-}507)$$

$$\left[\frac{v(x - x_0)}{U_0}\right]^{1/2}\kappa_i \to \kappa_i, \quad (5\text{-}508)$$

$$\frac{(x - x_0)v}{J_0 U_0}\varphi_{ij} \to \varphi_{ij}, \quad (5\text{-}509)$$

and the arrows are read "has been replaced by." It is of interest that φ_{11} and φ_{ii} are functions only of $(x - x_0)\tan\alpha/r_0$ and κ_i. Because of axial symmetry it is not necessary to obtain equations for φ_{22} and φ_{33}.

In order to integrate over wavenumber space, it is convenient to introduce spherical coordinates as follows:

$$\kappa_1 = \kappa\cos\theta, \qquad \kappa_2 = \kappa\cos\varphi\sin\theta, \qquad \kappa_3 = \kappa\sin\varphi\sin\theta. \quad (5\text{-}510)$$

Equation (5-304) then becomes, for $r = 0$,

$$\overline{u_i u_j} = \int_0^\infty \psi_{ij}\,d\kappa, \quad (5\text{-}511)$$

where

$$\psi_{ij} = \int_0^\pi \int_0^{2\pi} \varphi_{ij}\kappa^2 \sin\theta\,d\varphi\,d\theta. \quad (5\text{-}512)$$

The quantity ψ_{ij} is a function only of the magnitude of the wavenumber κ and represents an energy-spectrum tensor that has been integrated over all directions in wavenumber space.

The velocity variances $\overline{u_1^2}$, $\overline{u_i u_i}$, $\overline{u_2^2}$, and $\overline{u_3^2}$ are calculated from eqs. (5-504) through (5-512) as

$$\overline{u_1^2} = \frac{1}{24\sqrt{\pi}X^6}\left\{\frac{1}{2^{3/2}X^5\hbar g^{1/2}}\left(\frac{1 + X^6 - 2X^{12}}{\hbar} + \frac{X^{12}}{2f}\right) + \frac{1}{2^{5/2}X^5(1+\hbar)^{1/2}}\right.$$

$$\times\left[\frac{2X^6 - 3X^{12}}{\hbar} + \frac{X^{12}(3 + 2\hbar)}{1 + \hbar}\right] + \frac{3(1 - 4X^6 + 3X^{12})}{4X^4\hbar^2}H$$

$$\left. + \frac{15(1 - 2X^6 + X^{12})}{8X^6\hbar^3}(-\sqrt{2X}f^{1/2} + B)\right\}, \quad (5\text{-}513)$$

308 TURBULENT FLUID MOTION

$$\overline{u_i u_i} = -\frac{7^{3/2}}{96\sqrt{2\pi}X^2\hbar}(X-1)\left(\frac{X-1}{X^7-1}\right)^{1/2}\left[\frac{2-X^6-X^{12}}{X^7-1} + \frac{3(1-X^6)^2}{7\hbar X^6(X-1)}\right]$$
$$+\frac{1}{24\sqrt{\pi}X^6}\left[\frac{X(1+X^6)(3+2\hbar)}{2^{5/2}(1+\hbar)^{3/2}} + \frac{1+X^6-2X^{12}}{2^{5/2}X^5\hbar(1+\hbar)^{1/2}} + \frac{3(1-X^6)^2}{4X^4\hbar^2}H\right],$$
(5-514)

$$\overline{u_2^2} = \overline{u_3^2} = \frac{1}{2}\left(\overline{u_i u_i} - \overline{u_1^2}\right),$$
(5-515)

where

$$\frac{(x-x_0)^{5/2}v^{5/2}}{J_0 U_0^{5/2}}\overline{u_i u_j} \to \overline{u_i u_j},$$
(5-516)

$$\hbar = \frac{-6X^7 + 7X^6 - 1}{7X^6(X-1)},$$
(5-517)

$$f = \frac{X^7-1}{7X^6(X-1)},$$
(5-518)

$$\left.\begin{aligned}H &= \frac{1}{X\sqrt{2\hbar}}\ln\left(\sqrt{\hbar}+\sqrt{\hbar+1}\right) \quad \text{for } \hbar > 0,\\ H &= \frac{1}{X\sqrt{-2\hbar}}\sin^{-1}\sqrt{-\hbar} \quad \text{for } \hbar < 0,\end{aligned}\right\}$$
(5-519)

$$\left.\begin{aligned}B &= \frac{1}{\sqrt{2}}X\sqrt{\hbar+1} + \frac{X}{\sqrt{2\hbar}}\ln\left(\sqrt{\hbar}+\sqrt{\hbar+1}\right) \quad \text{for } \hbar > 0,\\ B &= \frac{1}{\sqrt{2}}X\sqrt{\hbar+1} + \frac{X}{\sqrt{-2\hbar}}\sin^{-1}\sqrt{-\hbar} \quad \text{for } \hbar > 0,\end{aligned}\right\}$$
(5-520)

where X has been rescaled by using eq. (5-507).

An important physical quantity related to the turbulence is the vorticity tensor $\overline{\omega_i \omega_j}$. The vorticity-spectrum tensor is given by eq. (5-348):

$$\Omega_{ij} = (\delta_{ij}\kappa^2 - \kappa_i\kappa_j)\varphi_{ll} - \kappa^2\varphi_{ij}.$$
(5-348)

By analogy with eq. (5-512) a directionally integrated vorticity-spectrum tensor can be defined as

$$\Lambda_{ij} = \int_0^\pi \int_0^{2\pi} \Omega_{ij}\kappa^2 \sin\theta\, d\varphi\, d\theta.$$
(5-521)

Then the vorticity tensor is given by

$$\overline{\omega_i \omega_j} = \int_0^\infty \Lambda_{ij}\, d\kappa.$$
(5-522)

One other quantity of considerable interest is the transfer term in eq. (5-490),

$$\frac{\partial (U_k)_m}{\partial (x_l)_m}\left(\kappa_k \frac{\partial \varphi_{ij}}{\partial \kappa_l} + \delta_{lk}\varphi_{ij}\right) = \beta_{ij}.$$
(5-523)

FOURIER ANALYSIS, SPECTRAL FORM OF THE CONTINUUM EQUATIONS **309**

This term is, of course, nonzero only in the presence of mean strain. The term is discussed in section 5-4-2-1 for the case of a shear flow. That it can be interpreted as a transfer term can be seen from eq. (5-489), where, if we let $r = 0$, we get

$$\int_{-\infty}^{\infty} \left(\kappa_k \frac{\partial \varphi_{ij}}{\partial \kappa_l} + \delta_{lk}\varphi_{ij} \right) d\kappa = 0. \tag{5-524}$$

That is, if the term is integrated over all of wavenumber space it gives zero contribution to the rate of change of $\overline{u_i u_j}$. (The quantity $(U_k)_m \partial \overline{u_i u_j}/\partial (x_k)_m$ in eq. (5-486) can be interpreted as a rate of change.) The term β_{ij}, however, can transfer energy between wavenumbers. Evidently the transfer takes place by the stretching or compressing of the vortex lines associated with the turbulence by the mean strain. This transfer is similar to that produced by triple correlations, except that in that case the stretching or compressing of the vortex lines is accomplished by the action of the turbulence on itself, rather than by a mean strain. For the present case of axially symmetric irrotational strain, eq. (5-523) becomes, for $i = j = 1$,

$$\beta_{11} = a \left(\kappa_1 \frac{\partial \varphi_{11}}{\partial \kappa_1} - \frac{1}{2}\kappa_2 \frac{\partial \varphi_{11}}{\partial \kappa_2} - \frac{1}{2}\kappa_3 \frac{\partial \varphi_{11}}{\partial \kappa_3} \right), \tag{5-525}$$

where φ_{11} is obtained from eq. (5-504). As in the case of φ_{ij} (eq. [5-512]), β_{ij} can be integrated over all directions in wavenumber space to give

$$T_{ij} = \int_0^\pi \int_0^{2\pi} \beta_{ij}\kappa^2 \sin\theta \, d\varphi \, d\theta. \tag{5-526}$$

Computed velocity variances, vorticities, and spectra are discussed next.

Turbulent velocity variances $\overline{u_1^2}$, $\overline{u_2^2}$, and $\overline{u_3^2}$, calculated using eqs. (5-513) through (5-515), are plotted in dimensionless form in Fig. 5-112. Also included is the curve

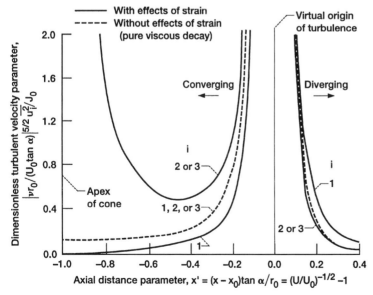

Figure 5-112 Variance of turbulent velocity components for flow through cone.

obtained by solving eq. (5-502) as though the effects of strain were absent by omitting the transfer, pressure, and production terms (terms two to six). The resulting equation is

$$\left|\frac{\nu r_0}{U_0 \tan \alpha}\right|^{5/2} \frac{\overline{u_1^2}}{J_0} = \frac{3}{16\sqrt{2\pi/3}} |X^3 - 1|^{-5/2}, \tag{5-527}$$

where eq. (5-507) is again used to rescale X. Negative values of x', the abscissa, correspond to a converging flow and positive values to a diverging flow. The virtual origin of the turbulence ($x' = 0$) is the point at which the energy of the turbulence is infinite.

In a converging flow, velocity fluctuations first decrease as one moves from the virtual origin toward the apex of the cone because of viscous dissipation. The distorting influence of the cone causes the longitudinal components to decrease more rapidly and the transverse components to decrease less rapidly than they would without the effects of strain. As the apex of the cone is approached, the longitudinal component continues to decrease rapidly toward zero. The transverse components, on the other hand, begin to increase as the effects of strain become greater than the effects of viscous dissipation. From eq. (5-500) it can be seen that the strain rate increases rapidly as the apex is approached. At the apex, where the mean velocity and strain rate approach infinity, the transverse velocity fluctuations also become infinite. In practice, of course, the tip of the cone must be removed in order to allow flow. It is of interest that this increase in transverse turbulent velocity component with mean velocity or contraction ratio at large velocity ratios is not observed in ref. [78], in which the velocity gradients are uniform.

For a diverging flow in a cone (positive x') the effects of strain are opposite to those for a converging flow. The longitudinal component decreases less rapidly than it would for no strain. Although the effect of strain is to increase the longitudinal component, that component continues to decrease as x' increases because, as shown in eq. (5-500), the strain rate decreases with x'. That is, the effect of viscous decay in this case is greater than the effect of strain.

Trends similar to those shown in Fig. 5-512 for converging flow have been observed experimentally in refs. [123] and [124]. A comparison between the present analysis and experimental results from ref. [123] for a four-to-one contraction is given in Fig. 5-113. The subscript a refers to conditions upstream of the contraction at the point at which the mean velocity begins to vary. (It should be noted that the contraction in the experiment evidently is not a true cone.) Values of U/U_a at the minimum point for the analytical curves are obtained by assuming that the value of U/U_a at the minimum point in the analytical curve for the transverse component ($i = 2$) corresponds to the value of U/U_a at the minimum point in the experimental curve for the same component. The overall change in the turbulent components produced by the contraction appears to be given reasonably well by the analysis, but the minimum in the experiment for $i = 2$ is sharper than that in the corresponding analytical curve.

To explain the trends shown in Fig. 5-112, turbulent vorticity variances are plotted in Fig. 5-114. The dashed curves in the figure, for no effects of strain, are obtained from the equation

$$\left|\frac{\nu r_0}{U_0 \tan \alpha}\right|^{7/2} \frac{\overline{\omega_1^2}}{J_0} = \frac{45}{64\sqrt{2\pi/3}} |X^3 - 1|^{-7/2}, \tag{5-528}$$

FOURIER ANALYSIS, SPECTRAL FORM OF THE CONTINUUM EQUATIONS **311**

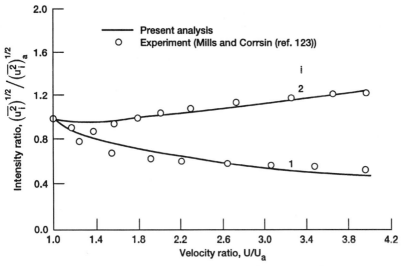

Figure 5-113 Present analysis compared with experiment for converging flow.

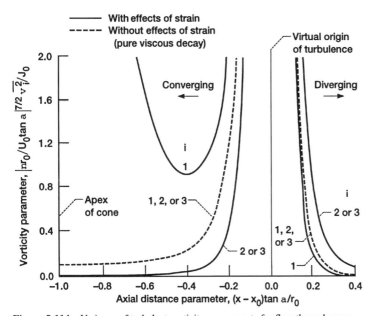

Figure 5-114 Variance of turbulent vorticity components for flow through cone.

where, as before, eq. (5-507) is used to rescale X. In general, the trends for the vorticity components are opposite to those for the velocities. For a converging flow, the longitudinal vortices are stretched and thus strengthened by the accelerating flow, and the transverse vortices are shortened along their axes and are thus weakened. The straining action also tends to turn the axes of the oblique vortices and to align them in the direction of the cone axis. Both of these effects tend to increase the longitudinal vorticity component and decrease the transverse components, as shown in Fig. 5-114. The opposite trends occur for a diverging flow; the strain in that case tends to decrease the longitudinal vorticity component and to increase the transverse components.

The relation of the vorticity to the fluctuating velocity components for a contracting flow has been pointed out by Prandtl [118] and Taylor [119]. If the vortex axes lie predominately in the direction of the mean flow, the longitudinal velocity fluctuations are small, whereas the transverse velocity fluctuations can be large. The opposite effects occur for a diverging flow. There is, however, an important difference between the two cases. This is illustrated in Fig. 5-115, in which $\overline{u_1^2}/\overline{u_2^2}$ and $\overline{\omega_1^2}/\overline{\omega_2^2}$ are plotted against U/U_0. (The relation between this abscissa and the one in the preceding plots is given by eq. (5-499). Although $\overline{u_1^2}/\overline{u_2^2}$ goes to zero as $\overline{\omega_1^2}/\overline{\omega_2^2}$ goes to infinity for a converging flow, the opposite limits are not approached for a diverging flow; that is, $\overline{u_1^2}/\overline{u_2^2}$ does not go to infinity as $\overline{\omega_1^2}/\overline{\omega_2^2}$ goes to zero but approaches a limiting value of 5. This occurs in the diverging flow because transverse as well as longitudinal velocity fluctuations are present if the vortex axes are aligned transversely; in converging flow, the longitudinal velocity fluctuations approach zero as the vortex axes become aligned longitudinally. Included in the plot (Fig. 5-115) for comparison is the curve for $\overline{u_1^2}/\overline{u_2^2}$ for a sudden contraction without viscosity as obtained from ref. [121].

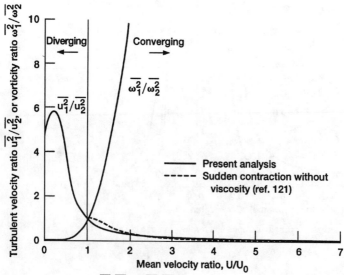

Figure 5-115 Plot of $\overline{u_1^2}/\overline{u_2^2}$ and $\overline{\omega_1^2}/\overline{\omega_2^2}$ against velocity ratio.

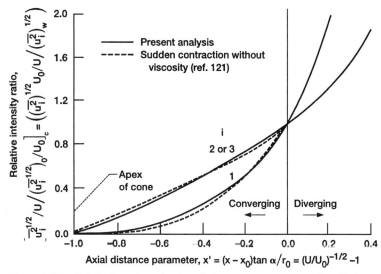

Figure 5-116 Relative intensity ratios for turbulent components corrected to eliminate decay.

Relative intensity ratios for turbulent components corrected to eliminate decay are plotted in Fig. 5-116. To obtain the ordinates in this figure, the ordinates for the solid curves in Fig. 5-112 are corrected for decay (as indicated by the subscript c) by dividing them by the ordinates for the dashed curves, which are for a pure viscous decay. (The subscript w means "without the effects of strain.") The result, after the square root has been extracted, is divided by U/U_0 to give intensity ratios relative to the local mean velocity. Curves obtained from ref. [121] for a sudden contraction and no viscosity are included in the plot in Fig. 5-116 and are similar to those for the present analysis.

The curve in Fig. 5-116 for $\{(\overline{u_2^2}^{1/2}/U)/[(\overline{u_2^2}^{1/2})_0/U_0]\}_c$ can be related approximately to the heat transfer between a cone wall and a fluid, which occurs, for instance, in a cooled rocket nozzle. The comparison is made by using the following argument, which is based on the momentum–heat transfer analogy: Except very close to the wall, the turbulent heat transfer is large compared with the molecular heat transfer, so that the total radial heat flow per unit area q is approximately $\rho c_p \varepsilon \, dT/dr$. The radial temperature gradient dT/dr at a particular radius is assumed proportional to $\Delta T/\delta$, where δ is the boundary layer thickness. The eddy diffusivity ε is replaced by the product of a transverse velocity fluctuation and a mixing length that is assumed proportional to δ, so that $\varepsilon \sim \overline{u_2^2}^{1/2} \delta$. Then the heat-transfer coefficient $h = q/\Delta T \sim \rho c_p \overline{u_2^2}^{1/2}$, or the Stanton number $St = h/(\rho U c_p) \sim \overline{u_2^2}^{1/2}/U$.

We assume that the change in turbulent intensity along the cone in the boundary layer is determined primarily by the normal strain rather than by the shear, as it might be if the strain is very rapid. (The initial turbulent intensity in the boundary layer at the entrance of the cone, of course, is determined by the shear in the upstream boundary layer.) By using the preceding relation for Stanton number, we can replace the ordinate of the curve for $i = 2$ in Fig. 5-116 by St/St_0. That curve, which has been corrected

314 TURBULENT FLUID MOTION

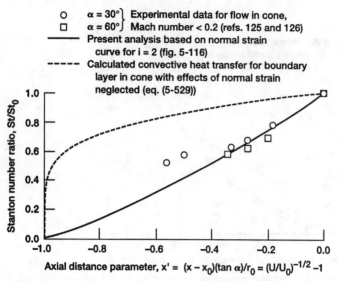

Figure 5-117 Approximate calculation of heat transfer between cone wall and fluid and comparison with experiment. Angle α (in degrees) in figure is angle between cone generator and flow direction.

to eliminate viscous decay, is used because the inhomogeneous shear that occurs in a boundary layer normally offsets the viscous decay. The curve, replotted in Fig. 5-117, shows that Stanton number decreases rapidly along a converging cone. If, by contrast, the Stanton number is calculated from local conditions in a boundary layer without considering normal strain, the decrease is much more gradual. In that case the Stanton number is roughly proportional to $(\rho U D)^{-0.2}$ (assuming that $\delta \sim D$), and

$$\frac{St}{St_0} = (1 + x')^{0.2}. \tag{5-529}$$

The curve for eq. (5-529) is the dashed curve in Fig. 5-117.

The heat transfer in the boundary layer of a rocket nozzle of course, may be more complicated than the case considered herein, where changes are assumed governed by the normal strain. The effects of shear and variable properties, as well as of normal strain, may not be negligible. However, experimental data for heat transfer in cooled conical nozzles [125–127] indicate trends very similar to those obtained herein. Data from refs. [125] and [126] for M < 0.2 in the conical portions of two nozzles are plotted in Fig. 5-117. Comparison of the data with the solid and dashed curves seems to indicate that changes in the Stanton numbers along the cone are more dependent on the normal strain than on the shear in the boundary layer. For plotting the data, it was assumed that the entrance of the cone corresponds to $x' = 0$. Although there is some uncertainty as to the point in the analysis that corresponds to the entrance of the cone in the experiment, it turns out that the results are insensitive to the point chosen. If, for instance, we chose $x' = -0.5$ instead of $x' = 0$ as the starting point, the results are nearly identical.

Spectra of $\overline{u_1^2}$ and $\overline{u_2^2}$ are plotted in Figs. 5-118 and 5-119 and show how contributions to $\overline{u_1^2}$ and $\overline{u_2^2}$ are distributed among dimensionless wavenumbers (or reciprocal eddy

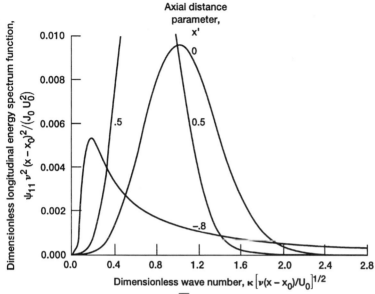

Figure 5-118 Spectra of dimensionless $\overline{u_1^2}$.

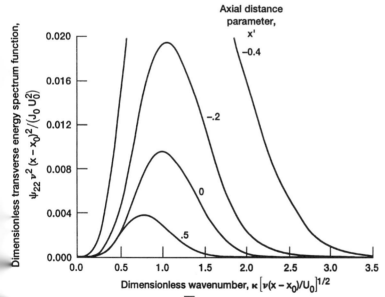

Figure 5-119 Spectra of dimensionless $\overline{u_2^2}$ (see Fig. 5-112 for definition of x').

sizes). Plotted in this way the curve for no convergence or divergence ($x' = 0$) does not change with x. Thus, comparison of the curves for various values of x' with the curve for $x' = 0$ shows how convergence or divergence affects the spectrum at a given position in comparison with the spectrum at the same position with no strain. For converging flow (negative x') the contributions to $\overline{u_1^2}$ occur at smaller wavenumbers (larger eddies) than they would for no convergence, whereas contributions to $\overline{u_2^2}$ move to higher wavenumbers. For diverging flow, by contrast, both ψ_{11} and ψ_{22} move to lower wavenumbers than they would for no divergence.

Of some theoretical interest is the extreme asymmetry of the spectrum of $\overline{u_1^2}$ for negatively large x'. This effect has been observed previously (section 5-4-2-1), but it is much more pronounced here, possibly because the strain rate increases sharply as x' becomes more negative. As in the previous cases the asymmetry is associated with a spectral-transfer term that depends on the mean strain rate. That the effect is associated with a transfer term (eq. [5-526]) was verified by solving the spectral equation with the transfer term omitted. The spectrum obtained was found to be nearly symmetrical.

A plot of the dimensionless transfer term associated with the mean strain is given in Fig. 5-120. The net area under each curve is zero, in agreement with eq. (5-524). The curve for negative x' is predominately negative at low wavenumbers and positive at higher ones, so that energy is in general transferred from low wavenumbers to higher ones. The curve also indicates that a small amount of reverse transfer to low wavenumbers takes place in the very low–wavenumber region.

The fact that the energy transfer in the longitudinal component of the turbulence is primarily from low to high wavenumbers may be surprising, because according to a

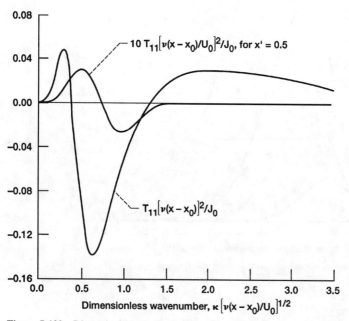

Figure 5-120 Dimensionless transfer spectra. x' is axial distance parameter (see Fig. 5-112).

simplified theory one might assume that the $\overline{u_1^2}$ component is produced by vortices that are aligned transversely. The axes of those vortices tend to be shortened in a converging flow. However, the proportion of vortices aligned in the transverse direction is small in a converging flow. Most of the contribution to $\overline{u_1^2}$ probably comes from oblique vortices, and in those vortices the energy transfer can be in the direction indicated in Fig. 5-120. For diverging flow (positive x') the energy transfer is in the opposite direction, that is, from high to low wavenumbers. In that case the effect of the energy transfer on the shape of the $\overline{u_1^2}$ spectrum seems to be small.

For the converging case, Fig. 5-120 shows that the positive area in the high-wavenumber region is spread out over a wide range of wavenumbers. As x' becomes more negative this range widens still more. This elongated positive area of energy transfer is responsible for the long tail on the spectrum of $\overline{u_1^2}$. To carry the effect to the extreme, the energy spectrum of dimensionless $\overline{u_1^2}$ for $x' = -0.99$ is plotted semilogarithmically in Fig. 5-121. Included also in the plot is the dissipation spectrum, which is proportional to $\kappa^2 \psi_{11}$. The energy and dissipation regions in this case show essentially complete separation and are separated by a pseudo-inertial subrange in which energy transfer is the dominating process. This inertial subrange is termed *pseudo-* because it occurs only in one component of the turbulence and because it is produced by inertial effects associated with the mean strain rather than with the triple correlations that are usually considered to be responsible for an inertial subrange [4]. Figure 5-121 is, however, a good illustration of a calculated case in which essentially complete separation of energy and dissipation regions occurs.

Figure 5-122 shows a log–log plot of the spectrum of dimensionless $\overline{u_1^2}$ for $x' = -0.99$. In this plot ψ_{11} is proportional to κ^{-1} over about four decades of κ. The region in

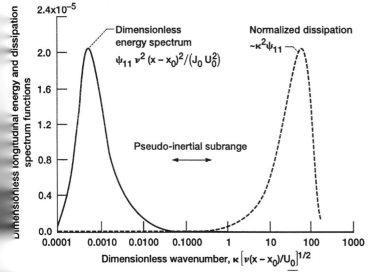

Figure 5-121 Energy and dissipation spectra for dimensionless $\overline{u_1^2}$ for $x' = -0.99$. Dissipation spectrum normalized to height of dimensionless ψ_{11} spectrum.

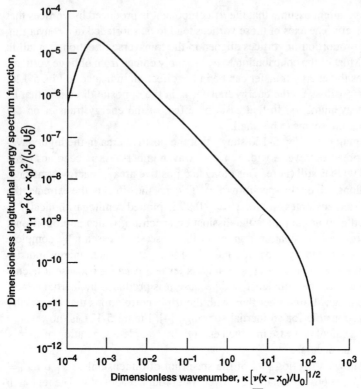

Figure 5-122 Log–log plot of spectrum of dimensionless $\overline{u_1^2}$ for $x' = -0.99$.

which the curve begins to fall off more rapidly than κ^{-1} is roughly the region in which dissipation effects become important. For $x' = -1$, the dissipation region is moved to infinity and ψ varies as κ^{-1} over the entire range of wavenumbers. The present results for a turbulence with large mean strain rates differ from the usual Kolmogorof $-5/3$-power spectrum that appears to apply at very high Reynolds numbers [128]. As shown by the present results, however, a $-5/3$-power region in an energy spectrum is not necessary for complete separation of the energy and dissipation regions. In fact, any power between 0 and -2 will do as well because the dissipation spectrum is obtained by multiplying the energy spectrum by κ^2.

Vorticity spectra also were calculated, and representative results for spectra of dimensionless $\overline{\omega_2^2}$ plotted in Fig. 5-123. As x' becomes more negative, the spectra move to higher dimensionless wavenumbers. Contributions to $\overline{\omega_2^2}$ also become spread out over a much wider range of wavenumbers or vortex sizes. As x' increases positively, the spectra move to lower dimensionless wavenumbers. The trends for the spectra of $\overline{\omega_1^2}$ (not shown) are similar to those shown in Fig. 5-123, with the exception that the shapes of the spectra do not change appreciably with strain. Thus, the vortices in converging flow tend to be smaller at a given x than they would be for no convergence, whereas they tend to be comparatively large for diverging flow.

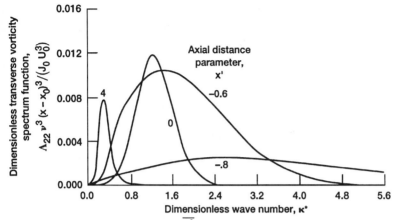

Figure 5-123 Spectra of dimensionless $\overline{\omega_2^2}$.

To summarize the results in this section, note that near the virtual origin of the turbulence, for both converging and diverging flow through a cone, all turbulence components decrease along the flow because of viscous dissipation. For a converging flow the effect of the distortion is to tend to align the turbulent vorticity in the direction of mean flow. This causes the longitudinal component of the velocity variance to decrease toward zero and the lateral component to increase. For a diverging flow the vorticity tends to be aligned in the transverse direction. In that case the ratio of longitudinal to transverse velocity variance ultimately approached a limit of 5. If the results for turbulent intensity are corrected to eliminate viscous decay and divided by local mean velocity, both longitudinal and transverse components decrease along a converging flow and increase along a diverging flow. The results are related approximately to heat transfer between a cone wall and a fluid and give trends very similar to those observed experimentally.

Turbulent vorticity spectra showed that the turbulent vortices in a converging flow tend to be smaller at a given location than they are for no convergence, whereas those in a diverging flow tend to be relatively large. A mean-strain transfer term in the spectral equation for the longitudinal component of the turbulence transferred energy in the high-wavenumber direction for a converging flow. This transfer causes the spectrum for the longitudinal component of the energy to become strongly asymmetric. Near the cone apex essentially complete separation of the longitudinal energy and dissipation spectra is obtained. A -1- rather than a $-5/3$-power spectrum that extended over a considerable range of wavenumbers was noted for the longitudinal energy components. Thus, a $-5/3$-power spectrum is not a necessary consequence of the separation of the energy and dissipation regions.

5-4-2-7 Turbulence and heat transfer with uniform normal strain. The analysis in the previous section is limited to turbulent flow through a cone. Here the study is extended to include turbulence and heat transfer with uniform normal strain [129]. That case has application, for example, to the heat transfer at a stagnation point.

320 TURBULENT FLUID MOTION

We consider the effect of uniform mean-velocity gradients dU_1/dx_1, dU_2/dx_2, dU_3/dx_3 on locally homogeneous turbulence and on longitudinal turbulent heat transfer. *Locally homogeneous*, as in the previous section, means that the intensity of the turbulence does not vary appreciably over a correlation or mixing length. Shear stresses are assumed to be absent, and the flow is considered incompressible and axisymmetric. The turbulence portion of this work (no heat transfer) has been considered by Pearson [78]. Additional results for the turbulence are given herein; Pearson gave results only for the accelerating case and did not include turbulent vorticities or spectra. Instead of a steady-state locally homogeneous but longitudinally varying turbulence, as considered herein, Pearson considered a homogeneous time-varying turbulence. The two treatments, however, give the same results.

The turbulence and turbulent heat transfer are assumed to be homogeneous in the transverse direction but only locally homogeneous in the longitudinal direction. The mean axial velocity is taken as uniform over a cross-section. The turbulence is initially isotropic but is allowed to become anisotropic under the distorting influence of the mean flow.

The equations for locally homogeneous turbulent heat transfer are considered first; the equations for the turbulence itself are obtained from section 5-4-2-6. By writing the incompressible Navier-Stokes and energy equations at two points P and P' in the turbulent fluid, we get

$$\frac{\partial \overline{\tau u'_j}}{\partial t} + U_k \frac{\partial \overline{\tau u'_j}}{\partial x_k} + \overline{u_k u'_j} \frac{\partial T}{\partial x_k} + \frac{\partial}{\partial x_k} \overline{\tau u_k u'_j} + \overline{\tau u'_k} \frac{\partial U'_j}{\partial x'_k} + U'_k \frac{\partial \overline{\tau u'_j}}{\partial x'_k} + \frac{\partial}{\partial x'_k} \overline{\tau u'_j u'_k}$$

$$= -\frac{1}{\rho} \frac{\partial \overline{\tau \sigma'}}{\partial x'_j} + \nu \frac{\partial^2 \overline{\tau u'_j}}{\partial x'_k \partial x'_k} + \alpha \frac{\partial^2 \overline{\tau u'_j}}{\partial x_k \partial x_k}. \qquad (5\text{-}530)$$

Equation (5-530) is the same as eq. (11) in ref. [108], in which the derivation is carried out in detail. The vector configuration for correlations between fluctuating quantities at points P and P' is shown in Fig. 5-110. The quantity τ is the fluctuating component of the temperature at P, u_k and u'_j are fluctuating velocity components at P and P', U_k and U'_j are mean velocity components, T is the mean temperature at P, x_k and x'_k are space coordinates, t is the time, ρ is the density, ν is the kinematic viscosity, α is the thermal diffusivity, and σ is the instantaneous (mechanical) pressure. The overbars designate corelations or averaged quantities. The subscripts can take on the values 1, 2, or 3, and a repeated subscript in a term indicates a summation. From eq. (4-148) we get,

$$\frac{1}{\rho} \frac{\partial^2 \overline{\tau \sigma'}}{\partial x'_j \partial x'_j} = -2 \frac{\partial U'_j}{\partial x'_k} \frac{\partial \overline{\tau u'_k}}{\partial x'_j} - \frac{\partial \overline{\tau u'_j u'_k}}{\partial x'_j \partial x'_k} \qquad (5\text{-}531)$$

Introducing the new independent variables $r_k \equiv x'_k - x_k$ and $(x_k)_m \equiv \frac{1}{2}(x_k + x'_k)$ in eqs. (5-530) and (5-531) (see Fig. 5-110) results in

$$\frac{\partial}{\partial t} \overline{\tau u'_j} + \frac{1}{2}(U_k + U'_k) \frac{\partial}{\partial (x_k)_m} \overline{\tau u'_j} + (U'_k - U_k) \frac{\partial}{\partial r_k} \overline{\tau u'_j} + \overline{u_k u'_j} \frac{\partial T}{\partial x_k}$$

$$+ \overline{\tau u'_k} \frac{\partial U'_j}{\partial x'_k} + \frac{1}{2} \frac{\partial}{\partial (x_k)_m} \left(\overline{\tau u_k u'_j} + \overline{\tau u'_j u'_k} \right) + \frac{\partial}{\partial r_k} \left(\overline{\tau u'_j u'_k} - \overline{\tau u_k u'_j} \right)$$

$$= -\frac{1}{2\rho} \frac{\partial \overline{\tau\sigma'}}{\partial (x_j)_m} - \frac{1}{\rho} \frac{\partial}{\partial r_j} \overline{\tau\sigma'} + \frac{1}{4}(\nu + \alpha) \frac{\partial^2 \overline{\tau u_j'}}{\partial (x_k)_m \partial (x_k)_m}$$

$$+ (\nu - \alpha) \frac{\partial^2 \overline{\tau u_j'}}{\partial (x_k)_m \partial r_k} + (\nu + \alpha) \frac{\partial^2 \overline{\tau u_j'}}{\partial r_k \partial r_k} \tag{5-532}$$

and

$$\frac{1}{\rho} \left[\frac{1}{4} \frac{\partial^2 \overline{\tau\sigma'}}{\partial (x_j)_m \partial (x_j)_m} + \frac{\partial^2 \overline{\tau\sigma'}}{\partial (x_j)_m \partial r_j} + \frac{\partial^2 \overline{\tau\sigma'}}{\partial r_j \partial r_j} \right] = -2 \frac{\partial U_j'}{\partial x_k'} \left[\frac{1}{2} \frac{\partial \overline{\tau u_k'}}{\partial (x_j)_m} + \frac{\partial \overline{\tau u_k'}}{\partial r_j} \right]$$

$$- \frac{1}{4} \frac{\partial^2 \overline{\tau u_j' u_k'}}{\partial (x_j)_m \partial (x_k)_m} - \frac{1}{2} \frac{\partial^2 \overline{\tau u_j' u_k'}}{\partial (x_j)_m \partial r_k} - \frac{1}{2} \frac{\partial^2 \overline{\tau u_j' u_k'}}{\partial (x_k)_m \partial r_j} - \frac{\partial^2 \overline{\tau u_j' u_k'}}{\partial r_j \partial r_k}. \tag{5-533}$$

For locally homogeneous turbulence and turbulent heat transfer, the turbulence is considered homogeneous over a correlation length, or the scale of the inhomogeniety is much greater than the scale of the turbulence. Thus, a quantity such as $\partial^2 \overline{\tau u_j'}/\partial (x_k)_m \partial (x_k)_m$ in eq. (5-532) is negligible compared with $\partial^2 \overline{\tau u_j'}/\partial r_k \partial r_k$. (A calculation for axially decaying turbulence without mean velocity gradients, Fig. 3 in ref. [7], implies that this is a good approximation except very close to the virtual origin of the turbulence.) In general, for locally homogeneous turbulence, $\partial/\partial (x_i)_m \ll \partial/\partial r_i$. Also, for that type of turbulence, the mean velocity and mean temperature vary linearly over distances for which correlations are appreciable so that $\partial U_j'/\partial x_k' = \partial U_j/\partial x_k = \partial (U_j)_m/\partial (x_k)_m$, $U_k' - U_k = r_l \partial (U_k)_m/\partial (x_l)_m$, $(U_k + U_k')/2 = (U_k)_m$, and $\partial T/\partial x_k = \partial T_m/\partial (x_k)_m$. Finally, in order to make the set of equations determinate, the turbulence is assumed to be weak enough or the mean gradients large enough to neglect terms containing triple correlations. The turbulence in a flow with large velocity or temperature gradients may not have to be as weak as that in a flow without mean gradients. The terms containing those gradients may be large compared with triple-correlation terms, even if the turbulence is moderately strong. Equations (5-532) and (5-333) become, for steady state at a fixed point,

$$\overline{\tau u_k'} \frac{\partial (U_j)_m}{\partial (x_k)_m} + (U_k)_m \frac{\partial}{\partial (x_k)_m} \overline{\tau u_j'} + r_l \frac{\partial (U_k)_m}{\partial (x_l)_m} \frac{\partial}{\partial r_k} \overline{\tau u_j'} + \overline{u_k u_j'} \frac{\partial T_m}{\partial (x_k)_m}$$

$$= -\frac{1}{\rho} \frac{\partial}{\partial r_j} \overline{\tau\sigma'} + (\nu + \alpha) \frac{\partial^2 \overline{\tau u_j'}}{\partial r_k \partial r_k} \tag{5-534}$$

and

$$\frac{1}{\rho} \frac{\partial^2 \overline{\tau\sigma'}}{\partial r_j \partial r_j} = -2 \frac{\partial (U_j)_m}{\partial (x_k)_m} \frac{\partial \overline{\tau u_k'}}{\partial r_j}. \tag{5-535}$$

The case of uniform axisymmetric strain with no shear and with mean temperature gradient in the longitudinal direction is considered herein. Equation (5-534) for $j = 1$

322 TURBULENT FLUID MOTION

and eq. (5-535) then become

$$(U_k)_m \frac{\partial}{\partial (x_k)_m}\overline{\tau u_l'} + a_{(ll)}r_l\frac{\partial}{\partial r_l}\overline{\tau u_l'} + b_{(1)}\overline{u_1 u_l'} + a_{(l1)}\overline{\tau u_l'} = -\frac{1}{\rho}\frac{\partial}{\partial r_1}\overline{\tau \sigma'} + (\nu+\alpha)\frac{\partial^2 \overline{\tau u_l'}}{\partial r_k \partial r_k} \quad (5\text{-}536)$$

and

$$\frac{1}{\rho}\frac{\partial^2 \overline{\tau\sigma'}}{\partial r_j \partial r_j} = -2a_{(ll)}\frac{\partial \overline{\tau u_l'}}{\partial r_l}, \quad (5\text{-}537)$$

where subscripts in parentheses are not strictly tensor subscripts. Equations (5-536) and (5-537) can be converted to spectral form by introducing the usual three-dimensional Fourier transforms defined as follows:

$$\overline{u_i u_j'} = \int_{-\infty}^{\infty} \varphi_{ij} e^{i\kappa \cdot r}\, d\kappa, \quad (5\text{-}538)$$

$$\overline{\tau u_j'} = \int_{-\infty}^{\infty} \gamma_j e^{i\kappa \cdot r}\, d\kappa, \quad (5\text{-}539)$$

and

$$\overline{\tau\sigma'} = \int_{-\infty}^{\infty} \zeta' e^{i\kappa \cdot r}\, d\kappa. \quad (5\text{-}540)$$

Then, using continuity and the inverse transform of eq. (5-539) gives

$$a_{(ll)}r_l\frac{\partial}{\partial r_l}\overline{\tau u_j'} = -\int_{-\infty}^{\infty} a_{(ll)}\kappa_l\frac{\partial \gamma_j}{\partial \kappa_l}e^{i\kappa\cdot r}\, d\kappa, \quad (5\text{-}541)$$

where κ is a wavevector having the dimension 1/length and $d\kappa = d\kappa_1 d\kappa_2 d\kappa_3$. Taking the Fourier transforms of eqs. (5-536) and (5-537) results in

$$(U_k)_m\frac{\partial \gamma_1}{\partial (x_k)_m} - a_{(ll)}\kappa_l\frac{\partial \gamma_1}{\partial \kappa_l} + b_1\varphi_{11} + a_{(11)}\gamma_1 = 2a_{(ll)}\frac{\kappa_1\kappa_l}{\kappa^2}\gamma_1 - (\alpha+\nu)\kappa^2\gamma_1, \quad (5\text{-}542)$$

where two equations have been combined into one by eliminating ζ'.

For axisymmetric strain $a_{22} = a_{33}$, and by continuity of the mean flow,

$$a_{22} = a_{33} = -\left(\frac{1}{2}\right)a_{11} \quad (5\text{-}543)$$

The turbulence also is assumed to be homogeneous in the transverse direction and changes only in the longitudinal or x_1 direction, so that

$$(U_k)_m\frac{\partial \gamma_1}{\partial (x_k)_m} = (U_1)_m\frac{\partial \gamma_1}{\partial (x_1)_m}. \quad (5\text{-}544)$$

To simplify the notation, let $(U_1)_m \equiv U$, $(x_1)_m \equiv x$, $a_{11} \equiv a$, and $b_1 \equiv b$ in the remainder of this section. Then eq. (5-542) becomes

$$U\frac{\partial \gamma_1}{\partial x} - a\kappa_1\frac{\partial \gamma_1}{\partial \kappa_1} + \frac{1}{2}a\kappa_2\frac{\partial \gamma_1}{\partial \kappa_2} + \frac{1}{2}a\kappa_3\frac{\partial \gamma_1}{\partial \kappa_3} = -b\varphi_{11} - a\gamma_1 + 3a\frac{\kappa_1^2}{\kappa^2}\gamma_1 - (\alpha+\nu)\kappa^2\gamma_1$$

(5-54

where use is made of the continuity relation in the form $\kappa_2\gamma_2 + \kappa_3\gamma_3 = -\kappa_1\gamma_1$. Corresponding expressions for φ_{11} and φ_{ii} are given as eqs. (5-497) and (5-498):

$$\frac{U}{a}\frac{\partial\varphi_{11}}{\partial x} - \kappa_1\frac{\partial\varphi_{11}}{\partial\kappa_1} + \frac{1}{2}\kappa_2\frac{\partial\varphi_{11}}{\partial\kappa_2} + \frac{1}{2}\kappa_3\frac{\partial\varphi_{11}}{\partial\kappa_3} = \varphi_{11}\left(6\frac{\kappa_1^2}{\kappa^2} - 2 - \frac{2\nu}{a}\kappa^2\right) \quad (5\text{-}497)$$

$$\frac{U}{a}\frac{\partial\varphi_{ii}}{\partial x} - \kappa_1\frac{\partial\varphi_{ii}}{\partial\kappa_1} + \frac{1}{2}\kappa_2\frac{\partial\varphi_{ii}}{\partial\kappa_2} + \frac{1}{2}\kappa_3\frac{\partial\varphi_{ii}}{\partial\kappa_3} = -3\varphi_{11} + \varphi_{ii} - 2\frac{\nu}{a}\kappa^2\varphi_{ii} \quad (5\text{-}498)$$

For uniform $a = dU/dx$,

$$c \equiv \frac{U}{U_0} = 1 + \frac{a(x - x_0)}{U_0}. \quad (5\text{-}546)$$

Equations (5-545), (5-497), and (5-498) can be written in terms of c as

$$c\frac{\partial\gamma_1}{\partial c} - \kappa_1\frac{\partial\gamma_1}{\partial\kappa_1} + \frac{1}{2}\kappa_2\frac{\partial\gamma_1}{\partial\kappa_2} + \frac{1}{2}\kappa_3\frac{\partial\gamma_1}{\partial\kappa_3} = -\frac{b}{a}\varphi_{11} + \gamma_1\left[3\frac{\kappa_1^2}{\kappa^2} - 1 - \frac{(\alpha+\nu)}{a}\kappa^2\right], \quad (5\text{-}547)$$

$$c\frac{\partial\varphi_{11}}{\partial c} - \kappa_1\frac{\partial\varphi_{11}}{\partial\kappa_1} + \frac{1}{2}\kappa_2\frac{\partial\varphi_{11}}{\partial\kappa_2} + \frac{1}{2}\kappa_3\frac{\partial\varphi_{11}}{\partial\kappa_3} = \varphi_{11}\left(6\frac{\kappa_1^2}{\kappa^2} - 2 - \frac{2\nu}{a}\kappa^2\right), \quad (5\text{-}548)$$

and

$$c\frac{\partial\varphi_{ii}}{\partial c} - \kappa_1\frac{\partial\varphi_{ii}}{\partial\kappa_1} + \frac{1}{2}\kappa_2\frac{\partial\varphi_{ii}}{\partial\kappa_2} + \frac{1}{2}\kappa_3\frac{\partial\varphi_{ii}}{\partial\kappa_3} = -3\varphi_{11} + \varphi_{ii}\left(1 - 2\frac{\nu}{a}\kappa^2\right), \quad (5\text{-}549)$$

For solving eqs. (5-547) through (5-549), the turbulence is assumed to be initially isotropic (at $c = 1$), and, as usual, eq. (5-333) is used:

$$(\varphi_{ij})_0 = \frac{J_0}{12\pi^2}\left(\kappa^2\delta_{ij} - \kappa_i\kappa_j\right), \quad (5\text{-}333)$$

where J_0 is a constant that depends on initial conditions. For the initial condition on γ_1 (at $c = 1$), it again is assumed that

$$(\gamma_1)_0 = 0. \quad (5\text{-}550)$$

That is, the turbulence-producing grid is assumed to be unheated, so that the temperature fluctuations are produced by the interaction of the mean longitudinal temperature gradient with the turbulence. Equation (5-333) appears to be the simplest condition that gives results that, at all values of x, reduce to those for isotropic turbulence as the mean strain goes to zero. The use of that initial condition for φ_{ij} implies that Pearson's parameter $\nu\kappa_0^2/a$ approaches ∞, where κ_0 is a characteristic initial wavenumber of the turbulence [78]. Thus, the present results should be applicable for large kinematic viscosity, small initial turbulence scale, or small strain rate. The case $\nu\kappa_0^2/a \to \infty$ was not considered by Pearson.

Equations (5-547) through (5-549) can be solved by methods similar to those given in ref. [79]. Solutions of these equations subject to the initial conditions given in eqs. (5-333) and (5-550) are, in dimensionless form,

$$\gamma_1 = -\frac{1}{12\pi^2}\frac{c^{-5}}{(c-1)}\frac{\kappa_2^2\kappa_3^2}{\kappa^2}(c^3\kappa_1^2+\kappa_2^2+\kappa_3^2)^2 \exp\left\{-\left(\frac{1}{Pr}+1\right)\left[\frac{1}{2}(c+1)\kappa_1^2+c^{-1}\right.\right.$$

$$\left.\left.\times(\kappa_2^2+\kappa_3^2)\right]\right\}\int_1^c\frac{\xi}{\kappa_1^2+c^{-3}\xi^3(\kappa_2^2+\kappa_3^2)}\exp\left\{\left(\frac{1}{Pr}-1\right)(c-1)^{-1}\right.$$

$$\left.\times\left[\frac{1}{2}\kappa_1^2\left(\frac{c}{\xi}\right)^2(\xi^2-1)+(\kappa_2^2+\kappa_3^2)c^{-1}(\xi-1)\right]\right\}d\xi, \qquad (5\text{-}551)$$

$$\varphi_{11} = \frac{1}{12\pi^2}(\kappa_2^2+\kappa_3^2)\frac{c^{-3}}{\kappa^4}(c^3\kappa_1^2+\kappa_2^2+\kappa_3^2)^2\exp\left\{-2\left[\frac{1}{2}(1+c)\kappa_1^2+c^{-1}(\kappa_2^2+\kappa_3^2)\right]\right\}, \qquad (5\text{-}552)$$

and

$$\varphi_{ii} = \frac{1}{12\pi^2}\left[\frac{c^{-3}}{\kappa^2}(c^3\kappa_1^2+\kappa_2^2+\kappa_3^2)^2+c^3\kappa_1^2+\kappa_2^2+\kappa_3^2\right]$$

$$\times \exp\left\{-2\left[\frac{1}{2}(1+c)\kappa_1^2+c^{-1}(\kappa_2^2+\kappa_3^2)\right]\right\}, \qquad (5\text{-}553)$$

where

$$\left[\frac{\nu(x-x_0)}{U_0}\right]^{1/2}\kappa_i \to \kappa_i, \qquad (5\text{-}554)$$

$$\frac{(x-x_0)\nu}{J_0 U_0}\varphi_{ij} \to \varphi_{ij}, \qquad (5\text{-}555)$$

$$\frac{\nu}{J_0 b}\gamma_1 \to \gamma_1, \qquad (5\text{-}556)$$

and

$$\frac{\nu}{\alpha} \to Pr. \qquad (5\text{-}557)$$

The arrows indicate "has been replaced by." It can be seen that γ_1, φ_{11}, and φ_{ii} are functions only of c, κ_i, and Prandtl number. For $Pr = 1$, eq. (5-551) can be integrated to give

$$\gamma_1 = \frac{1}{12\pi^2}(c-1)^{-1}c^{-3}\left(\frac{\kappa_2^2+\kappa_3^2}{\kappa_1^2}\right)^{1/3}\kappa^{-2}(c^3\kappa_1^2+\kappa_2^2+\kappa_3^2)^2$$

$$\times\left\{\frac{1}{6}\ln\frac{[(\kappa_2^2+\kappa_3^2)^{2/3}-\kappa_1^{2/3}(\kappa_2^2+\kappa_3^2)^{1/3}+\kappa_1^{4/3}][(\kappa_2^2+\kappa_3^2)^{1/3}c^{-1}+\kappa_1^{2/3}]^2}{[(\kappa_2^2+\kappa_3^2)^{2/3}c^{-2}-\kappa_1^{2/3}(\kappa_2^2+\kappa_3^2)^{1/3}c^{-1}+\kappa_1^{4/3}][(\kappa_2^2+\kappa_3^2)^{1/3}+\kappa_1^{2/3}]^2}\right.$$

$$\left.+\frac{1}{\sqrt{3}}\tan^{-1}\left[\frac{2(\kappa_2^2+\kappa_3^2)^{1/3}-\kappa_1^{2/3}}{\sqrt{3}\kappa_1^{2/3}}\right]-\frac{1}{\sqrt{3}}\tan^{-1}\left[\frac{2(\kappa_2^2+\kappa_3^2)^{1/3}c^{-1}-\kappa_1^{2/3}}{\sqrt{3}\kappa_1^{2/3}}\right]\right\}$$

$$\times \exp\left\{-2\left[\frac{1}{2}(1+c)\kappa_1^2+c^{-1}(\kappa_2^2+\kappa_3^2)\right]\right\}. \qquad (5\text{-}558)$$

In order to integrate over wavenumber space, spherical coordinates are introduced as follows:

$$\kappa_1 = \kappa \cos\theta, \tag{5-559a}$$

$$\kappa_2 = \kappa \cos\varphi \sin\theta, \tag{5-559b}$$

and

$$\kappa_3 = \kappa \sin\varphi \sin\theta. \tag{5-559c}$$

For $r = 0$, eqs. (5-538) and (5-539) then become

$$\overline{u_i u_j} = \int_0^\infty \psi_{ij}\, d\kappa \tag{5-560}$$

and

$$\overline{\tau u_j} = \int_0^\infty \Gamma_j\, d\kappa, \tag{5-561}$$

where

$$\psi_{ij} = \int_0^\pi \int_0^{2\pi} \varphi_{ij}\, \kappa^2 \sin\theta\, d\varphi\, d\theta \tag{5-562}$$

and

$$\Gamma_j = \int_0^\pi \int_0^{2\pi} \gamma_j\, \kappa^2 \sin\theta\, d\varphi\, d\theta. \tag{5-563}$$

The quantities φ_{ij} and Γ_j are functions only of the magnitude of the wavenumber and represent spectrum functions that have been integrated over all directions in wavenumber space.

The expressions for the velocity variances $\overline{u_1^2}$, $\overline{u_i u_i}$, $\overline{u_2^2}$, and $\overline{u_3^2}$ can be integrated as follows:

$$\overline{u_1^2} = \frac{1}{12\sqrt{\pi}c^3(c+1)^{1/2}} \left[\frac{3}{2(c+1)^2} + \frac{c(c^3-1)}{(c+1)(c^2+c-2)} + \frac{5}{2}\frac{c^2(c^3-1)^2}{(c^2+c-2)^2} \right.$$

$$+ \frac{3}{4}\frac{(c^2+c-2)}{c+1} + \frac{1}{4}\frac{c}{c+1}(2c^3+c^2+c-5) - \frac{3}{2}\frac{c^2(c^3-1)(c^3-3)}{(c^2+c-2)^2} \right]$$

$$+ \frac{1}{6\sqrt{\pi}c^3}\left\{\frac{3}{4}\frac{c^2(c^3-1)(c^3-3)}{(c^2+c-2)^2}H - \frac{15}{4}\frac{c^2(c^3-1)^2}{(c^2+c-2)^3}\left[\frac{c(c+1)^{1/2}}{2} - H\right]\right\}, \tag{5-564}$$

$$\overline{u_i u_i} = \frac{1}{6\sqrt{\pi}c^3(c+1)^{1/2}} \left[-\frac{c(2c^6-c^3-1)}{4(c+1)(c^2+c-2)} - \frac{3}{4}\frac{c^2(c^3-1)^2}{(c^2+c-2)^2} \right.$$

$$+ \frac{c(c^3+1)(c^2+c+1)}{8(c+1)} + \frac{c^2(c^3-1)(c^3+2)}{8(c^2+c-2)} + \frac{3}{4}\frac{c^2(c^3-1)^2(c+1)^{1/2}}{(c^2+c-2)^2}H \right], \tag{5-565}$$

and
$$\overline{u_2^2} = \overline{u_3^2} = \frac{1}{2}(\overline{u_i u_i} - \overline{u_1^2}), \tag{5-566}$$

where
$$\frac{(x-x_0)^{5/2} v^{5/2}}{J_0 U_0^{5/2}} \overline{u_i u_j} \to \overline{u_i u_j}, \tag{5-567}$$

$$H = \left(\frac{c}{c^2+c-2}\right)^{1/2} \ln\left\{\left(\frac{c^2+c-2}{2}\right)^{1/2} + \left[\frac{c(c+1)}{2}\right]^{1/2}\right\} \quad \text{for } c > 1, \tag{5-568a}$$

$$H = \left(\frac{-c}{c^2+c-2}\right)^{1/2} \sin^{-1}\left(\frac{c^2+c-2}{-2}\right)^{1/2} \quad \text{for } c < 1, \tag{5-568b}$$

and $c = U/U_0$.

Another quantity of importance is the turbulent vorticity tensor $\overline{\omega_i \omega_j}$. The vorticity-spectrum tensor is given by eq. (5-348), which we write as

$$\Omega_{ij} = (\delta_{ij}\kappa^2 - \kappa_i \kappa_j)\varphi_{ll} - \kappa^2 \varphi_{ij}. \tag{5-569}$$

As was the case for φ_{ij}, eq. (5-348) can be integrated over all directions in wavenumber space to give

$$\Lambda_{ij} = \int_0^\pi \int_0^{2\pi} \Omega_{ij} \kappa^2 \sin\theta \, d\varphi \, d\theta. \tag{5-570}$$

The vorticity tensor is then given by

$$\overline{\omega_i \omega_j} = \int_0^\infty \Lambda_{ij} \, d\kappa. \tag{5-571}$$

Calculated turbulent velocities, vorticities, temperature–velocity correlations, and spectra are considered next.

Figure 5-124 shows turbulent velocity variances $\overline{u_1^2}$, $\overline{u_2^2}$, and $\overline{u_3^2}$ plotted logarithmically in dimensionless form. Included in the plot is the curve obtained by solving eq. (5-548) as though the effects of strain are absent by omitting the second to sixth terms. This solution gives

$$\left|\frac{v}{a}\right|^{5/2} \frac{\overline{u_1^2}}{J_0} = \frac{\ln^{-5/2} c}{48\sqrt{2\pi}}. \tag{5-572}$$

For an accelerating flow with uniform strain, the longitudinal component $\overline{u_1^2}$ decreases more rapidly and the lateral components decrease less rapidly than they would if the effects of strain were absent. For large values of U/U_0, the lateral components reach a steady-state value as observed in Pearson's results [78]. This result differs from flow through a converging cone (see the previous section), where the increasing strain rate with distance causes the lateral components to increase without limit as the apex of the

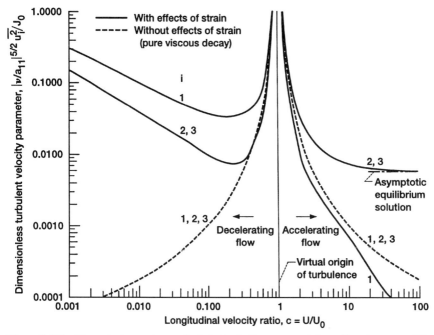

Figure 5-124 Dimensionless variance of turbulent velocity components for uniform incompressible strain.

cone is approached. The asymptotic equilibrium solution shown in Fig. 5-125 is given by

$$\overline{u_2^2} = \overline{u_3^2} = \frac{J_0}{96\sqrt{\pi}}\left(\frac{a}{\nu}\right)^{5/2}. \tag{5-573}$$

Thus, the solution represents a case in which the energy fed into the lateral components by straining action balances the energy dissipated by viscous forces. For decelerating flow near the virtual origin, both the longitudinal and the lateral components of the velocity fluctuations decrease in the direction of flow. For lower values of U/U_0, all components begin to increase as the effect of normal strain becomes greater than the effect of viscous dissipation. The region of increasing turbulent intensity in the decelerating case is not observed for the cone in the previous section, in which the strain rate a decreased sufficiently with distance to allow the turbulence to decay. As U approaches zero in the present case, the turbulence components tend to increase without limit. The assumption of local homogeneity, however, tends to break down in that region, and the turbulence components remain finite in a real situation. An increase in turbulent fluctuation in the decelerating flow near a stagnation point (in comparison with the free-stream fluctuation) has been observed experimentally in ref. [130].

The reasons for the trends observed in the turbulent velocity variances become clearer if the vorticity variances $\overline{\omega_i^2}$ plotted in Fig. 5-125 are considered. The dashed

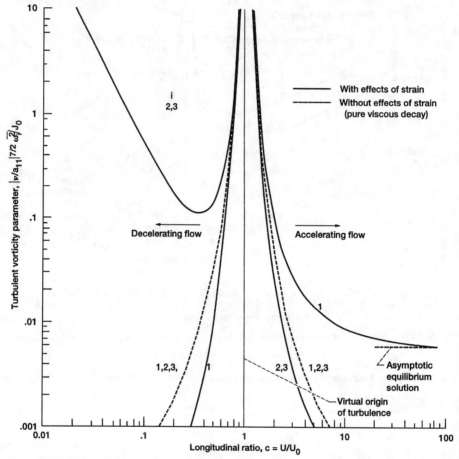

Figure 5-125 Dimensionless variance of turbulent vorticity components for uniform incompressible strain.

curves for no effects of strain are obtained from the equation

$$\left|\frac{\nu}{a}\right|^{7/2} \frac{\overline{\omega_1^2}}{J_0} = \frac{5}{192\sqrt{2\pi}} \ln^{-7/2} c. \tag{5-574}$$

Consider first the accelerating case. Here the trends are opposite to those for the velocity variances in Fig. 5-124; that is, the longitudinal vorticity component decays less rapidly and the lateral components decay more rapidly than they would if the effects of strain were absent. Thus, the turbulent vorticity tends to become aligned in the flow direction. That alignment occurs first because the longitudinal vortex filaments are strengthened by the stretching action of the mean flow, whereas the lateral filaments are shortened and thus weakened; and second because the mean strain rotates the vortex filaments that were originally oblique so that their axes tend to lie in the flow direction. The velocities associated with the turbulent vortex filaments then tend to lie in the transverse directions, in agreement with the curves for velocity variances in Fig. 5-124. As for the lateral components of the velocity variance, the longitudinal component of the vorticity variance approaches an equilibrium solution for large values of U/U_0 in which the vorticity generated by the mean strain balances that dissipated by viscous action. This solution is given by

$$\overline{\omega_1^2} = \frac{J_0}{96\sqrt{\pi}} \left(\frac{a}{\nu}\right)^{7/2}. \tag{5-575}$$

For decelerating flow at low values of U/U_0, the lateral components of the vorticity tend to increase, whereas the longitudinal component decreases more rapidly than it would if the effects of strain were absent. Thus, the vortex filaments tend to be aligned in the transverse directions. This alignment occurs because the lateral vortex filaments are strengthened by stretching, and the longitudinal components are weakened because they are shortened, and because the axes of vortex filaments that originally were oblique are rotated toward the transverse directions by the stretching action of the mean strain in the transverse directions. With the turbulent vortex filaments mostly aligned in the transverse directions, the velocities associated with them can be either in the longitudinal or the transverse directions. The explains why, for low values of U/U_0, the curves for both the longitudinal and the transverse components of the turbulent velocity variance in Fig. 5-124 increase in the flow direction, whereas in the curves for vorticity variance, only the lateral component can increase. The lateral stretching of the vortex filaments intensifies both the longitudinal and the transverse velocity fluctuations.

Relative intensity ratios for turbulence components corrected to eliminate viscous decay are plotted in Fig. 5-126. For obtaining the ordinates in this figure, values of turbulent velocity variance with the effects of strain included (solid curves in Fig. 5-124) are corrected to eliminate the effects of decay by dividing them by corresponding values for pure viscous decay (dashed curves in Fig. 5-124). The result (after taking the square root) is divided by U/U_0 to give intensity ratios that arc relative to the local mean velocity. In addition to the present results for uniform strain in an incompressible flow, results from the preceding section for flow through a cone and for uniform longitudinal strain in a compressible flow [131] are shown in the figure for comparison. The curves for uniform longitudinal strain in a compressible flow are obtained from eqs. (31) and (32) in ref. [131] by noting that $U_g = U_0 - ax_0$ and $U_g/\nu = U_0/\nu_0$ in those equations. The values for pure viscous decay are obtained by solving eq. (23) in ref. [133] with all but the first and last terms deleted, and again using $U_g/\nu = U_0/\nu_0$ and $U_g = U_0 - ax_0$. This solution gives $(U/U_0)^{-7/4}$ for the ordinate of the dot-dashed curve for $i = 1$ and $(1 + U/U_0)^{1/2}(U/U_0)^{-3/4}/\sqrt{2}$ for the ordinate of the dot-dashed curve for $i = 2, 3$.

The curves for the lateral components ($i = 2, 3$) for accelerating flow are of particular interest because, as shown in the previous section, the ordinates of those curves give approximately, for certain conditions, the Stanton-number ratio St/St_0 for the heat transfer between the fluid and a wall. In obtaining that relation, the normal strain is assumed to be so large that changes in the Stanton number along the flow are governed by normal strain rather than by shear.

The curves for the lateral components (and thus for the Stanton-number ratio) for accelerating flow in Fig. 5-126 indicate but a slight difference in the results for uniform incompressible strain and for flow in a cone. That is, if plotted in this way, the results at

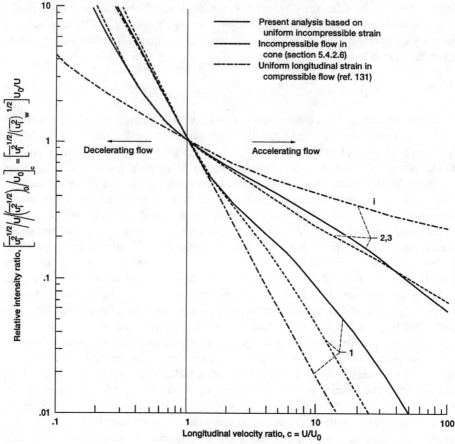

Figure 5-126 Comparison of relative intensity ratios for turbulent components corrected to eliminate decay for several situations. The subscript c means corrected. The subscript w means without strain.

a given U/U_0 for accelerating flow seem to depend but slightly on how the strain dU/dx varies along the flow. On the other hand, the results for uniform longitudinal strain in a compressible flow lie significantly above the others. These results evidently can be explained by the fact that the stretching of the vortex filaments is more intense in that case, because the lateral compressive strain is absent because of the lack of a radial flow. The dot-dashed curve might be related to heat transfer in a highly heated constant-area tube with fluid density changes along the length, whereas the other two curves are more closely related to nozzle heat transfer, in which the effects of compressibility are small (see Fig. 5-117).

Figure 5-127 shows the effect of uniform normal strain on dimensionless longitudinal turbulent heat transfer $\overline{\tau u_1}$. Because $\overline{\tau u_1}$ is divided by the temperature gradient b, the ordinates can be considered as representing the variation of longitudinal eddy conductivity with U/U_0. Results are given for Prandtl numbers of 0.01 (liquid metals), 1 and 0.7 (gases), and ∞. As Prandtl number decreases, the eddy conductivity decreases, apparently because a turbulent eddy in a high-conductivity fluid, such as a liquid metal

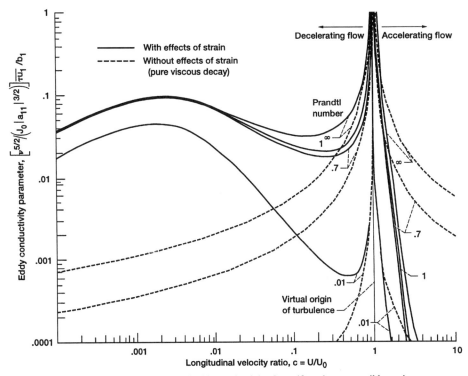

Figure 5-127 Dimensionless longitudinal eddy conductivity for uniform incompressible strain.

($Pr \sim 0.01$), gains or loses heat by conduction as it moves longitudinally and thus transfers heat with relatively low effectiveness (see Fig. 5-80). At a given Prandtl number, the trends in the curves are generally similar to those observed for the longitudinal component of the velocity variance $\overline{u_1^2}$ in Fig. 5-124. A notable exception, however, is that $\overline{\tau u_1}$ reaches a maximum at low values of U and then approaches zero at $U = 0$, whereas $\overline{u_1^2}$ becomes indefinitely large as U approaches zero. The ratio of maximum to minimum $\overline{\tau u_1}$ is greater at low Prandtl numbers.

The curves for decelerating flow in Fig. 5-127 illustrate the large increase that normal strain can produce in the longitudinal heat transfer between a body and a stream in the vicinity of a stagnation point if free-stream turbulence is present. The eddy conductivity decays to very low values (dashed curves) if the effects of strain are absent. (The time available for decay is quite large, because the fluid velocity becomes very small as the stagnation point is approached.) This increase in heat transfer is in agreement with the experiments in ref. [132] and the analysis of ref. [133]. The increase is evidently produced by the lateral stretching of vortex filaments, as assumed in ref. [133]. Reference [133], however, considered only transverse vortices, whereas the present analysis considers random vorticity in all directions. The present analysis does not, however, consider the damping effect of the wall at the stagnation point (effect of viscous diffusion), so that the increase in turbulent heat transfer due to normal strain probably is exaggerated here. In fact, an attempt to calculate the heat transfer near a stagnation point by assuming

332 TURBULENT FLUID MOTION

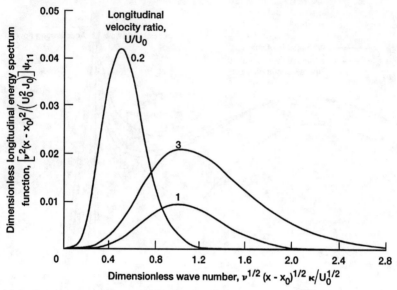

Figure 5-128 Spectra of dimensionless longitudinal velocity variance $\overline{u_1^2}$.

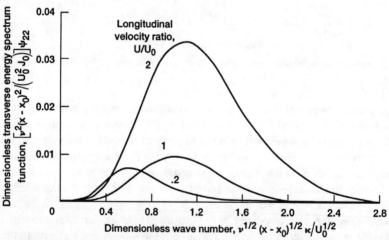

Figure 5-129 Spectra of dimensionless transverse velocity variance $\overline{u_2^2}$.

that the maximum in the curve in Fig. 5-527 ($Pr = 0.7$ for gases) corresponds to the maximum eddy conductivity in the boundary layer and that the minimum in the curve for $i = 1$ in Fig. 5-124 corresponds to the turbulence level in the undisturbed stream, gave (by dividing one ordinate by the other) increases in total heat transfer considerably higher than those observed experimentally. The results do indicate, however, that the combination of free-stream turbulence and normal strain (or lateral vortex stretching) can be an important factor in increasing the heat transfer in the vicinity of a stagnation point.

Dimensionless spectra of components of the velocity and vorticity variances are plotted in Figs. 5-128 through 5-130. The spectra show how contributions to the dimensionless

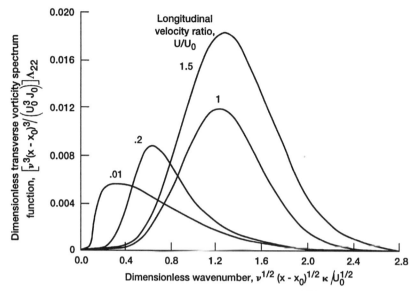

Figure 5-130 Spectra of dimensionless transverse vorticity variance $\overline{\omega_2^2}$.

mean velocity or vorticity fluctuations are distributed among dimensionless wavenumbers. Plotted in this way, the curves for no strain ($U/U_0 = 1$) reduce to a single curve that does not change with x. Thus, comparison of the curves for various values of U/U_0 shows how strain affects the spectrum at a given position in comparison with the spectrum at the same position with no strain. For instance, Fig. 5-130 shows that for decelerating flow ($U/U_0 < 1$) contributions to the transverse vorticity occur at smaller wavenumbers (larger vortices) than they would for no acceleration. This trend seems to be congruous with the observation in the analysis of ref. [133] that only the larger transverse vortices are amplified by stretching in the neighborhood of a stagnation point. Figures 5-128 and 5-129 show that contributions to components of the velocity variance also move to lower dimensionless wavenumbers as velocity ratio decreases. For $U/U_0 > 1$, the trends are opposite to those for decelerating flow, or contributions to the mean fluctuations move to higher dimensionless wavenumbers.

Figure 5-131 shows spectra corresponding to the asymptotic equilibrium solutions given by eqs. (5-573) and (5-575) for $U/U_0 \to \infty$. Note that the dimensionless quantities used in these spectra have been changed from those used in the preceding spectra in order to obtain finite dimensionless quantities for $U \to \infty$. The spectra in Fig. 5-131 are of interest because they show how contributions to the velocity and vorticity variances are distributed among wavenumbers for a case in which the energy fed into the turbulence by the mean strain exactly balances that dissipated by viscous action. Although the curves are in equilibrium at each wavenumber, there is not necessarily an equilibrium between production and dissipation at each wavenumber because energy can be transferred between wavenumbers by the stretching of the vortex filaments by the mean-velocity gradients, as we discussed for example in sections 5-4-2-1, 5-4-2-3, and 5-4-2-6.

Spectra of the dimensionless longitudinal eddy conductivity $\overline{\tau u_1}$ are plotted in Fig. 5-132 for Prandtl numbers of 0.7 and 0.01. The shifting of the curves to lower

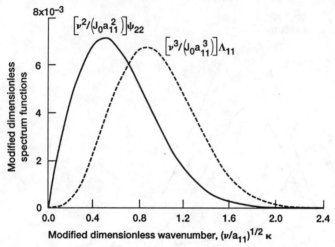

Figure 5-131 Asymptotic equilibrium spectra for longitudinal velocity ratio $U/U_0 \to \infty$.

dimensionless wavenumbers as U/U_0 decreases is similar to that for the spectra of the longitudinal velocity fluctuations shown in Fig. 5-128.

To summarize the results in this section, we note that for an incompressible accelerating flow with uniform strain, the longitudinal velocity fluctuations decrease more rapidly, the lateral fluctuations decrease less rapidly in the flow direction than they would if the effects of normal strain were absent. For large values of velocity ratio, the lateral components are found to reach a steady-state equilibrium value, as observed in Pearson's results. This result differs from flow through a converging cone, in which the increasing strain rate with distance cause the lateral components to increase without limit as the apex of the cone is approached.

For decelerating flow at low values of velocity ratio both the longitudinal and transverse velocity fluctuations increased in the flow direction as the effect of normal strain becomes greater than the effect of viscous dissipation. A somewhat similar increase in velocity fluctuation in the decelerating flow near a stagnation point has been observed experimentally. This region of increasing turbulent intensity in the decelerating case was not observed in the analysis of flow through a diverging cone, in which the strain rate decreased sufficiently with distance to allow the turbulence to decay.

Many of the trends observed for the velocity-fluctuation components can be explained by the analytical result that the vorticity becomes aligned in the longitudinal direction for accelerating flow and in the transverse directions for decelerating flow.

If the results for turbulent intensity are corrected for viscous decay and divided by local mean velocity, the transverse component for accelerating flow, which can be related to heat transfer between the fluid and a wall, is approximately the same for flow in a cone and for uniform incompressible strain. On the other hand the curve for uniform longitudinal compressible strain lies appreciably above the others, apparently because of the more intense vortex stretching for that case.

The results for longitudinal eddy conductivity in a decelerating flow show that normal strain can increase that quantity to values considerably above those which it

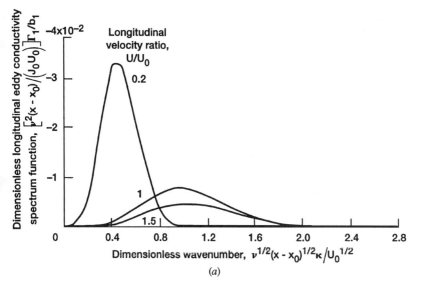

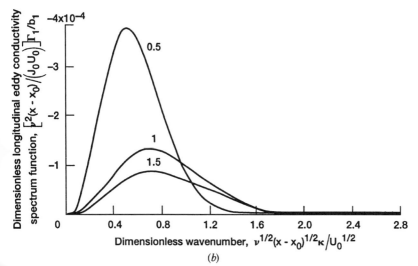

Figure 5-132 Spectra of dimensionless longitudinal eddy conductivity $\overline{\tau u_1}$. (a) Prandtl number, 0.70; (b) Prandtl number, 0.01.

would have if the effects of strain were absent. Thus, free-stream turbulence with normal strain (or lateral vortex stretching) can be an important factor in increasing the heat transfer in the vicinity of a stagnation point.

Turbulent vorticity spectra show that the turbulent vortexes in a decelerating flow tend to be larger at a given location than they would be for no deceleration. The eddies associated with the energy and with the longitudinal eddy conductivity also become comparatively lager. Spectra also are obtained for the energy and the vorticity for the asymptotic equilibrium solutions at large velocity ratios.

5-4-2-8 Turbulence and heat transfer with combined two-dimensional shear and normal strain.
The effects of uniform shear and of normal strain on turbulence have been analyzed separately, for example, in sections 5-4-2-1 and 5-4-2-7. There are important cases, however, in which shear and normal strain act simultaneously, as in the boundary layer of a fluid flowing through a contraction. Effects that are absent if one or the other type of strain acts by itself may be present if they act simultaneously. For instance, an apparent laminarization seems to occur in the boundary layers of certain accelerating flows [134–136].

Also, recall that there is a strong tendency for the energy to be drained out of a transverse component of a shear-flow turbulence, with a consequent ultimate decay of the total turbulent energy (see section 5-4-2-1 and Fig. 5-63). One way of maintaining the turbulence by preventing that drain, even in the absence of triple correlations, is to add a normal mean strain to the flow, as is discussed in section 5-4-2-9. (Another way of doing that is to introduce a destabilizing buoyancy, as discussed in section 5-4-2-5.)

In this section a simplified model is analyzed in an attempt to obtain some understanding of the effects of combined shear and normal strain on turbulence and turbulent heat transfer. Uniform shear and uniform normal velocity gradients, as well as a uniform transverse temperature gradient, are assumed to be acting on a field of initially isotropic turbulence. The turbulence quickly becomes anisotropic under the influence of the mean gradients. The turbulent field, although homogeneous in the transverse directions, is assumed to be only locally homogeneous in the longitudinal or flow direction; that is, the effects of changes in the intensity of the turbulence over a correlation or mixing length in the longitudinal direction are negligible. The normal strains in the present model correspond to a two-dimensional contraction with the transverse normal strains occurring in the same direction as the transverse shear and temperature gradients.

Two-point steady-state correlation equations for locally homogeneous turbulence have been obtained as eqs. (5-486) through (5-488) and (5-534) through (5-537). For uniform mean gradients one can write those equations as

$$\overline{u_k u'_j}\frac{\partial U_i}{\partial x_k} + \overline{u_i u'_k}\frac{\partial U_j}{\partial x_k} + \frac{\partial U_k}{\partial x_l} r_l \frac{\partial}{\partial r_k}\overline{u_i u'_j} + U_k \frac{\partial}{\partial x_k}\overline{u_i u'_j}$$

$$= -\frac{1}{\rho}\left(\frac{\partial}{\partial r_j}\overline{u_i \sigma'} - \frac{\partial}{\partial r_i}\overline{\sigma u'_j}\right) + 2\nu \frac{\partial^2 \overline{u_i u'_j}}{\partial r_k \partial r_k}, \qquad (5\text{-}576)$$

$$\frac{1}{\rho}\frac{\partial^2 \overline{u_i \sigma'}}{\partial r_j \partial r_j} = -2\frac{\partial U_j}{\partial x_k}\frac{\partial \overline{u_i u'_k}}{\partial r_j}, \qquad (5\text{-}577)$$

$$\frac{1}{\rho}\frac{\partial^2 \overline{\sigma u'_j}}{\partial r_i \partial r_i} = 2\frac{\partial U_i}{\partial x_k}\frac{\partial \overline{u_k u'_j}}{\partial r_i}, \qquad (5\text{-}578)$$

$$\overline{\tau u'_k}\frac{\partial U_j}{\partial x_k} + U_k \frac{\partial}{\partial x_k}\overline{\tau u'_j} + r_l \frac{\partial U_k}{\partial x_l}\frac{\partial}{\partial r_k}\overline{\tau u'_j} + \overline{u_k u'_j}\frac{\partial T}{\partial x_k} = -\frac{1}{\rho}\frac{\partial}{\partial r_j}\overline{\tau \sigma'} + (\nu+\alpha)\frac{\partial^2 \overline{\tau u'_j}}{\partial r_k \partial r_k}, \qquad (5\text{-}579)$$

and

$$\frac{1}{\rho}\frac{\partial^2 \overline{\tau\sigma'}}{\partial r_j \partial r_j} = -2\frac{\partial U_j}{\partial x_k}\frac{\overline{\partial \tau u'_k}}{\partial r_j}, \qquad (5\text{-}580)$$

where u_i and u'_j are fluctuating velocity components at the arbitrary points P and P', U_i is a mean velocity component, x_i is a space coordinate, r_i is a component of the vector extending from a point P to P', t is the time, ρ is the density, ν is the kinematic viscosity, σ is the instantaneous pressure, and τ is the temperature fluctuation. Bars over terms designate correlations or averaged quantities. The subscripts can take on the values 1, 2, or 3, and a repeated subscript in a term indicates a summation.

In obtaining eqs. (5-576) through (5-580), the instantaneous velocities and temperatures in the incompressible Navier-Stokes and energy equations are first broken into mean and fluctuating components. The resulting equations are then written at two points in the turbulent field, multiplied by appropriate temperatures or velocity components, and averaged. The equations for correlations involving pressures are obtained by taking the divergence of the Navier-Stokes equation and applying continuity. In order to make the locally homogeneous approximation, the turbulence is considered homogeneous over a correlation length, or the scale of the inhomogeneity is much greater than the scale of the turbulence. Thus, $\partial/\partial x_i \ll \partial/\partial r_i$, where the operators operate on two-point correlations. (A calculation for axially decaying turbulence without mean velocity gradients (Fig. 3 in ref. [7]) implies that this is a good approximation except in the region very close to the virtual origin of the turbulence.) Note also that, for locally homogeneous turbulence, the mean velocity and temperature may be considered to vary linearly over the small distances for which the correlations are appreciable, as assumed here. Finally, in order to make the set of equations determinate, the turbulence is assumed to be weak enough, or the mean gradients large enough, to neglect terms containing triple correlations. The turbulence in a flow with large velocity or temperature gradients may not have to be as weak as that in a flow without mean gradients. The terms containing those gradients may be large compared with triple-correlation terms, even if the turbulence is moderately strong.

Equations (5-576) through (5-580) can be converted to spectral form by introducing the usual three-dimensional Fourier transforms defined as follows:

$$\overline{u_i u'_j} = \int_{-\infty}^{\infty} \varphi_{ij} e^{i\kappa \cdot r} d\kappa, \qquad (5\text{-}581)$$

$$\overline{\sigma u'_j} = \int_{-\infty}^{\infty} \lambda_j e^{i\kappa \cdot r} d\kappa, \qquad (5\text{-}582)$$

$$\overline{u_i \sigma'} = \int_{-\infty}^{\infty} \lambda'_i e^{i\kappa \cdot r} d\kappa, \qquad (5\text{-}583)$$

$$\overline{\tau u'_j} = \int_{-\infty}^{\infty} \gamma_j e^{i\kappa \cdot r} d\kappa, \qquad (5\text{-}584)$$

and

$$\overline{\tau \sigma'} = \int_{-\infty}^{\infty} \zeta' e^{i\kappa \cdot r} d\kappa, \qquad (5\text{-}585)$$

where κ is a wavevector having the dimension 1/length and $d\kappa = d\kappa_1\, d\kappa_2\, d\kappa_3$. Taking the Fourier transforms of eqs. (5-576) through (5-580), eliminating the pressure–velocity and pressure–temperature terms, and using continuity result in

$$U_k \frac{\partial}{\partial x_k}\varphi_{ij} = \frac{\partial U_l}{\partial x_k}\left[\left(2\frac{\kappa_l\kappa_j}{\kappa^2} - \delta_{jl}\right)\varphi_{ik} + \left(2\frac{\kappa_l\kappa_i}{\kappa^2} - \delta_{il}\right)\varphi_{kj} + \kappa_l\frac{\partial \varphi_{ij}}{\partial \kappa_k}\right] - 2\nu\kappa^2\varphi_{ij} \tag{5-586}$$

and

$$U_k \frac{\partial \gamma_j}{\partial x_k} = \frac{\partial U_l}{\partial x_k}\left[\left(2\frac{\kappa_l\kappa_j}{\kappa^2} - \delta_{jl}\right)\gamma_k + \kappa_l\frac{\partial \gamma_j}{\partial \kappa_k}\right] - \frac{\partial T}{\partial x_k}\varphi_{kj} - (\alpha + \nu)\kappa^2\gamma_j, \tag{5-587}$$

where δ_{ij} is the Kronecker delta.

Equations (5-586) and (5-587) give contributions of various processes to the rates of change (with x_k) of spectral components of the turbulent energy tensor $\overline{u_i u_j}$ and of the turbulent heat transfer vector $\overline{\tau u_j}$. The terms in the equations that are proportional to $\partial/\partial \kappa_k$ are transfer terms that transfer activity into or out of a spectral component by the stretching or compressing of turbulent vortex filaments by the mean-velocity gradient, as discussed in sections 5-4-2-1, 5-4-2-6, and 5-4-2-7. The terms with κ^2 in the denominator are spectral components of pressure–velocity or pressure–temperature correlations and transfer activity between directional components (section 5-4-2-1). The last terms in the equations are dissipation terms, which dissipate activity by viscous or by conduction effects. The dissipation term in eq. (5-587) contains both viscous and conduction effects because it dissipates spectral components of velocity–temperature correlations. The remaining terms in the equations produce energy or activity by mean velocity- or temperature-gradient effects.

For the present model, a two-dimensional contraction with the throughflow in the x_1 direction and the contraction in the x_3 direction is considered. The shear and temperature gradients also occur in the x_3 direction. Thus, the mean gradients present in the flow are $\partial U_1/\partial x_1$, $\partial U_3/\partial x_3$, $\partial U_1/\partial x_3$, and $\partial T/\partial x_3$. These gradients all are taken to be independent of position. By continuity of the mean flow,

$$\frac{\partial U_1}{\partial x_1} = -\frac{\partial U_3}{\partial x_3} \equiv a_{11}. \tag{5-588}$$

Similarly, set

$$\frac{\partial U_1}{\partial x_3} \equiv a_{13}. \tag{5-589}$$

and

$$\frac{\partial T}{\partial x_3} \equiv b_3. \tag{5-590}$$

In addition, it is assumed that the turbulence is homogeneous in the transverse directions and that it changes only in the longitudinal or x_1 direction, so that

$$U_k \frac{\partial}{\partial x_k} = U_1 \frac{\partial}{\partial x_1}, \tag{5-591}$$

FOURIER ANALYSIS, SPECTRAL FORM OF THE CONTINUUM EQUATIONS 339

where the operators operate on the correlations or their Fourier transforms. For the model considered, then, eqs. (5-586) and (5-587) can be written as

$$U_1 \frac{\partial \varphi_{ij}}{\partial x_1} = a_{13}\left[\left(2\frac{\kappa_1\kappa_j}{\kappa^2} - \delta_{j1}\right)\varphi_{i3} + \left(2\frac{\kappa_1\kappa_i}{\kappa^2} - \delta_{i1}\right)\varphi_{3j} + \kappa_1 \frac{\partial \varphi_{ij}}{\partial \kappa_3}\right]$$

$$+ a_{11}\left[\left(2\frac{\kappa_1\kappa_j}{\kappa^2} - \delta_{j1}\right)\varphi_{i1} + \left(2\frac{\kappa_1\kappa_i}{\kappa^2} - \delta_{i1}\right)\varphi_{1j} - \left(2\frac{\kappa_3\kappa_j}{\kappa^2} - \delta_{j3}\right)\varphi_{i3}\right.$$

$$\left. - \left(2\frac{\kappa_3\kappa_i}{\kappa^2} - \delta_{i3}\right)\varphi_{3j} + \kappa_1 \frac{\partial \varphi_{ij}}{\partial \kappa_1} - \kappa_3 \frac{\partial \varphi_{ij}}{\partial \kappa_3}\right] - 2\nu\kappa^2 \varphi_{ij} \tag{5-592}$$

and

$$U_1 \frac{\partial \gamma_j}{\partial x_1} = a_{13}\left[\left(2\frac{\kappa_1\kappa_j}{\kappa^2} - \delta_{j1}\right)\gamma_3 + \kappa_1 \frac{\partial \gamma_j}{\partial \kappa_3}\right] + a_{11}\left[\left(2\frac{\kappa_1\kappa_j}{\kappa^2} - \delta_{j1}\right)\gamma_1\right.$$

$$\left. - \left(2\frac{\kappa_3\kappa_j}{\kappa^2} - \delta_{j3}\right)\gamma_3 + \kappa_1 \frac{\partial \gamma_j}{\partial \kappa_1} - \kappa_3 \frac{\partial \gamma_j}{\partial \kappa_3}\right] - b_3 \varphi_{3j} - (\alpha + \nu)\kappa^2 \gamma_j.$$

$$\tag{5-593}$$

In these equations the shear and normal strain terms are separated and written as the first and second bracketed terms on the right sides of the equations.

For solving eqs. (5-592) and (5-593) it is assumed that the turbulence is isotropic at $x_1 = (x_1)_0$. That condition is satisfied by the relation (5-333):

$$(\varphi_{ij})_0 = \frac{J_0}{12\pi^2}(\kappa^2 \delta_{ij} - \kappa_i \kappa_j),$$

where J_0, as before, is a constant that depends on initial conditions. For the initial condition on γ_i (at $x_1 = (x_1)_0$) it is assumed that

$$(\gamma_i)_0 = 0. \tag{5-594}$$

Thus, if the initial turbulence is produced by flow through a grid, that grid is unheated, and the temperature fluctuations are produced by the interaction of the mean temperature gradient with the turbulence.

Equations (5-592) and (5-593) are first-order partial-differential equations in the three independent variables x_1, κ_1, and κ_3. In solving the equations, it is convenient to introduce the velocity ratio c, which, for a uniform normal strain, is

$$c \equiv \frac{U_1}{(U_1)_0} = 1 + \frac{x_1 - (x_1)_0}{(U_1)_0} a_{11}. \tag{5-595}$$

Then,

$$U_1 \frac{\partial}{\partial x_1} = a_{11} c \frac{\partial}{\partial c}. \tag{5-596}$$

In order to reduce eqs. (5-592) and (5-593) to ordinary differential equations, the running variables ξ_1, ξ_3, and η are considered, of which κ_1, κ_3, and c are particular values such that $\xi_1 = \kappa_1$ and $\xi_3 = \kappa_3$ if $\eta = c$. If ξ_1, ξ_3, and η are introduced into the set of equations in place of κ_1, κ_3, and c, the resulting equations, of course, automatically satisfy the original set.

Equation (5-592) (and eq. [5-593] with φ_{ij} replaced by γ_j) then are of the form

$$-\xi_1 \frac{\partial \varphi_{ij}}{\partial \xi_1} + \eta \frac{\partial \varphi_{ij}}{\partial \eta} + \left(\xi_3 - \frac{a_{13}}{a_{11}}\xi_1\right)\frac{\partial \varphi_{ij}}{\partial \xi_3} = F(\xi_1, \xi_3, \varphi_{ij}, \kappa_2).$$

To determine under what conditions

$$-\xi_1 \frac{\partial \varphi_{ij}}{\partial \xi_1} + \eta \frac{\partial \varphi_{ij}}{\partial \eta} + \left(\xi_3 - \frac{a_{13}}{a_{11}}\xi_1\right)\frac{\partial \varphi_{ij}}{\partial \xi_3} = -\xi_1 \frac{d\varphi_{ij}}{d\xi_1}, \qquad (5\text{-}597)$$

note that φ_{ij} is a function of ξ_1, ξ_3, and η and κ_2, so that

$$-\xi_1 \frac{d\varphi_{ij}}{d\xi_1} = -\xi_1 \frac{\partial \varphi_{ij}}{\partial \xi_1} - \xi_1 \frac{\partial \varphi_{ij}}{\partial \xi_3}\frac{d\xi_3}{d\xi_1} - \xi_1 \frac{\partial \varphi_{ij}}{\partial \eta}\frac{d\eta}{d\xi_1}.$$

Comparison of this last equation with eq. (5-597) shows that they are equivalent if

$$-\xi_1 \frac{d\xi_3}{d\xi_1} = \xi_3 - \frac{a_{13}}{a_{11}}\xi_1$$

and

$$-\xi_1 \frac{d\eta}{d\xi_1} = \eta,$$

or

$$\xi_1\xi_3 - \frac{1}{2}\frac{a_{13}}{a_{11}}\xi_1^2 = (\text{constant})_1 = \kappa_1\kappa_3 - \frac{1}{2}\frac{a_{13}}{a_{11}}\kappa_1^2 \qquad (5\text{-}598)$$

and

$$\eta\xi_1 = (\text{constant})_2 = c\kappa_1. \qquad (5\text{-}599)$$

Thus, eq. (5-597) holds if $\xi_1\xi_3 - (1/2)(a_{13}/a_{11})\xi_1^2$ and $\eta\xi_1^2$ are constant during integration. With the introduction of eqs. (5-595) through (5-999), eqs. (5-592) and (5-593) become ordinary differential equations, components of which are

$$\frac{d\varphi_{11}(\xi_1)}{d\xi_1} = -\frac{2}{\xi_1}\left(2\frac{\xi_1^2}{h^2} - 1 - \frac{vh^2}{a_{11}}\right)\varphi_{11} - \frac{2}{\xi_1}\left[\frac{a_{13}}{a_{11}}\left(2\frac{\xi_1^2}{h^2} - 1\right) - 2\frac{\xi_1 f}{h^2}\right]\varphi_{13}, \qquad (5\text{-}600)$$

$$\frac{d\varphi_{13}(\xi_1)}{d\xi_1} = -2\frac{f}{h^2}\varphi_{11} - \frac{2}{\xi_1}\left[\frac{\xi_1 f}{h^2}\left(\xi_1^2 - f^2 + \frac{a_{13}}{a_{11}}\xi_1 f\right) - \frac{vh^2}{a_{11}}\right]\varphi_{13}$$

$$- \frac{1}{\xi_1}\left[-2\frac{\xi_1 f}{h^2} + \frac{a_{13}}{a_{11}}\left(2\frac{\xi_1^2}{h^2} - 1\right)\right]\varphi_{33}, \qquad (5\text{-}601)$$

$$\frac{d\varphi_{33}(\xi_1)}{d\xi_1} = -4\frac{f}{h^2}\varphi_{13} - \frac{2}{\xi_1}\left[-\left(2\frac{f^2}{h^2} - 1\right) + 2\frac{a_{13}}{a_{11}}\frac{\xi_1 f}{h^2} - \frac{vh^2}{a_{11}}\right]\varphi_{33}, \qquad (5\text{-}602)$$

$$\frac{d\varphi_{ii}(\xi_1)}{d\xi_1} = -\frac{2}{\xi_1}(\varphi_{33} - \varphi_{11}) + \frac{2}{\xi_1}\frac{a_{13}}{a_{11}}\varphi_{13} + \frac{2}{\xi_1}\frac{vh^2}{a_{11}}\varphi_{ii}, \qquad (5\text{-}603)$$

$$\frac{d\gamma_1(\xi_1)}{d\xi_1} = \frac{b_3}{\xi_1 a_{11}}\varphi_{13} - \frac{1}{\xi_1}\left[2\frac{\xi_1^2}{h^2} - 1 - (\alpha+v)\frac{h^2}{a_{11}}\right]\gamma_1 - \frac{1}{\xi_1}\left[\frac{a_{13}}{a_{11}}\left(2\frac{\xi_1^2}{h^2} - 1\right) - 2\frac{\xi_1 f}{h^2}\right]\gamma_3,$$

$$(5\text{-}604)$$

and

$$\frac{d\gamma_3(\xi_1)}{d\xi_1} = \frac{b_3}{\xi_1 a_{11}}\varphi_{33} - 2\frac{f}{h^2}\gamma_1 - \frac{1}{\xi_1}\left[1 - 2\frac{f^2}{h^2} + 2\frac{a_{13}}{a_{11}}\frac{\xi_1 f}{h^2} - (\alpha + \nu)\frac{h^2}{a_{11}}\right]\gamma_3, \quad (5\text{-}605)$$

where

$$f \equiv \frac{1}{\xi_1}\left[\frac{1}{2}\frac{a_{13}}{a_{11}}(\xi_1^2 - \kappa_1^2) + \kappa_1\kappa_3\right] \quad (5\text{-}606)$$

and

$$h^2 \equiv \xi_1^2 + \kappa_2^2 + \frac{1}{\xi_1^2}\left[\frac{1}{2}\frac{a_{13}}{a_{11}}(\xi_1^2 - \kappa_1^2) + \kappa_1\kappa_3\right]^2. \quad (5\text{-}607)$$

In these equations ξ_3 has been eliminated by eq. (5-598). The first three equations are independent of the remaining ones, but the converse is not true.

In order to apply initial conditions to the set of eqs. (5-600) through (5-605), let $\varphi_{ij}(\xi_1) = [\varphi_{ij}(\xi_1)]_0$ and $\gamma_i(\xi_1) = [\gamma_i(\xi_1)]_0$ if $\eta = 1$. These conditions then automatically satisfy the desired initial conditions that $\varphi_{ij}(\kappa_1) = [\varphi_{ij}(\kappa_1)]_0$ and $\gamma_i(\kappa_1) = [\gamma_i(\kappa_1)]_0$ if $c = 1$ (or $U_1 = [U_1]_0$) because, by definition, $\xi_1 = \kappa_1$ if $\eta = c$. Equation (5-599) shows that

$$(\xi_1)_0 = c\kappa_1. \quad (5\text{-}608)$$

Equation (5-608) gives the value of ξ_1 at which to start the integration for given values of κ_1 and c. In order to satisfy the initial conditons (5-333) and (5-594), let

$$\left.\begin{aligned}\varphi_{11}(\xi_1) &= \frac{J_0}{12\pi^2}(h^2 - \xi_1^2) \\ \varphi_{13}(\xi_1) &= -\frac{J_0}{12\pi^2}\xi_1 f \\ \varphi_{33}(\xi_1) &= \frac{J_0}{12\pi^2}(h^2 - f^2) \\ \varphi_{ii}(\xi_1) &= \frac{J_0}{6\pi^2}h^2 \\ \gamma_i &= 0\end{aligned}\right\} \text{ if } \xi_1 = (\xi_1)_0,$$

where f and h are again given by eqs. (5-606) and (5-607). The integration of eqs. (5-600) through (5-605) then goes from $(\xi_1)_0$ to $\xi_1 = \kappa_1$. We are mainly interested in the final values of φ_{ij} and γ_i, for which $\xi_1 = \kappa_1$ (and $\xi_2 = \kappa_2$ and $\eta = c$). The quantity ξ_1 can be considered as a dummy variable of integration.

In order to solve eqs. (5-600) through (5-605) numerically, it is convenient to convert them to dimensionless form by introducing the following dimensionless quantities:

$$\left[\frac{\nu(x-x_0)}{U_0}\right]^{1/2} \kappa_i \to \kappa_i, \qquad (5\text{-}609)$$

$$\left[\frac{\nu(x-x_0)}{U_0}\right]^{1/2} \xi_1 \to \xi_1, \qquad (5\text{-}610)$$

$$\left[\frac{(x-x_0)\nu}{J_0 U_0}\right] \varphi_{ij} \to \varphi_{ij}, \qquad (5\text{-}611)$$

$$\left(\frac{\nu}{J_0 b_3}\right)\gamma_i \to \gamma_i, \qquad (5\text{-}612)$$

$$\left[\frac{(x-x_0)}{U_0}\right] a_{13} \to a_{13}, \qquad (5\text{-}613)$$

and

$$\frac{\nu}{\alpha} \to Pr, \qquad (5\text{-}614)$$

and where the arrows mean "becomes."

In addition, spherical coordinates are introduced into the equations by using the transformations

$$\kappa_1 = \kappa \cos\varphi \sin\theta,$$
$$\kappa_2 = \kappa \sin\varphi \sin\theta, \qquad (5\text{-}615)$$
$$\kappa_3 = \kappa \cos\theta.$$

The integrations were carried out numerically on a high-speed coumputer for various fixed values of dimensionless κ, θ, φ, a_{13}, and c. Directionally integrated spectrum functions can be obtained (see sections 5-4-2-1 and 5-4-2-3) from

$$\begin{pmatrix}\psi_{ij}\\ \Gamma_i\\ \Lambda_{ij}\end{pmatrix} = \int_0^\pi \int_0^{2\pi} \begin{pmatrix}\varphi_{ij}\\ \gamma_i\\ \Omega_{ij}\end{pmatrix} \kappa^2 \sin\theta\, d\varphi\, d\theta. \qquad (5\text{-}616)$$

In this equation, Ω_{ij} is the vorticity spectrum tensor given by eq. (5-348):

$$\Omega_{ij} = (\delta_{ij}\kappa^2 - \kappa_i\kappa_j)\varphi_{ll} - \kappa^2\varphi_{ij}. \qquad (5\text{-}348)$$

The spectrum functions given by eq. (5-616) can be integrated over all wavenumbers to give

$$\begin{pmatrix}\overline{u_i u_j}\\ \overline{\tau u_i}\\ \overline{\omega_i \omega_j}\end{pmatrix} = \int_0^\infty \begin{pmatrix}\psi_{ij}\\ \Gamma_i\\ \Lambda_{ij}\end{pmatrix} d\kappa \qquad (5\text{-}617)$$

Thus, ψ_{ij}, Γ_i, and Λ_{ij} show how contributions to $\overline{u_i u_j}$, $\overline{\tau u_i}$, and $\overline{\omega_i \omega_j}$ are distributed among various wavenumbers or eddy sizes. Computed spectra and correlations are considered next. For the quantities that involve temperature gradients, the curves are given for a gas with a Prandtl number Pr of 0.7.

Calculated dimensionless energy spectra (spectra of dimensionless $\overline{u_i u_i}$) and dimensionless $\overline{\tau u_3}$ spectra are plotted in Figs. 5-133 and 5-134. The spectra are plotted for several values of the shear parameter and the normal strain parameter, which are, respectively, proportional to $\partial U_1/\partial x_3$ and $\partial U_1/\partial x_1$ (see Fig. 5-133). Both parameters are, in addition, proportional to longitudinal distance, so that increasing longitudinal distance has an effect similar to that of increasing the velocity gradients.

If plotted by using the similarity variables shown in Figs. 5-133 and 5-134, the dimensionless spectra for no shear and normal strain effects ($a_{13} = a_{11} = 0$) are the same for all values of x_1, although the turbulence itself decays. Comparison of the various curves indicates how normal strain and shear effects alter the spectra for a given position and initial mean velocity. If, for instance, a dimensionless spectrum lies above the curve for $a_{13} = a_{11} = 0$, the turbulent activity for that case is greater than it would be for no shear or normal strain effects.

The curves in Figs. 5-133 and 5-134 (as well as the succeeding ones) are all for positive values of a_{11} and correspond to an accelerating flow. The curves indicate that, in general, the effects of both shear and normal strain in an accelerating flow are to feed energy or activity into the turbulent field. The effect of shear on the spectra is greater at small values of a_{11} than at larger ones; that is, it is greater when the ratio a_{13}/a_{11} is large.

A turbulent velocity-component parameter $(\nu/a_{11})^{5/2} \overline{u_i^2}/J_0$, with $i = 1, 2,$ and 3, is plotted against logitudinal velocity ratio in Fig. 5-135. This parameter, in contrast to the spectral parameters in Figs. 5-133 and 5-134, does not contain $x_1 - (x_1)_0$, and thus can be used to show how $\overline{u_i^2}$ changes with longitudinal position (or velocity ratio) as well as with shear. Included in the plot is the curve obtained by solving eq. (5-592) with the effects of shear and normal strain absent. This solution gives

$$\left(\frac{\nu}{a_{11}}\right)^{5/2} \frac{\overline{u_i^2}}{J_0} = \frac{\ln^{-5/2} c}{48\sqrt{2\pi}} \tag{5-618}$$

Although the turbulence is taken to be initially isotropic, the results here show that the turbulence is already strongly anisotropic from the effects of shear and normal strain. As is the case for the spectra, these results show that $\overline{u_i^2}$ is increased by the shear and that the effect of shear is greatest at low values of velocity ratio or normal strain parameter. For low values of velocity ratio, all components decay because of the effects of viscosity. In that region the lateral components decrease less rapidly than they would if the effects of normal strain were neglected (compare with dashed curve) and, for $a_{13} = 0$, the longitudinal component decays more rapidly. At larger velocity ratios, the components in the x_2 and x_3 directions begin to increase as the effects of normal strain offset those of viscosity. The component in the x_1 direction continues to decrease, but at a slower rate than it would if the effects of normal strain were absent. In this way, the curves in Fig. 5-135, which are for a two-dimensional contraction, differ from those for the axially symmetric strains in ref. [78] and sections 5-4-2-6 and 5-4-2-7. For the axially symmetric strains, the longitudinal component decays more rapidly than it would for no effects of strain, whereas, in the present two-dimensional contraction, it decays less

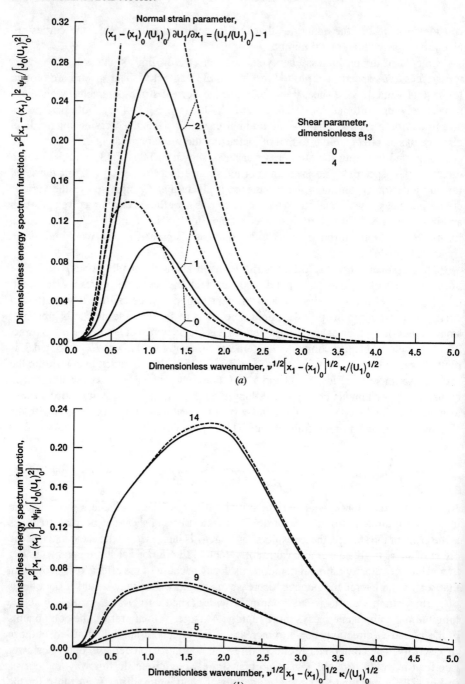

Figure 5-133 Effects of uniform shear and normal strain on spectra of dimensionless turbulent energy. (a) Low values of normal-strain parameter; (b) high values of normal-strain parameter.

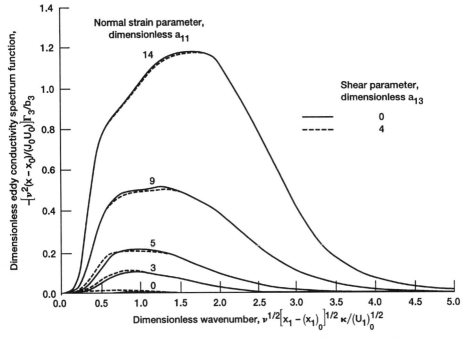

Figure 5-134 Effects of uniform shear and of normal strain on spectra of dimensionless eddy conductivity; Prandtl number, 0.7.

rapidly, except at small velocity ratios. Thus, in this case energy is fed into each of the three components of the turbulent energy by normal strain.

In an attempt to understand the trends shown in Fig. 5-135, the three components of the dimensionless turbulent vorticity $(v/a_{11})^{7/2}\overline{\omega_i^2}/J_0$ are plotted against velocity ratio in Fig. 5-136 for $a_{13} = 0$. The dashed curve for no effects of strain is obtained from the equation

$$\left(\frac{v}{a_{11}}\right)^{7/2}\frac{\overline{\omega_i^2}}{J_0} = \frac{5}{192\sqrt{2\pi}}\ln^{-7/2}c$$

The plot shows that the vorticity components in both the x_1 and x_2 directions decay less rapidly than they would if the effects of strain were absent; the x_3 component decays more rapidly. On the other hand, for the axially symmetric cases considered in sections 5-4-2-6 and 5-4-2-7, only the longitudinal component $\overline{\omega_1^2}$ decayed less rapidly. Thus, although in the axially symmetric case, the turbulent vortex filaments all tended to line up in the longitudinal direction, in the present two-dimensional contraction there is also a tendency (although less pronounced) for an alignment to occur in the x_2 direction (direction of no contraction). These trends are in agreement with the trends for velocity fluctuations shown in Fig. 5-135. The velocities associated with a vortex filament, of course, lie in planes normal to the direction of the filament. Thus, the vortex filaments aligned in the longitudinal direction tend to feed energy into the two lateral velocity

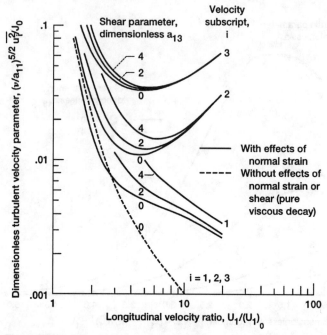

Figure 5-135 Effect of uniform shear and of normal strain (or velocity ratio) on dimensionless variance of turbulent velocity components.

components, while those aligned in the x_2 direction can give energy to the longitudinal velocity component, as well as to the x_3 component.

It may be of interest to compare the behavior of the components of turbulent energy at large velocity ratios for several types of mean strain. A qualitative comparison is given in the following table:

Type of mean normal strain	Behavior of turbulent energy components at large velocity ratios in accelerating flow
In compressible axisymmetric strain for flow in a cone (section 5-4-2-6)	Lateral components increase with longitudinal distance. Longitudinal component decreases faster than it would without effect of strain.
Uniform incompressible axisymmetric strain (section 5-4-2-7)	Lateral components approach steady state. Longitudinal component decreases faster with distance than it would without effects of strain.
Uniform compressible longitudinal axisymmetric strain (no lateral strain) [131]	Lateral components decrease less rapidly with distance than they would without effects of strain. Longitudinal component decreases more rapidly.
Uniform incompressible two-dimensional strain (present analysis)	Lateral components increase with longitudinal distance. Longitudinal component decreases, but a rate slower than it would without effect of strain.

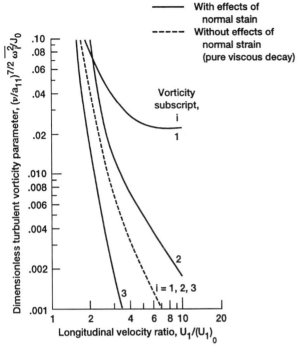

Figure 5-136 Effect of uniform normal strain (or velocity ratio) on dimensionless variance of turbulent vorticity components; shear parameter, 0.

Figure 5-137 shows the effect of uniform shear and normal strain (velocity ratio) on ratios of the turbulent energy components for accelerating flow. Both $\overline{u_3^2}/\overline{u_1^2}$ and $\overline{u_2^2}/\overline{u_1^2}$ tend to decrease with increasing shear parameter and to increase with increasing normal strain parameter; that is, the effect of shear is to make $\overline{u_3^2}$ and $\overline{u_2^2}$ less than $\overline{u_1^2}$, and normal strain tends to make those quantities greater than $\overline{u_1^2}$.

Dimensionless turbulent heat-transfer parameters $[\nu^{5/2}/(J_0 a_{11}^{3/2})]\overline{\tau u_i}/b_3$ with $i = 3$ and 1 are plotted in Fig. 5-138 as functions of velocity ratio and shear parameter. The trends shown here are qualitatively similar to those for the dimensionless velocity parameter shown in Fig. 5-135. It might seem surprising that there should be turbulent heat transfer in the longitudinal direction x_1, as given by the temperature–velocity correlation $\overline{\tau u_1}$, because there is no temperature gradient in the x_1 direction. However, because there is a correlation between τ and u_3 (because of the temperature gradient dT/dx_3) and a correlation between u_1 and u_3 (because of the velocity gradient dU_1/dx_3), it seems reasonable that there should be a correlation between τ and u_1, and thus a heat transfer in the x_1 direction.

Shear correlation coefficient $-\overline{u_1 u_3}/[(\overline{u_1^2})^{1/2}(\overline{u_3^2})^{1/2}]$ is plotted as a function of longitudinal velocity ratio and shear parameter in Fig. 5-139. The shear correlation is, of course, zero for zero shear and increases as dimensionless a_{13} increases. Except at small velocity ratios and large values of shear parameter, at which some increase in correlation

348 TURBULENT FLUID MOTION

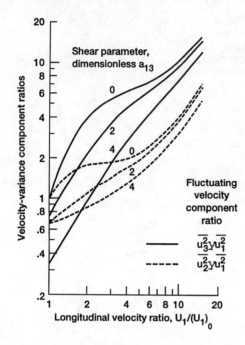

Figure 5-137 Effect of uniform shear and of normal strain (or velocity ratio) on velocity–variance component ratios.

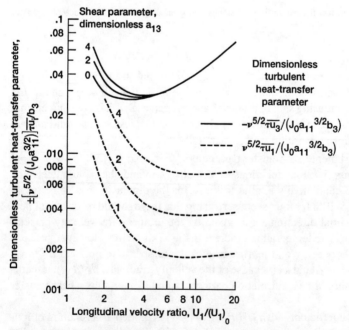

Figure 5-138 Effect of uniform shear and or normal strain (or velocity ratio) on dimensionless temperature–velocity correlations; Prandtl number, 0.7.

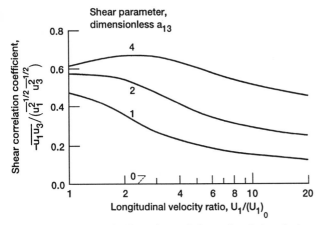

Figure 5-139 Effect of uniform shear and of normal strain (or velocity ratio) on shear correlation coefficient.

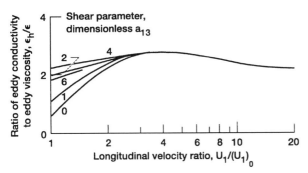

Figure 5-140 Effect of uniform shear and of normal strain (or velocity ratio) on ratio of eddy conductivity to eddy viscosity; Prandtl number, 0.7.

with increasing normal strain (velocity ratio) occurs, normal strain tends to destroy the shear correlation.

Figure 5-140 shows the ratio of eddy conductivity to eddy viscosity plotted as a function of velocity ratio and shear parameter. The eddy conductivity and eddy viscosity are defined by the relations

$$\varepsilon_h = -\frac{\overline{\tau u_3}}{dT/dx_3}$$

and

$$\varepsilon = -\frac{\overline{u_1 u_3}}{dU_1/dx_3}.$$

As the shear parameter, dimensionless a_{13}, increases, the ratio $\varepsilon_h/\varepsilon$ increases, reaches a maximum, and then decreases, although the trend is confined to moderately low values of velocity ratio. Results from section 5-4-2-3 for no normal strain show that $\varepsilon_h/\varepsilon$

ultimately approaches 1 as dimensionless a_{13} continues to increase. The results in Fig. 5-141 indicate that $\varepsilon_h/\varepsilon$ reaches a maximum with increasing velocity ratio, as well as with increasing dimensionless a_{13}.

As mentioned at the beginning of this section an apparent laminarization sometimes occurs in the turbulent boundary layers of accelerating flows. Some observed low heat-transfer values for flow in nozzles evidently can be explained by the fact that the mean velocity increases with distance in an accelerating flow (section 5-4-2-6). Some of the visual observations, however, seem to indicate that the turbulent energy itself decreases [134]. Back, Massier, and Gier [136] suggest that the effect may be caused by a normal strain term in the energy equation, which acts like a sink for turbulent energy. The term $2a_{11}(\overline{u_3^2} - \overline{u_1^2})$ corresponds to the second term in eq. (5-603), if that term is multiplied by $-\xi_1$ and integrated over all wavenumbers. The term can act like a sink only if $\overline{u_1^2}$ is greater than $\overline{u_3^2}$; otherwise, it acts like a normal-strain production term. The results in Fig. 5-137 for $\overline{u_3^2}/\overline{u_1^2}$ indicate that the normal-strain production term is negative at low values of velocity ratio and high values of shear parameter. On the other hand, the shear production term $-2a_{13}\overline{u_1 u_3}$ (which corresponds to the third term in eq. [5-603] multiplied by $-\xi_1$ and integrated over all wavenumbers) is always positive.

The ratio of the two production terms is shown in Fig. 5-141 as a function of longitudinal velocity ratio and shear parameter. The curves show that the normal-strain production term can be negative and thus act like a sink term for turbulent energy at

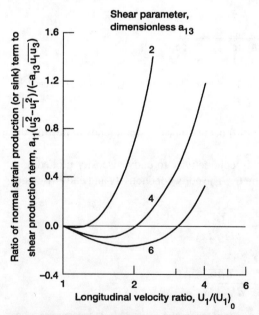

Figure 5-141 Ratio of normal-strain production (or sink) term to shear production term in turbulent energy equation (second and third terms in eq. [5-603] multiplied by ξ_1 and integrated over all wave numbers) as a function of shear and of normal strain (or velocity ratio).

low velocity ratios and high shear. However, in order for that term to offset the effect of the shear production term, the ratio of the two terms, of course, would have to be less than -1, and that does not occur for results in Fig. 5-141. It is possible that the ratio could be less than -1 at sufficiently large values of shear parameter. There appears to be a problem in making the normal-strain production term sufficiently negative to offset the effect of the shear production term. The shear must be large to make the normal-strain production term negative (by making $\overline{u_1^2} > \overline{u_3^2}$). In that case, however, the shear production term also is large. The curves in Fig. 5-141 show that as velocity ratio (or normal strain parameter) increases, the normal-strain production term becomes strongly positive, because the effect of normal strain is to make $\overline{u_1^2} < \overline{u_3^2}$.

To summarize the results of this section, note that, in general, both shear and normal strain in an accelerating flow increase the energy in the turbulent field in comparison with what would be present for no shear or normal strain. This increase occurs in spite of the normal-strain production term in the turbulent energy equation that can, under certain conditions of combined shear and normal strain, be negative and thus act as a turbulent energy sink. For the results computed, the shear production term more than offsets the effect of the sink term, and the net result is that the turbulent energy increases.

The present results for a two-dimensional contraction show that the lateral components of the turbulent energy increase with longitudinal distance at large mean velocity ratios. The longitudinal component decreases, but at a slower rate than it would if the effects of normal strain were absent. Thus, energy is fed into each of the three components of the turbulent energy by normal strain (and shear). This case differs from axially symmetric strain cases of accelerating flows, in which the longitudinal turbulence component decays faster with distance than it would if the effects of normal strain were absent. For the two-dimensional contraction, although most of the vortex filaments tend to line up in the longitudinal direction, there is also some tendency for them to align in the transverse direction of no normal strain.

The normal strain and shear both tend to produce anisotropy in the turbulence, but they work in opposite directions. The normal strain increases the ratios of the lateral components to the longitudinal component of the turbulent energy; shear decreases the ratio.

In general, the turbulent shear correlation tends to be destroyed by the normal strain. An exception occurs at small velocity ratios and large shear, at which some increase in correlation with increasing normal strain (velocity ratio) occurs.

As either the shear or normal-strain parameter increases, the ratio of eddy conductivity to eddy viscosity reaches a maximum and then decreases. In the presence of lateral mean-shear velocity gradients and lateral temperature gradients, turbulent heat transfer occurs in the longitudinal as well as in the lateral direction, even though there is no longitudinal temperature gradient.

5-4-2-9 Maintenance and growth of shear-flow turbulence. It is well known experimentally that turbulence can be maintained or caused to grow by a mean shear (e.g., in a boundary layer). However, it is not easy to explain that observation theoretically. It usually is assumed that the turbulence in a shear flow is maintained against viscous

352 TURBULENT FLUID MOTION

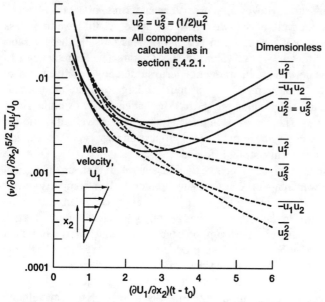

Figure 5-142 Growth due to shear of weak, locally homogeneous turbulence if $\overline{u_2^2} = \overline{u_3^2} = (1/2)\overline{u_1^2}$, and comparison with case for all components calculated as in section 5-4-2-1, in which energy is allowed to drain out of transverse components.

dissipation by work done on the Reynolds shear stress by the velocity gradient. Although that is no doubt true, it appears that the actual mechanism is slightly more subtle. For instance, although the turbulent energy may be maintained by work done on the Reynolds shear stress by the velocity gradient, it is not clear that the Reynolds shear stress itself does not decay.

Our study of homogeneous turbulence with uniform shear, which includes the interaction of the turbulence with the shear but neglects the turbulence self-interaction, shows that the turbulence ultimately decays (see section 5-4-2-1 and Fig. 5-63). This is shown by the dashed curves in Fig. 5-142, in which u_i is a velocity fluctuation component, ν is the kinematic viscosity, J_0 is a constant depending on initial conditions, U_1 is the mean velocity, x_2 is the coordinate in the direction of the velocity gradient dU_1/dx_2, t is the time, and the overbar indicates an averaged value. (This nomenclature differs slightly from that in the preceding section, in which the velocity gradient was in the x_3 direction, but is the same as that in section 5-4-2-1.) The turbulence is assumed isotropic at its virtual origin t_0. The decay occurs in spite of the fact that the model includes work done on the Reynolds stress by the velocity gradient. In section 5-4-2-1 it is shown that for certain initial conditions, the total energy at a point can grow for awhile. The transverse component of the turbulence in the direction of the mean gradient always decays, however, and eventually all of the components decay. On the basis of these results one might even be tempted to suppose that the Navier-Stokes equations are inadequate for investigating the possibility of a nondecaying turbulence; however, that supposition does not seem to be justified [137].

The equation for the rate of change of the two-point correlations (eq. [4-147]) for the present case can be written in abbreviated form as

$$\frac{\partial \overline{u_i u'_j}}{\partial t} = -\left(\delta_{i1}\overline{u_2 u'_j}\frac{dU_1}{dx_2} + \delta_{j1}\overline{u_i u'_2}\frac{dU'_1}{dx'_2}\right) + D_{ij} + P_{ij} + T_{ij} + G_{ij}, \quad (5\text{-}619)$$

where the primed and unprimed quantities are measured at the points P' and P, and i and j can take on the values 1, 2, or 3. The Kronecker delta δ_{ij} equals 1 for $i = j$ and 0 for $i \neq k$. The first term on the right-hand side of eq. (5-619) is the production term, and D_{ij}, P_{ij}, T_{ij}, and G_{ij} represent, respectively, dissipation, pressure, transfer, and diffusion terms.

It is argued in section 5-4-2-1 that $\overline{u_2^2}$ (or $\overline{u_2 u'_2}$), the turbulence component in the direction of the mean gradient, decays because the equation for the rate of change of that component does not contain a production term, and, in addition, the pressure–velocity correlations, for the model used, extract energy from that component. The decay of $\overline{u_2^2}$ then causes the Reynolds shear stress $\overline{u_1 u_2}$ to decay, because the latter contains u_2. There is then no mechanism for maintaining the turbulence, and all of the components ultimately decay because of viscous dissipation (see eq. [5-619]).

If the decay in the foregoing model of shear turbulence is due to the draining of energy out of the transverse component $\overline{u_2^2}$, as discussed above, then if that drain is prevented or counteracted, the turbulence should grow or at least be maintained. In the subsection on growth due to buoyancy of section 5-4-2-5 it is shown that this depletion can be counteracted by introducing destabilizing buoyancy forces in the direction of the mean-velocity gradient, and that all of the turbulence components then ultimately grow. However, it is not clear from that result just how much of the growth is due to the shear, because the buoyancy by itself can cause all three of the directional components of the turbulence to grow, although in that case the component in the direction of the buoyancy forces grows much faster than the others.

In order to determine whether shear by itself can cause turbulence to grow, we have prevented the energy drain from $\overline{u_2^2}$ by setting $\varphi_{22} = 1/2\varphi_{11}$ in the analysis of section 5-4-2-1, where φ_{ij} is the Fourier transform of $\overline{u_i u'_j}$. Because the equation for φ_{33}, like that for φ_{22}, does not contain a production term, we also set $\varphi_{33} = 1/2\varphi_{11}$ in order to prevent energy depletion in that component. However, the last assumption is unnecessary if we are interested only in the φ_{11}, φ_{22}, and φ_{12} components, because φ_{33} does not occur in the equations for those components. If we set $\varphi_{22} = \varphi_{33} = 1/2\varphi_{11}$, then $\overline{u_2^2} = \overline{u_3^2} = 1/2\overline{u_1^2}$. In making the calculations, we use the spectral equations of motion from section 5-4-2-1 for φ_{11} and φ_{12}, both of which contain production terms (see eq. [5-619]).

The results are shown by the solid curves in Fig. 5-142. In contrast with the dashed curves, where the energy was allowed to drain out of the transverse components and all of the components decayed, the components $\overline{u_1^2}$ and $\overline{u_1 u_2}$ (and thus $\overline{u_2^2}$ and $\overline{u_3^2}$, because $\overline{u_2^2} = \overline{u_3^2} = 1/2\overline{u_1^2}$) now grow at large times. That is, if we keep the energy from draining out of the transverse components (particularly $\overline{u_2^2}$), the mean shear can cause an ultimate growth of the turbulence. Although the assumption that $\overline{u_2^2} = \overline{u_3^2} = 1/2\overline{u_1^2}$ may be somewhat arbitrary, the calculation under that assumption is enlightening, in that before making it we had no assurance that turbulence could grow or be maintained by shear if the drain of energy out of the transverse components was prevented.

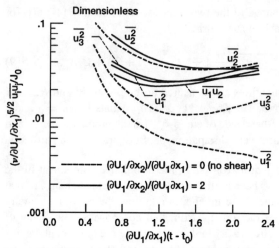

Figure 5-143 Growth due to shear of weak, locally homogeneous turbulence with normal strain.

The next question is how a severe imbalance between the directional components is prevented in an actual turbulent flow (other than a weak homogeneous turbulent flow). As discussed in section 5-4-2-1, if the turbulence is not weak or the shear is not large the interaction of triple correlations with the pressure–velocity correlations can cause the latter to have an equalizing effect on the directional components. Also, inhomogeneities in the turbulent field may have an effect [138]. Rotta [139] has discussed the directional redistribution of energy by the one-point pressure velocity-gradient correlations.

Another flow that sheds some light on the turbulence-producing mechanism is analyzed in the previous section. That is the case of combined two-dimensional shear and normal strain in locally homogeneous turbulence without turbulence self-interaction. The results given do not elucidate the turbulence-growth aspects of the flow. The results in Fig. 5-143, which show the evolution of the turbulence components for a given value of $(\partial U_1/\partial x_2)/(\partial U_1/\partial x_1)$, however, do show those aspects. The time increment $t - t_0$ is the time that would elaspse for an observer traveling with the accelerating fluid from the virtual origin of the turbulence at t_0 to the position x_1. For the results given, the fluid is being stretched in the direction to flow.

The dashed curves in Fig. 5-143 show that if the shear is zero, the positive longitudinal strain can cause $\overline{u_2^2}$ and $\overline{u_3^2}$ to grow, but that $\overline{u_1^2}$, the component in the direction of flow, decays. The shear component $\overline{u_1 u_2}$ is, of course, zero in that case. If a mean shear is applied (solid curves), all of the components ultimately grow. Here, it is clearer that the shear is having an important effect on the maintenance of turbulence than in the buoyancy case analyzed in section 5-4-2-5, where all of the directional components are maintained by the buoyancy alone. In the present case, the acceleration alone does not maintain the $\overline{u_1^2}$ component. (If $\partial U_1/\partial x_1 = 0$ but $\partial U_1/\partial x_2 \neq 0$, all of the components decay, as in the dashed curves of Fig. 5-142.) The growth of all of the components in the present case is again due to the equalization of the energy in the directional components (see Fig. 5-143). Here, the shear and the normal strain have opposite effects as far as the directional distribution of energy is concerned, so their combined effect is to keep the energy directionally distributed so that all components can grow.

It might seem surprising that in all of the cases considered here in which a severe imbalance of energy between the directional components is prevented, the turbulence continues to increase with time rather than level off. There are no boundaries on the flows considered here, however, so the effective Reynolds number of the mean flow is infinite. As the scale of the turbulence continues to grow, the eddies encounter larger and larger velocity differences, so that the effective driving forces on the disturbances continue to grow. The turbulence only grows, of course, at least in the case of uniform shear without acceleration, if sufficient energy is transferred into the transverse components, particularly into $\overline{u_2^2}$, as previously discussed. It is conceivable that the transfer of energy into $\overline{u_2^2}$ might in some cases be sufficient to prevent the turbulence from decaying, as it does in section 5-4-2-1, but insufficient to cause it to grow. That, in fact, seems to be the case in the experiments of Rose [93] and of Champagne et al. [94], in which a leveling off of intensity appears to occur, although the scale continues to grow. On the other hand, the work of Mulhearn and Luxton [95], in which the total strains are larger than those of refs. [93] and [94], indicates a growth of intensity at large times. In the present section we are mainly interested in whether, from a theoretical standpoint, the effect of a mean shear can be great enough to offset the effects of viscosity and keep a turbulent field from decaying, regardless of whether the turbulence ultimately grows or reaches a steady state.

5-4-3 Uniformly and Steadily Sheared Homogeneous Turbulence if Triple Correlations May Be Important

Thus far in section 5-4 we have considered only cases in which mean gradients are large or the turbulence is weak (see section 5-4-2). In those cases one generally can neglect triple-correlation terms in the correlation equations, those terms being small compared with other terms. For other cases those terms should be ratained.

In this section we obtain a numerical solution of the unaveraged Navier-Stokes equations for a uniformly and steadily sheared turbulence, as in ref. [140].[17] Conceptually, that is the simplest turbulent shear flow (although certainly not the simplest to produce experimentally—see, e.g., ref. [94]). A number of other significant numerical studies of that type of turbulence also have been made (see, e.g., ref. [141]). In those studies random initial conditions with a range of eddy sizes were used. That is in contrast to the present study, in which nonrandom initial conditions with a single length scale are used.

The numerical method used here is essentially the same as that in section 5-3-2-6; fourth-order-finite-spatial-differencing and a predictor–corrector time-differencing are used (a second-order leapfrog predictor and a third-order Adams-Moulton corrector [65]).

We use the unaveraged Navier-Stokes equations because, as discussed in the previous Chapter, the closure problem arises if nonlinear equations are averaged to obtain

---

[17]The results obtained here are qualitatively similar to those in ref. [140], although some of the latter were numerically underresolved.

correlation equations. The equations to be solved are the incompressible Navier-Stokes equations given by eq. (5-130),

$$\frac{\partial \tilde{u}_i}{\partial t} + \frac{\partial (\tilde{u}_i \tilde{u}_k)}{\partial x_k} = -\frac{1}{\rho}\frac{\partial \tilde{\sigma}}{\partial x_i} + \nu \frac{\partial^2 \tilde{u}_i}{\partial x_k \partial x_k}$$

where the mechanical pressure (see eq. [3-14]) is given by the Poisson equation (eq. [5-150]):

$$\frac{1}{\rho}\frac{\partial^2 \tilde{\sigma}}{\partial x_l \partial x_l} = -\frac{\partial^2 (\tilde{u}_l \tilde{u}_m)}{\partial x_l \partial x_m},$$

and where as usual, the subscripts can take on the values 1, 2, or 3, and a repeated subscript in a term indicates a summation. The quantities $\tilde{u}_i$ and $\tilde{u}'_j$ are instantaneous velocity components (which include the mean velocity components), x_i is a space coordinate, t is the time, ρ is the density, ν is the kinematic viscosity, and $\tilde{\sigma}$ is the instantaneous (mechanical) pressure. Equation (5-150) is obtained by taking the divergence of eq. (5-130) and using continuity (eq. [3-4]).

In the spirit of section 5-3-2-6, the present numerical study of uniformly sheared turbulence starts with simple determinate initial conditions that possess a single length scale. As in section 5-3-2-6, we can in this way study how the turbulence develops from nonturbulent initial conditions, as it does for experimental grid-generated turbulence. Again, much higher–Reynolds-number flows can be calculated with a given numerical grid if a single length scale is initially present, at least for early and moderate times.

As is seen subsequently, several interesting results that could not be obtained in the previous work on turbulent shear flow are obtained here. One of the significant findings is that the structure of the turbulence produced in the presence of shear is finer than that produced in its absence.

For the numerical solutions considered here, the inital velocity fluctuation is assumed to be given by

$$u_i = \sum_{n=1}^{3} a_i^n \cos \boldsymbol{q}^n \cdot \boldsymbol{x}. \tag{5-620}$$

Then, from eq. (4-14),

$$\tilde{u}_i = \sum_{n=1}^{3} a_i^n \cos \boldsymbol{q}^n \cdot \boldsymbol{x} + U_i. \tag{5-621}$$

The quantity a_i^n is an initial velocity amplitude or Fourier coefficient of the velocity fluctuation, $\boldsymbol{q}^n$ is an initial wavevector, and U_i is an initial mean-velocity component. In order to satisfy the continuity conditions, eqs. (4-10) and (4-21), we set

$$a_i^n q_i^n = 0. \tag{5-622}$$

For the present work let

$$\begin{aligned} a_i^1 = k(2, \pm 1, 1), \quad & a_i^2 = k(1, \pm 2, 1), \quad a_i^3 = k(1, \pm 1, 2), \\ q_i^1 = (-1, \pm 1, 1)/x_0, \quad & q_i^2 = (1, \mp 1, 1)/x_0, \quad q_i^3 = (1, \pm 1, -1)/x_0, \end{aligned} \tag{5-623}$$

where k has the dimensions of a velocity and determines the intensity of the initial velocity fluctuation. The quantity x_0 is the length scale of the initial velocity fluctuation. The quantities k and x_0, together with the kinematic viscosity ν and eq. (5-623), then determine the initial Reynolds number $(\overline{u_0^2})^{1/2} x_0/\nu$, because the square of eq. (5-620), averaged over a period, gives $\overline{u_0^2}$. In addition to satisfying the continuity eq. (5-622), eqs. (5-620) and (5-623) give

$$\overline{u_1^2} = \overline{u_2^2} = \overline{u_3^2} = \overline{u_0^2} \tag{5-624}$$

at the initial time. (The first three terms of eq. [5-624] apply at all times at which there are no mean gradients in the flow.) Thus eqs. (5-620), or (5-621), and (5-623) give a particularly simple initial condition, in that we need specify only one component of the mean-square velocity fluctuation. Moreover, for no mean shear, they give an isotropic turbulence at later times, as in section 5-3-2-6. Note that it is necessary to have at least three terms in the summation in eq. (5-620) or (5-621) to satisfy eq. (5-624). We do not specify an initial condition for the pressure because it is determined by the Poisson equation for the pressure (eq. [5-150]) and the initial velocities.

In order to carry out numerical solutions subject to the initial condition given by eqs. (5-620), or (5-621), and (5-623), we use a stationary cubical grid with a maximum of 128^3 points and with faces at $x_i^* = x_i/x_0 = 0$ and 2π. For boundary conditions we assume periodicity for the fluctuating quantities; we consider turbulence (or a turbulent-like flow) in a box with periodic walls. That is, let

$$(u_i)_{x_j^* = 2\pi + b_j^*} = (u_i)_{x_j^* = b_j^*} \tag{5-625}$$

and

$$\sigma_{x_j^* = 2\pi + b_j^*} = \sigma_{x_j^* = b_j^*}, \tag{5-626}$$

where $b_j^* = b_j/x_0$, $x_j^* = x_j/x_0$, and b_j is a variable length.

Using eqs. (4-14) and (4-15) these become

$$(\tilde{u}_i)_{x_j^* = 2\pi + b_j^*} = (\tilde{u}_i)_{x_j^* = b_j^*} + (U_i)_{x_j^* = 2\pi + b_j^*} - (U_i)_{x_j^* = b_j^*} \tag{5-627}$$

and

$$\tilde{\sigma}_{x_j^* = 2\pi + b_j^*} = \tilde{\sigma}_{x_j^* = b_j^*} + P_{x_j^* = 2\pi + b_j^*} - P_{x_j^* = b_j^*}. \tag{5-628}$$

In the present work we assume also that P, given by eq. (4-27) if buoyancy is absent, is periodic, so that

$$P_{x_j^* = 2\pi + b_j^*} = P_{x_j^* = b_j^*}, \tag{5-629}$$

and eq. (5-628) becomes

$$\tilde{\sigma}_{x_j^* = 2\pi + b_j^*} = \tilde{\sigma}_{x_j^* = b_j^*}. \tag{5-630}$$

These equations are used to calculate numerical derivatives at the boundaries of the computational grid.

Because we are considering a uniform shear, we let

$$U_i = \delta_{i1} \frac{dU_1}{dx_2} x_2 \tag{5-631}$$

in the initial condition (5-621) and

$$(U_i)_{x_j^*=2\pi+b_j^*} - (U_i)_{x_j^*=b_j^*} = \delta_{i1}\delta_{j2}2\pi\frac{dU_1}{dx_2} \tag{5-632}$$

in the boundary condition (5-627). Equation (5-631) applies, of course, at all times and all x_i. For the coefficients in eq. (5-621) we use eq. (5-623), where we choose the first set of signs. Equations (5-130) and (5-150) are written in terms of the total velocity $\tilde{u}_i$, but we can calculate the fluctuating component u_i from eq. (4-14). It should be emphasized that we do not consider here a sawtooth type of mean velocity profile, but rather a continuous profile in which the mean-velocity gradient is uniform at all points. Even with a uniform mean-velocity gradient, some local inhomogeneity is introduced into the fluctuations by the periodic boundary conditions. We shall not concern ourselves with that inhomogeneity, however, because we still can calculate products involving velocities or pressures averaged over a three-dimensional period. Those values are independent of the position of the boundaries of the cycle (see the paragraph containing eq. [4-4]).

There may be another related problem in using periodic boundary conditions (eqs. [5-625] through [5-630]) with a uniform mean-velocity gradient and a stationary (non-deforming) grid. For that case there is a tendency for singularities (discontinuities) to form at the boundaries of the computational box. This can be seen from eq. (4-22), which for our case can be written as

$$\frac{\partial u_i}{\partial t} = -\delta_{i1}\frac{dU_1}{dx_2}u_2 - \frac{dU_1}{dx_2}x_2\frac{\partial u_i}{\partial x_1} - \frac{\partial}{\partial x_k}(u_i u_k) - \frac{1}{\rho}\frac{\partial \sigma}{\partial x_i} + \nu\frac{\partial^2 u_i}{\partial x_k}. \tag{5-633}$$

As noted earlier, σ_e drops out of the equations of motion for $g_i = 0$ by virtue of the equation following (3-22).

By inspection of eq. (5-633) one sees that the term $(dU_1/dx_2)x_2\partial u_i/\partial x_1$ causes the fluctuation u_i to grow on the boundary at $x_2 = 2\pi x_0$ but not on the boundary at $x_2 = 0$. So with periodic boundary conditions there is a tendency for discontinuities, and thus numerical instabilities, to form on the boundaries for x_2. That tendency can be eliminated by using a deforming numerical grid, as in refs. [141] and [142]. In that case, however, the grid soon is distorted out of shape, and in addition the physical significance of the term $(dU_1/dx_2)x_2\partial u_i/dx_1$ is lost because that term is eliminated from the evolution equation for u_i. Fortunately the viscous term in eq. (5-633) tends to smooth out discontinuities, particularly if the Reynolds number is not high. Our results, which use a stationary grid, bear that out. However, before presenting results for the nonlinear case we discuss the simplified linearized equations. By doing that we may gain some insight into uniformly sheared turbulence, both with periodic boundary conditions and with boundary conditions at infinity, and into the relation of that turbulence to the term $(dU_1/dx_2)x_2\partial u_i/\partial x_1$.

5-4-3-1 Linearized problem. By using Reynolds decomposition, eq. (5-150) becomes, for our case (uniform mean shear and uniform mean pressure),

$$\frac{\partial^2 \sigma}{\partial x_l \partial x_l} = -\frac{\partial^2 (u_k u_l)}{\partial x_k \partial x_l} - 2\frac{\partial u_2}{\partial x_1}\frac{\partial U_1}{\partial x_2}, \tag{5-634}$$

where eqs. (4-14) and (4-15) are used. Equations (5-633) and (5-634) are linearized by neglecting the terms $-\partial(u_i u_k)/\partial x_k$ and $-\partial^2(u_k u_l)/\partial x_k \partial x_l$. The numerical solution, with initial and periodic boundary conditions given by eqs. (5-620), (5-623), (5-625), and (5-626), then proceeds as in the nonlinear case.

We first obtain an analytical solution for unbounded linearized fluctuations by using unbounded three-dimensional Fourier transforms. Instead of working with the averaged equations as in section 5-4-2-1, it is instructive to work with the unaveraged ones and use the initial condition given by eq. (5-620). In this case the Fourier transforms must be generalized functions (a series of δ functions) (section 5-2), but the method of solution is the same as that in the earlier work. Equation (5-633) for u_2 and eq. (5-634), if linearized, are independent of u_1 and u_3. The solution obtained by using the initial condition (5-620) is

$$u_2 = \sum_{n=1}^{3} U_2^n \cos(\mathbf{q}^n \cdot \mathbf{x} - a_{12} q_1^n t x_2), \qquad (5\text{-}635)$$

$$\sigma = \sum_{n=1}^{3} P^n \sin(\mathbf{q}^n \cdot \mathbf{x} - a_{12} q_1^n t x_2), \qquad (5\text{-}636)$$

where

$$U_2^n = \frac{a_2^n q^{n^2}}{q^{n^2} - 2a_{12} q_1^n q_2^n t + a_{12}^2 q_1^{n^2} t^2} \exp\left[-\nu t\left(q^{n^2} - a_{12} q_1^n q_2^n t + \frac{1}{3} a_{12}^2 q_1^{n^2} t^2\right)\right], \qquad (5\text{-}637)$$

$$P^n = \frac{-2\rho a_{12} a_2^n q_1^n q^{n^2}}{\left(q^{n^2} - 2a_{12} q_1^n q_2^n t + a_{12}^2 q_1^{n^2} t^2\right)^2} \exp\left[-\nu t\left(q^{n^2} - a_{12} q_1^n q_2^n t + \frac{1}{3} a_{12}^2 q_1^{n^2} t^2\right)\right], \qquad (5\text{-}638)$$

$a_{12} = dU_1/dx_2$, $q^{n^2} = q_1^{n^2} + q_2^{n^2} q_3^{n^2}$, and the a_i^n and q_i^n are given in the initial conditions (eqs. [5-620] and [5-623] with the first set of signs). Mean values are obtained by integrating over all space. For instance,

$$\overline{\sigma u_2} = \sum_{n=1}^{3} \frac{1}{2} P^n U_2^n. \qquad (5\text{-}639)$$

According to the linearized analytical solution given by eq. (5-635), the manufacture of small-scale fluctuations takes place only in the x_2 direction. Because of the analytical character of eq. (5-635) and the regularity of the initial condition, the fluctuation u_2 is nonrandom. Evidently, as in the case of no mean gradients, the only way one can have a linear trubulent solution is to put the turbulence in the initial conditions (section 5-4-2-1). The development of small-scale nonrandom structure is produced by the quantity $a_{12} q_1^n t x_2$ in the argument of the cosine in eq. (5-635) ($a_{12} = dU_1/dx_2$). That quantity arises from the term $-a_{12} x_2 \partial u_2/\partial x_1$ in eq. (5-633). Thus the term $a_{12} x_2 \partial u_2/\partial x_1$, or equivalently $a_{12} q_1^n t x_2$, acts like a chopper that breaks the flow into small-scale components.

For discussing the linearized case for constant periodic boundary conditions, it is convenient to convert eqs. (5-633) and (5-634) to a spectral form by taking their three-dimensional Fourier transforms (section 5-2). This gives for u_2, on neglecting nonlinear

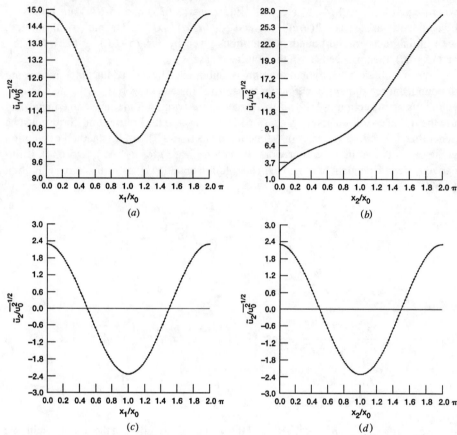

Figure 5-144 Instantaneous profiles of velocity components $\tilde{u}_i$. Initial Reynolds number $R_0 = \overline{u_0^2}^{1/2} x_0/\nu = 34.68$, $(x_0^2/\nu)dU_1/dx_2 = 138.7$. (a) $(\nu/x_0^2)t = 0$, $x_2/x_0 = x_3/x_0 = \pi$; (b) $(\nu/x_0^2)t = 0$, $x_1/x_0 = x_3/x_0 = \pi$; (c) $(\nu/x_0^2)t = 0$, $x_2/x_0 = x_3/x_0 = \pi$; (d) $(\nu/x_0^2)t = 0$, $x_1/x_0 = x_3/x_0 = \pi$.

terms,

$$\frac{\partial \varphi_2^n}{\partial t} = a_{12} q_1^n \sum_{\kappa_2'} \frac{1}{\kappa_2'} \varphi_2^n(\kappa_1, \kappa_2 - \kappa_2', \kappa_3) - \nu \left(q_1^{n^2} + \kappa_2^2 + q_3^{n^2} \right) \varphi_2^n + \frac{2 a_{12} q_1^n \kappa_2 \varphi_2^n}{q_1^{n^2} + \kappa_2^2 + q_3^{n^2}},$$

(5-640)

where

$$\varphi_2^n(\boldsymbol{\kappa}) = \frac{1}{8\pi^3} \int_{-\pi}^{\pi} dx_2 \int \int_{-\infty}^{\infty} u_2^n(x) e^{-i\boldsymbol{\kappa} \cdot \boldsymbol{x}} dx_1 dx_3, \quad (5\text{-}641)$$

$$u_2^n(x) = \sum_{\kappa_2 = -\infty}^{\infty} \int \int_{-\infty}^{\infty} \varphi_2^n(\boldsymbol{\kappa}) e^{i\boldsymbol{\kappa} \cdot \boldsymbol{x}} d\kappa_1 d\kappa_3, \quad (5\text{-}642)$$

$$u_2 = \sum_{n=-3}^{3} u_2^n, \qquad \varphi_2 = \sum_{n=-3}^{3} \varphi_2^n, \quad (5\text{-}643)$$

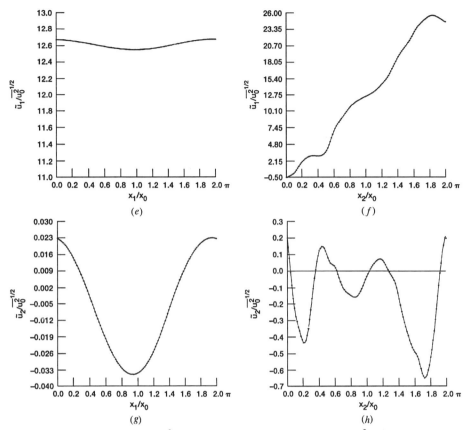

Figure 5-144 (*Continued*) (e) $(\nu/x_0^2)t = 0.1010$, $x_2/x_0 = x_3/x_0 = \pi$; (f) $(\nu/x_0^2)t = 0.1010$, $x_1/x_0 = x_3/x_0 = \pi$; (g) $(\nu/x_0^2)t = 0.1010$, $x_2/x_0 = x_3/x_0 = \pi$; (h) $(\nu/x_0^2)t = 0.1010$, $x_1/x_0 = x_3/x_0 = \pi$.

κ is the wavevector, and φ_2 is the Fourier transform of u_2. Note that a finite transform is used in the x_2 direction in order to satisfy periodic boundary conditions at $x_2/x_0 = -\pi, \pi$.

Strictly speaking, eq. (5-640) is for a sawtooth mean-velocity profile, whereas the numerical results are for a uniform mean-velocity gradient. Equation (5-640) still should apply, however, at least for the present discussion purposes, to points inside but not outside the computational grid.

For constant periodic boundary conditions for u_i, small-scale structure in the fluctuations or the transfer of energy between wavenumbers is produced by the term containing the summation over κ_2' in eq. (5-640). That term is the Fourier transform of $-a_{12}x_2 \partial u_2/\partial x_1$ (eq. [5-633]). From its form we see that it can produce a complicated interwavenumber interaction. The quantity φ_2^n at each κ_2 interacts with φ_2^n at every other allowable κ_2. A difference between the solutions for unbounded conditions and those for constant periodic conditions is that only fluctuations at intregal κ_2 are possible if periodic conditions are imposed, whereas for unbounded conditions, fluctuations are

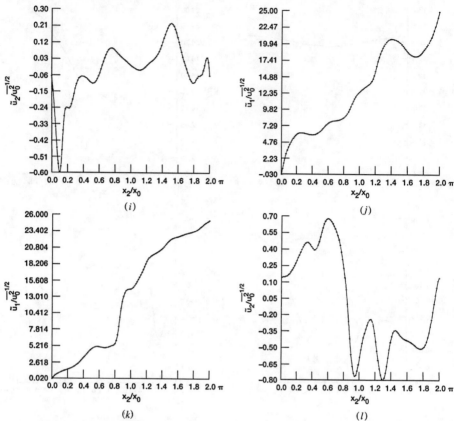

Figure 5-144 (*Continued*) (i) $(\nu/x_0^2)t = 0.1010$, $x_1/x_0 = (37/32)\pi$, $x_3/x_0 = (47/32)\pi$; (j) $(\nu/x_0^2)t = 0.310$, $x_1/x_0 = x_3/x_0 = \pi$; (k) $(\nu/x_0^2)t = 0.310$, $x_1/x_0 = (37/32)\pi$, $x_3/x_0 = (47/32)\pi$; (l) $(\nu/x_0^2)t = 0.310$, $x_1/x_0 = (37/32)\pi$, $x_3/x_0 = (47/32)\pi$.

possible at all values of κ_2. Thus spectra plotted against κ_2 are discrete for periodic boundary conditions, rather than continuous as they are for unbounded flows.

5-4-3-2 Nonlinear results. Figure 5-144 shows a numerically calculated development of instantaneous velocity profiles for uniformly sheared turbulence. The profiles, which are initially regular and given by eq. (5-621), soon take on a turbulent-like appearance. In particular, that is the case for profiles plotted in the x_2 direction. For the low initial Reynolds number shown, and 128^3 grid points, the profiles are well resolved, even though steep gradients characteristic of turbulent flow often occur. Note, however, that we have to use a much finer computational grid with sheared turbulence than we did for the unsheared turbulence in section 5-3-2-6, apparently to resolve singularities that may tend to form in our sheared case. The viscous terms in the instantaneous equations appear to do a good job of smoothing out any singularities that may tend to form (see discussion following eq. (5-633)). Calculated values at grid points are indicated by symbols.

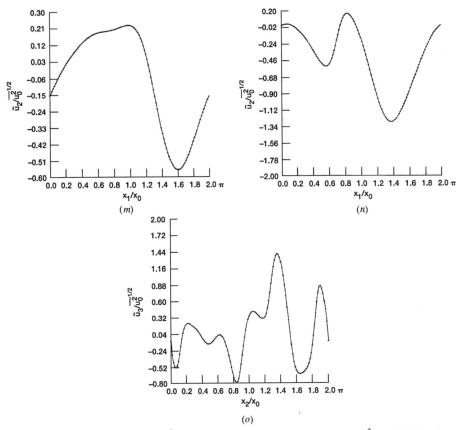

Figure 5-144 (*Continued*) (*m*) $(v/x_0^2)t = 0.310, x_2/x_0 = \pi, x_3/x_0 = \pi$; (*n*) $(v/x_0^2)t = 0.310, x_2/x_0 = (41/32)\pi, x_3/x_0 = (47/32)\pi$; (*o*) $(v/x_0^2)t = 0.310, x_1/x_0 = \pi, x_3/x_0 = \pi$.

As discussed in the previous section, the linear term $(dU_1/dx_2)x_2\partial u_i/\partial x_2$ in eq. (5-633) acts like a chopper that manufactures small-scale components but cannot produce by itself randomization. Only the nonlinear term $\partial(u_i u_k)/\partial x_k$ can do that by itself. But the linear term in combination with $\partial(u_i u_k)/\partial x_k$ can produce randomization by proliferation of harmonic components (one loses track of the individual components because of their sheer number) and by strange behavior (strange attractors in the nondecaying case), as is discussed in Chapter 6, on chaos and sensitive dependence on initial conditions.

The linear term $(dU_1/dx_2)x_2\partial u_i/\partial x_1$ in eq. (5-633) apparently produces small-scale temporal as well as small-scale spatial fluctuations. That is illustrated in Fig. 5-145. If the mean shear is removed from the flow, the small-scale temporal fluctuations die out, leaving only larger ones. Figure 5-145 shows, in a particularly graphic manner, the effectivenes of the term $(dU_1/dx_2)x_2\partial u_i/\partial x_1$ in producing small-scale turbulent structure. Note the correspondence between that term and the mean-gradient transfer term $T_{ij}''(\kappa)$ in eqs. (5-326) and (5-330).

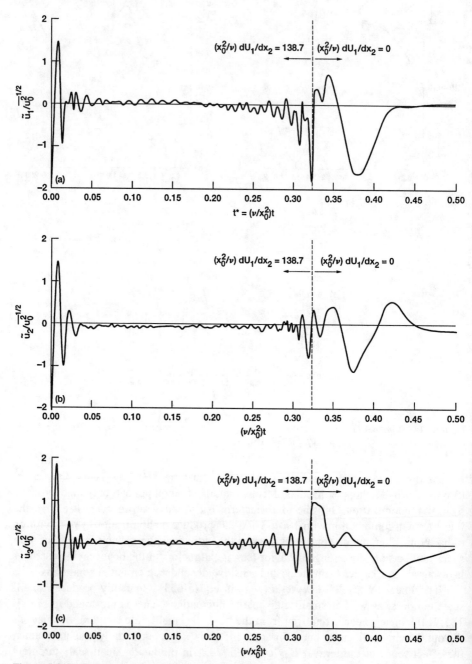

Figure 5-145 Evolution of instantaneous velocity components at center of computational grid. $R_0 = \overline{u_0^2}^{1/2} x_0/\nu = 34.68$. Mean shear removed at $t^* = 0.325$.

5-5 CLOSING REMARKS

This long chapter begins with a discussion of Fourier analysis; it is pointed out that Fourier analysis is a convenient way of studying the distribution and transfer of turbulent activity among scales of motion. In addition it simplifies the basic equations by replacing spatial derivatives by algebraic expressions. A distinction has to be made between equations containing only averaged quantities and those containing, in addition, instantaneous (unaveraged) quantities. The latter require a consideration of generalized functions (functions that do not exist in the ordinary sense). Basic continuum equations for turbulence are obtained in physical and in spectral space; these include the averaged and unaveraged (instantaneous) forms. Spectral transfer of turbulent activity is studied for both homogeneous and inhomogeneous turbulence.

The remainder of the chapter consists mostly of illustrative solutions of the basic turbulence equations, those solutions being obtained to study various turbulence processes. The solutions are divided into those with and without uniform mean gradients. The latter consider mainly spectral transfer between wavenumbers produced by nonlinear turbulence self-interaction and its interaction with turbulence dissipation. The former consider, on the other hand, the effect of mean gradients on spectral transfer, directional transfer, and production of turbulence by mean gradients. Linearized analytical solutions are obtained for weak turbulence without mean gradients and for stronger turbulence with large mean gradients; they are shown to agree quite well with available experimental data. In connection with directional transfer, it is shown that the presence of mean velocity gradients in the pressure terms of the turbulence equations causes the turbulence to become more anisotropic, particularly in the high-wavenumber region. That is, the mean gradients tend to prevent the development of local isotropy. It turns out that phenomenon has an important effect on the maintenance (or lack of maintenance) of shear-flow turbulence, because it drains turbulent energy out of the transverse component. That drain can be lessened if triple correlations are present and effective in the pressure terms, in which case the turbulence can be maintained. Solutions for stronger turbulence without mean gradients in which the correlation equations are closed by specification of sufficient random initial conditions also give realistic results in agreement with experiment. Those solutions involve the "gap" problem—that is, the problem or bridging the gap between the infinite amount of data theoretically required to specify the initial condition of the turbulence and the limited data generally available. Our solutions appear to successfully bridge the gap, in that the evolution of all of the quantities used to specify the initial turbulence are calculated. Also, nonlinear turbulent solutions with and without uniform mean-velocity gradients, and in which the initial conditions are nonrandom, are obtained numerically. The regular initial fluctuations quickly acquire a turbulent-like appearance. Moreover, the insertion of mean shear into the flow produces small-scale temporal fluctuations.

REFERENCES

1. Orszag, S.A., "Numerical Simulation of Turbulent Flows," in *Handbook of Turbulence. vol. 1*, edited by W. Frost and T.H. Moulden, pp. 281–313, Plenum Press, New York, 1977.

2. Farge, M., "Wavelet Transforms and Their Applications to Turbulence," in *Annual Review of Fluid Mechanics, vol. 24*, edited by J.L. Lumley, M. Van Dyke, and H.L. Reed, pp. 395–457, Annual Reviews, Palo Alto, CA, 1992.
3. Lumley, J.L., "Coherent Structures in Turbulence," in *Transition and Turbulence*, edited by R.E. Meyer, pp. 215–242, Academic Press, New York, 1981.
4. Batchelor, G.K., *The Theory of Homogeneous Turbulence*, Cambridge University Press, New York, 1953.
5. Tolstov, G.P., *Fourier Series*, Prentice-Hall, Englewood Cliffs, NJ, 1962.
6. Tranter, C.J., *Integral Transforms in Mathematical Physics*, John Wiley and Sons, New York, 1956.
7. Deissler, R.G., "Effects of Inhomogeneity and of Shear Flow in Weak Turbulent Fields," *Phys. Fluids*, vol. 4, no. 10, pp. 1187–1198, 1961.
8. Craya, A., "Contribution a l'Analyse de la Turbulence Associee a des Vitesses Moyennes," *Publ. Sci. Tech. Minist. Air Fr.*, no. 345, 1958.
9. Richardson, L.F., *Weather Prediction by Numerical Process*. Cambridge University Press, Cambridge, 1922.
10. Monin, A.S., and Yaglom, A.M., *Statistical Fluid Mechanics: Mechanics of Turbulence*, vol. 2, MIT Press, Cambridge, MA, 1975.
11. Gel'fand, I.M., and Shilov, G.E., *Generalized Functions, vol. 1, Properties and Operations*, Academic Press, New York, 1964.
12. Lightlhill, M.J., *Introduction to Fourier Analysis and Generalized Functions*, Cambridge University Press, New York, 1970.
13. Lumley, J.L., *Stochastic Tools in Turbulence*, Academic Press, New York, 1970.
14. Gradshteyn, I.S., and Ryzhik, I.M., *Table of Integrals, Series, and Products*, pp. 337, 496, 1034, Academic Press, New York, 1980.
15. Batchelor, G.K., and Proudman, I., "The Large-Scale Structure of Homogeneous Turbulence," *Philos. Trans. R. Soc. A*, vol. 248, no. 949, pp. 369–405, 1956.
16. Heisenberg, W., "Zur Statistischen Theorie der Turbulenz," *Z. Phys.*, vol. 124, pp. 628–657, 1948.
17. Kovaznay, L.S.G., "Spectrum of Locally Isotropic Turbulence," *J. Aeronaut. Sci.* vol. 15, no. 12, pp. 745–753, 1948.
18. Kraichnan, R.H., "Lagrangian-History Closure Approximation for Turbulence," *Phys. Fluids*, vol. 8, pp. 575–598, 1965.
19. Herring, J.R., "Statistical Turbulence Theory and Turbulence Phenomenology," in *Free Turbulent Shear Flows, vol. 1: Conference Proceedings*, pp. 41–66, NASA SP-321, 1973.
20. Herring, J.R., "An Introduction and Overview of Various Theoretical Approaches to Turbulence," in *Theoretical Approaches to Turbulence*, edited by D.L. Dwoyer, M.Y. Hussaini, and R.G. Voigt, Springer-Verlag, pp. 73–90, New York, 1985.
21. Deissler, R.G., "On the Decay of Homogeneous Turbulence Before the Final Period," *Phys. Fluids*, vol. 1, no. 2, pp. 111–121, 1958.
22. Deissler, R.G., "A Theory of Decaying Homogeneous Turbulence," *Phys. Fluids*, vol. 3, no. 2, pp. 176–187, 1960.
23. Kraichnan, R.H., "Relationships Among some Deductive Theories of Turbulence," in *Mechanics of Turbulence*, pp. 99–106, Gordon and Breach, New York, 1964.
24. Orszag, S.A., "Lectures on the Statistical Theory of Turbulence," in *Fluid Dynamics*, edited by R. Balian and J.L. Peube, pp. 235–374, Gordon and Breach, New York, 1985.
25. Leslie, D.C., *Developments in the Theory of Turbulence*, Clarendon, Oxford, England, 1973.
26. Kraichnan, R.H., "Dynamics of Nonlinear Stochastic Systems," *J. Math. Phys.*, vol. 2, pp. 124–148, 1961. [Erratum: vol. 3, 1962, p. 205].
27. von Kármán, T., and Howarth, L., "On the Statistical Theory of Isotropic Turbulence," *Proc. R. Soc. A.*, vol. 164, pp. 192–215, 1938.
28. Tatsumi, T., "The Theory of Decay Process of Incompressible Isotropic Turbulence," *Proc. R. Soc. London A*, vol. 230, pp. 16–145, 1957.
29. Saffman, P.G., "The Large-Scale Structure of Homogeneous Turbulence," *J. Fluid Mech.*, vol. 27, pp. 581–593, 1967.
30. Miller, W.L., and Gordon, A.R., "Numerical Evaluation of Infinite Series and Integrals Which Arise

in Certain Problems of Linear Heat Flow, Electrochemical Diffusion, etc." *J. Phys. Chem.*, vol. 35, pp. 2785–2884, 1931.
31. Loeffler, A.L., and Deissler R.G., "Decay of Temperature Fluctuations in Homogeneous Turbulence Before the Final Period," *Int. J. Heat Mass Transfer*, vol. 1, no. 4, pp. 312–324, 1961.
32. Batchelor, G.K., and Townsend, A.A., "Decay of Turbulence in the Final Period," *Proc. R. Soc. A*, vol. 194, pp. 527–543, 1948.
33. Clark, R.A., Ferziger, J.H., and Reynolds, W.C., "Evaluation of Subgrid-Scale Models Using an Accurately Simulated Turbulent Flow," *J. Fluid Mech.*, vol. 91, pp. 1–16, 1979.
34. Stewart, R.W., and Townsend, A.A., "Similarity and Self-Preservation in Isotropic Turbulence," *Phil. Trans. R. Soc. A*, vol. 243, no. 867, pp. 359–386, 1951.
35. Uberoi, M.S., "Energy Transfer in Isotropic Turbulence," *Phys. Fluids*, vol. 6, no. 8, pp. 1048–1056, 1963.
36. Van Atta, C.W., and Chen, W.Y., "Measurements of Spectral Energy Transfer in Grid Turbulence," *J. Fluid Mech.*, vol. 38, no. 4, pp. 743–763, 1969.
37. Ling, S.C., and Saad, A., "Experimental Study of the Structure of Isotropic Turbulence with Intermediate Range of Reynolds Number," *Phys. Fluids*, vol. 20, no. 11, pp. 1796–1799, 1977.
38. Ling, S.C., and Huang, T.T., "Decay of Weak Turbulence," *Phys. Fluids*, vol. 13, no. 12, pp. 2912–2924, 1970.
39. Kolmogorov, A.N., "The Local Structure of Turbulence in Incompressible Viscous Fluid for Very Large Reynolds Numbers." *Proc. R. Soc. London A*, vol. 434, pp. 9–13, 1991.
40. Obukhov, A.M., "Spectral Energy Distribution in a Turbulent Flow," *Dokl. Akad. Nauk SSSR*, vol. 32, no. 1, pp. 22–24, 1941.
41. Obukhov, A.M., "Structure of the Temperature Field in a Turbulent Flow," *Izv. Akad. Nauk SSSR, Ser. Geogr. i Geofiz.*, vol. 13, no. 1, pp. 58–69, 1949.
42. Hunt, J.C.R., and Vassilicos, J.C., "Kolmogorov's Contributions to the Physical and Geometrical Understanding of Small-Scale Turbulence and Recent Developments," *Proc. R. Soc. London A*, vol. 434, pp. 183–210, 1991.
43. Landahl, M.T., and Mollo-Christensen, E., *Turbulence and Random Processes in Fluid Mechanics*, p. 63, Cambridge University Press, New York, 1986.
44. Betchov, R., "On the Non-Gaussian Aspects of Turbulence," *Archiwum Mechaniki Stosowanej*, vol. 28, nos. 5–6, pp. 837–845, Warszawa, 1976.
45. Kraichnan, R.H., "The structure of Isotropic Turbulence at Very High Numbers," *J. Fluid Mech.* vol. 5, pp. 497–543, 1959.
46. Kraichnan, R.H., "Inertial-Range Transfer in Two- and Three-Dimensional Turbulence," *J. Fluid Mech.*, vol. 47, pp. 525–535, 1971.
47. Lii, K.S., Rosenblatt, M., and Van Atta, C., "Bispectral Measurements in Turbulence," *J. Fluid Mech.*, vol. 77, pp. 45–62, 1976.
48. Yeung, P.K., and Brasseur, J.G., "The Response of Isotropic Turbulence to Isotropic and Anisotropic Forcing at the Large Scales," *Phys. Fluids A*, vol. 3, no. 5, pp. 884–987, 1991.
49. Domaradzki, J.A., and Rogallo, R.S., "Local Energy Transfer and Nonlocal Interactions in Homogeneous, Isotropic Turbulence," *Phys. Fluids A*, vol. 2, no. 3, pp. 413–426, 1990.
50. Deissler, R.G., "On the Localness of the Spectral Energy Transfer in Turbulence," *Appl. Sci. Res.*, vol. 34, pp. 379–392, 1978.
51. Deissler, R.G., "Comparison of a Correlation Term-Discard Closure for Decaying Homogeneous Turbulence with Experiment," *Phys. Fluids*, vol. 22, no. 1, pp. 185–186, 1979.
52. Deissler, R.G., " Remarks on the Decay of Homogeneous Turbulence from a Given State," *Phys. Fluids*, vol. 17, no. 3, pp. 652–653, 1974.
53. Deissler, R.G., "Some Remarks on the Approximations for Moderately Weak Turbulence," *Phys. Fluids*, vol. 8, no. 11, pp. 2106–2107, 1965.
54. Kraichnan, R.H., "Convergents to Turbulence Functions," *J. Fluid Mech.*, vol. 41, pp. 189–217, 1970.
55. Kraichnan, R.H., "Invariance Principles and Approximation in Turbulence Dynamics," in *Dynamics of Fluids and Plasmas*, edited by S.I. Pai, pp. 239–255, Academic Press, San Diego, CA, 1966.
56. Erdélyi, A., *Asymptotic Expansions*, Dover Publications, Mineola, NY, 1956.

57. Deissler, R.G., Nonlinear Decay of a Disturbance in an Unbounded Viscous Fluid," *Appl. Sci. Res.*, vol. 21, no. 6, pp. 393–410, 1970.
58. Deissler, R.G., "Decay of Homogeneous Turbulence from a Given Initial State," *Phys. Fluids*, vol. 14, no. 8, pp. 1629–1638, 1971 [Corrected and expanded version in NASA TN D-6728, 1972].
59. Deissler, R.G., "Decay of Homogeneous Turbulence from a Given State at Higher Reynolds Number," *Phys. Fluids*, vol. 22, no. 10, pp. 1852–1856, 1979.
60. Hogge, H.D., and Meecham, W.C., "The Wiener-Hermite Expansion Applied to Decaying Isotropic Turbulence Using a Renormalized Time-Dependent Base," *J. Fluid Mech.*, vol. 85, pp. 325–347, 1978.
61. Graff, P., "An Exact Turbulence Theory for the Burgers Equation," *Z. Angew. Math. Phys.*, vol. 20, pp. 461–478, 1969.
62. Taylor, G.I., and Green, A.E., "Mechanism of the Production of Small Eddies from Large Ones," *Proc. R. Soc. London A*, vol. 158, no. 895, pp. 499–521, 1937.
63. McCormick, J.M., and Salvadore, M.G., *Numerical Methods in Fortran*, p. 38, Prentice-Hall, Englewood Cliffs, NJ, 1964.
64. Orszag, S.A., and Israeli, M., "Numerical Simulation of Viscous Incompressible Flows," in *Annual Review of Fluid Mechanics, vol. 6*, edited by M. Van Dyke, W.G. Vincenti, and J.V. Wehausen, pp. 281–318, Annual Reviews, Palo Alto, CA, 1974.
65. Ceschino, F., and Kuntzmann, J., *Numerical Solution of Initial Value Problems*, p. 141, 143, Prentice-Hall, Englewood Cliffs, 1966.
66. Deissler, R.G., "Turbulent Solutions of the Navier-Stokes Equations," *Phys. Fluids*, vol. 24, pp. 1595–1601, 1981.
67. Taylor, G.I., "Diffusion by Continuous Movements," *Proc. London Math. Soc.*, vol. 20, pp. 196–212, 1922.
68. Frenkiel, F.N., *Turbulent Diffusion: Mean Concentration Distribution in a Flow Field of Homogeneous Turbulence. Advances in Applied Mechanics*, vol. III, pp. 61–107, Academic Press, San Diego, CA, 1953.
69. Batchelor, G.K., and Townsend, A.A., "Turbulent Diffusion," in *Surveys in Mechanics*, edited by G.K. Batchelor and R.M. Davies, Cambridge University Press, Cambridge, UK, 1956.
70. Deissler, R.G., "Analysis of Multipoint-Multitime Correlations and Diffusion in Decaying Homogeneous Turbulence," National Aeronautics and Space Administration, NASA TR R-96, Washington, DC, 1961.
71. Burgers, J.M., "On Turbulent Fluid Motion," California Institute of Technology, Report E-34.1, Pasadena, CA, July 1951.
72. Baldwin, L.V., "Turbulent Diffusion in the Case of Fully Developed Pipe Flow," Ph.D. Diss., Case Institute of Technology, 1959.
73. Bass, J., *Space and Time Correlations in a Turbulent Fluid*, parts I and II: *University of California Publications in Statistics*, University of California Press, Berkeley, CA, 1954.
74. Lin, C.C., "Remarks on the spectrum of Turbulence," *Proceedings First Symposia of Applied Mathematics*, vol. 1, American Mathematical Society, Providence, RI, 1949.
75. Uberoi, M.S., and Corrsin, S., "Diffusion of Heat from a Line Source in Isotropic Turbulence," National Advisory Committee for Aeronautics, NACA Rept. 1142, Washington, DC, 1953.
76. Burgers, J.M., and Mitchner, M., "On Homogeneous Non-Isotropic Turbulence Connected With a Mean Motion Having a Constant Velocity Gradient I," *Koninklijke Nederlandse Akademie van Wetenschappen, Proc. B*, vol. 56, no. 3, pp. 228–235, 1953.
77. Burgers, J.M., and Mitchner, M., "On Homogeneous Non-Isotropic Turbulence Connected with a Mean Motion Having a Constant Velocity Gradient II," *Koninklijke Nederlandse Akademie van Wetenschappen, Proc. B*, vol. 56, no. 4, pp. 343–354, 1953.
78. Pearson, J.R.A., "The Effect of Uniform Distortion on Weak Homogeneous Turbulence," *J. Fluid Mech.*, vol. 5, pt. 2, pp. 274–288, 1959.
79. Hildebrand, F.B., *Advanced Calculus for Engineers*, p. 374, Prentice-Hall, Englewood Cliffs, NJ, 1948.
80. Fox, J., "Velocity Correlations in Weak Turbulent Shear Flow," *Phys. Fluids*, vol. 7, no. 4, pp. 562–564, 1964.
81. Sandborn, V.A., and Braun, W.H., "Turbulent Shear Spectra and Local Isotropy in the Low-Speed Boundary Layer," National Advisory Committee for Aeronautics, NACA TN-3761, Washington, DC, 1956.

82. Corrsin, S., "Local Isotropy in Turbulent Shear Flow," National Advisory Committee for Aeronautics, NACA RM 58B 11, Washington, DC, 1958.
83. Townsend, A.A.,"Local Isotropy in the Turbulent Wake of a Cylinder," *Aust. J. Sci. Research A*, vol. 1, no. 2, pp. 161–174, 1948.
84. Corrsin, S., and Uberoi, M.S., "Spectra and Diffusion in a Round Turbulent Jet," National Advisory Committee for Aeronautics, NACA Rept. 1040, Washington, DC, 1951.
85. Laufer, J., "Some Recent Measurements in a Two-Dimensional Turbulent Channel," *J. Aeronaut. Sci.*, vol. 17, pp. 277–287, 1950.
86. Townsend, A.A., "On the Find Scale Structure of Turbulence," *Proc. R. Soc. London A*, vol. 208, no. 1095, pp. 534–542, 1951.
87. Corrsin, S., *Some Current Problems in Turbulent Shear Flows: Symposium on Naval Hydrodynamics*. Academy of Sciences–National Research Council, Publication No. 515, pp. 373–407, Washington, 1957.
88. Taylor, G.I., "Production and Dissipation of Vorticity in a Turbulent Fluid," *Proc. R. Soc. London A*, vol. 164, no. 918, pp. 15–23, 1938.
89. Theodorsen, T., "The Mechanism of Turbulence." In *Proceedings of the Second Midwestern Conference on Fluid Mechanics*, pp. 1–18, Ohio State University, Columbus, OH, 1952.
90. Weske, J.R., and Plantholt, A.H., "Discrete Vortex Systems in the Transition Range of Fully Developed Flow in a Pine," *J. Aeronaut. Sci.*, vol. 20, pp. 717–718, 1953.
91. Hasen, E.M., "A Nonlinear Theory of Turbulence Onset in a Shear Flow," *J. Fluid Mech.*, vol. 29, pp. 721–729, 1967.
92. Deissler, R.G., "Effect of Initial Condition on Weak Homogeneous Turbulence with Uniform Shear," *Phys. Fluids*, vol. 13, no. 7, pp. 1868–1869, 1970.
93. Rose, W.G., "Results of an Attempt to Generate a Homogeneous Turbulent Shear Flow," *J. Fluid Mech.*, vol. 25, pp. 97–120, 1966.
94. Champagne, F.H., Harris, V.G., and Corrsin, S., "Experiments on Nearly Homogeneous Turbulent Shear Flow," *J. Fluid Mech.*, vol. 41, pp. 81–139, 1970.
95. Mulhearn, P.J., and Luxton, R.E., "The Development of Turbulence Structure in a Uniform Shear Flow," *J. Fluid Mech.*, vol. 68, pp. 577–590, 1975.
96. Lin, C.C., ed., *Turbulent Flows and Heat Transfer*, Princeton University Press, Princeton, NJ, 1959.
97. Hinze, J.O., *Turbulence*. McGraw-Hill, New York, 1959.
98. Knudsen, J.G., and Katz, D.L., *Fluid Dynamics and Heat Transfer*. McGraw-Hill, New York, 1958.
99. Jakob, M., *Heat Transfer, vol. II*. John Wiley and Sons, New York, 1957.
100. Eckert, E.R.G., and Drake, R.M., Jr., *Analysis of Heat and Mass Transfer*, McGraw-Hill, New York, 1972.
101. Burmeister, L.C., *Convective Heat Transfer*, John Wiley and Sons, New York, 1983.
102. Kays, W.M., *Convective Heat and Mass Transfer*, McGraw-Hill, New York, 1966.
103. Rohsenow, W.M., and Choi, H.Y., *Heat, Mass, and Momentum Transfer*, Prentice-Hall, Englewood Cliffs, NJ, 1961.
104. McAdams, W.H., *Heat Transmission*, McGraw-Hill, New York, 1954.
105. Kreith, F., *Principles of Heat Transfer*, ed. 4. Harper and Row, New York, 1986.
106. Corrsin, S., "Heat Transfer in Isotropic Turbulence," *J. Appl. Phys.*, vol. 23, pp. 113–118, 1952.
107. Dunn, D.W., and Reid, W.H., "Heat Transfer in Isotropic Turbulence during the Final Period of Decay," National Advisory Committee for Aeronautics, NACA TN-4168, Washington, DC, 1958.
108. Deissler, R.G., "Turbulent Heat Transfer and Temperature Fluctuations in a Field with Uniform Velocity and Temperature Gradients," *Int. J. Heat Mass Transfer*, vol. 6, pp. 257–270, 1963.
109. Deissler, R.G., "Analysis of Fully Developed Turbulent heat Transfer at Low Peclet Numbers in Smooth Tubes with Application to Liquid Metals," National Advisory Committee for Aeronautics, NACA RME52F05, Washington, DC, 1952.
110. Jenkins, R., "Variation of the Eddy Conductivity with Prandtl Modulus and Its Use in Prediction of Turbulent Heat Transfer Coefficients," in *Preprints of Papers, Heat Transfer and Fluid Mechanics Institute*, pp. 147–158, Stanford University Press, Stanford, CA, 1951.
111. Fox, J., "Turbulent Temperature Fluctuations and Two-Dimensional heat Transfer in a Uniform Shear Flow," *Int. J. Heat Mass Transfer*, vol. 8, pp. 467–480, 1965.

112. Loeffler, A.L., "Electric Field Modification of Turbulence in a Fluid Containing Space Charge," Grumman Research Report, RE-384, Grumman Aircraft Engineering, Bethpage, NY, 1970.
113. Deissler, R.G., "Turbulence in the Presence of a Vertical Body Force and Temperature Gradient," *J. Geophys. Res.*, vol. 67, no. 8, p. 3049, 1962.
114. Townsend, A.A., "Turbulent Flow in a Stably Stratified Atmosphere," *J. Fluid Mech.*, vol. 3, pp. 361–372, 1958.
115. Deissler, R.G., "Effects of Combined Buoyancy and Shear on Weak Homogeneous Turbulence," National Aeronautics and Space Administration, NASA TND-3999, Washington, DC, 1967.
116. Webster, C.A.G., "An Experimental Study of Turbulence in a Density-Stratified Shear Flow," *J. Fluid Mech.*, vol. 19, pp. 221–245, 1964.
117. Deissler, R.G., "Growth Due to Buoyancy of Weak Homogeneous Turbulence With Shear," *Z. Angew. Math. Phys.*, vol. 22, pp. 267–274, 1971.
118. Prandtl, L., "Attaining a Steady Air Stream in Wind Tunnels," National Advisory Committee for Aeronautics, NACA TM-726, Washington, DC, 1933.
119. Taylor, G.I., "Turbulence in Contracting Stream," *Z. Angew. Math. Mech.*, vol. 15, no. 1/2, pp. 91–96, 1935.
120. Ribner, H.S., and Tucker, M., "Spectrum of Turbulence in a Contracting Stream," National Advisory Committee for Aeronautics, NACA TN-2606, Washington, DC, 1952.
121. Batchelor, G.K., and Proudman, I., "The Effect of Rapid Distortion of a Fluid in Turbulent Motion," *Q. J. Mech. Appl. Math.*, vol. 7, pp. 83–103, 1954.
122. Deissler, R.G., "Weak Locally Homogeneous Turbulence in Idealized Flow Through a Cone," *J. Appl. Math. Phys. ZAMP*, vol. 18, pp. 165–183, 1967.
123. Mills, R.R., Jr., and Corrisin, S., "Effect of Contraction on Turbulence and Temperature Fluctuations Generated by a Warn Grid," National Aeronautics and Space Administration, NASA Memo 5-5-59W, Washington, DC, 1959.
124. Uberoi, S., Effect of Wind-Tunnel Contraction on Free-Stream Turbulence, *J. Aeron. Sci.*, vol. 23, no. 8, pp. 754–764, 1956.
125. Boldman, D.R., Schmidt, J.F., and Fortini, A., "Turbulence, Heat-Transfer, and Boundary Layer Measurements in a Conical Nozzle with a Controlled Inlet Velocity Profile," National Aeronautics and Space Administration, NASA TN D-3221, Washington, DC, 1966.
126. Boldman, D.R., Schmidt, J.F., and Ahlers, R.C., "Effect of Inlet Configuration on the Turbulent Boundary Layer and Heat Transfer a Conical Nozzle Operating with Air," American Society of Mechanical Engineers, New York, 1967.
127. Back, L.H., Massier, P.F., and Gier, H.L., "Convective Heat Transfer in a Convergent-Divergent Nozzle," *Int. J. Heat Mass Transfer*, vol. 7, no. 5, pp. 549–568, 1964.
128. Grant, H.L., Stewart, R.W., and Molliet, A., "Turbulence Spectra from a Tidal Channel," *J. Fluid Mech.*, vol. 12, pp. 241–268, 1962.
129. Deissler, R.G., "Weak Locally Homogeneous Turbulence and Heat Transfer with Uniform Normal Strain," *J. Appl. Math. Mech. ZAMM*, vol. 48, pp. 87–98, 1968.
130. Kuethe, A.M., Willmarth, W.W., and Crocker, G.H., "Stagnation Point Fluctuations on Bodies of Revolution with Hemispherical Noses," Rep. No. 02753-2-F (AFOSR TR 60-65), Michigan University, College of Engineering, June 1960.
131. Deissler, R.G., Effect of Uniform Longitudinal Strain Rate on Weak Homogeneous Turbulence in a Compressible Flow," National Aeronautics and Space Administration, NASA TN D-2800, Washington, DC, 1965.
132. Sutera, S.P., "Vorticity Amplification in Stagnation-Point Flow and Its Effect on Heat Transfer," *J. Fluid Mech.*, vol. 21, pp. 513–534, 1965.
133. Kestin, J., Maeder, P.F., and Sogin, H.H., "The Influence of Turbulence on the Transfer of Heat to Cylinders Near the Stagnation Point," *Z. Angew. Math. Phys.*, vol. 12, no. 2, pp. 115–132, 1961.
134. Kline, S.J., "Observed Structure Features in Turbulent and Transitional Boundary Layers." In *Fluid Mechanics of Internal Flow*, edited by G. Sovran, p. 39, Elsevier Publishing, New York, 1967.
135. Kays, W.M., *Convective Heat and Mass Transfer*, p. 96, McGraw-Hill, 1966.
136. Back, L.H., Massier, P.F., and Gier, H.L., "Convective Heat Transfer in a Convergent-Divergent Nozzle," *Int. J. Heat Mass Transfer*, vol. 7, no. 5, pp. 549–568, 1964.

137. Deissler, R.G., "Growth of Turbulence in the Presence of Shear," *Phys. Fluids*, vol. 15, no. 11, pp. 1918–1920, 1972.
138. Deissler, R.G., "Problem of Steady-State Shear-Flow Turbulence," *Phys. Fluids*, vol. 8, no. 3, pp. 391–398, 1965.
139. Rotta, J.C., "Turbulent Boundary Layers in Incompressible Flow," in *Progress in Aeronautical Science*, vol. 2, edited by A. Ferri, D. Kuchemann, and L.H.G. Sterne, pp. 1–219, Pergamon Press, Oxford, New York.
140. Deissler, R.G., "Turbulent Solutions of the Equations of Fluid Motion," *Rev. Mod. Phys.*, vol. 56, no. 2, pp. 223–254, 1984.
141. Rogallo, R.S., "Numerical Experiments in Homogeneous Turbulence," National Aeronautics and Space Administration, NASA TM-81315, Washington, DC, 1981.
142. Deissler, R.G., and Rosenbaum, B.M., "Nonlinear Evolution of a Disturbance in an Unbounded Viscous Fluid with Uniform Shear," National Aeronautics and Space Administration, NASA TN D-7284, Washington, DC, 1973.

CHAPTER
SIX

TURBULENCE, NONLINEAR DYNAMICS, AND DETERMINISTIC CHAOS

The unaveraged Navier-Stokes equations are used numerically in conjuction with tools and concepts from nonlinear dynamics, including time series, phase portraits, Poincaré sections, Liapunov exponents, power spectra, and strange attractors. Initially neighboring solutions for a low–Reynolds-number fully developed turbulence are compared, where the turbulence is sustained by a nonrandom time-independent external force. By reducing the Reynolds number (forcing), several nonturbulent solutions are also obtained and contrasted with the turbulent ones.

As is apparent from the preceding chapters, fluid turbulence is a many-faceted phenomenon. It has been characterized as random, nonlinear, multiscaled, dissipative, having a negative velocity-derivative skewness factor, transferring energy (mainly) to small-scale motions, being dissipated by small-scale motions, tending toward isotropy, and having an infinite number of components or degrees of freedom. Those descriptions appear in what might now be called the *classical* or *statistical* theory of turbulence [1–5, 28]. That theory is based mainly on averaged or moment equations obtained from the Navier-Stokes equations.

It is mentioned in the introductory remarks of Chapter 4 that the idea of using averaged equations, rather than the unaveraged Navier-Stokes equations, directly in an analysis has been adopted in the past mainly because it was thought that averaged, smoothly varying quantities should be easier to deal with than the haphazard motions occurring in the unaveraged equations. However, because of the nonlinearity of the Navier-Stokes equations, the averaging process introduces the closure problem of more unknowns than equations (see Chapter 4), so that it is not clear that averaging is advantageous as far as getting solutions is concerned.[1] The averaged or moment equations, however, are useful (if not necessary) for discussing the physical processes occurring in turbulence (see Chapters 4 and 5).

In recent years there have been attempts to utilize concepts from the theory of nonlinear dynamical systems in the analysis of turbulence [6–12]. There, in contrast to the statistical theory, the emphasis is on unaveraged rather than on averaged equations.

---
[1] Recall that parts of Chapters 4 and 5 are devoted to numerical solutions of the unaveraged equations.

374 TURBULENT FLUID MOTION

The use of unaveraged equations in which the velocities vary in a complicated way is made feasible by the advent of high-speed computers. By using ideas from nonlinear dynamics one might (as further evidence that turbulence is a many-faceted phenomenon) characterize turbulence as chaotic although deterministic, as aperiodic, as having sensitive dependence on initial conditions, as having time series without pattern, as having a positive Liapunov exponent, as having a phase portrait without pattern, as having Poincaré sections without pattern, as lying on a strange or chaotic attractor, and as having continuous time and spatial spectra.

Both the statistical (classical) theory and the newer nonlinear dynamics theory provide valid ways of looking at turbulence. The latter furnishes a number of new tools for probing the nature of turbulence (e.g., Liapunov exponents, Poincaré sections). As yet, however, it does not seem to provide a means of discussing such well-known aspects of turbulence as spectral and directional transfer of energy. Those aspects, however, can be considered within the framework of conventional turbulence theory (see Chapter 5).

Here we study the nature of Navier-Stokes turbulence (the turbulence obtained from solutions of the Navier-Stokes equations) by using concepts from nonlinear dynamics. Sensitive dependence on initial conditions and strange (chaotic) attractors are included. These are shown to occur in turbulence by obtaining and interpreting (mainly numerical) solutions of the Navier-Stokes equations.

In order to give a sharper characterization of turbulence, turbulent solutions are contrasted with periodic, quasiperiodic, and fixed-point solutions. Turbulent systems also are compared with those considered in the kinetic theory of gases. There is a certain suddenness inherent in turbulent mixing (section 4-3-2-3), as there is in molecular mixing by collision of gas particles.

Before investigating turbulence, we consider a low-order system. We see that solutions of that system have similarities to high- (infinite-) order turbulence.

6-1 LOW-ORDER NONLINEAR SYSTEM

An important (and surprising) result from nonlinear dynamics studies is that highly complex or chaotic motion can be obtained in systems with only a few (but not less than three) modes or degrees of freedom.[2] The best-known low-order example is the Lorenz system [12]:

$$\frac{d}{dt}\begin{bmatrix} x \\ y \\ z \end{bmatrix} = \begin{bmatrix} s(y-x) \\ rx - xz - y \\ xy - bz \end{bmatrix}, \tag{6-1}$$

where t is the time and x, y, and z are the three degrees of freedom of the system.

The variable (mode) y in eq. (6-1) is plotted against time t in Fig. 6-1, in which $b=1, r=26$, and $s=3$. The initial conditions are given by $x=0$, $y=1$, and

---

[2]It turns out that at least three modes are necessary if a chaotic solution is to be unique [7]. If, say, only two modes (or dimensions in phase or state space) were present, the curve representing a chaotic solution in phase space would intersect itself and thus the solution would not be unique at the points of intersection.

TURBULENCE, NONLINEAR DYNAMICS, AND DETERMINISTIC CHAOS 375

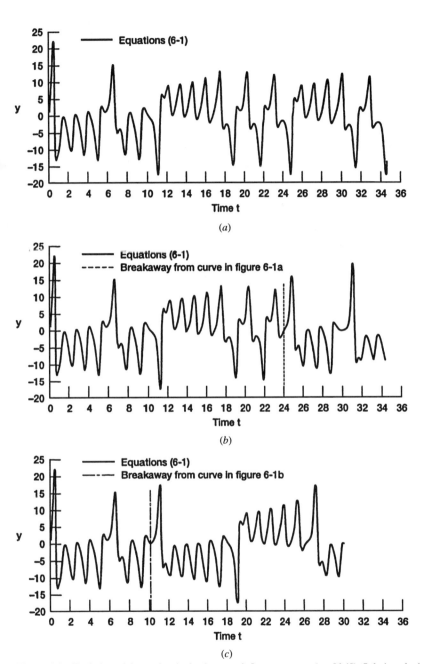

Figure 6-1 Evolution of the mode y in the three-mode Lorenz system (eq. [6-1]). Solution obtained numerically by improved Euler method. $b = 1, r = 26, s = 3$, initial $x = $ inital $z = 0$. (a) initial $y = 1.000$, Δt in numerical solution $= 0.002$; (b) initial $y = 1.001$, Δt in numerical solution $= 0.002$; (c) initial $y = 1.001$, Δt in numerical solution $= 0.01$.

$z = 0$ at $t = 0$. The quantity plotted in Fig. 6-1 has a random appearance, in spite of the fact that the system has only three degrees of freedom. In fact, the flow in Fig. 6-1 is somewhat like that in Fig. 1-3, in which a component of turbulence for a particular experimental flow is plotted against time; both flows have a random appearance. This is perhaps unexpected, because turbulence is described by partial differential equations (the Navier-Stokes equations) and thus has (theoretically) an infinite number of degrees of freedom. The random appearance of turbulence has been ascribed to the infinite number of degrees of freedom [13–15], but it now appears that as few as three are sufficient.

Comparison of Figs. 6-1a and b shows the effect of a small change in initial condition. A change in initial y of 0.1% produces a very large change in y at times beyond the indicated time of breakaway from the unperturbed flow, although there is no perceptible change at earlier times. The flow thus appears to be sensitively dependent on initial conditions—a necessary (although not sufficient) condition for a flow to be turbulent. This is another indication of the similarity of the low-order flow in Fig. 6-1 to a turbulent flow.

Finally, by comparing Figs. 6-1b and c we investigate the effect of time increment Δt in the numerical solution of the Lorenz system on the evolution of y. An improved Euler method [16] is used in the numerical solution. Increasing Δt from 0.002 in Fig. 6-1b to 0.01 in Fig. 6-1c (with everything else held constant) has no perceptible effect on the evolution of y for times smaller than the breakaway time indicated in Fig. 6-1c. Thus, the numerical calculations in the two graphs are accurate in the usual sense. For times greater than the breakaway time, however, the evolutions are completely different, apparently because of the inherent instability or chaoticity of the system for the parameters shown. Thus, the effect of increasing Δt is qualitatively similar to that of giving a small perturbation to the initial condition.

Following this brief study of a low-order chaotic system we go now to the Navier-Stokes system [17]. As in the preceding chapters the Navier-Stokes equations are used to describe turbulence in fluids.

6-2 BASIC EQUATIONS AND A LONG-TERM TURBULENT SOLUTION WITH STEADY FORCING

The incompressible Navier-Stokes equations, on which the present study is based, are

$$\frac{\partial u_i}{\partial t} = -\frac{\partial (u_i u_k)}{\partial x_k} - \frac{1}{\rho}\frac{\partial \sigma}{\partial x_i} + \nu \frac{\partial^2 u_i}{\partial x_k \partial x_k} + F_i, \tag{6-2}$$

together with a Poisson equation for the pressure

$$\frac{1}{\rho}\frac{\partial^2 \sigma}{\partial x_l \partial x_l} = -\frac{\partial^2 (u_l u_k)}{\partial x_l \partial x_k} + \frac{\partial F_l}{\partial x_l}. \tag{6-3}$$

The subscripts can have the values 1, 2, or 3, and a repeated subscript in a term indicates a summation, with the subscript successively taking on the values 1, 2, and 3. The quantity u_i is an instantaneous velocity component, x_i is a space coordinate, t is the time, ρ is the density, ν is the kinematic viscosity, σ is the instantaneous (mechanical) pressure, and F_i is a time-independent forcing term, or external force, which is taken as some fraction

χ of the negative of the initial viscous term at $t = 0$. That is,

$$F_i = -\chi \nu \left(\frac{\partial^2 u_i}{\partial x_k \partial x_k} \right)_{t=0}. \qquad (6\text{-}4)$$

The fraction χ controls the value of the asymptotic Reynolds number of the flow. Equations (6-2) and (6-3) are respectively the same as eqs. (3-19) and (3-21) if the forcing term F_i is replaced by the body force g_i. The initial nonrandom u_i in eq. (6-4) are given at $t = 0$ by

$$u_i = a_i \cos \boldsymbol{q} \cdot \boldsymbol{x} + b_i \cos \boldsymbol{r} \cdot \boldsymbol{x} + c_i \cos \boldsymbol{s} \cdot \boldsymbol{x}, \qquad (6\text{-}5)$$

where

$$a_i = k(2, 1, 1), \quad b_i = k(1, 2, 1), \quad c_i = k(1, 1, 2), \qquad (6\text{-}6)$$

$$q_i = (-1, 1, 1)/x_0, \quad r_i = (1, -1, 1)/x_0, \quad s_i = (1, 1, -1)/x_0,$$

k is a quantity that fixes the initial Reynolds number at $t = 0$, and x_0 is one over the magnitude of an initial wavenumber component. Through eq. (6-4), x_0 is also one over the magnitude of a wavenumber component of the forcing term F_i. Equations (6-5) and (6-6) satisfy continuity, and eqs. (6-2) through (6-4) ensure that continuity is maintained. Moreover, eqs. (6-4) through (6-6) give local values of F_i that are symmetric with respect to 90° rotations and translations of $2\pi x_0$. Then we find numerically that

$$\overline{u_1^2} \approx \overline{u_2^2} \approx \overline{u_3^2} \qquad (6\text{-}7a)$$

at all times, where the overbars indicate values averaged over space. After the initial transients have died out, the averages also may be taken over time, and the inexact equalities in eq. (6-7a) become equalities. Equation (6-7a) then becomes

$$\overline{\overline{u_1^2}} = \overline{\overline{u_2^2}} = \overline{\overline{u_3^2}}, \qquad (6\text{-}7b)$$

where the double bars indicate averages over space and time. The boundary conditions are periodic with a period of $2\pi x_0$. From eq. (6-4) and continuity, the last term in eq. (6-3) is zero for our system.

Equation (6-2) is a nonlinear dissipative equation for the evolution of the vector u_i. Although a Navier-Stokes fluid is linear (stress proportional to strain rate), a nonlinearity appears in eq. (6-2) as an effect of inertia. The equation is autonomous, because time does not appear explicitly on the right side, and deterministic because there are no random coefficients. Note that the equation would not be autonomous if the forcing term F_i were time-dependent. Equation (6-2), although three-dimensional in physical space, is infinite-dimensional in phase (or state) space, because it is a partial-differential equation. (The number of dimensions of the phase space of our system is the number of u_i required to specify the velocity field at a particular time. The pressure is not specified; it is calculated from equation [6-3].) The equation can be converted to an infinite system of ordinary differential equations by, for instance, introducing finite-difference representations of spatial derivatives (and letting grid spacing approach 0), or by taking the spatial Fourier transform of the equation. Because it is dissipative, the infinite system can be represented by a finite system of equations [11]. There should be a viscous cutoff, below which motion becomes unimportant as the scale of the motion decreases. Thus, a numerical solution should be possible, at least for low Reynolds numbers. Equation (6-2), together with eq. (6-3) for the pressure, eqs. (6-4) through (6-6) for the forcing term, eqs. (6-5) and (6-6) for the initial conditions, and periodic boundary conditions, can be considered a nonlinear, deterministic, autonomous, dissipative, dynamical system. The system is deterministic because there are no random elements in eqs. (6-2) through (6-6) or in the boundary conditions.

The numerical method used for the solution of eqs. (6-2) through (6-6) has been given in ref. [18] and in section 5-3-2-6. A cubical computational grid (32^3 grid points), fourth-order spatial differencing, and third-order predictor–corrector time differencing are used. In order to obtain numerical stability for the highest asymptotic Reynolds number (13.3), it is necessary to use about 50 time steps in each small fluctuation of velocity, so that the fluctuations with respect to time are well resolved. The spatial resolution is also good and is discussed later, in connection with Fig. 6-6.

It follows from eqs. (6-4) through (6-6) that the nonrandom initial condition on u_i applied at $t = 0$ is proportional to the steady forcing term F_i. The quantity $(u_i)_{t=0}$, or F_i, on an $x_j - x_k$ plane through the numerical grid center is plotted in Fig. 6-2. Figure 6-3 shows the magnitude of the vector $(u_i)_{t=0}$ or F_i. A high degree of spatial symmetry of $(u_i)_{t=0}$ and of F_i is apparent from these plots. Note that as a result of the symmetry, the subscript i can designate any component of the vectors, and that the $x_j - x_k$ plane can be any plane through the numerical grid center parallel to the grid axes. That is, i, j, and $k = 1, 2,$ or 3; $j \neq k$. Moreover, the symmetry allows the development of symmetric turbulence in a box, where the box has periodic walls.

Results for the evolution of $\overline{(u_1^2)}^{1/2}$ for χ in eq. (6-4) equal to 1 (asymptotic Reynolds number, 13.3) are given in Fig. 6-4 (see Fig. 6-4 for definition of the Reynolds number). The value of k in eq. (6-6) is 20, giving an initial Reynolds number at $t = 0$ of 34.6. The velocities have been divided by $(\overline{u_0^2})^{1/2}$, where the 0 again refers to $t = 0$. An asymptotic turbulent solution is obtained for $t^* > 5$. (The asterisk on t indicates that it has been nondimensionalized by x_0 and ν.)

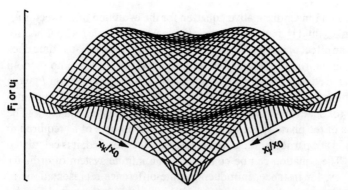

Figure 6-2 Plot of forcing term F_i in eq. (6-2) or of regular initial velocity component u_i on a plane through center of numerical grid. x_k and x_j are coordinates on the plane and x_0 is the reciprocal of a wavenumber component of the forcing term.

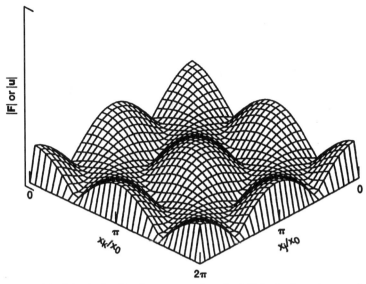

Figure 6-3 Magnitude of forcing vector or of regular initial velocity vector on a plane through numerical-grid center.

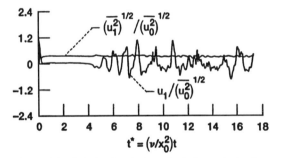

Figure 6-4 Calculated evolution of turbulent velocity fluctuations with a time-independent forcing term. Ordinates normalized by initial condition. Root-mean-square velocities (with a single bar) are spatially averaged. Developed Reynolds number $(\overline{\overline{u_1^2}})^{1/2} x_0/\nu = 13.3$, where the double bar indicates an average over space and time for $t^* > 5$, $\chi = 1$. $x_1^* = x_1/x_0 = 9\pi/8$, $x_2^* = 21\pi/16$, $x_3^* = 23\pi/16$, for unaveraged fluctuations. 32^3 spatial grid points.

A rather remarkable feature of turbulent flow is that a time-dependent haphazard flow can result if the applied exciting forces are steady (e.g., in a fully developed turbulent pipe flow with a steady applied pressure gradient). Figure 6-4 shows that the Navier-Stokes turbulence calculated here exhibits this feature, because a steady forcing term produces an apparently haphazard time-dependent motion. This is evidently the result of a kind of instability of the nonlinear Navier-Stokes equations except at very low Reynolds numbers, because initially neighboring solutions separate exponentially with time. That is, those solutions are chaotic [9]. It is seen subsequently that our steady forcing term also can produce time-dependent nonturbulent flow.

It should be mentioned that the symmetry present in the initial conditions (eqs. [6-5] and [6-6]) which, for instance, causes the three local velocity components to be equal for $x_1 = x_2 = x_3$ at $t = 0$, has been destroyed before $t^* = 5$, apparently by roundoff errors. This breaking of symmetry for local values indeed must occur in order for true turbulence to develop, and in fact the fluctuations eventually die out if the symmetry remains. Here the initial fluctuations are not strong enough to destroy the symmetry before the fluctuations become too small to be seen on the u_1 curve. The breaking of symmetry apparently occurs on the flat portion of the curve by the accumulation of roundoff errors. For higher initial Reynolds numbers (not shown) the initial fluctuations are strong enough to break the local symmetry earlier, and the flat portion of the u_1 curve is absent.

The mean skewness factor S of the velocity derivative of our Navier-Stokes turbulence in Fig. 6-4 is calculated to be

$$S = \overline{\left(\frac{\partial u_1}{\partial x_1}\right)^3} \Big/ \left[\overline{\left(\frac{\partial u_1}{\partial x_1}\right)^2}\right]^{3/2} = -0.52, \qquad (6\text{-}8)$$

where the skewness factor is averaged over time after the powers of the velocity derivative have been averaged over space. This value is close to those obtained experimentally for a variety of simple turbulent flows [19], in which the Reynolds numbers of the experiments are in the same range as that for the solution in Fig. 6-4.

Instantaneous (unaveraged) terms in the Navier-Stokes equation (equation [6-2]) for $i = 1$ at the numerical grid center are plotted in Fig. 6-5. These include the nonlinear convective term $-\partial(u_i u_k)/\partial x_k$, the steady forcing term F_i, the viscous term $\nu \partial^2 u_i/\partial x_k \partial x_k$, and the pressure term $-(1/\rho)\partial \sigma/\partial x_i$.

For the asymptotic or developed region (for $t^* > 5$) the viscous term is of the same order of magnitude as the steady forcing term. This is reasonable because the forcing term replenishes the energy lost by viscous action. On the other hand the nonlinear convective and pressure terms are much larger. (The pressure term is nonlinear through eq. [6-3].) It may seem surprising that a small forcing term can produce large convective and pressure terms; apparently those terms are amplified by the instability of the Navier-Stokes flow at the Reynolds number in Fig. 6-4. The tendency is even greater at higher Reynolds numbers (not shown). If we compare the nonlinear convective and pressure terms with the viscous term rather than with the forcing term, the trend is not surprising, because it is well known that the nonlinear terms become much greater than the viscous as the Reynolds number of a turbulent flow increases. As was mentioned before, the forcing term is of the same order of magnitude as the viscous.

Calculated spatial variations of velocity fluctuations are plotted in Fig. 6-6. Although the Reynolds number is low, there is some tendency for velocity gradients to become large in several nonadjacent regions, thus indicating the hydrodynamic instability of the flow and, in addition, the spatial intermittency of the flow in the smaller eddies [1]. (Note that steep gradients are associated with small eddies or large wavenumbers [20].) This tendency to form steep gradients is, of course, a well-known property of turbulent flows and evidently occurs as an effect of the nonlinear terms in the Navier-Stokes equations.

TURBULENCE, NONLINEAR DYNAMICS, AND DETERMINISTIC CHAOS 381

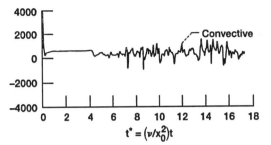
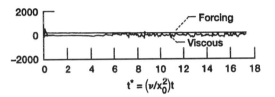
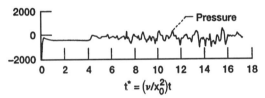

Figure 6-5 Calculated evolution of instantaneous terms in Navier-Stokes equation at grid center. $\chi = 1$. Developed Reynolds number = 13.3.

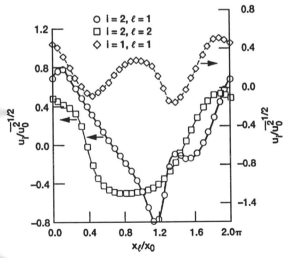

Figure 6-6 Calculated spatial variation of velocity fluctuations on a plane through grid center at $t^* = 13.74$. Symbols are at grid points.

In order to give an idea of the numerical resolution obtained, grid points are indicated by symbols; all of the scales of motion are well resolved.

The number of degrees of freedom or modes used in the present solution (32^3 grid points times three directional velocity components) is compared with the criteria for sufficient determining modes obtained by Constantin et al. [11]. Both on the basis of the ratio of the largest to smallest length scale and on the basis of Reynolds number, the number of determining modes used in the present solution is considerably larger than required for a qualitatively correct solution. So according to the criteria of ref. [11], there are plenty of determining modes for a qualitatively correct solution. That reference does not address the problem of a quantitatively correct solution.

After initial transients have died out (for $t^* > 5$), the flow considered in Figs. 6-2 through 6-8 lies on a strange attractor. This is because, as shown in ref. [9], the flow exhibits sensitive dependence on initial conditions, and because the Navier-Stokes equations represent a dissipative system, so that volumes in phase space, on the average, contract (for large times volumes in phase space approach zero) [7, 8, 21]. We also have shown that sensitive dependence on initial conditions occurs for moderately high–Reynolds-number decaying turbulence (22).

As a result of sensitive dependence on initial conditions it appears that one cannot obtain an analytical solution (at least in the usual sense), and we need not be apologetic about using a numerical solution for turbulent flows. Of course, one might use averaged, rather than instantaneous equations, but then the closure problem (more unknowns than equations) appears (see Chapter 4), so that a deductive solution cannot be obtained. (Averaged equations supposedly would not be chaotic.)

Figure 6-7 shows an instantaneous velocity vector field in the asymptotic (developed) region projected on the $x_1 - x_2$ plane through the numerical grid center. The time is $t^* = 13.28$. A few instantaneous streamlines also have been sketched in. The flow in Fig. 6-7 appears to be composed of randomly placed jets and whirls; other projections of the velocity vector field have a similar appearance, but with jets and whirls at different locations.

A three-dimensional representation of an instantaneous velocity field in the asymptotic region is given in Fig. 6-8. The magnitude of the velocity vector $|u|$ is plotted on the $x_1 - x_2$ plane through the numerical grid center. The time is again $t^* = 13.28$. Figure 6-8, as well as Fig. 6-7, illustrates the chaotic appearance of the velocity field. It is evident that the symmetry present in the nonrandom initial conditions in Figs. 6-2 and 6-3 has been broken for the developed flow in Figs. 6-7 and 6-8.

6-3 SOME COMPUTER ANIMATIONS OF A TURBULENT FLOW

Figures 6-2 through 6-8 give a good idea of the static appearance of the low–Reynolds-number turbulent flow. In order to illustrate the dynamic evolution of the same flow, we have made some computer animations and put them on the Internet. They can be accessed via the Worldwide Web by visiting the following address: http://www.lerc.nasa.gov/WWW/GVIS/Deissler.html

In particular, the animation designated "Long-term solution for turbulent velocity" appears to give a perspective not readily obtained from a static representation.

TURBULENCE, NONLINEAR DYNAMICS, AND DETERMINISTIC CHAOS **383**

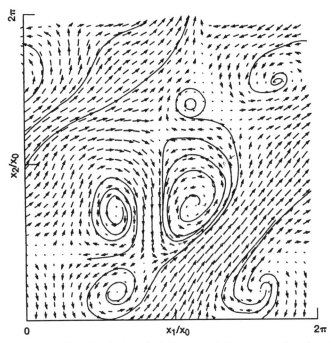

Figure 6-7 Plot of projection of velocity-vector field on $x_1 - x_2$ plane through grid center. Lengths of arrows are proportional to velocity magnitudes. Also shown are some streamlines. $\chi = 1$; $t^* = 13.28$.

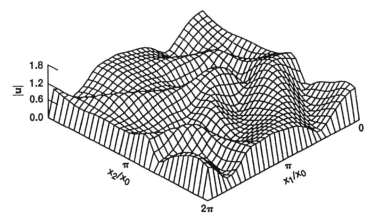

Figure 6-8 Magnitude of spatially chaotic initial velocity vector on plane through grid center. $t^* = 13.28$.

6-4 SOME TURBULENT AND NONTURBULENT NAVIER-STOKES FLOWS

In this section (except for one of the flows considered for illustrative purposes in Fig. 6-9) we use as initial conditions the spatially chaotic conditions in Figs. 6-7 and 6-8. These correspond to the flow in Fig. 6-4 at $t^* = 13.28$. As shown in ref. [9], that flow is chaotic

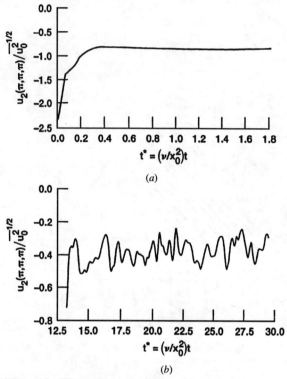

Figure 6-9 Effects of type of initial flow and of Reynolds number on asymptotic flow. (*a*) Initial flow, regular; initial Reynolds number = 17.3; asymptotic Reynolds number = 8.5. (*b*) Initial flow, chaotic; initial Reynolds number = 13.3; asymptotic Reynolds number = 7.4.

(the Liapunov characteristic exponent is positive). The use of chaotic initial conditions tends to assure that as many modes as possible are excited at a given Reynolds number. A chaotic initial condition, in fact, contains all modes; that is, it has a continuous spectrum.

The effectiveness of chaotic initial conditions in exciting unstable modes is illustrated in Fig. 6-9. The Reynolds numbers of both the nonchaotic initial conditions and of the asymptotic flow in Fig. 6-9*a* are higher than those in Fig. 6-9*b*, where the initial conditions are chaotic. Because the asymptotic flow in Fig. 6-9*a* is time-independent and that in Fig. 6-9*b* is chaotic, one sees that the character of these asymptotic flows is controlled by whether or not the initial conditions are chaotic, rather than by the Reynolds numbers. Of course if the initial Reynolds number is high enough, as in Fig. 6-4, the asymptotic flow may be chaotic even if the initial conditions are regular. At any rate it is clear from Fig. 6-9 that the use of chaotic initial conditions tends to make the asymptotic flow chaotic, if that is possible. It tends to insure that unstable modes are excited. But it is seen subsequently that, depending on the final Reynolds number, a variety of asymptotic flows can be obtained from chaotic initial conditions.

The procedure for the calculations in the remainder of this section is this: The initial conditions, which are spatially chaotic, are obtained from the chaotic flow in Fig. 6-4

TURBULENCE, NONLINEAR DYNAMICS, AND DETERMINISTIC CHAOS **385**

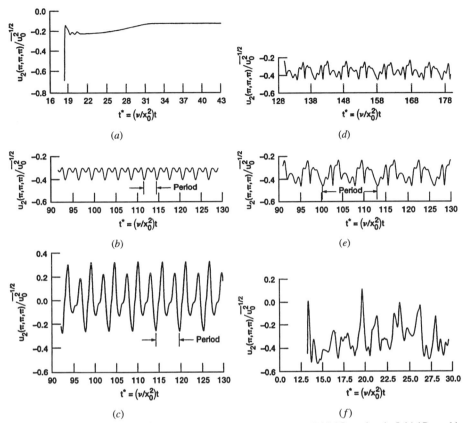

Figure 6-10 Calculated time series for evolution of velocity component. Initial flow, chaotic. Initial Reynolds number = 13.3. (a) Asymptotic Reynolds number $Re_a = 4.78$ or $\chi = 0.2$; fixed-point flow. (b) $Re_a = 6.24$ or $\chi = 0.3$; simple periodic flow. (c) $Re_a = 6.67$ or $\chi = 0.330$; period-two flow. (d) $Re_a = 6.72$ or $\chi = 0.338$. (e) $Re_a = 6.89$ or $\chi = 0.35$; complex periodic flow. (f) $Re_a = 6.93$ or $\chi = 0.4$.

for $t^* = 13.28$. (See also Figs. 6-7 and 6-8.) Using that initial condition, the asymptotic Reynolds number for each flow is fixed by setting the value of χ in the forcing term in eq. (6-4).

6-4-1 Time Series

Time series for seven different low–Reynolds-number flows are shown in Figs. 6-10 and 6-4. In Fig. 6-10a, in which the asymptotic Reynolds number Re_a is 4.78 ($\chi = 0.2$), the asymptotic (long-time) flow is time-independent. This happens although the initial conditions are chaotic. Thus, the asymptotic Reynolds number here appears not to be high enough to sustain a time-dependent chaotic or periodic flow; no modes are active. In phase space this type of flow is a fixed point, as is discussed in the next section.

For an asymptotic Reynolds number of 6.24 ($\chi = 0.3$) the longer-term solution (shown in Fig. 6-10b) is periodic in time. The curve has a rather simple shape, although

386 TURBULENT FLUID MOTION

it is not as simple as a sine wave. As discussed in the next section, this is a limit cycle in phase space.

A more complicated periodic flow is plotted in Fig. 6-10c, in which the asymptotic Reynolds number has been increased to 6.67 ($\chi = 0.330$). The period is about twice that of the flow in Fig. 6-10b. Thus, the solutions in Figs. 6-10b and c could be the first two members of a period-doubling sequence in a route to chaos [7]. Further characterization of the flows in Figs. 6-10b and c are given in the next section.

The asymptotic flow in Fig. 6-10d, which is for a Reynolds number of 6.72 ($\chi = 0.338$), has some parts that appear to repeat, but it is not periodic. Even after a very long time we are unable to obtain a complete repeating cycle. In order to see if roundoff errors could produce that result, we increased those errors by several orders of magnitude, but the results were unchanged. Figure 6-10d by itself does not provide enough information to satisfactorily characterize the flow in that figure. After we have calculated phase portraits, Poincaré sections, and Liapunov exponents, we will be in a better position to characterize the flow.

Consider next the asymptotic flow in Fig. 6-10e, where the Reynolds number is 6.89 ($\chi = 0.35$). At first glance this flow might appear chaotic because of its complexity. It is, however, periodic, although the velocity variation within each period is quite complicated. This complex periodic flow has a period close to four times that of the simple periodic flow in Fig. 6-10b. But it is not a third member of the period-doubling sequence of which the solutions in Figs. 6-10b and c appear to be the first two members. That is so because of the existence of the aperiodic flow in Fig. 6-10d; it breaks up the sequence. Further discussion of the periodic flow in Fig. 6-10e is given in following sections.

Finally, by increasing the asymptotic Reynolds number to 6.93 ($\chi = 0.4$) we get in Fig. 6-10f what appears to be a chaotic flow, because it has no apparent pattern. The flow has an appearance similar to that in Fig. 6-4 ($\chi = 1$), which already is shown to be chaotic [9].

In summarizing the information obtained from the time series for the various asymptotic flows, we note that the only flows that could be identified with reasonable certainty from the time series alone were the time-independent flow (Fig. 6-10a) and the periodic flows in Figs. 6-10b, c and e. We are able to get a better understanding even of those flows from representations yet to be considered.

6-4-2 Phase Portraits

The term *phase portrait* as used here refers to a solution trajectory in the phase space of a flow. Because one cannot readily visualize a space of more than three dimensions, our representations are projections of the higher-dimensional portraits onto two-dimensional planes or three-dimensional volumes in phase space.

The trajectory in Fig. 6-11a, which corresponds to the time series in Fig. 6-10a, shows an initial transient that ends at a stable fixed point in phase space. The arrow indicates the direction of increasing time (the direction of motion of the phase point). Because the velocity components at all points in physical space are time-independent for large times, the phase point occupies the same position in phase space for all large

TURBULENCE, NONLINEAR DYNAMICS, AND DETERMINISTIC CHAOS

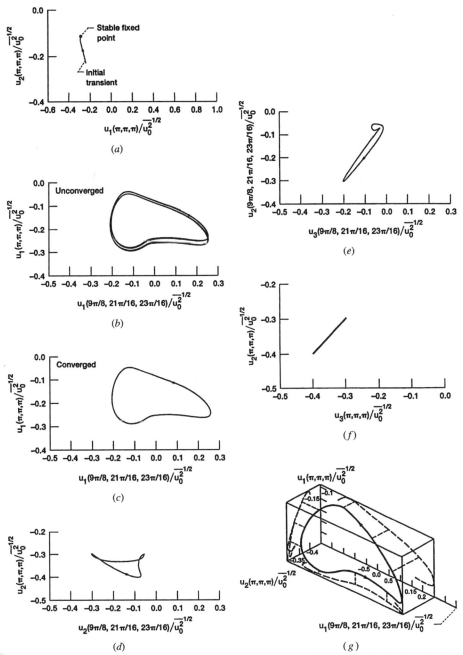

Figure 6-11 Projected phase portraits. (a) $Re_a = 4.78$ or $\chi = 0.2$; transient and fixed point flow; $t^* > 21$. (b) $Re_a = 6.24$ or $\chi = 0.3$; simple periodic flow. (c) $Re_a = 6.24$ or $\chi = 0.3$. (d) $Re_a = 6.24$ or $\chi = 0.3$. (e) $Re_a = 6.24$ or $\chi = 0.3$; simple periodic flow. (f) $Re_a = 6.24$ or $\chi = 0.3$. (g) $Re_a = 6.24$ or $\chi = 0.3$.

388 TURBULENT FLUID MOTION

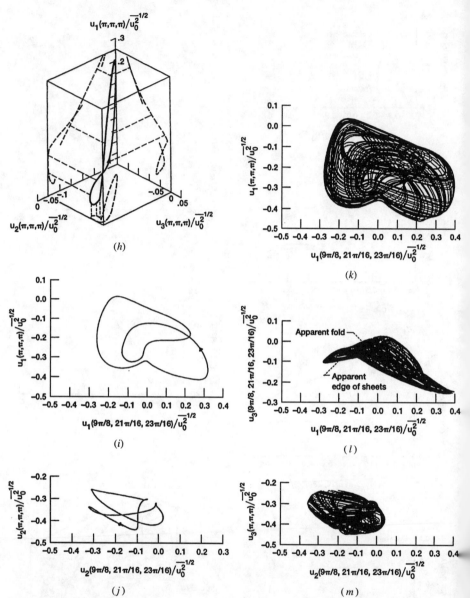

Figure 6-11(*Continued*)   (*h*) $Re_a = 6.24$ or $\chi = 0.3$; simple periodic flow. (*i*) $Re_a = 6.67$ or $\chi = 0.330$; period-two flow. (*j*) $Re_a = 6.67$ or $\chi = 0.330$; period-two flow. (*k*) $Re_a = 6.72$ or $\chi = 0.338$. (*l*) $Re_a = 6.72$ or $\chi = 0.338$. (*m*) $Re_a = 6.72$ or $\chi = 0.338$.

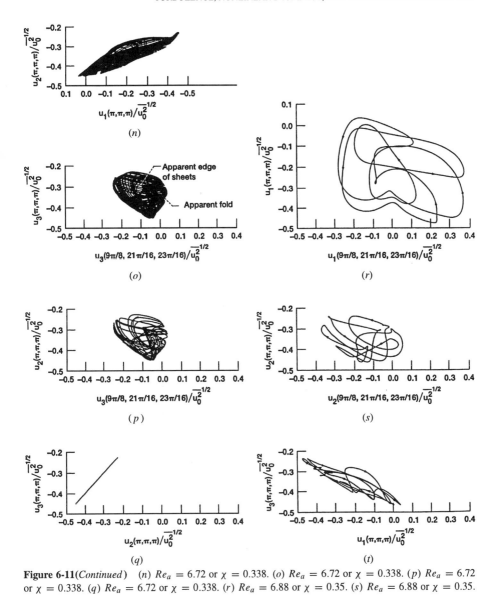

Figure 6-11(*Continued*) (n) $Re_a = 6.72$ or $\chi = 0.338$. (o) $Re_a = 6.72$ or $\chi = 0.338$. (p) $Re_a = 6.72$ or $\chi = 0.338$. (q) $Re_a = 6.72$ or $\chi = 0.338$. (r) $Re_a = 6.88$ or $\chi = 0.35$. (s) $Re_a = 6.88$ or $\chi = 0.35$. (t) $Re_a = 6.88$ or $\chi = 0.35$.

times. The projection in Fig. 6-11a is onto a $u_1(\pi, \pi, \pi) - u_2(\pi, \pi, \pi)$ plane; other projections are similar. This is the simplest example of an attractor, the trajectory in phase space being attracted to a single stable point. Once the phase point arrives there it does not leave. As mentioned earlier, volumes in phase space contract, on the average, in a dissipative system [7, 21]. In this case the volumes shrink down to a zero-volume zero-dimensional point. Motion in physical space of course does not cease but becomes time-independent.

390 TURBULENT FLUID MOTION

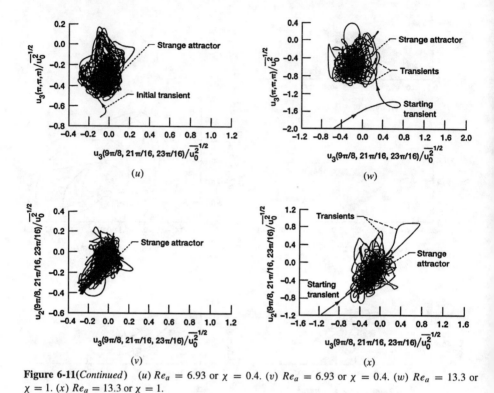

Figure 6-11(*Continued*) (*u*) $Re_a = 6.93$ or $\chi = 0.4$. (*v*) $Re_a = 6.93$ or $\chi = 0.4$. (*w*) $Re_a = 13.3$ or $\chi = 1$. (*x*) $Re_a = 13.3$ or $\chi = 1$.

Consider next the periodic phase portrait corresponding to the time series in Figure 6-10*b* (see Figs. 6-11*b* through *h*). Figures 6-11*b* and *c* show trajectories projected onto a $u_1(9\pi/8, 21\pi/16, 23\pi/16) - u_1(\pi, \pi, \pi)$ plane. Comparison of the unconverged orbit in Fig. 6-11*b* with the converged one in Fig. 6-11*c* shows that the unconverged curve wobbles around (on both sides of) the converged curve until it finally settles down on it. Thus, the trajectory is attracted to a stable limit cycle or periodic attractor. The fact that the phase point traces the same curve over and over (after convergence) confirms the periodicity of the orbit.

This formation of the stable limit cycle appears to be an example of order being born out of chaos (self-organization), because the flow initially is chaotic. For that to occur, it is only necessary to reduce the Reynolds number (or forcing) below its initial value by an appropriate amount.[3]

The contraction of volumes in phase space for a dissipative system again manifests itself here. Whereas in Fig. 6-11*a* the volumes shrink down to a zero-volume

---

[3]There may be an analogy here with the formation of the universe according to the presently accepted big-bang theory. According to that theory ordered structures (e.g., atoms, galaxies, stars, etc.) arose from an initial formless chaos (radiation). The structures could form when a parameter with an initially enormous magnitude, say the temperature, had decreased sufficiently. A nontechnical account of the theory is given in ref. [23].

zero-dimensional point, for the periodic attractor considered here they shrink down to a zero-volume one-dimensional closed curve. The coordinate axis used to plot the curve has the same shape as the curve itself. Thus, although the curve itself is one-dimensional, the one-dimensional coordinate system, or the basis function, may require many orthogonal dimensions to represent it. The curve is, strictly speaking, one-dimensional only if used with its own optimum one-dimensional coordinate system or basis function. Although the curve does not cross itself in its optimum coordinate system, it may cross if projected onto a two-dimensional orthogonal coordinate system (see Fig. 6-11d).

Additional projections of the periodic attractor onto planes in phase space are shown in Figs. 6-11e and f in order to give an idea of the variety of curve shapes that can be obtained. Note, that in Fig. 6-11f, part of the symmetry present for $t = 0$ (see Fig. 6-2) has returned. (This symmetry is absent in the fully chaotic flows.) Projections of the orbit onto three-dimensional volumes in phase space are plotted in Figs. 6-11g and h.

Projections of the period-two trajectory (corresponding to the time series in Fig. 6-10c are plotted in Figs. 6-11i and j. Comparison of those plots with the period-one plots in Figs. 6-11c and d shows that the single-cycle flow (with one large loop) has undergone a bifurcation to a two-cycle flow so that period doubling has occurred.

Phase-portrait projections corresponding to the time series in Fig. 6-10d are plotted in Figs. 6-11k through q. This portrait differs qualitatively from the others shown so far, because it tends to fill a region of space in most of the two-dimensional projections. It is found that the longer the running time, the blacker the portrait for the projections in Figs. 6-11k through p. Thus, the trajectory is clearly not periodic, because if it were it would be a closed curve in all projections. If it were quasiperiodic (with two independent frequencies), the phase portrait would lie on a torus. Figure 6-11m resembles a torus in some respects but is more complicated. In particular, it has a knob in the central region.

The projections in Figs. 6-11l, n, and o appear to show a sheet-like structure. Whereas for the periodic attractor of Figs. 6-11c through h, phase-space volumes shrink down to a zero-volume line, here they appear to shrink down to a zero-volume sheet (or sheets). The notch in the projection in Fig. 6-11n is probably the result of a superposition of sheets. Sheet-like structures with folds are generic in strange attractors [7]. Because in a chaotic flow there is stretching in at least one direction in phase space, there must be folding in order to keep the flow bounded. There appear to be some folds in the projections in Figs. 6-11l, m, and o, thus indicating that chaos is probable. The confused appearance of the trajectories in Figs. 6-11k and p also is indicative of chaos. Further evidence relative to the classification of this hard-to-classify flow is considered in succeeding sections.

Projections of the periodic trajectory corresponding to the time series in Fig. 6-10e are plotted in Figs. 6-11r to t. Initial transients have died out. Because of the very complicated appearance of the trajectory a cursory look might lead one to guess that it is chaotic (see also Fig. 6-10e). It is not chaotic, however, because it is not space-filling. No matter how long a time the solution is continued, there is no blackening of the phase portrait; the same closed curve is traced over and over, indicating periodicity of the orbit. Because initial deviations or transients present in the flow (not shown) die out as the flow is attracted to a limiting curve, the long-term solution trajectory is a periodic attractor or limit cycle. The flow appears to have a remarkable memory in being able to repeat such a complicated orbit. The fact that such a complicated curve can be retraced also is indicative of the accuracy of the numerical method. As is the case for the simpler periodic attractors considered previously, the present periodic attractor, although much more complicated, shows the shrinking of phase-space volumes to a zero-volume one-dimensional curve. The discussion given there concerning the sense in which the curve is one-dimensional also applies here.

Increasing the Reynolds numbers to those in Figs. 6-10f and 6-4 we again get (as for Figs. 6-11k through o) space-filling attractors. Projections of these are plotted in Figs. 6-11u through x. After transients have died out, the trajectories are attracted to the black regions in the plots. These look like astrophysical black holes. Indeed, these attractors are similar to black holes in that for large times points cannot leave. A possible difference is that for somewhat earlier times, the phase points can cross over the attractors, leaving them momentarily. However, that situation is temporary. After initial transients have completely died out, the phase points must remain forever on the attractors. These trajectories appear to be even more chaotic (have less of a pattern) than those in Figs. 6-11k through o. However sheets and folding are less apparent than in the attractors for the lower Reynolds number in Figs. 6-11k through o, probably because of the higher dimensionality of the attractors for the higher Reynolds numbers. More is said about that in section 6-4-7.

6-4-3 Poincaré Sections

Poincaré sections are obtained by plotting the points at which the phase point of a trajectory pierces (with increasing time) one side of a plane in phase space. The resulting plot has a dimension one less than that of the corresponding phase portrait. The lower-dimensional Poincaré section is sometimes easier to interpret. Here the pierced plane (Poincaré plane) is taken as a $u_1(\pi, \pi, \pi) - u_2(\pi, \pi, \pi)$ plane, and points are plotted if $u_1(9\pi/8, 21\pi/16, 23\pi/16)$ changes from positive to negative or from negative to positive. (Figure 6-11g may aid in visualizing the operation, at least for the simple periodic case.)

For the fixed-point attractor in Fig. 6-11a a Poincaré section does not exist, except in a trivial sense, because the phase point does not pass through a plane as time increases. So we go on to the simple periodic attractor of Figs. 6-11c through h. For that attractor the Poincaré sections are points. Figure 6-12a shows two Poincaré sections, one for $u_1(9\pi/8, 21\pi/16, 23\pi/16)$ changing from positive to negative and one for that coordinate changing from negative to positive as the phase point passes through a $u_1(\pi, \pi, \pi) - u_2(\pi, \pi, \pi)$ plane. (See also Fig. 6-11g, which plots the three coordinates.) Even after the phase point has pierced the Poincaré plane a large number of times (eight to ten), each section consists of a single point.

Consider next some Poincaré sections of the phase portrait for Figs. 6-11k through q ($\chi = 0.338$). These are plotted in Figs. 12b and c. Some portions of the plots appear to be lines; that tends to indicate quasiperiodicity of the flow (with two independent frequencies). However, in other parts of the plots the points are scattered somewhat randomly with no apparent pattern; that tends to indicate chaos. Thus, the flow has

TURBULENCE, NONLINEAR DYNAMICS, AND DETERMINISTIC CHAOS   393

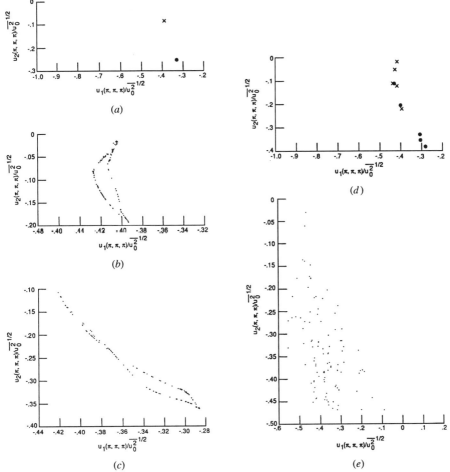

Figure 6-12 Poincaré sections of attractors. (a) $Re_a = 6.24$ or $\chi = 0.3$; simple periodic flow; ×, Poincaré plane pierced from positive side; •, plane pierced from negative side. (b) Plane pierced from positive side, $Re_a = 6.72$ or $\chi = 0.338$. (c) Plane pierced from negative side, $Re_a = 6.72$ or $\chi = 0.338$. (d) $Re_a = 6.88$ or $\chi = 0.35$; complex periodic flow; × plane pierced from positive side; •, plane pierced from negative side. (e) Plane pierced from negative side; $Re_a = 6.93$ or $\chi = 0.4$.

both chaotic and quasiperiodic features. It is not periodic because longer running times produce more points on the Poincaré section.

Two Poincaré sections for the complex periodic attractor of Figs. 11r through t are plotted in Fig. 12d. These sections are similar to those in Fig. 12a, but because of the complexity of the attractor of Figs. 11r through t, each section consists of five points instead of one. As is the case for the simpler periodic attractor, the number of points does not increase with increasing running time.

Finally, in Figs. 12e through h, we consider Poincaré sections for our two highest–Reynolds-number flows ($\chi = 0.4$ and 1). Phase portraits for these flows are considered

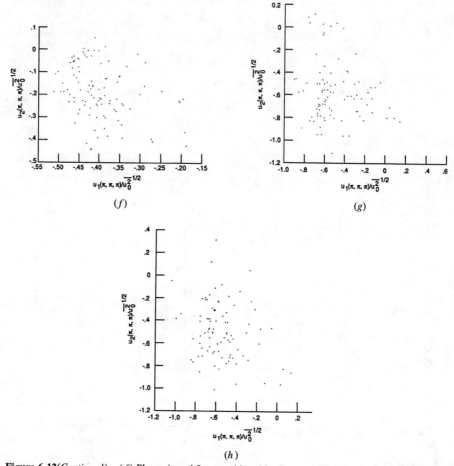

Figure 6-12(*Continued*) (*f*) Plane pierced from positive side; $Re_a = 6.93$ or $\chi = 0.4$. (*g*) Plane pierced from negative side. $Re_a = 13.3$ or $\chi = 1$. (*h*) Plane pierce from positive side; $Re_a = 13.3$ or $\chi = 1$.

in Figs. 11*u* through *x*. These Poincaré sections are similar to those in Figs. 12*b* and *c* insofar as longer running times produce more plotted points. However, they are qualitatively different, because there are no regions in which the points lie along a curve. They tend to fill a region of space in an apparently random fashion; there is no evident pattern.

6-4-4 Liapunov Exponent

The Liapunov characteristic exponent (or largest Liapunov exponent if a spectrum of exponents is considered) provides a definitive way of determining whether or not a flow is chaotic. A positive Liapunov exponent indicates sensitive dependence on initial conditions, which in turn often is considered as synonymous with chaoticity.

The method used here to determine the sensitivity of our solutions to small changes in initial conditions, and to determine Liapunov exponents, is similar to one we used previously [9]. The values of u_i at a time after initial transients have died out are perturbed by small spatially random numbers R, where $-10^{-6} < R < 10^{-6}$ or $-10^{-4} < R < 10^{-4}$. The perturbations are applied at each spatial grid point at one time. The distance between the perturbed and unperturbed solutions at various times is then calculated from

$$D = \left(\sum_{i,j} [u_{i,\text{perturbed}}(x_j, t) - u_{i,\text{unperturbed}}(x_j, t)]^2 \right)^{1/2} \tag{6-9}$$

where i, which can have values from 1 to 3, indicates different directional velocity components, and j, which can go from 1 to some number M, indicates different points in physical space. Then D represents a distance or norm in a 3M-dimensional space. For M equal to the number of grid points, D is the distance in the phase space of the discretized system. (Note that the distance D has the dimensions of a velocity.)

In ref. [9], D is represented by embedding it in one-, three-, six-, and twelve-dimensional space. It is found that increasing the embedding dimension from 3 to 12 had little or no effect on the calculated value of the Liapunov exponent. Here we adopt six dimensions as giving a sufficiently good representation of D. That is, we use three velocity components at each of two points in physical space as the dimensions ($M = 2$).

Thus, embedding the distance between perturbed and unperturbed solutions in a six-dimensional space and plotting $\log (D/u_0^2)^{-1/2}$ against dimensionless time, we obtain Figs. 6-13a through c for $\chi = 0.338, 0.4$, and 1. The values of $\log D$, on the average, increase linearly with time, indicating that D increases exponentially. That is, initially neighboring solutions diverge exponentially on the average. Thus it appears that we can characterize these three flows as chaotic.

The fact that the mean slopes of the distance-evolution curves are constant over a considerable range also allows us to use our results to obtain an estimate of the Liapunov characteristic exponent. The Liapunov characteristic exponent σ (for times after initial transients have died out) is defined as [7]

$$\sigma = \lim_{\substack{t \to \infty \\ D(0) \to 0}} \left(\frac{1}{t} \right) \ln \frac{D(t)}{D(0)}, \tag{6-10}$$

where the $D(t)$ are values of distance between initially neighboring solutions that might be obtained from Fig. 6-13. However, if the values of D were obtained from the wavy curves in Fig. 6-13, we would have to go to very large times in order to obtain a reasonable estimate for σ. This would take us out of the region of exponential growth of D, unless $D(0)$ were very small (probably below the computer noise level). One way of getting around this difficulty is to use a renormalization procedure [7].

For our purposes it seems that, because the mean slopes of the distance evolution curves in Fig. 6-13 are constant over a considerable range, the best procedure is to replace the wavy curves by straight lines through them. Then eq. (6-10) is replaced by

$$\sigma = [\ln(D_m/D_a)]/(t - t_a), \tag{6-11}$$

396 TURBULENT FLUID MOTION

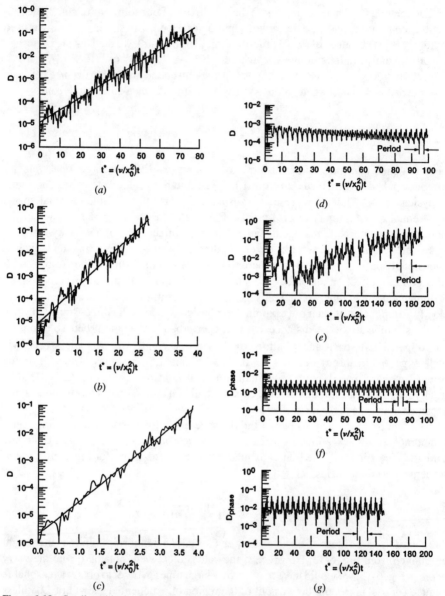

Figure 6-13 Semilogarithmic plots showing evolution of distance D between initially neighboring solutions. (a) $Re_a = 6.72$ or $\chi = 0.338$; $-10^{-6} < R < 10^{-6}$; chaotic flow. (b) $Re_a = 6.93$ or $\chi = 0.4$; $-10^{-6} < R < 10^{-6}$; chaotic flow. (c) $Re_a = 13.3$ or $\chi = 1$; $-10^{-6} < R < 10^{-6}$; chaotic flow. (d) $Re_a = 6.24$ or $\chi = 0.3$; $-10^{-4} < R < 10^{-4}$; simple periodic flow. (e) $Re_a = 6.88$ or $\chi = 0.35$; complex periodic flow. (f) Separation distance if phase point is perturbed only along trajectory; $Re_a = 6.24$ or $\chi = 0.3$; simple periodic flow. (g) Separation distance if phase point is perturbed only along trajectory; $Re_a = 6.88$ or $\chi = 0.35$; cimplex periodic flow.

where the values of D_m and D_a are read from the straight line in each figure at times t and t_a, respectively. The straight line in each figure is drawn so that its mean square deviation from the wavy curve is a minimum; this procedure should give a good estimate for σ. The values of dimensionless σ so obtained for Figs. 6-13a through c are, respectively,

$$(x_0^2/\nu)\sigma \approx 0.12, \quad 0.35, \quad \text{and} \quad 2.7. \tag{6-12}$$

The value 2.7 agrees with that obtained for the same flow (but for a different time of perturbation and different embedding dimension) in ref. [9]. The Liapunov exponents in eq. (6-12) give us a measure of the mean exponential rate of divergence of two initially neighboring solutions, or of the chaoticity of the flows. The important point is that σ is positive, indicating that these three flows are chaotic [7]. It is noted that as the Reynolds number increases σ increases (for constant x_0 and ν), or the flows become more chaotic.

Plots of dimensionless D versus t^* for our two periodic flows are given in Figs. 6-13d and e. (Note two lost-data gaps in the Fig. 13e curve.) These plots are qualitatively different from those for chaotic flows. If they were not, of course, our method for calculating Liapunov exponents would be in error. Whereas D for chaotic flow increases exponentially (on the average) for about four orders of magnitude until it is of the same order as u_i, D for the periodic flows, on the average, shows no tendency to increase exponentially. Thus the Liapunov exponent does not show a tendency to be positive, as of course it should not, because the flow is not chaotic. Theoretically the largest Liapunov exponent, the one associated with perturbations along a trajectory, should be zero for a periodic attractor [24].

The following simple argument shows that the largest Liapunov exponent for a limit cycle is zero: A limit cycle is stable, so the flow must return to the same periodic attractor after a perturbation. That is, the trajectory, a long time after perturbation, must occupy the same points in phase space as it did before perturbation. So the only possible difference between the perturbed and unperturbed trajectories is that there may be a phase difference; although the trajectory, a long time after perturbation, must occupy the same points in phase space as does the unperturbed trajectory, it may do so at different times. A phase difference is allowable because our dynamical system is autonomous; time does not appear on the right side of eq. (6-2). Because the velocity components are all periodic in time, D is periodic, as in Figs. 6-13f and g. There the limit cycle is perturbed along its trajectory by introducing a small phase difference Δt; the distance between neighboring solutions is calculated from

$$D_{\text{phase}} = \left(\sum_{i,j} [u_i(x_j, t + \Delta t) - u_i(x_j, t)]^2 \right)^{1/2} \tag{6-13}$$

in place of eq. (6-9). Thus the average D over a long time has zero slope, so that for a periodic flow, the largest Liapunov exponent (associated with perturbations along the trajectory) is zero. Other Liapunov exponents (associated with perturbations normal to the trajectory) are negative, because the flow is attracted to the limit cycle. Note that Figs. 6-13f and g do not by themselves, without the rest of the above argument, show that the largest Liapunov exponent is zero. However, the wavy curves in Figs. 6-13d and e do approach those in Figs. 6-13f and g respectively for very long times. In particular the wavy-curve shape in Fig. 6-13g is nearly identical with that near the end of the curve in Fig. 13e. So the use of eq. (6-13) is a way of producing the asymptotic D immediately if it is known that the asymptotic D are the result of a phase difference, or of a perturbation along the trajectory. The effects of perturbations normal to the trajectory are absent in Figs. 13f and g.

6-4-5 Ergodic Theory Interpretations

It may be worthwhile to look at our results in the light of modern ergodic theory (see refs. [7] and [25] and section 4-1-1). According to that theory there is a hierarchy of random systems (Fig. 6-14). At the bottom of the hierarchy are ergodic systems (those with equivalence of time, space, and ensemble averages); those systems embody the weakest notion of randomness. The so-called mixing systems (those that approach equilibrium) have a stronger notion of randomness than do those that are only ergodic, and systems that exhibit sensitive dependence on initial conditions, or chaoticity, have a stronger notion of randomness than do those that are only ergodic, or only ergodic and mixing. Mixing implies ergodicity, and chaoticity implies both ergodicity and mixing, but the converse is not true. At the top of the hierarchy are the most random systems, those that, although deterministic, may appear in a certain sense to behave as randomly as the numbers produced by a roulette wheel (Bernoulli systems) [25].

Recall that our flows with asymptotic Reynolds numbers Re_a of 6.72, 6.93, and 13.3 show sensitive dependence on initial conditions (chaoticity), because of their positive Liapunov exponents (Figs. 6-13a through c). Thus, according to the hierarchy of randomness, they both also must be ergodic and mixing. The flows for the two higher asymptotic Reynolds numbers ($Re_a = 6.93$ and 13.3), in addition to being chaotic, mixing, and ergodic, have a Poincaré section without apparent pattern, in contrast to the lower–Reynolds-number flow ($Re_a = 6.72$), where there was pattern in the Poincaré section. (Similar differences are observed in plots showing projections of the attractors for the two Reynolds numbers onto planes in phase space.) Thus the higher–Reynolds-number flows have a higher degree of randomness than that required for chaoticity, and so

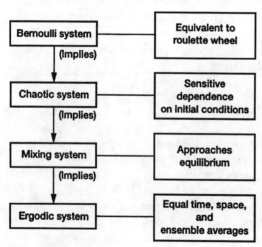

Figure 6-14 Randomness hierarchy.

are higher in the randomness hierarchy than is the lower–Reynolds-number flow. That is, the higher–Reynolds-number flows seem to have a stronger notion of randomness than chaoticity. In fact, the points on its Poincaré section appear to be placed about as randomly as the numbers produced by Roulette wheels.

For example, consider two Roulette wheels, one of which produces numbers corresponding to the abscissas of plotted points, and the other of which produces numbers corresponding to their ordinates. (Of course, one wheel could alternately be used for abscissas and ordinates.) The plot so obtained from the spins of Roulette wheels would be similar to those in Figs. 12e through h (no apparent pattern), so that our higher Reynolds number flows may be close to a Bernoulli system. A possible explanation for the differences in the randomness exhibited by our two flows is that the higher–Reynolds-number flow has an attractor of higher dimension (>3) and thus a more confused (random) appearance.

6-4-5-1 Chaotic versus turbulent flows. This leads us to a possible distinction between flows that are chaotic and those that, in addition, might be called turbulent. Perhaps one should reserve the term *turbulent* for flows that have both a positive Liapunov exponent and Poincaré sections without apparent pattern, as have those for Reynolds numbers Re_a of 6.93 and 13.3. Most flows called turbulent appear to be more random than required for a flow to be chaotic, although they are certainly chaotic, as well as ergodic and mixing.

Another characteristic that often is given as indicative of turbulence is a negative skewness factor S of the velocity derivative, where usually $-1 < S < 0$ [19]. However, for the time-dependent flows considered here, both turbulent and nonturbulent, the skewness factor does not vary significantly from that given in eq. (6-8). Even for the fixed-point flow (Figs. 6-10a and 6-11a) the value of S was about -0.25. Thus, although a negative S is necessary for the presence of turbulence, it certainly is not a sufficient indicator. A negative S in fact seems to be more an indicator of nonlinearity than of turbulence. All of the flows here are highly nonlinear.

6-4-6 Power Spectra

Power spectra give the distribution with frequency of the energy in a flow. We obtain the spectra by computing the fast Fourier transforms of the time series for the velocity components. The squares of the absolute values of those transforms are then plotted against dimensionless frequency. The results are given in Fig. 6-15.

Two types of spectra are indicated—discrete for the periodic flows and continuous for the chaotic ones. However, the spectra do not appear able to distinguish qualitatively between the weakly chaotic (Fig. 6-15b) and the fully chaotic (Figs. 6-15d and e) flows. In that respect they are less sensitive indicators than are the Poincaré sections. If one considers the discrete and continuous spectra separately, then higher frequency components become excited as the Reynolds number increases (as χ increases). In the case of the discrete spectra, the simple periodic flow (Fig. 6-15a) requires only four spectral components to represent u_2, whereas the much more complex periodic flow (Fig. 6-15c) requires 36 nonnegligible components. In both cases the frequencies of the components are related to one another as ratios of integers (one fundamental frequency in each case).

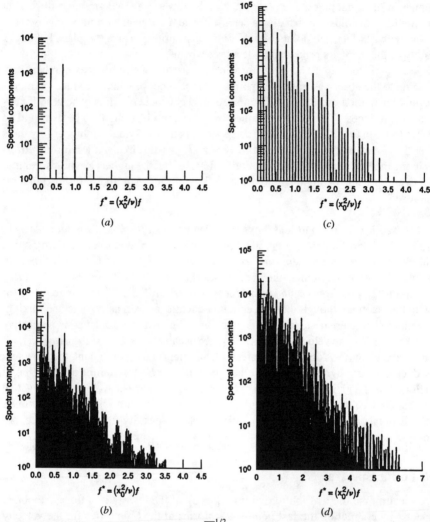

Figure 6-15 Power spectra of $u_2(\pi, \pi, \pi) / \overline{u_0^2}^{1/2}$. (a) $Re_a = 6.24$ or $\chi = 0.3$; simple periodic flow, (b) $Re_a = 6.72$ or $\chi = 0.338$; weakly chaotic flow. (c) $Re_a = 6.88$ or $\chi = 0.35$; complex periodic flow; (d) $Re_a = 6.93$ or $\chi = 0.4$; chaotic flow.

6-4-7 Dimensions of the Attractors

As a final characterization of our Navier-Stokes flows, we consider the dimensions of the attractors on which the flows reside. The dimension of a space gives, in general, the number of quantities required to specify the position of a point in the space; for example, one, two, or three coordinates are respectively required to specify a point in a one-, two-, or three-dimensional physical space. The same applies to an n-dimensional phase space, or to an attractor that is a portion of the phase space. The attractor is generally of lower

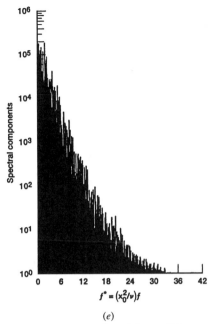

Figure 6-15(*Continued*)   (*e*) $Re_a = 13.3$ or $\chi = 1$; chaotic flow.

dimension than that of the phase space because of the shrinking of volumes in the phase space of a dissipative system. It is partly this possibility of a decreased dimension of the attractor, and consequent simplification of the problem (in principle), that makes calculation of dimension an interesting pursuit. The dimension can be considered the lower bound on the number of essential variables needed to describe the dynamics of a system [26]. Unfortunately it usually is difficult to obtain reliable estimates of that quantity.

As mentioned in section 6-4-2, the dimensions of our fixed-point and periodic attractors are respectively zero and one; a point in any space is zero-dimensional, and a closed curve, no matter how complicated its shape, is one-dimensional if the optimum coordinate system or basis function is used (see discussion in section 6-4-2).

One might question why more than one spectral component is required in Figs. 6-15*a* and *c* for the representation of one-dimensional periodic attractors. However, the need for more than one component in those representations means only that the basis functions used there, sines and cosines, are not optimum for those cases. In the case of our complex periodic flow (Fig. 6-15*c*) it is necessary to use an extremely complicated basis function for one-spectral-component representation—most likely a basis function represented numerically rather than by an analytical function.

We also attempted to calculate the pointwise dimensions of our chaotic or strange attractors [26, 27]. In that attempt we have not been able to obtain a long enough time series for the dimension to become independent of time-series length. Thus, all we can say with certainty is that the dimension must be greater than 2; if it were not, trajectories for our chaotic flows would cross in phase space. They cannot cross for an autonomous system because if they did, there would be more than one trajectory for the same conditions (at the point at which the trajectories cross), and the problem would not be deterministic.

One might expect that for our (weakly) chaotic flow the dimension would be only slightly greater than 2 because apparent folding can be seen in the phase portrait (Figs. 6-11l and o); if the attractor were many-dimensional, stretching and folding would occur in many directions and, because of the resulting confusion, could not be discerned in a two-dimensional plot. That is apparently what happens for the turbulent flows (Figs. 6-11u through x). There the dimension must be significantly greater than 2; stretching and folding, although certainly present, is many dimensional, so that the result is a confused appearance of the phase portrait. However, even there the dimension of the attractor should be limited by the overall shrinkage of volumes in phase space.

6-5 CLOSING REMARKS

Navier-Stokes turbulence is a chaotic phenomenon. Our long-term solutions with steady forcing show that the calculated turbulence has a positive Liapunov exponent, which in turn means that it is sensitively dependent on initial conditions.

Turbulence has, for a long time, been assumed to be random [1], or at least to have the appearance of randomness. Sensitive dependence on initial conditions provides an explanation for the occurrence of apparent randomness in turbulence. But in spite of its random appearance turbulence has a deterministic element, inasmuch as the Navier-Stokes equations that describe it are fully deterministic. The phrase "deterministic chaos" might therefore provide a fitting description for turbulence. Although turbulence is time-dependent and random in appearance, our solutions show that it can form with no time-dependent or random input. This again is a result of sensitive dependence of the solutions on initial conditions.

It may not be, however, a sufficiently complete description of turbulence to say that it is chaotic. Some of our low–Reynolds-number flows have a positive Liapunov exponent, and thus are chaotic, but their Poincaré sections show a pattern in some of their parts. On the other hand, solutions at somewhat higher Reynolds numbers show a complete lack of pattern. Perhaps we should reserve the term *turbulent* for flows that have a positive Liapunov exponent and, in addition, have Poincaré sections without pattern. Interpreting our results in the light of modern ergodic theory, turbulence is more random than required for a system to be chaotic; its randomness approaches that of a Bernoulli system, an example of which is a Roulette wheel.

Turbulence is also aperiodic or nonperiodic. As examples of flows that contrast with turbulence, we were able to obtain some periodic and fixed-point solutions. Whereas the fixed-point (in phase space) flows are time-independent, and the periodic flows are closed curves in phase space (points on Poincaré sections), the turbulent flows are time-dependent and fill a portion of phase space. The turbulent, periodic, and fixed-point flows are all attracted to lower-dimensional regions of phase space called attractors. The turbulent flows lie on strange or chaotic attractors.

Another requirement that often is given for flows to be turbulent is that they have negative velocity-derivative skewness factors. However, our periodic and fixed-point solutions have skewness factors that do not vary greatly from those for turbulent flows.

A negative skewness factor seems to be more an indication of nonlinearity (all of our forced flows are highly nonlinear) than of turbulence.

REFERENCES

1. Batchelor, G.K., *The Theory of Homogeneous Turbulence*, Cambridge University Press, New York, 1953.
2. Herring, J.R., "An Introduction and Overview of Various Theoretical Approaches to Turbulence," in *Theoretical Approaches to Turbulence*, edited by D.L. Dwoyer, M.Y. Hussaini, and R.G. Voigt, pp. 73–90, Springer-Verlag, New York, 1985.
3. Kraichnan, R.H., "Decimated Amplitude Equations in Turbulence Dynamics," in *Theoretical Approaches to Turbulence*, edited by D.L. Dwoyer, M.Y. Hussaini, and R.G. Voigt, pp. 91–136, Springer-Verlag, New York, 1985.
4. Deissler, R.G., "Turbulence Processes and Simple Closure Schemes," in *Handbook of Turbulence, vol. 1*, edited by W. Frost and T.H. Moulden, pp. 165–186, Plenum Press, New York, 1977.
5. Lin, C.C., and Reid, W.H., "Turbulent Flow, Theoretical Aspects," in *Handbuch der Physik, vol. VIII/2*, edited by S. Flügge, and C. Truesdell, Springer-Verlag, New York, 1963.
6. Ruelle, D., and Takens, F., "On the Nature of Turbulence," *Commun. Math. Phys.*, vol. 20, pp. 167–192, 1971.
7. Lichtenberg, A.J., and Lieberman, M.A., *Regular and Stochastic Motion*, Springer-Verlag, New York, 1983.
8. Deissler, R.G., "Turbulent Solutions of the Equations of Fluid Motion," *Rev. Mod. Phys.* vol. 56, pp. 223–254, 1984.
9. Deissler, R.G., "Is Navier-Stokes Turbulence Chaotic?" *Phys. Fluids*, vol. 29, pp. 1453–1457, 1986.
10. Deissler, R.J., "External Noise and the Origin and Dynamics of Structure in Convectively Unstable Systems," *J. Stat. Phys.*, vol. 54, nos. 5/6, pp. 1459–1488, 1989.
11. Constantin, P., Foias, C., Manley, O.P., and Teman, R., "Determining Modes and Fractal Dimension of Turbulent Flows." *J. Fluid Mech.*, vol. 150, pp. 427–440, 1985.
12. Lorenz, E.N., "Deterministic Nonperiodic Flow," *J. Atmos. Sci.*, vol. 20, pp. 130–141, 1963.
13. Landau, L.D., "On the Problem of Turbulence." *C.R. Acad. Sci. USSR*, vol. 44, p. 311, 1944.
14. Landau, L.D., and Lifshitz, E.M., *Fluid Mechanics*, chapter III. Pergamon Press, New York, 1959.
15. Deissler, R.G., "Nonlinear Decay of a Disturbance in an Unbounded Viscous Fluid," *Appl. Sci. Res.*, vol. 21, no. 6, pp. 393–410, 1970.
16. Goldstein, M.E., and Braun, W.H., "Advanced Methods for the Solution of Differential Equations," National Aeronautics and Space Administration, NASA SP-316, Washington, DC, 1973.
17. Deissler, R.G., "On the Nature of Navier-Stokes Turbulence," National Aeronautics and space Administration, NASA TM-101983, Washington, DC, 1989.
18. Clark, R.A., Ferziger, J.H., and Reynolds, W.C., "Evaluation of Subgrid-Scale Models Using an Accurately Simulated Turbulent Flow," *J. Fluid Mech.*, vol. 91, pp. 1–16, 1979.
19. Tavoularis, S., Bennett, J.C., and Corrsin, S., "Velocity-Derivative Skewness in Small Reynolds Number, Nearly Isotropic Turbulence," *J. Fluid Mech.*, vol. 88, pp. 63–69, 1978.
20. Sokolnikoff, I.S., *Advanced Calculus*, pp. 391, 392, McGraw-Hill, New York, 1939.
21. Constantin, P., Foias, C., and Temam, R., "Attractors Representing Turbulent Flows," *Memoirs, Amer. Math. Soc.*, vol. 53, no. 314, p. 57, 1985.
22. Deissler, R.G., and Molls, F.B., "Effect of Spatial Resolution on Apparent Sensitivity to Initial Conditions of a Decaying Flow as it Becomes Turbulent," *J. Comp. Phys.*, vol. 100, no. 2, pp. 430–432, 1992.
23. Jastrow, R., *God and the Astronomers*, W.W. Norton, New York, 1978.
24. Haken, H., *Advanced Synergetics*, Springer-Verlag, New York, 1983.
25. Lebowitz, J.L., and Penrose, O., "Modern Ergodic Theory," *Phys. Today*, vol. 26, no. 2, pp. 23–29, 1973.
26. Farmer, J.D., Ott, E., and Yorke, J.A., "The Dimension of Chaotic Attractors," *Physica D*, vol. 7D, pp. 153–180, 1983.
27. Guckenheimer, J., and Buzyna, G., "Dimension Measurements for Geostrophic Turbulence," *Phys. Rev. Lett.*, vol. 51, no. 16, pp. 1438–1441, 1983.
28. McComb, W.D., *The Physics of Fluid Turbulence*, Oxford University Press, New York, 1990.

AFTERWORD

Advances in solving the turbulence problem continue to be made along theoretical, computational, and experimental lines. Considerable progress in understanding turbulence mechanisms, particularly spectral energy transfer and the interaction between energy transfer and dissipation, has been made by using theoretical and computational methods. Much of the recent activity in turbulence research is related to advances in high-speed computation. For example, deductive–computational solutions of the unaveraged Navier-Stokes equations are now available, at least for turbulent flows at low and moderate Reynolds numbers. A Reynolds number higher than those that can be handled by available computers and computational schemes, of course, can always be picked. However, at least from a research standpoint, high–Reynolds-number turbulence does not differ qualitatively from that at lower Reynolds numbers; the turbulent energy is just spread out over a wider range of wavenumbers, there being no bifurcations in going from low to high Reynolds numbers except possibly in the transition region between laminar and turbulent flow. The mathematical and computational methods appropriate for low and for high Reynolds numbers of course may differ.

Another area of progress is the application of nonlinear-dynamics and chaos theory to turbulence research. Here again the research could not get very far without the use of high-speed computation. Nonlinear-dynamics and chaos theory have given us the means of interpreting numerical results by providing new investigative tools. Turbulence, although chaotic, is shown to be more random than is required for a chaotic flow.

From an engineering standpoint quantitative results at high Reynolds numbers are required. There, modeling, or the use of information in addition to that provided by the Navier-Stokes equations, still has a place in turbulence calculations.

Advances in solving the turbulence problem no doubt will continue. For example, faster computers should become available and make possible turbulent solutions of the unaveraged equations at higher Reynolds numbers; those solutions could then be interpreted by the use of nonlinear-dynamics and chaos theory. At the same time, advances in the use of modeling in the solution of averaged Navier-Stokes equations should give engineering results at still higher Reynolds numbers. Eventually, as still faster computers become available, results from the two approaches (deductive–computational and modeling) hopefully will merge.

INDEX

Alternating tensor, 25
Averages:
    kinds of, 49–50
    properties of, 51–52

Bernoulli's equation, 7
Boundary layer:
    turbulent, (illus.) 5
    moderately short, highly accelerated, 80–91, 92
Buoyancy:
    effect on turbulence, 263–277
Buoyancy and shear combined:
    effect on turbulence, 278–301

Chaos:
    deterministic, 4, 373–403
    and turbulence, 373–03
Chaotic velocity, 383
Chaotic vs. turbulent flows, 399
Choaoticity, 4
Closure:
    by specification of sufficient random initial conditions, 164–198
Closure assumption, 71, 79–80
Closure problem, 55–106
Computer animation of turbulent flow, 382
Cone:
    turbulent flow through, 301–319
Continuum approach:
    justification of for turbulence, 31
Contraction, 27

Delta function:

Dirac, 116
Dimensional analysis:
    rule for, 45–47
Dimensions of attractors, 400–402
Directed-line segment, 21
Direct numerical simulation:
    triple correlations important, 73, 94–98, 200–215
Dot product, 27
Dummy subscript, 20

Earth:
    turbulent, (illus.) 10
Eddy conductivity, 58
Eddy conductivity, eddy viscosity ratio:
    computation of, 260–263
Eddy viscosity, 58
Equations of motion:
    dimensionless form of, 41
    of viscous fluid, 38–41
Ergodic theory, 51, 398
Eulerian-time correlation, 216

Final period turbulence, 131–134
Fluid-flow equations:
    dimensionless form of, 41–42
Forcing term:
    time independent, 376–377
Fourier analysis, 109–131
    averaged quantities and equations, 110–115
    unaveraged quantities and equations, 116–131
Fourier coefficients, 110
Fourier integral, 111
Fourier series, 110

Fourier transform, 111
  finite, 112
Friction factor data:
  dimensionless correlation of, 41–42
Fully developed flow and heat transfer, 65–78

Gap problem, 195
Generalized functions, 116
Growth due to buoyancy:
  in turbulence with shear, 294–301

Heat transfer, convective, 76–77, (illus.) 78,
  dimensionless correlation of, 44–47
  partial differential equation of, 42–44
Heat transfer and temperature fluctuations:
  in uniformly sheared turbulence, 250–263
Heat transfer vector:
  turbulent, 56–57
Heisenberg, W., 10
Homogeneous turbulence and heat transfer:
  with uniform mean gradients, 220–355
Homogeneous turbulence:
  without mean gradients, 122–221

Initial condition, effect of on uniformly sheared
  turbulence, 239–242
Inner product, 27, 28
Invariant, 27

Jet:
  turbulent, (illus.) 11, (illus.) 12

Kampé De Feriet, J., 10
Kármán constant, 70, 80
Kolmogorov, A. N., 10, 147
Kolmogorov-Obukov spectrum, 147
Kraichnan, R. H., 367
Kronecker delta, 20, 24

Lagrangian correlations, 216
Law of the wall, 66–68
Liapunov exponent, 394–398
Local isotropy, 231
Locally homogeneous turbulence, 320
Loeffler, A. L., 269
Logarithmic law, 69–71
Lorenz system, 374–376

Maintenance and growth of shear-flow turbulence,
  351–355
Mixing length, 59–60
Mixing, turbulent:
  nonuniformity of, 60–62

Momentum:
  conservation of, 34–41
Multitime-multipoint correlations, 215

Nonlinear dynamics and turbulence, 373–403
Nontensor, 28
Normal strain:
  with turbulence and heat transfer, 319–335
Normal strain and shear combined:
  with turbulence and heat transfer, 336–351
Numerical solution:
  triple correlations important, 73, 94–98, 200–215

One-point correlation equations, 91–93,
  physical interpretation of terms in, 93–94
Orszag, S. A., 18, 365, 366
Outer product, 22

Phase portraits, 386–392
Pipe flow, 75, 104
Poincaré sections, 392–394
Power spectra, 399–401
Prandtl, L., 10
Pressure:
  mechanical, 37
  thermodynamic, 39, 40
Processes in isotropic turbulence, 213–215

Quotient law, 23, 26

Reynolds decomposition, 53
Reynolds, O., 1, 8
Reynolds number, 1
Reynolds stress tensor, 55–56

Sensitive dependence on initial conditions, 4,
  373–403
Sheared homogeneous turbulence with turbulence
  self-interaction, 355–364,
  numerical results, 362–364
Skewness factor, 380
Spectra:
  discussion of computed in uniformly sheared
    turbulence, 227, 254
Spectral equations:
  interpretation of terms in, 224–225
  solutions of, 225–227
Spectral energy transfer and dissipation:
  interaction between, 145–146
Spectrum for infinite Reynolds number, 146–147
Statistical or classical theory of turbulence, 373
Strange attractors, (illus.) 390, 392
Stress tensor, 23
  viscous, 34–38

Summation convection:
  Einstein, 20

Taylor, G. I., 8, 215
Tensor:
  cartesian, 20n
  isotropic, 24
  of order zero, 19, 27
  of order one, 19, 21
  of order two, 19, 22
  two-point isotropic, 226n
Theoretical physics:
  fluid dynamics and turbulence, as part of, 8
Transfer, spectral:
  by mean gradients, 113–115, 232–234
  by self-interaction, 113–115, 156
  mode of occurrence, 148–159
  poetry, 115, 156
Turbulence:
  astrophysical, (illus.) 16
  before final period, 137
  description of, 3–4
  in clouds, (illus.) 9
  in final period, 131–134
  grid generated, (illus.) 15
  in smoke, (illus.) 6
  in a space shuttle launch, (illus.) 8
  specification of, 164
  ubiquity of, 4–6
  in a volcano, (illus.) 7
Turbulent diffusion, 215–219
Turbulent flow:
  in a passage, (illus.) 33, (illus.) 73
Turbulent processes with uniform shear, 234–235
Turbulent shear stress and heat transfer near wall:
  conditions satisfied by, 62–65
Two-point correlation equations, 98–102

Uniformly sheared turbulence, 223–236

Vector:
  as first-order tensor, 21, 22
  position or displacement, 22
Velocity correlations, 19
Velocity-defect law, 68, 69
Van Karman, T., 10, 79
Vorticity, 25
  direction of maximum in turbulent shear flow, 236–239

Wake:
  turbulent, (illus.) 13